# Test File

## *to accompany*

Life The Science of Biology

**SEVENTH EDITION**

*PURVES • SADAVA • ORIANS • HELLER*

**Catherine Ueckert** *Northern Arizona University*

**Chris Romero** *Front Range Community College*

**Betty McGuire** *Smith College*

**Paula Mabee** *University of South Dakota*

**Erica Bergquist** *Holyoke Community College*

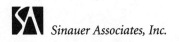 *Sinauer Associates, Inc.*

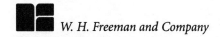 *W. H. Freeman and Company*

Cover photo © Steve Bloom/stevebloom.com.

**Test File** to accompany *Life: The Science of Biology,* **Seventh Edition**

**Address editorial correspondence to:**
Sinauer Associates, Inc.
23 Plumtree Road
Sunderland, MA, 01375 U.S.A.
Fax: 413-549-1118
Internet: www.sinauer.com; publish@sinauer.com

**Address orders to:**
VHPS/W. H. Freeman & Co. Order Department
16365 James Madison Highway,
U.S. Route 15, Gordonsville, VA, 22942 U.S.A.
Internet: www.whfreeman.com

ISBN 0-7167-5815-6
Printed in the U.S.A.

# To the Instructor

This Test File has been prepared to aid you in the preparation of exams for the course for which you have adopted the Seventh Edition of *Life: The Science of Biology*. This edition of the Test File has been revised and updated to be consistent with all the changes made to the Seventh Edition of the textbook. The authors of these test questions have endeavored to provide you with a broad range of questions from which to draw. For each chapter of the Seventh Edition of *Life*, this Test File has a comprehensive set of questions that cover the full range of material presented in the chapter. Due to variation in chapter length and the amount and complexity of material covered in each chapter, the number of questions per chapter in the Test File is necessarily variable.

To aid you in selecting a range of questions, we have employed a three-tiered system for categorizing the complexity/difficulty of each question:

1. The bulk of the questions are designed to test the student's mastery of the content of the chapter. These are mostly factual recall in nature, and are not marked in any way.
2. Questions that the authors deem more difficult are marked with a star (★).
3. Questions that are more conceptual in nature (those that require thinking beyond the bare facts) are marked with a square (■).

(Some question are marked with both a star and a square.)

## Electronic Versions of This Test File

This printed Test File is meant to serve as your reference guide to the test questions that are available. To make it as easy as possible for you to incorporate these questions into your exams, we have provided these questions in two electronic formats:

1. On the Seventh Edition Instructor's Media Library, we have provided Microsoft® Word® files for each of the 58 chapters of the Test File. To use these files, simply open them in Microsoft Word, and copy the questions you want into your own document.
2. We also have a Computerized Test Bank CD available to adopters. This CD includes the powerful Brownstone® Diploma® test creation software. When you run the installation from the CD, the software and all the test questions will be installed onto your hard drive. You can then use Diploma to create your exams, and take advantage of all the features it has to offer.

## Note to Instructors Who Have Used Previous Editions

In this edition, we did not include the extra questions from other sources (Study Guide, Student Website, etc.) in the *printed* Test File. These questions *are* included in the electronic versions of the Test File. Both the Word files on the Instructor's Media Library and the Computerized Test Bank CD include many additional questions taken from the Student Website Online Quizzes, the printed Study Guide, and the textbook end-of-chapter Self-Quizzes. If you would like a copy of either the Instructor's Media Library (ISBN 0-7167-5814-8) or the Computerized Test Bank CD (ISBN 0-7167-5816-4), please contact your W. H. Freeman representative.

# Contents

## Part 5  The Evolution of Diversity

## Part 6  The Biology of Flowering Plants

## Part 7  The Biology of Animals

## Part 8  Ecology and Biogeography

# 1 An Evolutionary Framework for Biology

## Fill in the Blank

1. _____ noticed the similarities of the limbs of different mammals.
   *Answer: Buffon*

2. All scientific study begins with observations and the formation of testable _____.
   *Answer: hypotheses*

3. Under present conditions on Earth, cells do not arise from _____ material.
   *Answer: noncellular*

4. In contrast to eukaryotic cells, bacteria lack _____.
   *Answer: membrane-enclosed compartments*

5. According to _____ theory, if humans continue to use their brains, their brains will become larger and more developed.
   *Answer: Lamarck's*

6. _____ and _____ were the first to suggest that organisms change gradually through the natural selection of variable characteristics.
   *Answer: Darwin; Wallace*

7. Single-celled organisms that lack a nucleus belong to the two kingdoms called _____ and _____.
   *Answer: Archaea; Bacteria*

8. Eukaryotes can be described as _____ within a cell.
   *Answer: cells*

9. As many as _____ species inhabit Earth.
   *Answer: 30 million*

10. Multicellular organisms that are photosynthetic belong to the kingdom called _____.
    *Answer: Plantae*

11. Nonphotosynthetic multicellular organisms that digest their food outside their body's cells and absorb the products belong to the kingdom called _____.
    *Answer: Animalia*

12. About _____ years ago, prokaryotes acquired the ability to photosynthesize.
    *Answer: 2.5 billion*

13. Currently, scientists agree with the estimate that life first appeared approximately _____ years ago.
    *Answer: 4 billion*

14. Organisms maintain a consistent internal environment despite changing external conditions. The term for this process is _____.
    *Answer: homeostasis*

15. There are three domains used to categorize life forms that have evolved separately for about a billion years: _____, _____, and _____.
    *Answer: Archaea; Bacteria; Eukarya*

16. The _____ hypothesis states that no difference exists due to the variable under investigation.
    *Answer: null*

17. Fungi and animals are both _____.
    *Answer: heterotrophs*

18. An _____ shows the evolutionary relationships among species.
    *Answer: evolutionary tree*

19. A National Science Foundation project to determine the evolutionary relationships among all species on Earth is known as _____.
    *Answer: Assembling the Tree of Life (or ATOL)*

20. _____ is an organized genetic unit capable of metabolism, reproduction, and evolution.
    *Answer: Life*

21. The total chemical activity of a living organism is its _____.
    *Answer: metabolism*

22. _____ is a change in the genetic composition of a population of organisms over time.
    *Answer: Biological evolution*

23. _____ are divided into four groups: Protista, Fungi, Plantae, and Animalia.
    *Answer: Eukarya*

24. A Pacific tree frog has the scientific nomenclature of *Hyla regilla*. This particular tree frog belongs to the species _____.
    *Answer: regilla*

## Multiple Choice

1. Before Darwin, one scientist who wrote extensively about evolution was
   a. Hooke.
   b. Leeuwenhoek.
   c. Lamarck.
   d. Pasteur.
   e. Virchow.
   *Answer: c*

2. Count de Buffon thought that pigs have small function-less toes because
   a. they are evolving toward having functioning toes but have not yet reached the goal.
   b. they have defective toe-producing information in their DNA.
   c. they evolved from ancestors that had functioning toes.
   d. constant parasitization over generations caused the loss of their toes.
   e. All of the above
   *Answer: c*

3. Darwin noted that all populations have the potential to grow _____, but that in nature most populations _____ over time.
   a. linearly; are stable
   b. exponentially; grow more slowly
   c. linearly; fluctuate unpredictably
   d. exponentially; are stable
   e. linearly; decrease slowly
   *Answer: d*

4. Plants are
   a. eukaryotic, multicellular photosynthesizers.
   b. eukaryotic, unicellular autotrophs.
   c. eukaryotic, multicellular heterotrophs.
   d. prokaryotic, multicellular autotrophs.
   e. prokaryotic, unicellular heterotrophs.
   *Answer: a*

5. What distinguishes living organisms from nonliving matter?
   a. Only living organisms are characterized by the processes of metabolism, reproduction, and evolution.
   b. Only living organisms change in response to their environment.
   c. Only living organisms are composed of molecules.
   d. Living organisms do not obey physical and chemical laws, whereas nonliving matter does.
   e. Only living organisms increase in size.
   *Answer: a*

6. Metabolism is
   a. the consumption of energy.
   b. the release of energy.
   c. all conversions of matter and energy taking place in an organism.
   d. the production of heat by chemical reactions.
   e. the exchange of nutrients and waste products with the environment.
   *Answer: c*

7. All living organisms obtain their energy from
   a. food.
   b. external sources.
   c. sunlight.
   d. heterotrophs.
   e. autotrophs.
   *Answer: b*

8. The smallest chemical units of matter are
   a. cells.
   b. lipids.
   c. molecules.
   d. hydrogen.
   e. atoms.
   *Answer: e*

9. Sexual reproduction enhances chances for adaptation by means of
   a. mutation.
   b. recombination.
   c. random distribution.
   d. gene loss.
   e. gene flow.
   *Answer: b*

10. Heterotrophs cannot obtain their energy from
    a. food.
    b. autotrophs.
    c. other heterotrophs.
    d. complex chemical substances.
    e. sunlight.
    *Answer: e*

11. The hierarchical order of the units of life, from simple to complex is
    a. atom, molecule, cell, tissue, organ, organism, population, community.
    b. cell, molecule, atom, tissue, organ, organism, population, community.
    c. molecule, cell, organ, atom, tissue, organism, population, community.
    d. atom, molecule, cell, tissue, organ, population, organism, community.
    e. atom, molecule, tissue, organ, cell, population, organism, community.
    *Answer: a*

12. Earth is approximately _____ years old.
    a. 4 billion
    b. 4 trillion
    c. 4 million
    d. 6,000
    e. 40 trillion
    *Answer: a*

13. The fundamental unit of life is the
    a. aggregate.
    b. organelle.
    c. organism.
    d. membrane.
    e. cell.
    *Answer: e*

14. A species is
    a. all the organisms that live together in a particular area.
    b. a group of similar organisms that cannot interbreed.
    c. a group of similar organisms capable of interbreeding.
    d. an adult organism and all of its offspring.
    e. a group of similar organisms that live in the same area.
    *Answer: c*

15. Which of the following is *not* a major stage of the hypothetico-deductive method?
    a. Controlling an environment
    b. Making an observation
    c. Forming a hypothesis
    d. Making a prediction
    e. Testing a prediction
    *Answer: a*

■16. Which of the following questions *cannot* be answered using the hypothetico-deductive method?
    a. Are bees more attracted to red roses than to yellow roses?
    b. Are red roses more beautiful than yellow roses?
    c. Why are red roses red?
    d. Do red roses bloom earlier than yellow roses?
    e. Are red roses more susceptible to mildew than yellow roses?
    *Answer: b*

■17. After observing that fish live in clean water but not in polluted water, you make the statement, "polluted water kills fish." Your statement is an example of
    a. scientific inquiry.
    b. biological evolution.
    c. a prediction.
    d. a hypothesis.
    e. a theory.
    *Answer: d*

18. Which of the following is *not* a feature of scientific hypotheses?
    a. They are true.
    b. They make predictions.
    c. They are based on observations.
    d. They can be tested by experimentation.
    e. They can be tested by observational analysis.
    *Answer: a*

■19. Based on the large numbers of offspring produced by many organisms, Darwin proposed that mortality was high and only a few individuals survived to reproduce. He called the differential reproductive success of individuals with particular variations
    a. evolution.
    b. artificial selection.
    c. the cell theory.
    d. natural selection.
    e. inheritance of acquired characteristics.
    *Answer: d*

20. Members of the kingdom Animalia obtain their energy from
    a. decomposing organic matter.
    b. photosynthesis.
    c. other organisms.
    d. sunlight.
    e. inorganic molecules.
    *Answer: c*

21. A typical cell
    a. can be composed of many types of tissues.
    b. is found only in plants and animals.
    c. is the smallest entity studied by biologists.
    d. is the simplest biological structure capable of independent existence and reproduction.
    e. All of the above
    *Answer: d*

22. Which kingdom contains eukaryotic, single-celled organisms?
    a. Plantae
    b. Archaea
    c. Animalia
    d. Protista
    e. None of the above
    *Answer: d*

23. The key purpose of experimentation is to
    a. obtain accurate quantitative measurements.
    b. prove unambiguously that a particular hypothesis is correct.
    c. avoid comparative analysis.
    d. answer as many key questions in one experiment as possible.
    e. control factors that might affect the result.
    *Answer: e*

24. A key point in Darwin's explanation of evolution is that
    a. the biological structures most likely inherited are those that have become better suited to the environment through constant use.
    b. mutations that occur are those that will help future generations fit into their environments.
    c. slight variations among individuals significantly affect the chance that a given individual will survive in its environment and be able to reproduce.
    d. genes change in order to help organisms cope with problems encountered within their environments.
    e. extinction is nature's way of weeding out undeserving organisms.
    *Answer: c*

25. The smallest entities studied by biologists are
    a. cells.
    b. tissues.
    c. organelles.
    d. molecules.
    e. membranes.
    *Answer: d*

26. It is thought that some prokaryotes were consumed by, then integrated into other prokaryotes _____ years ago.
    a. about 4,000 years ago.
    b. about 10,000 years ago.
    c. more than one million years ago.
    d. more than one billion years ago.
    e. more than one trillion years ago.
    *Answer: d*

27. All _____ must obtain their energy from the sun.
    a. plants
    b. autotrophs
    c. organisms
    d. heterotrophs
    e. bacteria
    *Answer: b*

28. Heterotrophs obtain their energy from
    a. fungi.
    b. water.
    c. other organisms.
    d. vitamins.
    e. heat.
    *Answer: c*

29. The two important developments necessary for the existence of multicellular life forms were
    a. differentiation and cell clumping.
    b. symmetry and asymmetry.
    c. gene regulation and the evolution of the nucleus.
    d. sexual reproduction and recombination.
    e. isolation and compartmentalization.
    *Answer: a*

30. The initial accumulation of oxygen in the atmosphere was the result of photosynthesis from an organism most like modern
    a. cyanobacteria.
    b. algae.
    c. mosses.
    d. kelp.
    e. eukaryotes.
    *Answer: a*

31. A prerequisite for the survival of life on land was the accumulation of
    a. $O_2$.
    b. $CO_2$.
    c. water vapor.
    d. $O_3$.
    e. bacteria in the soil.
    *Answer: d*

32. The chemical formula for ozone is
    a. O.
    b. $O_2$.
    c. $H_2O_2$.
    d. $O_3$.
    e. None of the above
    *Answer: d*

33. Ozone is important to life on Earth because it
    a. is toxic to all forms of life.
    b. can be used in place of oxygen.
    c. blocks much ultraviolet radiation.
    d. provides energy to some basic forms of life.
    e. disinfects.
    *Answer: c*

34. When biologists organize species into groups, they attempt to do so based on
    a. physical similarities.
    b. ecological niches.
    c. chronological order.
    d. evolutionary relationships.
    e. All of the above
    *Answer: d*

35. When attempting to group species, scientists use
    a. fossils.
    b. physical structures.
    c. gene similarities.
    d. All of the above
    e. None of the above
    *Answer: d*

36. Eukarya include
    a. Protista.
    b. Plantae.
    c. Fungi.
    d. Animalia.
    e. All of the above
    *Answer: e*

37. Which of the following is *not* a characteristic of most multicellular organisms?
    a. Cells change structure and function during development.
    b. Cells stick together after division.
    c. Cells specialize.
    d. Certain cells specialize for purposes of sexual reproduction.
    e. Cells can grow without regulation.
    *Answer: e*

38. The advantage of controlled experiments is that
    a. all factors are controlled except one.
    b. the hypothesis is proven right.
    c. patterns can be predicted.
    d. investigations can be done in the field.
    e. a massive amount of data can be synthesized.
    *Answer: a*

39. A worldwide decline of amphibians is due to
    a. the parasite *Ribeiroia*.
    b. UV-B susceptibility.
    c. agricultural pesticides.
    d. atmospheric contamination.
    e. no single factor.
    *Answer: e*

40. A study has been done to test the hypothesis that pesticides from agricultural lands are contributing to the decline of amphibian populations. The study
    a. shows that amphibian species are declining in areas exposed to agricultural pesticides.
    b. indicates that more studies are needed before a conclusion can be made.
    c. shows that some species are declining while others are unaffected.
    d. proves that the use of agricultural pesticides must be stopped.
    e. is an example of an experimental study.
    *Answer: c*

41. Which of the following is a true statement?
    a. The diversity of life is dependent upon a stable environment.
    b. The nature and diversity of life have changed over time.
    c. Earth has existed and changed over a few thousand years.
    d. Ancestral forms of life were very similar to the organisms that currently exist.
    e. Giraffes have a long neck because of a need to reach the taller branches of a tree.
    *Answer: b*

42. An evolutionary tree
    a. shows evolutionary relationships.
    b. places the most closely related groups on the same branch of the tree.
    c. places the organisms that share a common ancestor on the same branch of the tree.
    d. shows the order in which populations split and evolved into new species.
    e. All of the above
    *Answer: e*

43. Data for assembling the Tree of Life (ATOL) is obtained
    from
    a. fossils.
    b. DNA sequencing.
    c. information technology.
    d. Both a and c
    e. a, b and c
    *Answer: e*

**\*\***44. From the diagram  below it can be determined that

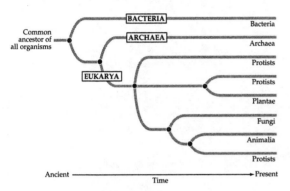

a. Eukarya are divided into three groups.
b. protists share a common ancestor with bacteria.
c. plants and animals share a common ancestor.
d. plants, fungi, and animals are descendants of different
   protist ancestors.
e. Archaea and Bacteria share a common ancestor.
*Answer: d*

# 2 Life and Chemistry: Small Molecules

## Fill in the Blank

1. Every atom except _____ has one or more neutrons in its nucleus.
   *Answer: hydrogen*

2. The sum of the atomic weights in any given molecule is called its _____.
   *Answer: molecular weight*

3. _____ occurs when one atom, such as $^{14}C$, is transformed into another atom, such as $^{14}N$, with an accompanying emission of energy.
   *Answer: Radioactive decay*

4. The nutritionist's Calorie, which biologists call a kilocalorie, is the equivalent of _____ heat-energy calories.
   *Answer: 1,000*

5. The water strider skates along the surface of water due to a property of liquids called _____.
   *Answer: surface tension*

6. The chemical properties of an element are determined by the number of _____ its atoms contain.
   *Answer: electrons*

7. The attraction between a slight positive charge on a hydrogen atom and the slight negative charge of a nearby atom is a _____.
   *Answer: hydrogen bond*

8. A chemical reaction that can proceed in either direction is called a _____.
   *Answer: reversible reaction*

9. A _____ is two or more atoms linked by chemical bonds.
   *Answer: molecule*

10. A _____ is the amount of heat needed to raise the temperature of 1g of pure water from 14.5°C to 15.5°C.
    *Answer: calorie*

## Multiple Choice

1. A 1.0 M solution of HCl has a pH of
   a. 1.0.
   b. 7.0.
   c. 14.0.
   d. 11.2.
   e. 14
   *Answer: a*

2. The part of the atom that determines how the atom behaves chemically is the
   a. proton.
   b. electron.
   c. neutron.
   d. innermost shell.
   e. nucleus.
   *Answer: b*

3. Hydrogen, deuterium, and tritium all have the same
   a. atomic weight.
   b. atomic number.
   c. mass.
   d. density.
   e. nuclear composition.
   *Answer: b*

4. Which component of an atom does *not* significantly add to the mass of an atom?
   a. Proton
   b. Neutron
   c. Electron
   d. Both a and b
   e. a, b, and c
   *Answer: c*

5. Which of the following describes an unusual property of water?
   a. Water will not react with other atoms.
   b. Water's solid state is more dense than its liquid state.
   c. Energy is not required to change water from a solid to a liquid.
   d. Little heat energy is needed to raise the temperature of water.
   e. The hydrogen bonds between water molecules continually form and break.
   *Answer: e*

6. In addition to covalent and ionic bonds, which of the following interactions are important in biological systems?
   a. van der Waals interactions
   b. Hydrogen bonds
   c. Hydrophobic interactions
   d. Both a and b
   e. a, b, and c
   *Answer: e*

■7. What is the difference between an atom and an element?
   a. An atom is made of protons, electrons, and sometimes neutrons; an element is a substance composed of only one kind of atom.
   b. An element is made of protons, electrons, and sometimes neutrons; an atom is a substance composed of only one kind of element.
   c. An atom does not contain electrons, whereas an element does.
   d. An atom contains protons and electrons, whereas an element contains protons, electrons, and neutrons.
   e. None of the above
   *Answer: a*

8. The number of protons in an atom equals the number of
   a. neutrons.
   b. electrons.
   c. electrons plus neutrons.
   d. neutrons minus electrons.
   e. isotopes.
   *Answer: b*

9. Which of the following statements about atoms is true?
   a. An electron has a more negative charge than a proton has positive charge, which is why there are usually more protons than electrons in an atom.
   b. Neutrons simply add mass to an atom without ever influencing other properties.
   c. When protons equal electrons, an atom has a neutral charge.
   d. Atoms of an element are radioactive whenever they vary in their number of neutrons.
   e. The energy level of electrons is higher in shells close to the nucleus.
   *Answer: c*

10. Because atoms can have the same number of protons but a different number of neutrons, elements have
    a. isotopes.
    b. more than one atomic mass listed on the periodic table.
    c. more than one atomic number.
    d. various means of forming chemical bonds.
    e. isomers.
    *Answer: a*

11. An element has a weight of 131.3. The reason the number includes 0.3 is that
    a. the atomic weight includes the weight of electrons.
    b. atomic weight is the average of the mass numbers of all its isotopes.
    c. the neutrons do not have a single unit weight.
    d. the atomic weight does not include the weight of protons.
    e. the number of electrons may vary.
    *Answer: b*

12. The atomic number of an element is the same as the number of _____ in each atom.
    a. neutrons
    b. protons plus electrons
    c. protons
    d. neutrons plus protons
    e. neutrons plus electrons
    *Answer: c*

13. The atomic weight of an element is the same as the number of _____ in each atom.
    a. electrons
    b. protons
    c. neutrons
    d. protons plus neutrons
    e. electrons plus neutrons
    *Answer: d*

14. When $^7C$ decays by releasing a particle, it becomes
    a. significantly lighter.
    b. nitrogen, by gaining a proton.
    c. boron, by losing a proton.
    d. nitrogen, by losing a proton.
    e. $^{13}C$, by losing a neutron.
    *Answer: b*

15. The best reference for the atomic number and mass number for elements is
    a. a good chemistry text.
    b. a dictionary.
    c. the periodic table.
    d. a general physics book.
    e. a good biology text.
    Answer: c

*16. All except _____ follow the octet rule.
    a. sodium
    b. chlorine
    c. carbon
    d. hydrogen
    e. nitrogen
    *Answer: d*

17. In a hydrogen molecule, the two atoms are held together by
    a. hydrogen bonds.
    b. a shared pair of electrons.
    c. van der Waals forces.
    d. ionic attractions.
    e. gravity.
    *Answer: b*

18. Hydrogen bonds
    a. form between two hydrogen atoms.
    b. form between hydrogen and oxygen atoms within a molecule.
    c. form between different molecules.
    d. involve sharing of electrons.
    e. are the strongest bonds because of their length.
    *Answer: c*

19. What determines if a molecule is polar, nonpolar, or ionic?
    a. The number of protons
    b. The bond distances
    c. The differences in the electronegativities of the atoms
    d. The ionic charges
    e. The distance of the electrons from the nucleus
    *Answer: c*

20. Which pair has similar chemical properties?
    a. $^1H$ and $^{22}Na$
    b. $^{12}C$ and $^{28}Si$
    c. $^{16}O$ and $^8S$
    d. $^{12}C$ and $^{14}C$
    e. $^8H$ and $^2He$
    *Answer: d*

*21. $^{31}_{15}P$ and $^{32}_{15}P$ have virtually identical chemical and biological properties because they have the same
 a. half-life.
 b. number of neutrons.
 c. atomic weight.
 d. mass number.
 e. number of electrons.
 *Answer: e*

22. The number of different elements found in the universe is closest to
 a. 12.
 b. 24.
 c. 48.
 d. 100.
 e. 192.
 *Answer: d*

*23. Why is the atomic weight of hydrogen 1.008 and not exactly its mass number, 1.000?
 a. Atomic weight does not take into account the weight of the rare isotopes of an element.
 b. Atomic weight is the average of the mass numbers of a representative sample of the element, including all its isotopes.
 c. The atomic weight includes the weight of the electrons.
 d. The atomic weight does not include the weight of the protons.
 e. The mass number of an element is always lower than its atomic weight.
 *Answer: b*

24. Which of the following elements is the most chemically reactive?
 a. Hydrogen
 b. Helium
 c. Neon
 d. Argon
 e. They all have the same chemical reactivity.
 *Answer: a*

25. A single covalent chemical bond represents the sharing of how many electrons?
 a. One
 b. Two
 c. Three
 d. Four
 e. Six
 *Answer: b*

26. The molecular weight of water is 18.0154. One mole of water weighs exactly _____ grams.
 a. 9
 b. 18
 c. 18.0154
 d. 36.031
 e. $6.023 \times 10^{23}$
 *Answer: c*

27. Ionic bonds are
 a. attractions between oppositely charged ions.
 b. the result of electron sharing.
 c. the strongest of the chemical bonds.
 d. caused by partial electrical charges.
 e. dependent upon hydrophobic interactions.
 *Answer: a*

*28. What is the difference between covalent and ionic bonds?
 a. Covalent bonds are the sharing of neutrons, whereas ionic bonds are the sharing of electrons.
 b. Ionic bonds are the sharing of electrons between atoms, whereas covalent bonds are the electric attraction between two atoms.
 c. Covalent bonds are the sharing of protons between atoms, whereas ionic bonds are the electric attraction between two atoms.
 d. Covalent bonds are the sharing of protons between atoms, whereas ionic bonds are the sharing of electrons between two atoms.
 e. Covalent bonds are the sharing of electrons between atoms, whereas ionic bonds are the transfer of electrons from one atom to another.
 *Answer: e*

29. Which contains more molecules, a mole of hydrogen or a mole of carbon?
 a. A mole of carbon
 b. A mole of hydrogen
 c. Both contain the same number of molecules.
 d. Inadequate information is provided.
 e. It depends on atmospheric pressure.
 *Answer: c*

30. Sweating is a useful cooling device for humans because water
 a. takes up a great deal of heat in changing from its liquid state to its gaseous state.
 b. takes up a great deal of heat in changing from its solid state to its liquid state.
 c. can exist in three states at temperatures common on Earth.
 d. is an outstanding solvent.
 e. ionizes readily.
 *Answer: a*

31. When exposed to extreme heat, the human body relies on _____ to absorb excess calories of heat and maintain normal body temperature.
 a. evaporation
 b. condensation
 c. respiration
 d. transpiration
 e. All of the above
 *Answer: a*

*32. Two characteristics of water make it different from most other compounds. Its solid state is _____ its liquid state, and it takes up _____ heat to change to its gaseous state.
 a. less dense than; large amounts of
 b. more dense than; small amounts of
 c. less dense than; small amounts of
 d. more dense than; large amounts of
 e. just as dense as; no
 *Answer: a*

33. Specific heat is defined as
 a. the heat released from ice during freezing.
 b. the heat needed to raise the temperature of water.
 c. one calorie.
 d. the calories needed to raise a gram 1°C.
 e. 4.184 joules.
 *Answer: d*

■34. Which characteristic of water contributes to the relatively constant temperatures of the oceans?
   a. Water ionizes only slightly.
   b. It takes a small amount of heat energy to raise the temperature of water.
   c. Water can contain large amounts of salt.
   d. Water has the ability to ionize readily.
   e. It takes a large amount of heat energy to raise the temperature of water.
   *Answer: e*

35. Ice floats because
   a. the crystal takes up more space than the liquid.
   b. substances expand when cooled.
   c. heat is released and heat makes water expand.
   d. hydrogen bonds must break.
   e. heat is absorbed.
   *Answer: a*

36. Ice is used in beverages because
   a. it is composed only of water.
   b. it is cold.
   c. more water is needed in the drink.
   d. people like to chew it.
   e. it absorbs a lot of heat when it melts.
   *Answer: e*

■37. If you place a paper towel in a dish of water, the water will move up the towel by capillary action. Which property of water gives rise to capillary action?
   a. Water molecules ionize.
   b. Water is a good solvent.
   c. Water molecules have hydrophobic interactions.
   d. Water can form hydrogen bonds.
   e. Water takes up large amounts of heat when it vaporizes.
   *Answer: d*

38. An alkaline solution contains
   a. more $OH^-$ ions than $H^+$ ions.
   b. more $H^+$ ions than $OH^-$ ions.
   c. the same number of $OH^-$ ions as $H^+$ ions.
   d. no $OH^-$ ions.
   e. None of the above
   *Answer: a*

39. The pH 6.0 contains
   a. $10^6$ hydrogen ions.
   b. $6^{10}$ hydrogen ions.
   c. $6^{10}$ moles of hydrogen ions.
   d. more $OH^-$ than $H^+$.
   e. $10^{-6}$ moles of hydrogen ions.
   *Answer: e*

40. Solutions that contain buffers tend to resist pH changes because buffers
   a. are bases.
   b. change from ionic to non-ionic in solution.
   c. change from non-ionic to ionic in response to changes in pH and release or absorb $H^+$.
   d. are weak acids or bases.
   e. are ionic, polar molecules that add or absorb $H^+$ in solutions.
   *Answer: c*

■41. What is the difference between an acid and a base?
   a. An acid undergoes a reversible reaction, whereas a base does not.
   b. An acid releases $OH^-$ ions in solution, whereas a base accepts $OH^-$ ions.
   c. An acid releases $H^+$ ions in solution, whereas a base releases $OH^-$ ions.
   d. An acid releases $OH^-$ ions in solution, whereas a base releases $H^+$ ions.
   e. An acid releases $H^+$ ions in solution, whereas a base accepts $H^+$ ions.
   *Answer: e*

42. All of the following are nonpolar *except*
   a. $O_2$.
   b. $N_2$.
   c. $CH_4$.
   d. NaCl.
   e. $H_2$.
   *Answer: d*

43. Polar molecules
   a. have an overall negative electric charge.
   b. have an equal distribution of electric charge.
   c. have an overall positive electric charge.
   d. have an unequal distribution of electric charge.
   e. are ions.
   *Answer: d*

44. The electron cloud of a water molecule
   a. is equally dense throughout the molecule.
   b. is most dense near the hydrogen atoms.
   c. is most dense near the oxygen atom.
   d. covers only the positive portion of the molecule.
   e. is separated by an angle of 104.5°.
   *Answer: c*

45. All of the following are consequences of hydrogen bonding *except*
   a. the attraction between water molecules.
   b. surface tension.
   c. cohesion.
   d. capillary action.
   e. Avogadro's number.
   *Answer: e*

46 A van der Waals interaction is an attraction between
   a. the electrons of one molecule and the nucleus of a nearby molecule.
   b. the electrons of one molecule and the nucleus of the same molecule.
   c. the electrons of one molecule and the electrons of a nearby molecule.
   d. nonpolar molecules, due to the exclusion of water.
   e. nonpolar molecules, because they are surrounded by water molecules.
   *Answer: a*

47. Oil remains as a droplet in water because of
   a. the van der Waals interactions of the nonpolar oil molecules.
   b. the hydrophobic interactions of the nonpolar oil molecules.
   c. the hydrogen bonds formed between the nonpolar molecules of the oil and the water molecules.
   d. the covalent bonds formed between the nonpolar molecules of the oil.
   e. the covalent bonds formed between the nonpolar molecules of the oil and the water molecules.
   *Answer: b*

▪48. A drop of oil in water disperses when detergent is added because
 a. the nonpolar parts of the detergent molecules associate with the oil and the polar parts of the detergent molecules associate with the water.
 b. the polar parts of the detergent molecules associate with the oil and the nonpolar parts of the detergent molecules associate with the water.
 c. the nonpolar parts of the detergent molecules associate with the oil and with the water.
 d. the polar parts of the detergent molecules associate with the oil and with the water.
 e. the detergent lowers the surface tension of the water.
 *Answer: a*

49. When potassium hydroxide (KOH) is added to water, it ionizes, releasing hydroxide ions. The resulting solution is
 a. acidic.
 b. basic.
 c. neutral.
 d. molar.
 e. a buffer.
 *Answer: b*

50. $H_2SO_4$ can ionize to yield two $H^+$ ions and one $SO_4^{2-}$ ion. $H_2SO_4$ is
 a. molar.
 b. a base.
 c. a buffer.
 d. a solution.
 e. an acid.
 *Answer: e*

51. Acid rain is a serious environmental problem. A sample of rainwater collected in the Adirondack Mountains had an $H^+$ concentration of $10^{-4}$ mol/L. The pH of this sample was
 a. .0001.
 b. −4.
 c. 4.
 d. 0.
 e. 10,000.
 *Answer: c*

▪52. Carbonic acid and sodium bicarbonate act as buffers in the blood. When a small amount of acid is added to this buffer, the $H^+$ ions are used up as they combine with the bicarbonate ions. When this happens, the pH of the blood
 a. becomes basic.
 b. becomes acidic.
 c. doesn't change.
 d. is reversible.
 e. ionizes.
 *Answer: c*

53. Cholesterol is composed primarily of carbon and hydrogen atoms. Which property would you expect cholesterol to have?
 a. It is insoluble in water.
 b. It is soluble in water.
 c. It is a base.
 d. It is an acid.
 e. It is a buffer.
 *Answer: a*

54. The compound inositol has six hydroxyl groups attached to a six-carbon backbone. Thus inositol can be classified as a(n)
 a. amine.
 b. acid.
 c. ketone.
 d. alcohol.
 e. buffer.
 *Answer: d*

55. Butane and isobutane have the same chemical formula but different arrangements of atoms. These two compounds are called
 a. ionic.
 b. alcohols.
 c. functional groups.
 d. amines.
 e. isomers.
 *Answer: e*

56. One dalton is the same as the mass of one
 a. electron.
 b. carbon atom.
 c. proton.
 d. gram.
 e. charge unit.
 *Answer: c*

57. Oxygen and carbon are defined as different elements because they have atoms with a different
 a. number of electrons.
 b. number of protons.
 c. number of neutrons.
 d. mass.
 e. charge.
 *Answer: b*

58. The mass of an atom is determined primarily by the _____ it contains.
 a. number of electrons
 b. number of protons
 c. sum of the number of protons and electrons
 d. sum of the number of protons and neutrons
 e. number of charges
 *Answer: d*

*59. Two atoms are held together in four covalent bonds because of forces between the
 a. electrons and protons.
 b. electrons.
 c. protons and neutrons.
 d. protons.
 e. electrons and neutrons.
 *Answer: b*

60. Two carbon atoms held together in a double covalent bond share _____ electron(s).
 a. one
 b. two
 c. four
 d. six
 e. eight
 *Answer: c*

61. Particles having a net negative charge are called
    a. electronegative.
    b. cations.
    c. anions.
    d. acids.
    e. bases.
    *Answer: c*

62. Avogadro's number is the number of
    a. grams in a mole.
    b. molecules in a mole.
    c. moles in gram molecular weight.
    d. molecules in a gram.
    e. moles in a gram.
    *Answer: b*

63. A mole is
    a. $6.02 \times 10^{23}$ molecules.
    b. the molecular weight of a compound.
    c. an abbreviation for molecule.
    d. $6.02 \times 10^{-23}$ atoms.
    e. None of the above
    *Answer: a*

64. To determine the number of molecules in a teaspoon of sugar you need to know
    a. the density of the sugar.
    b. the weight of the sugar.
    c. the molecular weight of the sugar.
    d. Avogadro's number.
    e. the weight and molecular weight of the sugar, and Avogadro's number.
    *Answer: e*

**65. How would you make 100 ml of an aqueous solution with a 0.25 M concentration of a compound that has a molecular weight of 200 daltons?
    a. Add 0.25 grams of the compound to 100 ml of water.
    b. Add 250 grams of the compound to 100 ml of water.
    c. Take 250 grams of the compound and add water until the volume equals 100 ml.
    d. Take 50 grams of the compound and add water until the volume equals 100 ml.
    e. Take 5 grams of the compound and add water until the volume equals 100 ml.
    *Answer: e*

66. Of the following compounds containing $^1H$, $^{12}C$, and $^{16}O$, the one with the greatest number of molecules in a sample with mass 2 grams would be
    a. CO.
    b. $CO_2$.
    c. HCOOH.
    d. $C_2H_5OH$.
    e. $C_6H_{12}O_6$.
    *Answer: a*

67. Which of the following molecules is held together primarily by ionic bonds?
    a. $H_2O$
    b. $C_6H_{12}O_6$
    c. NaCl
    d. $H_2$
    e. $NH_3$
    *Answer: c*

68. The hydrogen bond between two water molecules arises because water is
    a. polar.
    b. nonpolar.
    c. a liquid.
    d. a small molecule.
    e. hydrophobic.
    *Answer: a*

69. The functional group written as —COOH is called the _____ group.
    a. hydroxyl
    b. carbonyl
    c. amino
    d. ketone
    e. carboxyl
    *Answer: e*

70. The functional group diagramed =O is a(n)
    a. carbonyl.
    b. carboxyl.
    c. aldehyde.
    d. amino.
    e. hydroxyl.
    *Answer: a*

71. The more acidic of two solutions has
    a. more hydroxyl ions.
    b. more hydrogen acceptors.
    c. more $H^+$ ions per liter.
    d. a higher pH.
    e. None of the above
    *Answer: c*

72. Which of the following atoms usually has the greatest number of covalent bonds with other atoms?
    a. Carbon
    b. Oxygen
    c. Sulfur
    d. Hydrogen
    e. Nitrogen
    *Answer: a*

73. Of the following atomic configurations, the one that has an atomic mass of 14 is the atom with
    a. 14 neutrons.
    b. 14 electrons.
    c. 7 neutrons and 7 electrons.
    d. 7 protons and 7 electrons.
    e. 6 protons and 8 neutrons.
    *Answer: e*

74. An atom that is neutrally charged contains
    a. only neutrons.
    b. the same number of neutrons as electrons.
    c. the same number of neutrons as protons.
    d. the same number of positive particles as negative particles.
    e. no charged particles.
    *Answer: d*

75. The four elements most common in organisms are
    a. calcium, iron, hydrogen, and oxygen.
    b. water, carbon, hydrogen, and oxygen.
    c. carbon, oxygen, hydrogen, and nitrogen.
    d. nitrogen, carbon, iron, and hydrogen.
    e. phosphorus, water, carbon, and oxygen.
    *Answer: c*

76. The notation [H⁺] refers to the
    a. number of H⁺ ions present in a solution.
    b. number of protons in an H⁺ ion.
    c. charge of an H⁺ ion.
    d. concentration of H⁺ ions in moles per liter.
    e. chemical reactivity of H⁺ ions.
    *Answer: d*

77. Of the following types of chemical bonds, the strongest bond in biological systems is the _____ bond.
    a. hydrogen
    b. van der Waals
    c. ionic
    d. acidic
    e. covalent
    *Answer: e*

▪78. Select the statement that describes a difference between ionic bonds and covalent bonds.
    a. An ionic bond is stronger.
    b. Electron sharing is more equal in the covalent bond.
    c. An ionic bond occurs more often in aqueous solutions.
    d. An ionic bond occurs only in acids.
    e. A covalent bond occurs only in nonpolar molecules.
    *Answer: b*

79. Oxygen forms _____ bond(s), carbon forms _____, and hydrogen forms _____.
    a. one; four; one
    b. four; four; four
    c. two; four; none
    d. two; four; one
    e. two; two; two
    *Answer: d*

80. Of the following types of molecules, the one that always contains nitrogen is
    a. thiol.
    b. sugar.
    c. hydrocarbon.
    d. alcohol.
    e. amino acid.
    *Answer: e*

81. A typical distance between two atoms held together by a covalent bond is
    a. 1 millimeter.
    b. 2 micrometers.
    c. 108 meters.
    d. 0.2 nanometer.
    e. $10^{-12}$ meter.
    *Answer: d*

82. Surface tension and capillary action occur in water because it
    a. is wet.
    b. is dense.
    c. has hydrogen bonds.
    d. is nonpolar.
    e. has ionic bonds.
    *Answer: c*

83. Which of the following is the correct order for the relative strengths of chemical bonds in decreasing order?
    a. Covalent, ionic, hydrogen, van der Waals forces
    b. Ionic, covalent, hydrogen, van der Waals forces
    c. van der Waals forces, covalent, ionic, hydrogen
    d. Hydrogen, covalent, van der Waals forces, ionic
    e. Ionic, covalent, van der Waals forces, hydrogen
    *Answer: a*

84. Probes indicating that there is a dry lake bed and trapped water beneath the poles on Mars are of major significance because they suggest that
    a. life may exist on Mars.
    b. climate patterns change with time.
    c. water is a moderator of temperature changes.
    d. Earth began forming 4.6 billion years ago.
    e. life originated in water.
    *Answer: a*

85. The ability of atoms to combine with other atoms is determined by
    a. the atom's atomic weight.
    b. the number and distribution of electrons.
    c. the atom's ability to form isomers.
    d. the atom's nucleus.
    e. pH.
    *Answer: b*

86. Which two functional groups are present in amino acids?
    a. Aldehyde and ketone
    b. Ketone and alcohol
    c. Alcohol and carboxyl
    d. Carboxyl and amine
    e. Amine and aldehyde
    *Answer: d*

87. Chemical bonds formed by electrical attractions are
    a. covalent.
    b. ionic.
    c. hydrogen bonds.
    d. van der Waals forces.
    e. Both b and c
    *Answer: e*

88. Phosphorus has an atomic number of 15 and an atomic weight of 30.974. How many neutrons does phosphorus have?
    a. 5
    b. 16
    c. 30
    d. 31
    e. 47
    *Answer: b*

89. From the information given in the previous question, it can be determined that this element
    a. has isotopes.
    b. forms isomers.
    c. has a pH of 7.
    d. is an ion.
    e. is radioactive.
    *Answer: a*

90. Which of the following has the greatest concentration of hydrogen ions?
    a. Household ammonia at pH 11
    b. Baking soda at pH 9
    c. Human blood at pH 7
    d. Black coffee at pH 5
    e. Cola at pH 3
    *Answer: e*

91. Molecular behavior is influenced by
    a. the densely packed nucleus.
    b. the presence of functional groups.
    c. the existence of isotopes.
    d. buffers.
    e. All of the above
    *Answer: b*

92. The optimum pH for growing strawberries is 6.5, whereas the optimum pH for growing blueberries is 4.5. Therefore, the number of hydrogen ions needed to grow strawberries is _____ the number needed for blueberries.
    a. 2 times
    b. 10 times
    c. 100 times
    d. 1000 times
    e. 1,000,000 times
    *Answer: c*

# 3 Life and Chemistry: Large Molecules

## Fill in the Blank

1. Fluidity and melting point of fatty acids are partially determined by the number of _____ bonds.
**Answer: unsaturated (or carbon double)**

2. Many monosaccharides like fructose, mannose, and galactose have the same chemical formula as glucose ($C_6H_{12}O_6$), but the atoms are combined differently to yield different structural arrangements. These varying forms of the same chemical formula are called _____.
**Answer: isomers**

3. The highly branched polysaccharide that stores glucose in the muscle and the liver of animals is _____.
**Answer: glycogen**

4. In proteins, amino acids are linked together by _____ bonds.
**Answer: peptide**

5. The only amino acid that has no stereoisomer is _____.
**Answer: glycine**

6. The amino acid that most limits rotation around the α carbon is _____.
**Answer: proline**

7. The bonds between the units in a carbohydrate polymer are called _____ bonds.
**Answer: glycosidic**

8. The linear arrangement of amino acids in the polypeptide chain is referred to as the _____ structure of the protein.
**Answer: primary**

9. Starch is a polymer of glucose subunits. The subunits of any polymer are called _____.
**Answer: monomers**

10. Fatty acids with more than one carbon–carbon double bond are called _____.
**Answer: polyunsaturated**

11. Cholesterol, vitamin D, and testosterone all have a multiple-ring structure and are members of a family of lipids known as _____.
**Answer: steroids**

12. Carbohydrates made up of two simple sugars are called _____.
**Answer: disaccharides**

13. The covalent bond forces between the sulfur atoms of two cysteine side chains is called a _____.
**Answer: disulfide bridge**

14. Disulfide bonds can form between _____ residues in proteins.
**Answer: cysteine**

15. All amino acids have a hydrogen atom, a carboxyl group, and an amino group attached to a carbon atom. The variability in the 20 different amino acids lies in the structure of their _____.
**Answer: R group**

16. A(n) _____ linkage connects the fatty acid molecule to glycerol.
**Answer: ester**

17. The _____ molecules found in humans are the same as those found in tomato plants.
**Answer: glucose**

18. The bonds that link sugar monomers together in a starch molecule are _____ bonds.
**Answer: glycosidic**

19. Cholesterol is classified as a(n) _____.
**Answer: lipid**

★20. The reaction $A — H + B — OH \rightarrow A — B + H_2O$ represents a _____.
**Answer: condensation reaction**

## Multiple Choice

1. The major classes of biologically significant large molecules include all of the following *except*
   a. proteins.
   b. nucleic acids.
   c. carbohydrates.
   d. lipids.
   e. triglycerides.
   **Answer: e**

2. Lipids are
   a. insoluble in water.
   b. readily soluble in organic solvents.
   c. characterized by their solubility.
   d. important constituents of biological membranes.
   e. All of the above
   **Answer: e**

3. Which of the following is characteristic of proteins?
   a. They are insoluble in water.
   b. They are the structural units of glycogen.
   c. They possess glycosidic linkages between amino acids.
   d. Some function as enzymes.
   e. a, b, and c
   *Answer: d*

4. Molecules with molecular weights greater than 1,000 daltons are usually called
   a. proteins.
   b. polymers.
   c. nucleic acids.
   d. macromolecules.
   e. monomers.
   *Answer: d*

5. Polymerization reactions in which proteins are synthesized from amino acids
   a. require the formation of phosphodiester bonds between the amino acids.
   b. occur in the nucleus of the cell.
   c. are hydrolysis reactions.
   d. depend upon van der Waals forces to hold the amino acids together.
   e. result in the formation of water.
   *Answer: e*

6. In condensation reactions, the atoms that make up a water molecule are derived from
   a. oxygen.
   b. only one of the reactants.
   c. both of the reactants.
   d. carbohydrates.
   e. enzymes.
   *Answer: c*

7. Which of the following is *not* a macromolecule?
   a. RNA
   b. DNA
   c. An enzyme
   d. A protein
   e. Salt
   *Answer: e*

8. The bonds that form between the units of polymeric macromolecules are _____ bonds.
   a. hydrogen
   b. peptide
   c. disulfide
   d. covalent
   e. ionic
   *Answer: d*

9. Which of the following is *not* a characteristic of lipids?
   a. They are readily soluble in water.
   b. They are soluble in organic solvents.
   c. They release large amounts of energy when broken down.
   d. They form two layers when mixed with water.
   e. They act as an energy storehouse.
   *Answer: a*

■10. You have isolated an unidentified liquid from a sample of beans. You add the liquid to a beaker of water and shake vigorously. After a few minutes, the water and the other liquid separate into two layers. To which class of large biological molecules does the unknown liquid most likely belong?
   a. Carbohydrates
   b. Lipids
   c. Proteins
   d. Enzymes
   e. Nucleic acids
   *Answer: b*

■■11. Lipids form the barriers surrounding various compartments within an organism. Which property of lipids makes them a good barrier?
   a. Many biologically important molecules are not soluble in lipids.
   b. Lipids are polymers.
   c. Lipids store energy.
   d. Triglycerides are lipids.
   e. Lipids release large amounts of energy when broken down.
   *Answer: a*

■■12. You look at the label on a container of shortening and see "hydrogenated vegetable oil." This means that during processing the number of carbon–carbon double bonds in the oil was decreased. What is the result of decreasing the number of double bonds?
   a. The oil now has a lower melting point.
   b. The oil is now a solid at room temperature.
   c. There are more "kinks" in the fatty acid chains.
   d. The oil is now a derivative carbohydrate.
   e. The fatty acid is now a triglyceride.
   *Answer: b*

13. The portion of a phospholipid that contains the phosphorous group has one or more electric charges. That makes this region of the molecule
   a. hydrophobic.
   b. hydrophilic.
   c. nonpolar.
   d. unsaturated.
   e. saturated.
   *Answer: b*

■14. Cholesterol is soluble in ether, an organic solvent, but it is not soluble in water. Based on this information, what class of biological macromolecules does cholesterol belong to?
   a. Nucleic acids
   b. Carbohydrates
   c. Proteins
   d. Enzymes
   e. Lipids
   *Answer: e*

15. Which of the following is *not* a function in which lipids play an important role?
   a. Vision
   b. Storing energy
   c. Membrane structure
   d. Storing genetic information
   e. Chemical signaling
   *Answer: d*

16. In a biological membrane, the phospholipids are arranged with the fatty acid chains facing the interior of the membrane. As a result, the interior of the membrane is
   a. hydrophobic.
   b. hydrophilic.

c. charged.
d. polar.
e. filled with water.
*Answer: a*

17. The monomers that make up polymeric carbohydrates like starch are called
    a. nucleotides.
    b. trisaccharides.
    c. monosaccharides.
    d. nucleosides.
    e. fatty acids.
    *Answer: c*

18. The atoms that make up carbohydrates are
    a. C, H, and N.
    b. C and H.
    c. C, H, and P.
    d. C, H, and O.
    e. C, H, O, and N.
    *Answer: d*

19. Glucose and fructose both have the formula $C_6H_{12}O_6$, but the atoms in these two compounds are arranged differently. Glucose and fructose are known as
    a. isomers.
    b. polysaccharides.
    c. oligosaccharides.
    d. pentoses.
    e. steroids.
    *Answer: a*

20. A nucleotide contains a pentose, a phosphate, and a(n)
    a. lipid.
    b. acid.
    c. nitrogen-containing base.
    d. amino acid.
    e. glycerol.
    *Answer: c*

21. A simple sugar with the formula $C_5H_{10}O_5$ can be classified as a
    a. hexose.
    b. polysaccharide.
    c. disaccharide.
    d. pentose.
    e. lipid.
    *Answer: d*

22. Lactose, or milk sugar, is composed of one glucose unit and one galactose unit. It can be classified as a
    a. disaccharide.
    b. hexose.
    c. pentose.
    d. polysaccharide.
    e. monosaccharide.
    *Answer: a*

23. Two important polysaccharides made up of glucose monomers are
    a. guanine and cytosine.
    b. RNA and DNA.
    c. sucrose and lactose.
    d. cellulose and starch.
    e. testosterone and cortisone.
    *Answer: d*

24. Polysaccharides that serve as energy storage molecules tend to have _____ linkages.

a. α-1,4
b. α-2,3
c. β-1,4
d. β-2,3
e. Both a and c
*Answer: e*

25. Cellulose is the most abundant organic compound on Earth. Its main function is
    a. to store genetic information.
    b. as a storage compound for energy in plant cells.
    c. as a storage compound for energy in animal cells.
    d. as a component of biological membranes.
    e. to provide mechanical strength to plant cell walls.
    *Answer: e*

26. In animals, glucose is stored in the compound
    a. cellulose.
    b. amylose.
    c. glycogen.
    d. fructose.
    e. cellobiose.
    *Answer: c*

27. Which of the following monomer/polymer pairs is *not* correct?
    a. Monosaccharide/polysaccharide
    b. Amino acid/protein
    c. Triglyceride/lipid
    d. Nucleotide/DNA
    e. Nucleotide/RNA
    *Answer: c*

28. Amino acids can be classified by the
    a. number of monosaccharides they contain.
    b. number of carbon–carbon double bonds in their fatty acids.
    c. number of peptide bonds they can form.
    d. number of disulfide bridges they can form.
    e. characteristics of their side chains.
    *Answer: e*

29. During the formation of a peptide linkage, which of the following occurs?
    a. A molecule of water is formed.
    b. A disulfide bridge is formed.
    c. A hydrophobic bond is formed.
    d. A hydrophilic bond is formed.
    e. An ionic bond is formed.
    *Answer: a*

30. The side chain of leucine is a hydrocarbon. In a folded protein, where would you expect to find leucine?
    a. In the interior of a cytoplasmic enzyme
    b. On the exterior of a protein embedded in a membrane
    c. On the exterior of a cytoplasmic enzyme
    d. Both a and b
    e. Both a and c
    *Answer: d*

31. What is the theoretical number of different proteins that you could make from 50 amino acids?
    a. $50^{20}$
    b. $20 \times 50$
    c. $20^{50}$
    d. $10^{50}$
    e. $2^{50}$
    *Answer: c*

32. The shape of a folded protein is often determined by
    a. its tertiary structure.
    b. the sequence of its amino acids.
    c. whether the peptide bonds have α or β linkages.
    d. the number of peptide bonds.
    e. the base-pairing rules.
    *Answer: b*

33. The amino acids of the protein keratin are arranged in a helix. This secondary structure is stabilized by
    a. covalent bonds.
    b. peptide bonds.
    c. glycosidic linkages.
    d. polar bonds.
    e. hydrogen bonds.
    *Answer: e*

34. What is the nucleotide sequence of the complementary strand of this DNA molecule: A A T G C G A?
    a. T T A C G C T
    b. A A T G C G A
    c. G G C A T A G
    d. C C G T T A T
    e. A G C G T A A
    *Answer: a*

35. Which of the following is *not* a difference between DNA and RNA?
    a. DNA has thymine, whereas RNA has uracil.
    b. DNA usually has two polynucleotide strands, whereas RNA usually has one strand.
    c. DNA has deoxyribose sugar, whereas RNA has ribose sugar.
    d. DNA is a polymer, whereas RNA is a monomer.
    e. In DNA, A pairs with T, whereas in RNA, A pairs with U.
    *Answer: d*

36. DNA molecules that carry different genetic information can be distinguished by looking at
    a. the number of strands in the helix.
    b. how much uracil is present.
    c. the sequence of nucleotide bases.
    d. differences in the base-pairing rules.
    e. the shape of the helix.
    *Answer: c*

37. The "backbone" of nucleic acid molecules is made of
    a. nitrogenous bases.
    b. alternating sugars and phosphate groups.
    c. purines.
    d. pyrimidines.
    e. nucleosides.
    *Answer: b*

38. According to the base-pairing rules for nucleic acids, purines always pair with
    a. deoxyribose sugars.
    b. uracil.
    c. pyrimidines.
    d. adenine.
    e. guanine.
    *Answer: c*

39. What type of amino acid side chains would you expect to find on the surface of a protein embedded in a cell membrane?
    a. Cysteine

b. Hydrophobic
c. Hydrophilic
d. Charged
e. Polar, but not charged
*Answer: b*

40. A molecule with the formula $C_{16}H_{32}O_2$ is a
    a. hydrocarbon.
    b. carbohydrate.
    c. lipid.
    d. protein.
    e. nucleic acid.
    *Answer: c*

41. A molecule with the formula $C_{16}H_{30}O_{15}$ is a
    a. hydrocarbon.
    b. carbohydrate.
    c. lipid.
    d. protein.
    e. nucleic acid.
    *Answer: b*

42. Fatty acids are molecules that
    a. contain fats bonded to a glycerol.
    b. are composed of hydrogen, carbon, and a carboxyl group.
    c. are carbohydrates linked to a hydrocarbon chain.
    d. contain glycerol and a carboxyl group.
    e. are always saturated.
    *Answer: b*

43. Sucrose is a
    a. hexose.
    b. lipid.
    c. disaccharide.
    d. glucose.
    e. simple sugar.
    *Answer: c*

44. DNA and RNA contain
    a. pentoses.
    b. hexoses.
    c. fructoses.
    d. maltoses.
    e. amyloses.
    *Answer: a*

45. The 20 different common amino acids have different
    a. amino groups.
    b. R groups.
    c. acid groups.
    d. peptide linkages.
    e. primary structures.
    *Answer: b*

46. The primary structure of a protein is determined by its
    a. disulfide bridges.
    b. α helix structure.
    c. sequence of amino acids.
    d. branching.
    e. three-dimensional structure.
    *Answer: c*

47. When a protein loses its three-dimensional structure and becomes nonfunctional, it is
    a. permanent.
    b. reversible.
    c. denatured.
    d. hydrolyzed.

e. environmentalized.

*Answer: c*

*48. A β pleated sheet organization in a polypeptide chain is an example of _____ structure.
   a. primary
   b. secondary
   c. tertiary
   d. quaternary
   e. coiled

*Answer: b*

49. A protein can best be defined as a polymer
   a. of amino acids.
   b. containing one or more polypeptide chains.
   c. containing 20 amino acids.
   d. containing 20 peptide linkages.
   e. containing double helices.

*Answer: a*

50. The four nitrogenous bases of RNA are abbreviated as
   a. A, G, C, and T.
   b. A, G, T, and N.
   c. G, C, U, and N.
   d. A, G, U, and T.
   e. A, G, C, and U.

*Answer: e*

51. Polysaccharides, polypeptides, and polynucleotides all
   a. contain simple sugars.
   b. are formed in condensation reactions.
   c. are found in cell membranes.
   d. contain nitrogen.
   e. have molecular weights less than 30,000 daltons.

*Answer: b*

52. DNA carries genetic information in its
   a. helical form.
   b. sequence of bases.
   c. tertiary sequence.
   d. sequence of amino acids.
   e. phosphate groups.

*Answer: b*

53. A molecule often spoken of as having a head and tail is a(n)
   a. phospholipid.
   b. oligosaccharide.
   c. RNA.
   d. steroid.
   e. triglyceride.

*Answer: a*

54. A molecule that has an important role in limiting what gets into and out of cells is
   a. glucose.
   b. maltose.
   c. phospholipid.
   d. fat.
   e. phosphohexose.

*Answer: c*

55. Waxes are formed by
   a. adding water to fatty acids.
   b. removing water from fatty acids.
   c. combining fatty acids with alcohol.
   d. condensing fatty acids with glycerol.
   e. vitamin P.

*Answer: c*

56. A molecule that has an important role in long-term storage of energy is
   a. a steroid.
   b. RNA.
   c. glycogen.
   d. an amino acid.
   e. hexose.

*Answer: c*

57. A peptide linkage (peptide bond) holds together two _____ molecules.
   a. protein
   b. amino acid
   c. sugar
   d. fatty acid
   e. phospholipid

*Answer: b*

58. In DNA molecules,
   a. purines pair with pyrimidines.
   b. A pairs with C.
   c. G pairs with A.
   d. purines pair with purines.
   e. C pairs with T.

*Answer: a*

59. A type of molecule very often drawn with a single six-sided ring structure is
   a. sucrose.
   b. an amino acid.
   c. glucose.
   d. a fatty acid.
   e. a steroid.

*Answer: c*

60. Maltose and lactose are similar in that they both are
   a. simple sugars.
   b. amino acids.
   c. insoluble in water.
   d. disaccharides.
   e. hexoses.

*Answer: d*

61. Starch and glycogen are different in that only one of them
   a. is a polymer of glucose.
   b. contains ribose.
   c. is made in plants.
   d. is an energy storage molecule.
   e. can be digested by humans.

*Answer: c*

62. Enzymes are
   a. DNA.
   b. lipids.
   c. carbohydrates.
   d. proteins.
   e. amino acids.

*Answer: d*

63. The type of bond that holds two amino acids together in a polypeptide chain is a(n)
   a. ionic bond.
   b. disulfide bridge.
   c. hydrogen bond.
   d. peptide linkage.
   e. dehydration bond.

*Answer: d*

64. Peptides have _____ and _____ end.
    a. a start; a stop
    b. a +; a –
    c. an N terminus; a C terminus
    d. a 5'; a 3'
    e. an A; a Z
    *Answer: c*

65. The _____ structure of a protein concerns the way separate polypeptides are assembled together.
    a. primary
    b. secondary
    c. tertiary
    d. quaternary
    e. helical
    *Answer: d*

66. Quaternary structure is found in proteins
    a. composed of subunits.
    b. of membranes.
    c. of the quadruple complex.
    d. that change over time.
    e. None of the above
    *Answer: a*

67. In DNA, A hydrogen bonds with T and G with C; these are examples of a specific type of reaction called
    a. complementary base pairing.
    b. a dehydration reaction.
    c. a reduction reaction.
    d. a hydrophobic reaction.
    e. a purine–purine reaction.
    *Answer: a*

68. A fat contains fatty acids and
    a. glycerol.
    b. a base.
    c. an amino acid.
    d. a phosphate.
    e. None of the above
    *Answer: a*

69. There are _____ different types of tripeptides (molecules with three amino acids linked together) that can exist using the 20 common amino acids.
    a. 3
    b. 20
    c. 60
    d. 900
    e. 8,000
    *Answer: e*

70. Chitin is a polymer of
    a. galactosamine.
    b. glucose.
    c. glucosamine.
    d. glycine.
    e. All of the above
    *Answer: c*

71. A type of protein that functions by helping other proteins to fold correctly is called
    a. foldzyme.
    b. renaturing protein.
    c. chaperonin.
    d. hemoglobin.
    e. denaturing protein.
    *Answer: c*

72. The composition of a protein is
    a. its three-dimensional structure.
    b. the number and kinds of amino acids present.
    c. the particular side chain or R group present.
    d. similar in every protein because every protein contains 20 different amino acids.
    e. universal because all proteins have an equal number of amino acids.
    *Answer: b*

73. Why does bread become hard and stale?
    a. Cellulose molecules aggregate in the absence of water.
    b. In the absence of water, unbranched starch forms hydrogen bonds between polysaccharides, which then aggregate.
    c. The release of carbon dioxide causes the bread to harden.
    d. Water and heat cause the polysaccharide chains to bind together.
    e. Mold growth interferes with α linkages, causing the bread to harden.
    *Answer: b*

74. The difference between α- and β- glucose is
    a. in the number of covalent bonds present.
    b. in the placement of OH and H atoms.
    c. in the type of R group attached to the terminal carbon.
    d. that α-glucose is polar, whereas β-glucose is nonpolar.
    e. that α-glucose is a pentose, whereas β-glucose is a hexose.
    *Answer: b*

75. Examination of meterorites has revealed that they contain the chemistry of life. All of the following have been found *except*
    a. purines and pyrimidines.
    b. amino acids.
    c. glucose.
    d. magnetite.
    e. polycyclic aromatic hydrocarbons.
    *Answer: c*

76. Examination of meteorites suggests that
    a. life is not limited to Earth.
    b. life originated in outer space from nonliving matter.
    c. comets brought Earth most of its water.
    d. meteorites brought life to Earth.
    e. they are responsible for the Earth's magnetic field.
    *Answer: a*

77. Miller-Urey showed that in any environment with conditions similar to those of Earth,
    a. inorganic molecules would react to form organic molecules.
    b. RNA would self-replicate.
    c. organic molecules would form primitive cells.
    d. an oxygen atmosphere would develop.
    e. DNA would be synthesized.
    *Answer: a*

78. Earth's early atmosphere contained all of the following gases *except*
    a. carbon dioxide.
    b. sulfur dioxide.
    c. hydrogen sulfide.
    d. oxygen.
    e. methane.
    *Answer d*

79. Incorrect folding of a protein can have serious consequences. For instance, an accumulation of misfolded proteins in the brain is a characteristic of
    a. rickets.
    b. Alzheimer's disease.
    c. hemophilia.
    d. excessively dry skin.
    e. night blindness.
    *Answer b*

80. It is important to know the exact shape of a protein because the knowledge allows scientists to
    a. create specific proteins to block the action of another protein.
    b. create multi-protein machines to synthesize RNA.
    c. synthesize antibodies that kill viruses.
    d. predict whether the protein will form an α or β chain.
    e. predict the energy required to break the protein's bonds.
    *Answer a*

81. The double helix structure of DNA is due to
    a. complementary base pairings.
    b. purines bonding with pyrimidines.
    c. the phosphodiester bonds between deoxyribose and phosphate.
    d. hydrogen bonding of the two complementary polynucleotide strands.
    e. ionic bonding of base pairs.
    *Answer d*

82. Spontaneous generation was disproved by
    a. Miller.
    b. Urey.
    c. Redi and Pasteur.
    d. Allan Hills.
    e. van der Waal.
    *Answer c*

83. RNA molecules that act as catalysts are called
    a. ribozymes.
    b. proteases.
    c. chaperonins.
    d. disulfide bridges.
    e. triglycerides.
    *Answer a*

# 4 Cells: The Basic Units of Life

## Fill in the Blank

1. _____ are a model of how cells may have originated.
   *Answer: Protobionts*

2. In biology, we call the basic unit of life the _____.
   *Answer: cell*

3. Photosynthetic membrane systems and mesosomes are internal membrane components of certain organisms termed _____.
   *Answer: prokaryotes*

4. The light microscope has glass lenses that focus light (photons) for imaging, whereas the electron microscope has _____ that focus electrons for imaging.
   *Answer: magnets*

5. Membranous compartments with distinctive shapes and functions are termed _____.
   *Answer: organelles*

6. _____ is the process whereby light energy is converted into chemical bonds.
   *Answer: Photosynthesis*

7. The _____ is the organelle with many folds called cristae.
   *Answer: Mitochondrion*

8. The _____ is an organelle that serves as a sort of "postal depot" where some of the proteins synthesized on ribosomes and rough ER are processed.
   *Answer: Golgi apparatus*

9. RNA carries information for protein synthesis from the DNA in the nucleus to the ribosomes in the cytoplasm. To get from the nucleoplasm to the cytoplasm, RNA must pass through _____.
   *Answer: nuclear pores*

10. All organisms are composed of cells; all cells come from preexisting cells. These statements are called _____.
    *Answer: the cell theory*

11. When you cut an orange in half, you _____ the surface area-to-volume ratio.
    *Answer: increase*

12. The DNA in a prokaryotic cell can be found in the _____ region.
    *Answer: nucleoid*

13. The _____ of some bacteria help them avoid being detected by the human immune system.
    *Answer: capsules*

14. The meshwork of intermediate filaments found on the interior surface of the nuclear membrane is called the _____.
    *Answer: nuclear lamina*

15. Steroids, fatty acids, phospholipids, and carbohydrates are synthesized in the _____.
    *Answer: smooth ER*

16. The side of the Golgi facing the ER is the _____ face.
    *Answer: cis*

17. The substances that enter the Golgi come from the _____.
    *Answer: ER*

18. Toxic peroxides that are formed unavoidably as side products of important cellular reactions are found and neutralized in _____.
    *Answer: peroxisomes*

19. The _____ is the cytoskeletal component with the smallest diameter.
    *Answer: actin (or microfilament)*

20. Keratin is classified as an _____ type of filament.
    *Answer: intermediate*

## Multiple Choice

*1. The surface area-to-volume ratio of an object can be decreased by
   a. cutting it into smaller pieces.
   b. flattening it.
   c. stretching it.
   d. making it spherical.
   e. All of the above
   *Answer: d*

2. What must cells do in order to survive?
   a. Obtain and process energy
   b. Convert genetic information into proteins
   c. Keep certain biochemical reactions separate from one another
   d. Both a and b
   e. a, b, and c
   *Answer: e*

3. Cholesterol is synthesized by
   a. chloroplasts.
   b. lysosomes.
   c. the SER.
   d. the Golgi.
   e. mitochondria.
   *Answer: c*

4. Examples of cellular "appendages" include
   a. the Golgi apparatus.
   b. cilia.
   c. flagella.
   d. pili.
   e. b, c, and d
   *Answer: e*

5. Roles of biological membranes in eukaryotic cells include which of the functions listed below?
   a. Separating a cell from its environment
   b. Selecting what goes into and out of the cell
   c. Maintaining a constant internal environment
   d. Communicating with adjacent cells
   e. All of the above
   *Answer: e*

6. The utilization of "food" in the mitochondria, with the associated formation of ATP, is termed
   a. cellular respiration.
   b. metabolic rate.
   c. diffusion.
   d. metabolic processing of fuels.
   e. catabolism.
   *Answer: a*

7. The DNA of mitochondria
   a. is needed to hydrolyze monomers.
   b. is used to make proteins needed for cellular respiration.
   c. directs photosynthesis.
   d. controls the cell's activities.
   e. synthesizes polysaccharides for the plant cell wall.
   *Answer: b*

8. The DNA of a chloroplast is located in the
   a. intermembrane space.
   b. matrix.
   c. cristae.
   d. stroma.
   e. granum.
   *Answer: d*

9. Components of chloroplasts include
   a. grana and thylakoids.
   b. chromatin and nucleoplasm.
   c. cristae and matrix.
   d. a *trans* region and a *cis* region.
   e. lysosomes and phagosomes.
   *Answer: a*

■10. The cell is the basic unit of function and reproduction because
   a. subcellular components cannot regenerate whole cells.
   b. cells are totipotent.
   c. single cells sometimes can produce an entire organism.
   d. cells can come only from preexisting cells.
   e. a cell can arise by the fusion of two cells.
   *Answer: a*

■11. What is the major distinction between prokaryotic and eukaryotic cells?
   a. A prokaryotic cell does not have a nucleus, whereas a eukaryotic cell does.
   b. A prokaryotic cell does not have DNA, whereas a eukaryotic cell does.
   c. Prokaryotic cells are smaller than eukaryotic cells.
   d. Prokaryotic cells have not prospered, whereas eukaryotic cells are evolutionary "successes."
   e. Prokaryotic cells cannot obtain energy from their environment.
   *Answer: a*

12. Which of the following is *not* a characteristic of a prokaryotic cell?
   a. A plasma membrane
   b. A nuclear envelope
   c. A nucleoid
   d. Ribosomes
   e. Enzymes
   *Answer: b*

13. Members of the domains Bacteria and Archaea
   a. have nuclei.
   b. have chloroplasts.
   c. are multicellular.
   d. are prokaryotes.
   e. have flagella.
   *Answer: d*

14. Ribosomes are made up of
   a. DNA and RNA.
   b. DNA and proteins.
   c. RNA and proteins.
   d. proteins.
   e. DNA.
   *Answer: c*

15. Which of the following is (are) found in prokaryotic cells?
   a. Mitochondria
   b. Chloroplasts
   c. Nuclear membrane
   d. Ribosomes
   e. Endoplasmic reticulum
   *Answer: d*

16. In some prokaryotic organisms the plasma membrane folds to form an internal membrane system that is able to
   a. carry on photosynthesis.
   b. engulf and phagocytize bacteria.
   c. synthesize proteins.
   d. propel the cell.
   e. hydrolyze carbohydrates to ATP.
   *Answer: a*

17. The DNA of prokaryotic cells is found in the
   a. plasma membrane.
   b. nucleus.
   c. ribosome.
   d. nucleoid region.
   e. mitochondria.
   *Answer: d*

18. Which structure supports the plant cell and determines its shape?
   a. Capsule
   b. Flagellum

c. Cell wall
d. Cytosol
e. Cytoplasm
*Answer: c*

19. Some bacteria are able to propel themselves through liquid by means of a structure called the
    a. flagellum.
    b. pilus.
    c. cytoplasm.
    d. cell wall.
    e. peptidoglycan molecule.
    *Answer: a*

■20. If you removed the pili from a bacterial cell, which of the following would you expect to happen?
    a. The bacterium would no longer be able to swim.
    b. The bacterium would not adhere to other cells as well.
    c. The bacterium would no longer be able to regulate the movement of molecules into and out of the cell.
    d. The bacterium would dry out.
    e. The shape of the bacterium would change.
    *Answer: b*

21. Ribosomes are not visible under a light microscope, but they can be seen with an electron microscope because
    a. electron beams have more energy than light beams.
    b. electron microscopes focus light with magnets.
    c. electron microscopes have more resolving power than light microscopes.
    d. electrons have such high energy that they pass through biological samples.
    e. living cells can be observed under the electron microscope.
    *Answer: c*

■22. Using a light microscope, it is possible to view cytoplasm streaming around the central vacuole in cells of the green alga *Nitella*. Why would you use a light microscope instead of an electron microscope to study this process?
    a. Electron microscopes have less resolving power than light microscopes.
    b. Structures inside the cell cannot be seen with the electron microscope.
    c. Whole cells cannot be viewed with the electron microscope.
    d. The electron microscope cannot be used to observe living cells.
    e. The central vacuole is too small to be seen with a scanning electron microscope.
    *Answer: d*

23. Which of the following is a general function of all cellular membranes?
    a. They regulate which materials can cross the membrane.
    b. They support the cell and determine its shape.
    c. They produce energy for the cell.
    d. They produce proteins for the cell.
    e. They move the cell.
    *Answer: a*

24. Which statement about the nuclear envelope is true?
    a. It contains pores for the passage of large molecules.
    b. It is composed of two membranes.
    c. It contains ribosomes on the inner surface.
    d. Both a and b

e. All of the above
*Answer: d*

25. What is the purpose of the folds of the inner mitochondrial membrane?
    a. They increase the volume of the mitochondrial matrix.
    b. They create new membrane-enclosed compartments within the mitochondrion.
    c. They increase the surface area for the exchange of substances across the membrane.
    d. They anchor the mitochondrial DNA.
    e. The folds have no known purpose.
    *Answer: c*

26. Which type of organelle is found in plants but *not* in animals?
    a. Ribosomes
    b. Mitochondria
    c. Nuclei
    d. Plastids
    e. None of the above
    *Answer: d*

27. Where in the cell do you *not* find DNA?
    a. Mitochondrial matrix
    b. Chloroplast stroma
    c. Cell cytosol
    d. Cell nucleus
    e. You find DNA in all of the above.
    *Answer: c*

28. Which of the following statements about cells is true?
    a. Animal cells do not produce chloroplasts.
    b. Animal cells do not have mitochondria.
    c. All plant cells contain chloroplasts.
    d. Plant cells do not have plastids.
    e. None of the above
    *Answer: a*

■29. Which of the following is *not* an argument for the endosymbiotic theory?
    a. Mitochondria and chloroplasts have double membranes.
    b. Mitochondria and chloroplasts cannot be grown in culture free of a host cell.
    c. Mitochondria and chloroplasts have DNA and ribosomes.
    d. Mitochondrial ribosomes are similar to bacterial ribosomes.
    e. All of the above are arguements for the endosymbiotic theory.
    *Answer: b*

■30. What is the difference between "free" and "attached" ribosomes?
    a. Free ribosomes are in the cytoplasm, whereas attached ribosomes are anchored to the endoplasmic reticulum.
    b. Free ribosomes produce proteins in the cytosol, whereas attached ribosomes produce proteins that are inserted into the ER.
    c. Free ribosomes produce proteins that are exported from the cell, whereas attached ribosomes make proteins for mitochondria and chloroplasts.
    d. Both a and b
    e. Both a and c
    *Answer: d*

31. The carotenoid pigments that give ripe tomatoes their red color are contained in organelles called
    a. chloroplasts.
    b. proplastids.
    c. protoplasts.
    d. leucoplasts.
    e. chromoplasts.
    *Answer: e*

32. Which is a function of a plant cell vacuole?
    a. Storage of wastes
    b. Support for the cell
    c. Excretion of wastes
    d. Both a and b
    e. Both b and c
    *Answer: d*

33. Microtubules are made of
    a. actin, and they function in locomotion.
    b. tubulin, and they are found in cilia.
    c. tubulin, and they are found in microvilli.
    d. actin, and they function to change cell shape.
    e. polysaccharides, and they function in locomotion.
    *Answer: b*

34. The size of the smallest structure that can be seen clearly through a light microscope is
    a. 1 millimeter.
    b. 0.1 millimeter.
    c. 10 micrometers.
    d. 2 micrometers.
    e. 0.2 micrometer.
    *Answer: e*

35. The two major types of cells are
    a. human and nonhuman.
    b. prokaryotic and eukaryotic.
    c. blood and muscle.
    d. plant and animal.
    e. warm-blooded and cold-blooded.
    *Answer: b*

36. The one type of cell always lacking a cell wall is the _____ cell.
    a. bacterial
    b. plant
    c. animal
    d. fungal
    e. prokaryotic
    *Answer: c*

37. An organelle found only in plant cells is the
    a. cilium.
    b. nucleus.
    c. mitochondrion.
    d. glyoxysome.
    e. peroxisome.
    *Answer: d*

38. An organelle found in all eukaryotic cells during some portion of their lives is the
    a. chloroplast.
    b. nucleus.
    c. flagellum.
    d. vacuole.
    e. centriole.
    *Answer: b*

39. Ribosomes are the structures in which
    a. chemical energy is stored by making ATP.
    b. cell division is controlled.
    c. genetic information is used to make proteins.
    d. sunlight energy is converted into chemical energy.
    e. new organelles are made.
    *Answer: c*

40. Chloroplasts are the structures in which
    a. chemical energy is stored by making ATP.
    b. cell division is controlled.
    c. genetic information is used to make proteins.
    d. sunlight energy is converted into chemical energy.
    e. new organelles are made.
    *Answer: d*

41. An organelle consisting of a series of flattened sacks stacked somewhat like pancakes is the
    a. mitochondrion.
    b. chloroplast.
    c. Golgi apparatus.
    d. rough endoplasmic reticulum.
    e. flagellum.
    *Answer: c*

42. An organelle with an internal cross section showing a characteristic "9 + 2" morphology is the
    a. mitochondrion.
    b. vacuole.
    c. Golgi apparatus.
    d. flagellum.
    e. cytoskeleton.
    *Answer: d*

43. An organelle continuous with the endoplasmic reticulum is the
    a. nucleus.
    b. Golgi apparatus.
    c. mitochondria.
    d. flagellum.
    e. lysosome.
    *Answer: a*

44. Chromatin is a series of entangled threads composed of
    a. microtubules.
    b. DNA and protein.
    c. fibrous proteins.
    d. cytoskeleton.
    e. membranes.
    *Answer: b*

45. Ribosomes are *not* found in
    a. the mitochondrion.
    b. a chloroplast.
    c. the rough endoplasmic reticulum.
    d. a prokaryotic cell.
    e. the Golgi apparatus.
    *Answer: e*

46. The overall shape of a cell is determined by its
    a. cell membrane.
    b. cytoskeleton.
    c. nucleus.
    d. cytosol.
    e. endoplasmic reticulum.
    *Answer: b*

47. Of the following structures of an animal cell, the one with the largest volume is the
    a. cilium.

b. mitochondrion.
c. lysosome.
d. nucleus.
e. ribosome.
*Answer: d*

48. Of the following structures of a plant cell, the one that most often has the greatest volume is the
    a. glyoxysome.
    b. lysosome.
    c. chromosome.
    d. ribosome.
    e. vacuole.
    *Answer: e*

49. Of the following structures, the one that contains both a matrix and cristae is the
    a. plastid.
    b. lysosome.
    c. Golgi apparatus.
    d. mitochondrion.
    e. chromatin.
    *Answer: d*

50. Light energy for conversion to chemical energy is trapped in the
    a. mitochondrion.
    b. chromoplast.
    c. thylakoid.
    d. endoplasmic reticulum.
    e. Golgi apparatus.
    *Answer: c*

51. Proteins that will function outside of the cytosol are made by
    a. the Golgi apparatus.
    b. ribosomes within the mitochondrion.
    c. the smooth endoplasmic reticulum.
    d. ribosomes on the rough endoplasmic reticulum.
    e. ribosomes within the nucleus.
    *Answer: d*

52. Cilia contain
    a. microtubules.
    b. microfilaments.
    c. intermediate filaments.
    d. ribosomes.
    e. plasmodesmata.
    *Answer: a*

53. You would *not* expect to find RNA in which of the following structures?
    a. Nucleus
    b. Mitochondrion
    c. Vacuole
    d. Ribosome
    e. Prokaryotic cell
    *Answer: c*

54. Which of the following structures is (are) involved with the movement of organelles within a cell?
    a. Golgi apparatus
    b. Endoplasmic reticulum
    c. Mitochondrion
    d. Microfilaments
    e. Intermediate filaments
    *Answer: d*

55. Chloroplasts are a kind of
    a. leucoplast.
    b. endoplasmic reticulum.
    c. chromoplast.
    d. Golgi apparatus.
    e. plastid.
    *Answer: e*

56. Which of the following is *not* a component of the endomembrane system?
    a. Rough endoplasmic reticulum
    b. Smooth endoplasmic reticulum
    c. Golgi apparatus
    d. Lysosomes
    e. Plastids
    *Answer: e*

57. A prokaryotic cell does *not* have a _____ or _____.
    a. nucleus; organelles
    b. nucleus; DNA
    c. nucleus; ribosomes
    d. nucleus; membranes
    e. cell wall; membranes
    *Answer: a*

58. The pores found in the nuclear membrane are composed of
    a. one large protein.
    b. eight large protein granules.
    c. keratin.
    d. intermediate filaments.
    e. lipids.
    *Answer: b*

59. Starch molecules are stored inside
    a. chromoplasts.
    b. granularplasts.
    c. chloroplasts.
    d. potatoplasts.
    e. leucoplasts.
    *Answer: e*

60. Some organelles in eukaryotic cells are thought to have
    a. originated from extracellular symbiotic relationships.
    b. their own endoplasmic reticulum.
    c. their own mitochondria.
    d. originated from endosymbiotic relationships.
    e. the ability to live free from the host cell.
    *Answer: d*

61. The membranes of the endoplasmic reticulum are continuous with the membranes of the
    a. nucleus.
    b. Golgi apparatus.
    c. nucleolus.
    d. plasma membrane.
    e. mitochondria.
    *Answer: a*

62. The rough ER is the portion of the ER that
    a. lacks ribosomes.
    b. is older and was once the smooth ER.
    c. has ribosomes attached to it.
    d. is connected to the Golgi apparatus.
    e. is the site of steroid synthesis.
    *Answer: c*

▪63. The difference between the Golgi of plants, protists, and fungi when compared to that of vertebrates, is that the vertebrates' Golgi
   a. forms a large apparatus from a few stacked sacks.
   b. forms small, widely distributed sacks.
   c. forms a single large sack.
   d. connects directly to the ER.
   e. lacks integral membrane proteins.
   *Answer: a*

64. Materials that enter and leave the Golgi
   a. are transported by proteins.
   b. are packaged on or in vesicles.
   c. are destined for export from the cell.
   d. require a docking protein.
   e. originated in the nucleus.
   *Answer: b*

65. Proteins from the Golgi are transported to the correct location due to
   a. signals found on the packaged proteins.
   b. the direction in which all vesicles travel within the cell.
   c. the control provided by the nucleus.
   d. motor proteins.
   e. microtubules.
   *Answer: a*

66. A secondary lysosome is a lysosome that
   a. provides a backup to the primary lysosomes.
   b. is smaller than a primary lysosome.
   c. will become a primary lysosome after it fuses with a phagosome.
   d. is a primary lysosome that has fused with a phagosome.
   e. has exocytosed.
   *Answer: d*

67. Lysosomes are important to eukaryotic cells because they contain
   a. photosynthetic pigments.
   b. starch molecules for energy storage.
   c. their own DNA molecules.
   d. the cells' waste materials.
   e. digestive enzymes.
   *Answer: e*

68. Which of the following cellular components is (are) most important for stabilizing the shape of an animal cell?
   a. Golgi apparatus
   b. Nuclear lamina
   c. Microfilaments
   d. Microtubules
   e. Cell wall
   *Answer: c*

69. The surface area of some eukaryotic cells is greatly increased by
   a. microtubules.
   b. pili.
   c. thylakoid membranes.
   d. myosin.
   e. microvilli.
   *Answer: e*

70. Microvilli are created by projections of
   a. microtubules.
   b. actin.
   c. myosin.

   d. intermediate filaments.
   e. None of the above
   *Answer: b*

71. Hair and intermediate filaments are composed of
   a. microtubules.
   b. microfilaments.
   c. collagen.
   d. hydroxyapatite.
   e. keratin.
   *Answer: e*

72. Microtubules are composed of subunits of
   a. $\alpha$- and $\beta$-tubulin.
   b. d- and l- actin.
   c. r- and s- myosin.
   d. kappa tubules.
   e. kappa actinomin.
   *Answer: a*

73. In cells possessing them, microvilli are useful
   a. as aids in their locomotion.
   b. to help concentrate food particles.
   c. for intracellular trafficking of molecules.
   d. to greatly increase their surface area.
   e. for intercellular communications.
   *Answer: d*

74. Which of the following is a cellular component found at the base of each cilium?
   a. Centriole
   b. Basal body
   c. Nucleolus
   d. Flagellum
   e. Microvillus
   *Answer: b*

75. The cellular structures that are most like centrioles are
   a. basal bodies.
   b. microbodies.
   c. chromoplasts.
   d. microfilaments.
   e. centromeres.
   *Answer: a*

76. Intermediate filaments
   a. hold organelles in place within the cell.
   b. strengthen cellular structures.
   c. provide movement in animal cell division.
   d. are involved in the structure and function of cilia and flagella.
   e. are responsible for pseudopod extensions.
   *Answer: a*

77. A specialized structure found in some prokaryotes is the
   a. cell wall.
   b. ribosome.
   c. cytosol.
   d. mitochondrion.
   e. chloroplast.
   *Answer: a*

78. The cytoplasm
   a. is a static region of the cell.
   b. contains DNA.
   c. is composed largely of water.
   d. supports the cell and determines its shape.
   e. chemically modifies proteins and other molecules.
   *Answer c*

79. The membrane surrounding each organelle
    a. is composed of hydrophobic proteins.
    b. regulates traffic into and out of the cell.
    c. is studded with ribosomes.
    d. allows for interactions among molecules.
    e. is perforated with pores.
    *Answer: b*

80. Which of the following organelles were once independent prokaryote organisms?
    a. Mitochondria and lysosomes
    b. Mitochondria and chloroplasts
    c. Chloroplasts and Golgi apparatus
    d. Golgi apparatus and ribosomes
    e. Ribosomes and lysosomes
    *Answer: b*

81. The extracellular matrix of animal cells
    a. is composed of cellulose.
    b. contains plasmodesmata.
    c. limits the cell volume by remaining rigid.
    d. helps orient cell movements during embryonic development.
    e. acts as a barrier to disease-causing fungi.
    *Answer: d*

82. Which of the following organelles assembles ribosomes?
    a. Chloroplast
    b. Smooth endoplasmic reticulum
    c. Mitochondrion
    d. Nucleolus
    e. Golgi apparatus
    *Answer: d*

83. How does the surface area-to-volume ratio of a 1 mm cube compare to the surface area-to-volume ratio of a 3 mm cube?
    a. The 3 mm cube has a higher ratio.
    b. The ratio increases as the cube becomes larger.
    c. Increasing the volume increases the ratio.
    d. The ratio decreases as the cube becomes larger.
    e. The ratio does not change.
    *Answer: d*

84. Ribosomes are found in all of the following *except*
    a. the Golgi apparatus.
    b. chloroplasts.
    c. mitochondria.
    d. the rough endoplasmic reticulum.
    e. None of the above
    *Answer: a*

85. Which of the following best describes the role of the Golgi apparatus?
    a. Detoxifies peroxides
    b. Lipid biosynthesis
    c. Modification of proteins to be secreted
    d. Synthesis of ribosomal subunits
    e. Stores poisonous substances
    *Answer: c*

86. Which of the following is true of lysosomes?
    a. Malfunctioning lysosomes may result in Tay-Sachs Disease.
    b. Lysosomes provide turgor in plant cells.
    c. Lysosomes may contain anthocyanins that aid in pollination.
    d. Lysosomes are found only in plants.
    e. Lysosomes may have arisen through endosymbiosis.
    *Answer: a*

87. Why are prokaryotic cells generally smaller than eukaryotic cells?
    a. Prokaryotes have more diverse energy sources.
    b. Prokaryotes have a capsule that limits cell growth.
    c. The rigid cell wall found in prokaryotes limits cell size.
    d. Prokaryotes lack the genetic material needed for protein synthesis.
    e. Eukaryotes have compartmentalization, which allows for specialization.
    *Answer: e*

# 5 Cellular Membranes

## Fill in the Blank

1. The _____ molecules of membranes act as barriers to the passage of many materials and serve to maintain the membrane's physical integrity.
   *Answer: lipid*

2. In a complex solution, the _____ of each substance is independent of that of the other substances.
   *Answer: diffusion*

3. Lipids and proteins can move _____ but not across biological membranes.
   *Answer: laterally*

4. _____ involves coated pits, clathrin, and coated vesicles.
   *Answer: Receptor-mediated endocytosis*

5. The ability of some materials to move through biological membranes more readily than others is called selective _____.
   *Answer: permeability*

6. The major lipids in biological membranes are called _____.
   *Answer: phospholipids*

7. If a cell placed within a solution shrinks, the solution is _____ relative to the cell.
   *Answer: hypertonic*

8. The cells of the intestinal epithelium are joined to one another by _____ that prevent substances from passing between the cells of this tissue.
   *Answer: tight junctions*

9. The coupled transport system by which glucose and sodium ions enter intestinal epithelial cells is called _____.
   *Answer: symport*

10. Mammalian embryos have protein complexes that couple cells, allowing communication between cells by small molecules. These complexes of proteins are called _____.
    *Answer: gap junctions*

11. Diffusion occurs _____ a concentration gradient.
    *Answer: down*

12. Three factors that influence the diffusion rate of a molecule are _____, _____, and _____.
    *Answer: the size of the molecule; the temperature; the concentration gradient*

13. The force that increases inside a plant cell when it is placed in water is called _____ pressure.
    *Answer: turgor*

14. The sodium–potassium pump of cell membranes is a(n) _____ active transport system.
    *Answer: antiport*

15. Biological membranes are organized in the manner described by the _____ model.
    *Answer: fluid mosaic*

16. Membrane proteins with attached carbohydrates are called _____.
    *Answer: glycoproteins*

17. Membrane lipids with attached carbohydrates are called _____.
    *Answer: glycolipids*

18. The amount of protein in the cell membranes of the myelin sheath is (more/less) _____ than the amount of protein in the membranes of the mitochondria.
    *Answer: more*

19. The cell adhesion molecule of sponges is a _____.
    *Answer: glycoprotein*

20. When cell adhesion molecules are of the same type, they are called _____.
    *Answer: homotypic*

21. The type of cell adhesion molecule found in sponges is _____.
    *Answer: homotypic*

22. In a comparison of carrier and channel proteins, _____ proteins were found to be faster, and they do not saturate.
    *Answer: channel*

23. Membrane synthesized on the ER moves to other points of the cell as _____.
    *Answer: vesicles*

24. The process of one cell engulfing another is called _____.
    *Answer: phagocytosis*

## Multiple Choice

1. Peripheral membrane proteins have
   a. hydrophobic regions within the lipid portion of the bilayer.
   b. hydrophilic regions that protrude in aqueous environments on either side of the membrane.
   c. lateral but not vertical movement within the bilayer.
   d. control of diffusion.
   e. polar regions that interact with similar regions of integral membrane proteins.
   *Answer: e*

2. Which of these are *not* specialized cell junctions?
   a. Gap junctions
   b. Tight junctions
   c. Desmosomes
   d. Lipid rafts
   e. Both a and b
   *Answer: d*

3. Substances move through biological membranes against concentration gradients via
   a. simple osmosis.
   b. active transport.
   c. reverse osmosis.
   d. simple diffusion.
   e. None of the above
   *Answer: b*

4. Because the sodium–potassium pump imports $K^+$ ions while exporting $Na^+$ ions, it is a coupled transport system called a(n)
   a. symport.
   b. antiport.
   c. secondary active transporter.
   d. facilitated transport.
   e. diffusion mechanism.
   *Answer: b*

5. Whether a membrane protein can be integral and can traverse the membrane depends on the presence of _____ R groups.
   a. primary
   b. secondary
   c. tertiary
   d. quaternary
   e. hydrophobic
   *Answer: e*

■6. When placed in water, wilted plants lose their limpness because of
   a. active transport of salts from the water into the plant.
   b. active transport of salts into the water from the plant.
   c. osmosis of water into the plant cells.
   d. osmosis of water from the plant cells.
   e. diffusion of water from the plant cells.
   *Answer: c*

★■7. Houseplants adapted to indoor temperatures may die when accidentally left outdoors in the cold because their
   a. DNA cannot function.
   b. membranes lack adequate fluidity.
   c. photosynthesis is impaired.
   d. chloroplasts malfunction.
   e. membranes need more cholesterol.
   *Answer: b*

8. The compounds in biological membranes that form a barrier to the movement of materials across the membrane are
   a. integral membrane proteins.
   b. carbohydrates.
   c. lipids.
   d. nucleic acids.
   e. peripheral membrane proteins.
   *Answer: c*

9. The interior of the phospholipid bilayer is
   a. hydrophilic.
   b. hydrophobic.
   c. aqueous.
   d. solid.
   e. charged.
   *Answer: b*

10. In biological membranes, the phospholipids are arranged in a
    a. bilayer, with the fatty acids pointing toward each other.
    b. bilayer, with the fatty acids facing outward.
    c. single layer, with the fatty acids facing the interior of the cell.
    d. single layer, with the phosphorus-containing region facing the interior of the cell.
    e. bilayer, with the phosphorus groups in the interior of the membrane.
    *Answer: a*

11. A protein that forms an ion channel through a membrane is most likely to be
    a. a peripheral protein.
    b. an integral protein.
    c. a phospholipid.
    d. an enzyme.
    e. entirely outside the phospholipid bilayer.
    *Answer: b*

■12. When a mouse cell and a human cell are fused, the membrane proteins of the two cells become uniformly distributed over the surface of the hybrid cell. This occurs because
    a. many proteins can move around within the bilayer.
    b. all proteins are anchored within the membrane.
    c. proteins are asymmetrically distributed within the membrane.
    d. all proteins in the plasma membrane are peripheral.
    e. different membranes contain different proteins.
    *Answer: a*

13. The hydrophilic regions of a membrane protein are most likely to be found
    a. only in muscle cell membranes.
    b. associated with the fatty acid region of the lipids.
    c. in the interior of the membrane.
    d. exposed on the surface of the membrane.
    e. either on the surface or inserted into the interior of the membrane.
    *Answer: d*

14. Biological membranes are composed of
    a. nucleotides and nucleosides.
    b. enzymes, electron acceptors, and electron donors.
    c. fatty acids.
    d. monosaccharides.

e. lipids, proteins, and carbohydrates.
*Answer: e*

■15. The LDL receptor is an integral protein that crosses the plasma membrane, with portions of the protein extending both outside and into the interior of the cell. The amino acid side chains in the region of the protein that crosses the membrane are most likely to be
a. charged.
b. hydrophilic.
c. hydrophobic.
d. carbohydrates.
e. lipids.
*Answer: c*

16. Which of the following functions as a recognition signal between cells?
a. RNA
b. Phospholipids
c. Cholesterol
d. Fatty acids
e. Glycolipids
*Answer: e*

17. When a membrane is prepared by freeze-fracture and examined under the electron microscope, the exposed interior of the membrane bilayer appears to be covered with bumps. These bumps are
a. integral membrane proteins.
b. ice crystals.
c. platinum.
d. organelles.
e. vesicles.
*Answer: a*

★18. Structures that contain networks of keratin fibers and hold adjacent cells together are called
a. extracellular matrices.
b. glycoproteins.
c. gap junctions.
d. desmosomes.
e. phospholipid bilayers.
*Answer: d*

19. The electric signal for contraction passes rapidly from one muscle cell to the next by way of
a. tight junctions.
b. desmosomes.
c. gap junctions.
d. integral membrane proteins.
e. freeze-fractures.
*Answer: c*

20. Tight junctions serve an important function in epithelial cell layers by
a. restricting the extracellular movement of molecules between the adjacent cells.
b. allowing nerve impulses to move from one cell to the next.
c. providing cytoplasmic channels between adjacent cells.
d. providing channels between the cytoplasm and the extracellular environment.
e. acting as recognition sites for foreign substances.
*Answer: a*

■21. You fill a shallow pan with water and place a drop of red ink in one end of the pan and a drop of green ink in the other end. Which of the following is true at equilibrium?
a. The red ink is uniformly distributed in one-half of the pan, and the green ink is uniformly distributed in the other half of the pan.
b. The red and green inks are uniformly distributed throughout the pan.
c. Each ink is moving down its concentration gradient.
d. The concentration of each ink is higher at one end of the pan than at the other end.
e. No predictions can be made without knowing the molecular weights of the pigment molecules.
*Answer: b*

22. Which of the following does *not* affect the rate of diffusion of a substance?
a. Temperature
b. Concentration gradient
c. Electrical charge of the diffusing material
d. Presence of other substances in the solution
e. Molecular diameter of the diffusing material
*Answer: d*

23. For cells in which carbon dioxide crosses the plasma membrane by simple diffusion, what determines the rate at which carbon dioxide enters the cell?
a. The concentration of carbon dioxide on each side of the membrane
b. The amount of ATP produced by the cell
c. The number of carrier proteins in the membrane
d. The amount of energy available
e. The concentration of hydrogen ions on each side of the membrane
*Answer: a*

24. Plant cells transport sucrose across the vacuole membrane against its concentration gradient by a process known as
a. simple diffusion.
b. active transport.
c. passive transport.
d. facilitated diffusion.
e. cellular respiration.
*Answer: b*

25. When placed in a hypertonic solution, plant cells
a. shrink.
b. swell.
c. burst.
d. transport water out.
e. concentrate.
*Answer: a*

★■26. You place cells in a solution of glucose and measure the rate at which glucose enters the cells. As you increase the concentration of the glucose solution, the rate increases. However, when the glucose concentration of the solution is increased above 10 M, the rate no longer increases. Which of the following is the most likely mechanism for glucose transport into the cell?
a. Facilitated diffusion via a carrier protein
b. Facilitated diffusion via a channel protein
c. Pinocytosis
d. Secondary active transport
e. Symport
*Answer: a*

27. Transporting substances across a membrane from an area of lower concentration to an area of higher concentration requires
    a. phospholipids.
    b. diffusion.
    c. gap junctions.
    d. facilitated diffusion.
    e. energy.
    *Answer: e*

28. In the parietal cells of the stomach, the uptake of chloride ions is coupled to the transport of bicarbonate ions out of the cell. This type of transport system is called
    a. a uniport.
    b. a symport.
    c. an exchange channel.
    d. diffusion.
    e. an antiport.
    *Answer: e*

29. When a red blood cell is placed in an isotonic solution, which of the following will occur?
    a. The cell will shrivel.
    b. The cell will swell and burst.
    c. The cell will shrivel, and then return to normal.
    d. The cell will swell, and then return to normal.
    e. Water moves into and out of the cell at an equal rate.
    *Answer: e*

30. When a plant cell is placed in a hypotonic solution, which of the following occurs?
    a. The cell takes up water until the osmotic potential equals the pressure potential of the cell wall.
    b. The cell takes up water and eventually bursts.
    c. The cell shrinks away from the cell wall.
    d. There is no movement of water into or out of the cell.
    e. Water moves out of the cell.
    *Answer: a*

■31. When vesicles from the Golgi apparatus deliver their contents to the exterior of the cell, they add their membranes to the plasma membrane. Why doesn't the plasma membrane increase in size?
    a. Some vesicles from the Golgi apparatus fuse with the lysosomes.
    b. Membrane vesicles carry proteins from the endoplasmic reticulum to the Golgi apparatus.
    c. Membrane is continually being lost from the plasma membrane by endocytosis.
    d. New phospholipids are synthesized in the endoplasmic reticulum.
    e. The phospholipids become more tightly packed together in the membrane.
    *Answer: c*

32. Receptor-mediated endocytosis is the mechanism for transport of
    a. clathrin.
    b. all macromolecules.
    c. ions.
    d. specific macromolecules.
    e. integral membrane proteins.
    *Answer: d*

33. During the formation of muscle, the association of individual muscle cells to form a tissue requires specific membrane proteins called
    a. coated vesicles.
    b. cell adhesion molecules.
    c. glycolipids.
    d. carrier molecules.
    e. transport proteins.
    *Answer: b*

■34. The neurotransmitter acetylcholine can activate a muscle cell and cause it to contract, even though the acetylcholine molecule never enters the cell. How is this possible?
    a. The acetylcholine receptor protein is a peripheral protein.
    b. Acetylcholine can bind to all proteins in the plasma membrane.
    c. The acetylcholine receptor protein spans the plasma membrane.
    d. Acetylcholine is hydrophobic.
    e. Acetylcholine enters the cell by receptor-mediated endocytosis.
    *Answer: c*

★35. Insulin is a protein secreted by cells of the pancreas. What is the pathway for the synthesis and secretion of insulin?
    a. Rough ER, Golgi apparatus, vesicle, plasma membrane
    b. Golgi apparatus, rough ER, lysosome
    c. Lysosome, vesicle, plasma membrane
    d. Plasma membrane, coated vesicle, lysosome
    e. Rough ER, cytoplasm, plasma membrane
    *Answer: a*

36. Carbohydrates associated with cellular membranes are typically associated with
    a. proteins inside cells.
    b. lipids inside cells.
    c. proteins outside cells.
    d. proteins of the internal organelles.
    e. the nuclear membrane.
    *Answer: c*

★★37. You are studying how low-density lipoproteins (LDL) enter cells. When you examine cells that have taken up LDL, you find that the LDL is inside clathrin-coated vesicles. What is the most likely mechanism for the uptake of LDL?
    a. Facilitated diffusion
    b. Proton antiport
    c. Receptor-mediated endocytosis
    d. Gap junctions
    e. Ion channels
    *Answer: c*

■38. If you compare the proteins of the plasma membrane and the proteins of the inner mitochondrial membrane, which of the following will be true?
    a. Both membranes will have only peripheral proteins.
    b. Only the mitochondrial membrane will have integral proteins.
    c. Only the mitochondrial membrane will have peripheral proteins.
    d. All of the proteins from both membranes will be hydrophilic.
    e. The proteins from the two membranes will be different.
    *Answer: e*

39. The molecules in a membrane that limit its permeability are the
    a. carbohydrates.
    b. phospholipids.

c. proteins.
d. negative ions.
e. water.
*Answer: b*

40. The plasma membrane of animals contains carbohydrates
    a. on the side of the membrane facing the cytosol.
    b. on the side of the membrane facing away from the cell.
    c. on both sides of the membrane.
    d. on neither side of the membrane.
    e. within the membrane.
    *Answer: b*

41. Integral membrane proteins tend to
    a. be of small molecular weight.
    b. be very soluble in water.
    c. have hydrophobic amino acids on much of their intramembrane surface.
    d. be rare.
    e. contain much cholesterol.
    *Answer: c*

*42. An important function of certain integral proteins of a eukaryotic cell's plasma membrane is
    a. movement of the cell.
    b. binding with signals in the cell's environment.
    c. usage of genetic information.
    d. digestion of food molecules.
    e. generation of ATP.
    *Answer: b*

43. Cholesterol molecules act to
    a. help hold a membrane together.
    b. transport ions across membranes.
    c. attach to carbohydrates.
    d. disrupt membrane function.
    e. alter the fluidity of the membrane.
    *Answer: e*

■44. You have observed that the rate of facilitated diffusion of a specific molecule across a membrane does not continue to increase as the concentration difference of the molecule across the membrane increases. Why?
    a. Facilitated diffusion requires the use of ATP.
    b. As the concentration difference increases, molecules interfere with one another.
    c. The transport protein must be of the carrier type.
    d. The transport protein must be of the channel type.
    e. The diffusion constant depends on the concentration difference.
    *Answer: c*

45. Active transport is important because it can move molecules
    a. from their high concentration to a lower concentration.
    b. from their low concentration to a higher concentration.
    c. that resist osmosis across the membrane.
    d. with less ATP than might otherwise be needed to move the molecules.
    e. by increasing their diffusion coefficient.
    *Answer: b*

46. Active transport usually moves molecules
    a. in the same direction in which diffusion moves them.
    b. in the opposite direction from the one in which diffusion moves them.
    c. in a direction that tends to bring about equilibrium.

d. toward higher pH.
e. toward higher osmotic potential.
*Answer: b*

47. Osmosis is a specific form of
    a. diffusion.
    b. facilitated transport.
    c. active transport.
    d. secondary active transport.
    e. movement of water by carrier proteins.
    *Answer: a*

48. Osmosis moves water from a region of
    a. high concentration of dissolved material to a region of low concentration.
    b. low concentration of dissolved material to a region of high concentration.
    c. hypertonic solution to a region of hypotonic solution.
    d. negative osmotic potential to a region of positive osmotic potential.
    e. low concentration of water to a region of high concentration of water.
    *Answer: b*

49. Clathrin-coated pits are structures associated with
    a. active transport.
    b. phagocytosis.
    c. flagellar movement.
    d. receptor-mediated endocytosis.
    e. secretory vesicles.
    *Answer: d*

■50. Which of the following molecules is the most likely to diffuse across a cell membrane?
    a. Glucose
    b. $Na^+$
    c. A steroid
    d. A protein common in blood
    e. A peripheral protein
    *Answer: c*

■51. Cell growth can involve movement of membrane material from
    a. the cell membrane to the vesicles.
    b. the Golgi apparatus to the cell membrane.
    c. the smooth ER to the rough ER.
    d. coated pits to the inside of the cell.
    e. lysosomes to the cell membrane.
    *Answer: b*

52. An important function of specialized membranes found in certain organelles is to
    a. help the organelles move.
    b. protect the organelles from increased temperatures.
    c. transform energy.
    d. make use of the cells' internal genetic information.
    e. destroy cellular waste products.
    *Answer: c*

53. Membrane molecules that help individual cells organize themselves into tissues are known as
    a. peripheral proteins.
    b. cell adhesion molecules.
    c. clathrins.
    d. secondary active transport proteins.
    e. symports.
    *Answer: b*

54. For each molecule of ATP consumed during active transport of sodium and potassium,
    a. two Na$^+$ ions are imported and three K$^+$ ions are exported.
    b. two Na$^+$ ions are imported and one K$^+$ ion is exported.
    c. one K$^+$ ion is imported and three Na$^+$ ions are exported.
    d. two K$^+$ ions are imported and three Na$^+$ ions are exported.
    e. three K$^+$ ions are imported and two Na$^+$ ions are exported.
    *Answer: d*

55. Materials can be limited from moving through the spaces between cells by
    a. coated pits on the surfaces of the cells.
    b. desmosomes between the cells.
    c. carbohydrates on the surfaces of the cells.
    d. the extracellular matrix around the cells.
    e. tight junctions between the cells.
    *Answer: e*

56. Keratin is a protein found in
    a. plasmodesmata.
    b. desmosomes.
    c. gap junctions.
    d. clathrin pits.
    e. microfilaments.
    *Answer: b*

57. A concentration gradient of glucose across a membrane means that
    a. there are more moles of glucose on one side of the membrane than on the other.
    b. glucose molecules are more crowded on one side of the membrane than on the other.
    c. there is less water on one side of the membrane than on the other.
    d. the glucose molecules are chemically more tightly bonded on one side than on the other.
    e. there are more glucose molecules within the membrane than outside of the membrane.
    *Answer: b*

58. Connexons occur in
    a. the cytoskeleton.
    b. tight junctions.
    c. desmosomes.
    d. plasmodesmata.
    e. gap junctions.
    *Answer: e*

59. Transport proteins that simultaneously move two molecules across a membrane in the same direction are called
    a. uniports.
    b. symports.
    c. antiports.
    d. active transporters.
    e. diffusive ports.
    *Answer: b*

60. The only process that can bring glucose molecules into cells and does *not* involve the metabolic energy of ATP is
    a. phagocytosis.
    b. pinocytosis.
    c. active transport.
    d. diffusion.
    e. osmosis.
    *Answer: d*

61. The functional roles for different proteins found in membranes include all *except* which of the following?
    a. Allowing movement of molecules that would otherwise be excluded by the lipid components of the membrane
    b. Transferring signals from outside the cell to inside the cell
    c. Maintaining the shape of the cell
    d. Facilitating the transport of macromolecules across the membrane
    e. Stabilizing the lipid bilayer
    *Answer: e*

62. The tensile strength of connections between adjacent cells in tissues comes from
    a. gap junctions.
    b. tight junctions.
    c. desmosomes.
    d. slip junctions.
    e. weld junctions.
    *Answer: c*

63. Desmosomes include or associate with
    a. dense plaquelike regions.
    b. keratin fibers.
    c. external cell adhesion molecules.
    d. a and b only
    e. All of the above
    *Answer: e*

64. Secondary active transport involves all the following *except*
    a. the direct use of ATP.
    b. coupling to another transport system.
    c. use of regained energy from an existing gradient.
    d. the requirement of energy.
    e. the ability to concentrate the transported molecule.
    *Answer: a*

★65. How do amino acids get into cells against their concentration gradients?
    a. Simple diffusion
    b. Facilitated diffusion
    c. Primary active transport
    d. Secondary active transport
    e. Antiport transport
    *Answer: d*

66. Receptor-mediated endocytosis involves
    a. clathrin.
    b. coatomer.
    c. vesiclease.
    d. LDH.
    e. cholesterol.
    *Answer: a*

67. Stephen Hawking's muscle cells do not respond to stimulation by nerve cells. This condition is due to
    a. a failure of the plasma membrane's protein channels to open.
    b. a deficiency of lipids in the plasma membrane.
    c. a high level of cholesterol in the plasma membrane.
    d. the rapid passage of hydrophilic materials through the plasma membrane.
    e. poor diet and lack of exercise.
    *Answer: a*

68. A characteristic of plasma membranes that helps them to fuse during vesicle formation and phagocytosis is
    a. the ratio of 1 protein molecule for every 25 phospholipid molecules.
    b. the capacity of lipids to associate and maintain a bilayer organization.
    c. the constant fatty acid chain length and degree of saturation.
    d. the ability of phospholipid molecules to flip over and trade places with other phospholipid molecules.
    e. the asymmetrical distribution of membrane proteins.
    *Answer: b*

69. The plasma membranes of winter wheat are able to remain fluid when it is extremely cold by
    a. increasing the number of cholesterol molecules present.
    b. closing protein channels.
    c. decreasing the number of hydrophobic proteins present.
    d. replacing saturated fatty acids with unsaturated fatty acids.
    e. using fatty acids with longer tails.
    *Answer: d*

70. Protein movement within a membrane may be restricted by
    a. glycolipids and glycoproteins.
    b. closure of gated channels.
    c. the cytoskeleton and lipid rafts.
    d. cell adhesion.
    e. tight junctions and desmosomes.
    *Answer: c*

71. The difference between tight junctions, desmosomes, and gap junctions is that
    a. desmosomes and gap junctions contain keratin, whereas tight junctions have collagen.
    b. gap junctions and tight junctions have specialized protein channels called connexons; desmosomes do not.
    c. tight junctions and desmosomes have mechanical roles, whereas gap junctions facilitate communication between cells.
    d. desmosomes and gap junctions are found in epithelial tissue, whereas tight junctions are found in nerve cells.
    e. they all have different functions; however, their structure is the same.
    *Answer: c*

72. How does an ion channel exert its specificity for one ion and not another?
    a. It is a simple matter of charge and ionic size.
    b. The ion channel hydrates ions as they pass through.
    c. The ion channel makes use of aquaporins.
    d. There are recognition sites in the ion channel.
    e. The ion lets go of its water and is attracted to a channel pore protein.
    *Answer: e*

73. Which of the following processes is highly specific?
    a. Phagocytosis
    b. Receptor-mediated endocytosis
    c. Pinocytosis
    d. Diffusion
    e. Osmosis
    *Answer: b*

74. A patient is brought to the hospital severely dehydrated. An IV of normal saline is started immediately. Why doesn't the doctor order an IV of distilled water instead?
    a. An IV of distilled water would cause water to leave the cells, causing them to collapse.
    b. The patient needs the nutrients available in normal saline.
    c. An IV of distilled water would cause blood cells to swell and eventually burst.
    d. Normal saline is more economical than pure water.
    e. The distilled water may be contaminated by bacteria.
    *Answer: c*

75. Which of these is *not* a true statement regarding diffusion?
    a. Diffusion depends on the intrinsic kinetic energy of molecules.
    b. Diffusion continues until the concentrations are in equilibrium.
    c. In diffusion, molecules move from areas of greater concentration to areas of lesser concentration.
    d. Diffusion is a random process.
    e. Simple diffusion depends upon specific carrier proteins.
    *Answer: e*

76. Which type of membrane protein would you expect to remove most easily in a laboratory experiment?
    a. Integral proteins
    b. Channel proteins
    c. Peripheral proteins
    d. Transmembrane proteins
    e. Gated channels
    *Answer: c*

# 6 Energy, Enzymes, and Metabolism

## Fill in the Blank

1. Cells cannot create energy because _____.
   *Answer: energy cannot be created or destroyed*

\*2. Variations of enzymes that allow organisms to adapt to changing environments are termed _____.
   *FAnswer: isozymes*

3. Although some enzymes consist entirely of one or more polypeptide chains, others possess a tightly bound nonprotein portion called a _____.
   *Answer: prosthetic group*

4. Cells mostly use _____ as an immediate source of energy to drive reactions.
   *Answer: ATP*

\*5. A _____ reaction, in which one reaction is used to drive another, is the major means of carrying out energy-requiring reactions within cells.
   *Answer: coupled*

6. The second law of thermodynamics states that the _____, or disorder, of the universe is constantly increasing.
   *Answer: entropy*

7. When a drop of ink is added to a beaker of water, the dye molecules become randomly dispersed throughout the water. This is an example of an increase in _____.
   *Answer: entropy*

8. For a reaction to be spontaneous, the change in free energy of the reaction, $\Delta G$, must be _____.
   *Answer: negative*

9. The enzyme phosphoglucoisomerase catalyzes the conversion of glucose 6-phosphate to fructose 6-phosphate. The region on phosphoglucoisomerase where glucose 6-phosphate binds is called the _____.
   *Answer: active site*

10. The $\Delta G$ of a spontaneous reaction is negative, indicating that the reaction releases free energy. Such a reaction is _____.
   *Answer: exergonic*

11. Enzymes are biological _____.
   *Answer: catalysts*

12. The zinc ion in the active site of the enzyme thermolysin is called a _____.
   *Answer: prosthetic group*

13. When an enzyme is heated until its three-dimensional structure is destroyed, the enzyme is said to be _____.
   *Answer: denatured*

14. The temperature of water above a waterfall is probably _____ than the temperature where the water falls.
   *Answer: colder*

15. _____ is the term used for all the chemical activity of a living organism.
   *Answer: Metabolism*

16. Heat, light, electricity, and motion are all examples of _____ energy.
   *Answer: kinetic*

17. The energy in a system that exists due to position is _____ energy.
   *Answer: potential*

18. Potential energy can be converted to _____ energy, which does work.
   *Answer: kinetic*

19. The building up of molecules in a living system is _____; the breaking down of molecules in a living system is _____.
   *Answer: anabolism; catabolism*

20. The first law of thermodynamics is that _____.
   *Answer: energy is neither created nor destroyed*

## Multiple Choice

1. Water held back by a dam represents what kind of energy?
   a. Hydroelectric
   b. Irrigation
   c. Potential
   d. Kinetic
   e. At times, all of the above
   *Answer: c*

2. The change in free energy is related to a
   a. change in heat.
   b. change in entropy.
   c. change in pressure.
   d. Both a and b
   e. a, b, and c
   *Answer: d*

3. Enzymes are sensitive to
   a. temperature.
   b. pH.
   c. irreversible inhibitors such as DIPF.
   d. allosteric effectors.
   e. All of the above
   *Answer: e*

4. End products of biosynthetic pathways often act to block the initial step in that pathway. This phenomenon is called
   a. allosteric inhibition.
   b. denaturation.
   c. branch pathway inhibition.
   d. feedback inhibition.
   e. binary inhibition.
   *Answer: d*

\*5. Which of the following identifies a group of enzymes that is important in fine-tuning the metabolic activities of cells?
   a. Isozymes
   b. Alloenzymes
   c. Allosteric enzymes
   d. Both a and c
   e. a, b, and c
   *Answer: d*

■6 Competitive and noncompetitive enzyme inhibitors differ with respect to
   a. the precise location on the enzyme to which they bind.
   b. their pH.
   c. their binding affinities.
   d. their energies of activation.
   e. None of the above
   *Answer: a*

■7. During photosynthesis, plants use light energy to synthesize glucose from carbon dioxide. However, plants do not use up energy during photosynthesis; they merely convert it from light energy to chemical energy. This is an illustration of
   a. increasing entropy.
   b. chemical equilibrium.
   c. the first law of thermodynamics.
   d. the second law of thermodynamics.
   e. a spontaneous reaction.
   *Answer: c*

■8. The standard free energy change for the hydrolysis of ATP to ADP + $P_i$ is –7.3 kcal/mol. What can you conclude from this information?
   a. The reaction will never reach equilibrium.
   b. The free energy of ADP and phosphate is higher than the free energy of ATP.
   c. The reaction requires energy.
   d. The reaction is endergonic.
   e. The reaction is exergonic.
   *Answer: e*

■9. The hydrolysis of maltose to glucose is an exergonic reaction. Which of the following statements is true?
   a. The reaction requires the input of free energy.
   b. The free energy of glucose is larger than the free energy of maltose.
   c. The reaction is not spontaneous.
   d. The reaction releases free energy.
   e. At equilibrium, the concentration of maltose is higher than the concentration of glucose.
   *Answer: d*

10. The first law of thermodynamics states that the total energy in the universe is
    a. decreasing.
    b. increasing.
    c. constant.
    d. being converted to free energy.
    e. being converted to matter.
    *Answer: c*

■11. If the enzyme phosphohexosisomerase is added to a 0.3 M solution of fructose 6-phosphate, and the reaction is allowed to proceed to equilibrium, the final concentrations are 0.2 M glucose 6-phosphate and 0.1 M fructose 6-phosphate. These data give an equilibrium constant of 2. What is the equilibrium constant if the initial concentration of fructose 6-phosphate is 3 M?
    a. 2
    b. 3
    c. 5
    d. 10
    e. 20
    *Answer: a*

■12. You are studying the effects of temperature on the rate of a particular enzyme-catalyzed reaction. When you increase the temperature from 40°C to 70°C, what effect will this have on the rate of the reaction?
    a. It will increase.
    b. It will decrease.
    c. It will decrease to zero because the enzyme denatures.
    d. It will increase and then decrease.
    e. This cannot be answered without more information.
    *Answer: e*

■13. If $\Delta G$ of a chemical reaction is negative and the change in entropy is positive, what can you conclude about the reaction?
    a. It requires energy.
    b. It is endergonic.
    c. It is exergonic.
    d. It will not reach equilibrium.
    e. It decreases the disorder in the system.
    *Answer: c*

14. Which of the following determines the rate of a reaction?
    a. $\Delta S$
    b. $\Delta G$
    c. $\Delta H$
    d. The activation energy
    e. The overall change in free energy
    *Answer: d*

15. In a chemical reaction, transition-state species have free energies
    a. lower than either the reactants or the products.
    b. higher than either the reactants or the products.
    c. lower than the reactants, but higher than the products.
    d. higher than the reactants, but lower than the products.
    e. lower than the reactants, but the same as the products.
    *Answer: b*

■16. The hydrolysis of sucrose to glucose and fructose is exergonic. However, if you dissolve sucrose in water and keep the solution overnight at room temperature, there is no detectable conversion to glucose and fructose. Why?
    a. The change in free energy of the reaction is positive.
    b. The activation energy of the reaction is high.
    c. The change in free energy of the reaction is negative.

d. This is a condensation reaction.

e. The free energy of the products is higher than the free energy of the reactants.

*Answer: b*

17. The enzyme α-amylase increases the rate at which starch is broken down into smaller oligosaccharides. It does this by

a. decreasing the equilibrium constant of the reaction.

b. increasing the change in free energy of the reaction.

c. decreasing the change in free energy of the reaction.

d. increasing the change in entropy of the reaction.

e. lowering the activation energy of the reaction.

*Answer: e*

18. The enzyme glyceraldehyde 3-phosphate dehydrogenase catalyzes the reaction glyceraldehyde 3-phosphate → 1,3-diphosphoglycerate. The region of the enzyme where glyceraldehyde 3-phosphate binds is called the

a. transition state.

b. groove.

c. catalyst.

d. active site.

e. energy barrier.

*Answer: d*

*19. The enzyme glucose oxidase binds the six-carbon sugar glucose and catalyzes its conversion to glucono-1,4-actone. Mannose is also a six-carbon sugar, but glucose oxidase cannot bind mannose. The specificity of glucose oxidase is based on the

a. free energy of the transition state.

b. activation energy of the reaction.

c. change in free energy of the reaction.

d. three-dimensional shape and structure of the active site.

e. rate constant of the reaction.

*Answer: d*

■20. In the presence of alcohol dehydrogenase, the rate of reduction of acetaldehyde to ethanol increases as you increase the concentration of acetaldehyde. Eventually, the rate of the reaction reaches a maximum, at which point further increases in the concentration of acetaldehyde have no effect. Why?

a. All of the alcohol dehydrogenase molecules are bound to acetaldehyde molecules.

b. At high concentrations of acetaldehyde, the activation energy of the reaction increases.

c. At high concentrations of acetaldehyde, the activation energy of the reaction decreases.

d. The enzyme is no longer specific for acetaldehyde.

e. At high concentrations of acetaldehyde, the change in free energy of the reaction decreases.

*Answer: a*

21. When an enzyme catalyzes both an exergonic reaction and an endergonic reaction, the two reactions are said to be

a. substrates.

b. endergonic.

c. kinetic.

d. activated.

e. coupled.

*Answer: e*

■22. In glycolysis, the exergonic reaction 1,3-diphosphoglycerate → 3-phosphoglycerate is coupled to the reaction ADP + $P_i$ → ATP. Which of the following is most likely to be true about the reaction ADP + $P_i$ → ATP?

a. The reaction never reaches equilibrium.

b. The reaction is spontaneous.

c. There is a large decrease in free energy.

d. The reaction is endergonic.

e. Temperature will not affect the rate constant of the reaction.

*Answer: d*

■23. Trypsin and elastase are both enzymes that catalyze hydrolysis of peptide bonds. But trypsin only cuts next to lysine and elastase only cuts next to alanine. Why?

a. Trypsin is a protein, and elastase is not.

b. ΔG for the two reactions is different.

c. The shape of the active site for the two enzymes is different.

d. One of the reactions is endergonic, and the other is exergonic.

e. Hydrolysis of lysine bonds requires water; hydrolysis of alanine bonds does not.

*Answer: c*

24. The enzyme catalase has a ferric ion tightly bound to the active site. The ferric ion is called a(n)

a. side chain.

b. enzyme.

c. coupled reaction.

d. prosthetic group.

e. substrate.

*Answer: d*

■■25. The addition of the competitive inhibitor mevinolin slows the reaction HMG-CoA → mevalonate, which is catalyzed by the enzyme HMG-CoA reductase. How could you overcome the effects of mevinolin and increase the rate of the reaction?

a. Add more mevalonate

b. Add more HMG-CoA

c. Lower the temperature of the reaction

d. Add a prosthetic group

e. Lower the rate constant of the reaction

*Answer: b*

■■26. How does a noncompetitive inhibitor inhibit binding of a substrate to an enzyme?

a. It binds to the substrate.

b. It binds to the active site.

c. It lowers the activation energy.

d. It increases the ΔG of the reaction.

e. It changes the shape of the active site.

*Answer: e*

27. Which type of inhibitor can be overcome completely by the addition of more substrate?

a. Irreversible

b. Noncompetitive

c. Competitive

d. Prosthetic

e. Isotonic

*Answer: c*

28. Binding of substrate to the active site of an enzyme is

a. reversible.

b. irreversible.

c. noncompetitive.

d. coupled.

e. allosteric.

*Answer: a*

29–30. Consider the following metabolic pathway. Reactants and products are designated by capital letters; enzymes are designated by numbers.

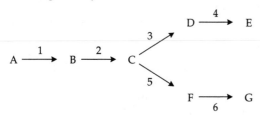

29. Which enzyme is end product G most likely to inhibit?
    a. 1
    b. 2
    c. 3
    d. 5
    e. 6
    *Answer: d*

30. Assume that end product E is a negative feedback regulator of enzyme 1. What happens to a cell when it is grown in the presence of large amounts of E?
    a. The cell can't make G.
    b. The cell can't make A.
    c. The cell makes too much G.
    d. The cell makes too much E.
    e. The cell makes too much D.
    *Answer: a*

31. An allosteric inhibitor
    a. decreases the concentration of inactive enzyme.
    b. decreases the concentration of active enzyme.
    c. increases the concentration of product.
    d. decreases the concentration of substrate.
    e. increases the concentration of enzyme–substrate complex.
    *Answer: b*

32. An RNA molecule that has enzyme activity is called
    a. RNAse.
    b. ribonuclease.
    c. an allosteric enzyme.
    d. a regulatory enzyme.
    e. a ribozyme.
    *Answer: e*

33. The presence in a system of energy that is unusable for the purpose of doing work is related to the system's
    a. temperature.
    b. entropy.
    c. work.
    d. thermodynamics.
    e. equilibrium.
    *Answer: b*

34. The molecules that are acted on by an enzyme are called
    a. products.
    b. substrates.
    c. carriers.
    d. prosthetics.
    e. effectors.
    *Answer: b*

35. The sum total of all the chemical reactions in a living structure is called its
    a. energetics.
    b. activity.
    c. digestive power.
    d. entropy.
    e. metabolism.
    *Answer: e*

36. The rate of a chemical reaction in a cell is the measure of how
    a. often the reaction occurs.
    b. quickly the reaction reaches equilibrium.
    c. much energy must be added to have the reaction occur.
    d. much activation energy is required to have the reaction occur.
    e. easily the reaction is inhibited.
    *Answer: b*

37. The statement "Enzymes are highly specific" means that certain
    a. enzymes are found in certain cells.
    b. reactions involving certain substrates are catalyzed by certain enzymes.
    c. enzymes require certain concentrations of substrates.
    d. reactions with certain activation energies are catalyzed by certain enzymes.
    e. concentrations of substrates work with certain enzymes.
    *Answer: b*

38. An active site is
    a. the part of the substrate that binds with an enzyme.
    b. the part of the enzyme that binds with a substrate.
    c. the site where energy is added to an enzyme catalyst.
    d. the site where enzymes are found in cells.
    e. None of the above
    *Answer: b*

■39. Enzymatic reactions can become saturated as substrate concentration increases because
    a. enzymes have the maximum possible number of hydrogen atoms attached to them.
    b. the concentration of substrate cannot increase any higher.
    c. substrates are inhibitors of enzymes.
    d. the activation energy of the reaction cannot be further lowered.
    e. there are a limited number of the enzyme molecules present.
    *Answer: e*

40. Competitive inhibitors of enzymes work by
    a. fitting into the active site.
    b. fitting into a site other than the active site.
    c. altering the shape of the enzyme.
    d. changing the enzyme into an inactive form.
    e. increasing the activation energy of the enzyme-catalyzed reaction.
    *Answer: a*

41. Allosteric inhibitors act by
    a. decreasing the amount of enzyme molecules.
    b. increasing the amount of enzyme molecules.
    c. decreasing the amount of the inactive form of the enzyme.
    d. decreasing the amount of the active form of the enzyme.
    e. increasing the amounts of substrate.
    *Answer: d*

42. Negative feedback in a sequence of chemical reactions involves a chemical that appears
    a. late in the sequence and inhibits an earlier reaction.
    b. early in the sequence and inhibits a later reaction.
    c. early in the sequence and activates a later reaction.
    d. late in the sequence and activates an earlier reaction.
    e. late in the sequence and inhibits a later reaction.
    *Answer: a*

43. Denatured enzymes are the same as
    a. ribozymes.
    b. abzymes.
    c. isozymes.
    d. destroyed enzymes.
    e. coenzymes.
    *Answer: d*

44. The inhibition of enzyme activity by noncompetitive inhibitors can be reduced
    a. by decreasing the concentration of allosteric enzymes.
    b. by decreasing the concentration of substrate.
    c. by increasing the concentration a competitive inhibitor.
    d. by increasing the concentration of substrate.
    e. only when they become unbound.
    *Answer: e*

45. The concentration of a substrate in a reaction at equilibrium depends most strongly on the concentration(s) of
    a. enzyme.
    b. the active form of the enzyme.
    c. the activator.
    d. the other substrates and products.
    e. the products.
    *Answer: e*

46. An allosteric site of an enzyme is the place where a(n) _____ may bind.
    a. competitive inhibitor
    b. substrate
    c. prosthetic group
    d. activator
    e. coenzyme
    *Answer: d*

47. In some cases, a substrate–enzyme complex is stabilized by
    a. hydrogen bonds.
    b. covalent bonds.
    c. ionic attractions.
    d. hydrophobic interactions.
    e. All of the above
    *Answer: e*

48. Once a spontaneous reaction is initiated, the speed with which it reaches equilibrium without a catalyst is influenced by
    a. the equilibrium constant.
    b. a change in free energy.
    c. a change in entropy.
    d. activation energy.
    e. standard free energy change.
    *Answer: b*

49. Factors that can either activate or inhibit allosteric enzymes are called
    a. proteins.
    b. coenzymes.
    c. sites.
    d. effectors.
    e. competitors.
    *Answer: d*

50. The diversity of chemical reactions occurring in a cell depends mostly on which molecules in the cell?
    a. Isozymes
    b. Coenzymes
    c. Ribozymes
    d. Abzymes
    e. Enzymes
    *Answer: e*

*51. When organisms move from one environment to another, they sometimes synthesize variations of existing enzymes, which are called
    a. coenzymes.
    b. abzymes.
    c. isozymes.
    d. effectors.
    e. activators.
    *Answer: c*

52. A type of enzyme inhibitor that binds within the enzyme's active site is termed
    a. allosteric.
    b. noncompetitive.
    c. competitive.
    d. extracompetitive.
    e. None of the above
    *Answer: c*

*53. The catalysis mechanism used by lysozyme is
    a. acid-base catalysis.
    b. covalent catalysis.
    c. metal cofactor redox catalysis.
    d. induced strain.
    e. unknown.
    *Answer: d*

54. The ability of an enzyme's active site to sometimes bind inhibitors that are larger than the substrate is called
    a. induced fit.
    b. enzyme flex.
    c. the lock and key paradox.
    d. substrate-induced active site shaping.
    e. enzyme retrofit.
    *Answer: a*

55. The process that involves an end product acting as an inhibitor of an earlier step in a metabolic pathway is called
    a. feedback activation.
    b. feedback inhibition.
    c. positive feedback.
    d. concerted activation.
    e. competitive inhibition.
    *Answer: b*

56. What can never be created or destroyed?
    a. Entropy
    b. Energy
    c. Free energy
    d. Thermal energy
    e. Potential energy
    *Answer: b*

57. The maximum possible rate of an enzyme reaction influenced by a competitive inhibitor depends on the concentration of
    a. inhibitor.
    b. substrate.
    c. product.
    d. enzyme.
    e. free energy.
    *Answer: b*

*58. Enzymes of the acid-base catalysis type contain
    a. a metal ion bound to a side chain.
    b. a prosthetic group.
    c. a coenzyme.
    d. an acid or base amino acid residue in the active site.
    e. a covalent-activated active site.
    *Answer: d*

59. Which of the following is true regarding allosteric regulators?
    a. Positive regulators stabilize the active form of the enzyme.
    b. All enzymes are allosterically regulated.
    c. Enzymes that are allosterically regulated are made of multiple polypeptide subunits.
    d. Both the active site and the regulatory site are present on the same subunit.
    e. Allosteric regulators bind to the active site, blocking enzyme function.
    *Answer: a*

60. When ATP loses a phosphate to form ADP,
    a. free energy is released by the loss of a phosphate.
    b. energy is consumed.
    c. the reaction ends.
    d. chemical energy is converted to light energy.
    e. ribose loses an oxygen to become deoxyribose.
    *Answer: a*

61. Fireflies
    a. release a considerable amount of energy as heat.
    b. light up to signal danger.
    c. use ATP to begin luciferin oxidation.
    d. are constantly converting light energy into chemical energy.
    e. have a short life cycle due to rapid depletion of ATP.
    *Answer: c*

62. How does the second law of thermodynamics apply to organisms?
    a. As energy transformations occur, free energy increases and unusable energy decreases.
    b. To maintain order, life requires a constant input of energy.
    c. The potential energy of ATP is converted to kinetic energy such as muscle contractions.
    d. Reactions occur only with an input of energy.
    e. It doesn't; the complexity of organisms is in disagreement with the second law.
    *Answer: b*

63. Which of the following is true of ATP?
    a. The hydrolysis of ATP is exergonic.
    b. ATP consists of adenine bonded to deoxyribose.
    c. ATP releases a relatively small amount of energy when hydrolyzed.
    d. An active cell requires a hundred or so molecules of ATP per second.

    e. On average, ATP is consumed within a second of its formation.
    *Answer: e*

64. Aspirin reduces swelling and pain by
    a. adding hydroxyl groups to amino acids.
    b. hindering the conversion of linear fatty acids to a ring structure.
    c. blocking a substrate's access to enzymes.
    d. accelerating the formation of a fatty acid ring structure.
    e. repairing damage to the stomach wall and aiding in blood clotting.
    *Answer: b*

65. Prescription drugs are designed to
    a. block specific chemical transformations by inhibiting specific enzymes.
    b. break chemical bonds, thereby inhibiting metabolism.
    c. phosphorylate many different molecules.
    d. block the energy coupling cycle of ATP.
    e. block acetylcholinesterase and subsequent nerve impulse transmission.
    *Answer: a*

66. Which of the following statements about enzymes is *false*?
    a. An enzyme changes shape when it binds to a substrate.
    b. Enzymes lower the activation energy.
    c. Enzymes are highly specific.
    d. An enzyme may orient substrates, induce strain, or temporarily add chemical groups.
    e. Most enzymes are much smaller than their substrates.
    *Answer: e*

67. The enzyme sucrase increases the rate at which sucrose is broken down into glucose and fructose. Sucrase works by
    a. increasing the amount of free energy of the reaction.
    b. lowering the activation energy of the reaction.
    c. decreasing the equilibrium constant of the reaction.
    d. supplying energy to speed up the reaction.
    e. changing the shape of the active site.
    *Answer: b*

68. Enzymes are highly sensitive to pH and temperature because
    a. changes in the environment raise their activation energy.
    b. changes in temperature and pH readily break their hydrogen bonds.
    c. of their three-dimensional structure and side chains.
    d. at extreme temperatures and pH levels, coenzymes add chemical groups to the substrate.
    e. extremes of temperature and pH level change the ionization rate.
    *Answer: c*

69. How do competitive and noncompetitive enzyme inhibitors differ?
    a. Competitive inhibitors bind to the active site, whereas noncompetitive inhibitors change the shape of the active site.
    b. Competitive inhibitors have a higher energy of activation than noncompetitive inhibitors have.
    c. They function at different pH values.
    d. Noncompetitive enzyme inhibitors contain magnesium, whereas competitive inhibitors contain iron.
    e. Noncompetitive enzyme inhibitors are reversible, whereas competitive inhibitors are irreversible.
    *Answer: a*

# 7 Cellular Pathways that Harvest Chemical Energy

## Fill in the Blank

1. Oxidation and _____ occur together.
   *Answer: reduction*

2. Due to its ability to carry electrons and free energy, _____ is the most common electron carrier in cells.
   *Answer: NAD*

3. Pyruvate is _____ to form acetate.
   *Answer: oxidized*

4. The chemiosmotic formation of ATP during the operation of the respiratory chain is called _____.
   *Answer: oxidative phosphorylation*

5. The loss of an electron by a ferrous ion ($Fe^{2+}$) to yield a ferric ion ($Fe^{3+}$) is called _____.
   *Answer: oxidation*

6. In a redox reaction, the reactant that becomes oxidized is called a _____.
   *Answer: reducing agent*

7. A chemical reaction resulting in the transfer of electrons or hydrogen atoms is called a _____ reaction.
   *Answer: redox*

8. The pathway for the oxidation of glucose to pyruvate is called _____.
   *Answer: glycolysis*

9. The conversion of glucose to lactic acid is a form of

   _____.
   *Answer: fermentation*

10. Fatty acids must be converted to _____ before they can be used for respiratory ATP production.
    *Answer: acetyl CoA*

11. During alcoholic fermentation, $NAD^+$ is regenerated by the reduction of acetaldehyde to _____.
    *Answer: ethanol*

12. NAD is an abbreviation for _____.
    *Answer: nicotinamide adenine dinucleotide*

13. An enzyme that transfers a phosphate group from ATP to another protein is called a _____.
    *Answer: kinase*

## Multiple Choice

1. ATP is
   a. a short-term energy-storage compound.
   b. the cell's principle compound for energy transfers.
   c. synthesized within mitochondria.
   d. the molecule all living cells rely on to do work.
   e. All of the above
   *Answer: e*

2. When a molecule loses hydrogen atoms (not hydrogen ions), it becomes
   a. reduced.
   b. oxidized.
   c. redoxed.
   d. hydrogenated.
   e. hydrolyzed.
   *Answer: b*

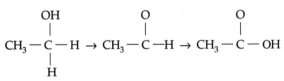

*3. In the diagram shown above, the first reaction is _____, and the second reaction is _____.
   a. oxidation; oxidation
   b. oxidation; reduction
   c. reduction; reduction
   d. reduction; oxidation
   e. oxiduction; oxiduction
   *Answer: a*

4. The end product of glycolysis is
   a. pyruvate.
   b. the starting point for the citric acid cycle.
   c. the starting point for the fermentation pathway.
   d. Both a and b
   e. a, b, and c
   *Answer: e*

5. In the conversion of succinate to fumarate, hydrogen atoms are transferred to FAD. The conversion of succinate and FAD to fumarate and $FADH_2$ is an example of
   a. hydrolysis.
   b. an allosteric reaction.
   c. a metabolic pathway.
   d. an aerobic reaction.
   e. a redox reaction.
   *Answer: e*

6. The oxidation of malate to oxaloacetate is coupled to the reduction of NAD$^+$ to NADH + H$^+$. NAD$^+$ is a(n)
   a. reducing agent.
   b. oxidizing agent.
   c. vitamin.
   d. phosphate ester.
   e. phosphorylating agent.
   *Answer: b*

7. When NADH donates two electrons to ubiquinone during respiration, ubiquinone is
   a. reduced.
   b. oxidized.
   c. phosphorylated.
   d. aerobic.
   e. hydrolyzed.
   *Answer: a*

8. Which of the following oxidizes other compounds by gaining free energy and hydrogen atoms and reduces other compounds by giving up free energy and hydrogen atoms?
   a. Vitamins
   b. Adenine
   c. ATP
   d. NAD
   e. Riboflavin
   *Answer: d*

9. Within the cell, isocitrate dehydrogenase, an enzyme of the citric acid cycle, is located in the
   a. thylakoids.
   b. cytoplasm.
   c. chloroplast.
   d. mitochondrial matrix.
   e. plasma membrane.
   *Answer: d*

10. In the first reaction of glycolysis, glucose receives a phosphate group from ATP. This reaction is
    a. respiration.
    b. a redox reaction.
    c. exergonic.
    d. endergonic.
    e. fermentation.
    *Answer: d*

■■11. How does the reduction of pyruvate to lactic acid during fermentation allow glycolysis to continue in the absence of oxygen?
    a. Water is formed during this reaction.
    b. This reaction is a kinase reaction.
    c. This reaction is coupled to the oxidation of NADH to NAD$^+$.
    d. This reaction is coupled to the formation of ATP.
    e. This reaction is coupled to the reduction of NAD$^+$ to NADH.
    *Answer: c*

12. During glycolysis, for each mole of glucose oxidized to pyruvate
    a. 6 moles of ATP are produced.
    b. 4 moles of ATP are used, and 2 moles of ATP are produced.
    c. 2 moles of ATP are used, and 4 moles of ATP are produced.

d. 2 moles of NAD$^+$ are produced.
e. no ATP is produced.
*Answer: c*

■■13. In steps 6 through 10 of glycolysis, the conversion of one mole of glyceraldehyde 3-phosphate to pyruvate yields 2 moles of ATP. But the oxidation of glucose to pyruvate produces a total of 4 moles of ATP. Where do the remaining 2 moles of ATP come from?
    a. One mole of glucose yields 2 moles of glyceraldehyde 3-phosphate.
    b. Two moles of ATP are used during the conversion of glucose to glyceraldehyde 3-phosphate.
    c. Glycolysis produces 2 moles of NADH.
    d. Fermentation of pyruvate to lactic acid yields 2 moles of ATP.
    e. Fermentation of pyruvate to lactic acid yields 2 moles of NAD$^+$.
    *Answer: a*

14. For glycolysis to continue, all cells require
    a. a respiratory chain.
    b. oxygen.
    c. mitochondria.
    d. chloroplasts.
    e. NAD$^+$.
    *Answer: e*

*15. The free energy released during the oxidation of glyceraldehyde 3-phosphate to 1,3 bisphosphoglycerate is
    a. used to oxidize NADH.
    b. lost as heat.
    c. used to synthesize ATP.
    d. used to reduce NAD$^+$.
    e. stored in lactic acid.
    *Answer: d*

16. The oxidation of pyruvate to carbon dioxide is called
    a. fermentation.
    b. the citric acid cycle.
    c. glycolysis.
    d. oxidative phosphorylation.
    e. the respiratory chain.
    *Answer: b*

*17. During the energy-priming portion of glycolysis, the phosphates from ATP molecules are
    a. added to the first and sixth carbons.
    b. added to the second and fourth carbons.
    c. wasted, as an energy investment.
    d. used to make pyruvate.
    e. used to make lactate.
    *Answer: a*

18. Which of the following is produced during the citric acid cycle?
    a. FAD
    b. Pyruvate
    c. Reduced electron carriers
    d. Lactic acid
    e. Water
    *Answer: c*

■19. Some of the free energy released by oxidation of pyruvate to acetate is stored in acetyl CoA. How does acetyl CoA store free energy?
    a. Acetyl CoA has a higher free energy than acetate.
    b. Acetyl CoA is an electron carrier.

c. Acetyl CoA is a phosphate donor.
d. Acetate + CoA → acetyl CoA is an exergonic reaction.
e. Reduction of acetyl CoA is coupled to ATP synthesis.
*Answer: a*

20. The oxidizing agent at the end of the respiratory chain is
a. $O_2$.
b. $NAD^+$.
c. ATP.
d. FAD.
e. ubiquinone.
*Answer: a*

21. During the citric acid cycle, energy stored in acetyl CoA is used to
a. create a proton gradient.
b. drive the reaction $ADP + P_i \rightarrow ATP$.
c. reduce $NAD^+$ to NADH.
d. drive the reaction oxaloacetate → citric acid.
e. reduce FAD to $FADH_2$.
*Answer: d*

22. During the citric acid cycle, oxidative steps are coupled to
a. oxidative phosphorylation.
b. the oxidation of water.
c. the oxidation of electron carriers.
d. the hydrolysis of ATP.
e. the reduction of electron carriers.
*Answer: e*

*23. Animals breathe in air containing oxygen and breathe out air with less oxygen and more carbon dioxide. The carbon dioxide comes from
a. hydrocarbons and the air.
b. the citric acid cycle.
c. glycolysis.
d. waste products.
e. All of the above
*Answer: b*

**■24. The drug 2,4-dinitrophenol (DNP) destroys the proton gradient across the inner mitochondrial membrane. What would you expect to be the effect of incubating isolated mitochondria in a solution of DNP?
a. Oxygen would no longer be reduced to water.
b. No ATP would be made during transport of electrons down the respiratory chain.
c. Mitochondria would show a burst of increased ATP synthesis.
d. Glycolysis would stop.
e. Mitochondria would switch from glycolysis to fermentation.
*Answer: b*

25. Electron transport within NADH-Q reductase, cytochrome reductase, and cytochrome oxidase can be coupled to proton transport from the mitochondrial matrix to the space between the inner and outer mitochondrial membranes because those protein complexes are
a. in the mitochondrial matrix.
b. within the inner mitochondrial membrane.
c. in the space between the inner and outer mitochondrial membranes.
d. in the cytoplasm.
e. loosely attached to the inner mitochondrial membrane.
*Answer: b*

26. According to the chemiosmotic theory, the energy for the synthesis of ATP during the flow of electrons down the respiratory chain is provided directly by the
a. hydrolysis of GTP.
b. reduction of $NAD^+$.
c. diffusion of protons.
d. reduction of FAD.
e. hydrolysis of ATP.
*Answer: c*

27. In the absence of oxygen, cells capable of fermentation
a. accumulate glucose.
b. no longer produce ATP.
c. accumulate pyruvate.
d. oxidize FAD.
e. oxidize NADH to produce $NAD^+$.
*Answer: e*

28. For bacteria to continue growing rapidly when they are shifted from an environment containing oxygen to an anaerobic environment, they must
a. increase the rate of the citric acid cycle.
b. produce more ATP per mole of glucose during glycolysis.
c. produce ATP during the oxidation of NADH.
d. increase the rate of transport of electrons down the respiratory chain.
e. increase the rate of the glycolytic reactions.
*Answer: e*

29. In alcoholic fermentation, $NAD^+$ is produced during the
a. oxidation of pyruvate to acetyl CoA.
b. reduction of pyruvate to lactic acid.
c. reduction of acetaldehyde to ethanol.
d. hydrolysis of ATP to ADP.
e. oxidation of glucose.
*Answer: c*

30. During the fermentation of 1 molecule of glucose, the net production of ATP is _____ molecule(s).
a. 1
b. 2
c. 3
d. 6
e. 8
*Answer: b*

31. The portion of aerobic respiration that produces the most ATP per mole of glucose is
a. oxidative phosphorylation.
b. the citric acid cycle.
c. glycolysis.
d. lactic acid fermentation.
e. alcoholic fermentation.
*Answer: a*

■32. More free energy is released during the citric acid cycle than during glycolysis, but only 1 mole of ATP is produced for each mole of acetyl CoA that enters the cycle. What happens to most of the remaining free energy that is produced during the citric acid cycle?
a. It is used to synthesize GTP.
b. It is used to reduce electron carriers.
c. It is lost as heat.
d. It is used to reduce pyruvate.
e. It is converted to kinetic energy.
*Answer: b*

*33. Animals inhale air containing oxygen and exhale air with less oxygen and more carbon dioxide. After inhalation, the extra oxygen from the air will mostly be found in
a. the carbon dioxide that is exhaled.
b. water.
c. organic molecules.
d. ethanol.
e. lactate.
*Answer: b*

34. When the supply of acetyl CoA being produced exceeds the demands of the citric acid cycle, some of the acetyl CoA is diverted to the synthesis of
a. pyruvate.
b. NAD.
c. proteins.
d. fatty acids.
e. lactic acid.
*Answer: d*

35. In the cell, the site of oxygen utilization is the
a. nucleus.
b. chloroplast.
c. endoplasmic reticulum.
d. mitochondrion.
e. cytosol.
*Answer: d*

36. Before starch can be used for respiratory ATP production, it must be hydrolyzed to
a. pyruvate.
b. fatty acids.
c. amino acids.
d. glucose.
e. oxaloacetate.
*Answer: d*

37. When yeast cells are switched from aerobic to anaerobic growth conditions, the rate of glycolysis increases. The rate of glycolysis is regulated by the concentration of _____ in the cells.
a. ATP
b. acetyl CoA
c. oxaloacetate
d. FAD
e. protein
*Answer: a*

38. When acetyl CoA builds up in the cell, it increases the activity of the enzyme that synthesizes oxaloacetate from pyruvate and carbon dioxide. Acetyl CoA is acting as a(n)
a. electron carrier.
b. substrate.
c. allosteric activator.
d. acetate donor.
e. proton pump.
*Answer: c*

39. In yeast, if the citric acid cycle is shut down because of a lack of oxygen, glycolysis will probably
a. shut down.
b. increase.
c. produce more ATP per mole of glucose.
d. produce more NADH per mole of glucose.
e. produce acetyl CoA for fatty acid synthesis.
*Answer: b*

40. The function of $NAD^+$ is to
a. cause the release of energy to adjacent cells when energy is needed in aerobic conditions.
b. hasten the release of energy when the cell has been deprived of oxygen.
c. carry hydrogen atoms and free energy from compounds being oxidized and to give hydrogen atoms and free energy to compounds being reduced.
d. block the release of energy to adjacent cells.
e. None of the above
*Answer: c*

41. The end result of glycolysis is the
a. creation of 38 molecules of ATP.
b. reduction of 8 molecules of NAD.
c. formation of 2 molecules of pyruvate.
d. conversion of 1 molecule of glucose to lactic acid.
e. None of the above
*Answer: c*

*42. The first five reactions of the glycolytic pathway result in
a. the addition of phosphates, modification of sugars, and formation of G3P.
b. oxidative steps, proton pumping, and reactions with oxygen.
c. the oxidation of pyruvate and formation of acetyl CoA.
d. the removal of hydrogen and protons from glucose.
e. None of the above
*Answer: a*

*43. For the citric acid cycle to proceed, it is necessary for
a. pyruvate to bind to oxaloacetate.
b. carbon dioxide to bind to oxaloacetate.
c. an acetyl group to bind to oxaloacetate.
d. water to be oxidized.
e. None of the above
*Answer: c*

44. Which of the following events occurs in the respiratory chain?
a. Carbon dioxide is released.
b. Carbon dioxide is reduced.
c. Cytochromes, FADH, and NADH are oxidized.
d. Only $NAD^+$ is reduced.
e. None of the above
*Answer: c*

45. The respiratory chain contains three large enzymes: NADH-Q reductase, cytochrome reductase, and cytochrome oxidase. The function of these enzymes is to
a. transport electrons.
b. ensure the production of water and oxygen.
c. regulate the passage of water through the chain.
d. oxidize NADH.
e. None of the above
*Answer: a*

46. Oxygen is used by
a. glycolysis.
b. the citric acid cycle.
c. the electron transport chain.
d. substrate-level phosphorylation.
e. ATP synthase.
*Answer: c*

47. Pyruvate oxidation generates
a. acetate.
b. NADH + $H^+$ from $NAD^+$.
c. a change in free energy.

d. a capture of energy.

e. All of the above

*Answer: e*

*48. Water is a by-product of cellular respiration. The water is produced as a result of the

a. combining of carbon dioxide with protons.

b. conversion of pyruvate to acetyl CoA.

c. degradation of glucose to pyruvate.

d. reduction of oxygen at the end of the electron transport chain.

e. None of the above

*Answer: d*

49. The formation of ethanol from pyruvate is an example of

a. an exergonic reaction.

b. an extra source of energy as the result of glycolysis.

c. a fermentation process that takes place in the absence of oxygen.

d. cellular respiration.

e. None of the above

*Answer: c*

50. Regardless of the electron or hydrogen acceptor employed, fermentation always produces

a. AMP.

b. DNA.

c. $P_i$.

d. $NAD^+$.

e. None of the above

*Answer: d*

■51. Yeast cells tend to create anaerobic conditions because they use oxygen more quickly than it can be replaced by diffusion through the cell membrane. For this reason, yeast cells

a. exhibit a red pigment.

b. exhibit a green pigment.

c. die.

d. produce ethanol.

e. None of the above

*Answer: d*

52. In human muscle cells, the fermentation process produces

a. lactic acid.

b. 12 moles of ATP.

c. pyruvic acid.

d. an excessive amount of energy.

e. None of the above

*Answer: a*

53. If a cell has an abundant supply of ATP, acetyl CoA may be used

a. to enhance fermentation.

b. to enhance oxidative metabolism.

c. for fatty acid synthesis.

d. to convert glucose to glycogen.

e. None of the above

*Answer: c*

54. In order for glucose to be used as an energy source, it is necessary that

a. glucose be formed from fructose.

b. glucose phosphate be formed from fructose phosphate.

c. glucose be degraded to carbon dioxide.

d. two ATP molecules be invested in the system.

e. None of the above

*Answer: d*

55. Many species derive their energy from fermentation. The function of fermentation is to

a. reduce $NAD^+$.

b. oxidize carbon dioxide.

c. oxidize $NADH + H^+$, ensuring a continued supply of ATP.

d. produce acetyl CoA.

e. None of the above

*Answer: c*

56. The chemiosmotic generation of ATP is driven by

a. osmotic movement of water into an area of high solute concentration.

b. the addition of protons to ADP and phosphate via enzymes.

c. oxidative phosphorylation.

d. a difference in $H^+$ concentration on both sides of a membrane.

e. None of the above

*Answer: d*

57. When a cell needs energy, cellular respiration is regulated by isocitrate dehydrogenase, an enzyme of the citric acid cycle. This enzyme is stimulated by

a. $H^+$.

b. heat.

c. oxygen.

d. ADP.

e. None of the above

*Answer: d*

58. Substrate-level phosphorylation is the transfer of a(n)

a. phosphate to a protein.

b. phosphate to a substrate.

c. phosphate to an ADP.

d. ATP to a protein.

e. phosphate from ATP to a substrate.

*Answer: c*

59. The proton-motive force is

a. ATP synthase.

b. the proton concentration gradient and electric charge difference.

c. a metabolic pathway.

d. a redox reaction.

e. None of the above

*Answer: b*

60. Most ATP produced in our bodies is made

a. by glycolysis.

b. in the citric acid cycle.

c. using ATP synthase.

d. from photosynthesis.

e. by burning fat.

*Answer: c*

61. In some mammals, such as new born humans and hibernating animals, body temperature is raised by means of

a. the uncoupling of respiration by the protein thermogenin.

b. an increase in the rate of glycolysis.

c. shivering.

d. leakage of hydrogen ions across the cell's plasma membrane.

e. cytochrome reductase.

*Answer: a*

62. ATP can be used to drive nonspontaneous reactions because
    a. nonspontaneous reactions are exergonic.
    b. the breakdown of ATP to ADP is exergonic.
    c. the breakdown of ATP to ADP is endergonic.
    d. when ATP is broken down to ADP, $P_i$ is released.
    e. ADP possesses more free energy than ATP does.
    *Answer: b*

63. The main control mechanism in glycolysis is the
    a enzyme isocitrate dehydrogenase.
    b. negative feedback of citrate accumulation.
    c. presence or absence of oxygen.
    d the enzyme phosphofructokinase.
    e. supply of NAD.
    *Answer: d*

64. In all cells, glucose metabolism begins with
    a. glycolysis.
    b. fermentation.
    c. pyruvate oxidation.
    d. the citric acid cycle.
    e. chemiosmosis.
    *Answer: a*

65. The citric acid cycle begins with
    a. glucose.
    b. pyruvate.
    c. acetyl CoA.
    d. NADH + H$^+$.
    e. ATP synthase.
    *Answer: c*

66. NAD
    a. is a key electron carrier in redox reactions.
    b. requires oxygen to function.
    c. is found only in prokaryotes.
    d. binds with an acetyl group to form acetyl CoA.
    e. detoxifies hydrogen peroxide.
    *Answer: a*

67. Which of the following is true of metabolic pathways?
    a. Complex chemical transformations in the cell occur in a single reaction.
    b. Each reaction in a pathway requires oxygen.
    c. Metabolic pathways in eukaryotes occur in the cytoplasm.
    d. Metabolic pathways vary from organism to organism.
    e. Each metabolic pathway is regulated by specific enzymes.
    *Answer: e*

68. Which process converts glucose to pyruvate, generating a small amount of ATP but no carbon dioxide?
    a. Pyruvate oxidation
    b. Glycolysis
    c. The citric acid cycle
    d. Respiratory chain
    e. Gluconeogenesis
    *Answer: b*

69. The hydrogen ion gradient is maintained by
    a. electron transport and proton pumping.
    b. the splitting of water.
    c. the ionization of glucose.
    d. ATP synthase.
    e. acetyl CoA.
    *Answer: a*

70. Which of these is *not* a characteristic of metabolic pathways?
    a. The product of one reaction becomes the reactant for the next reaction.
    b. They are a series of enzyme-catalyzed reactions.
    c. Almost all metabolic pathways are anabolic.
    d. Metabolic pathways are similar in all organisms.
    e. Many metabolic pathways are compartmentalized in eukaryotes.
    *Answer: c*

71. Which of the following is (are) true of the respiratory chain?
    a. Electrons are received from NADH and FADH$_2$.
    b. Electrons are passed from donor to recipient carrier molecules in a series of oxidation-reduction reactions.
    c. Usually the terminal electron acceptor is oxygen.
    d. Most of the enzymes are part of the inner mitochondrial membrane.
    e. All of the above
    *Answer: e*

72. When hydrogen ions are pumped from the mitochondrial matrix across the inner membrane into the intermembraneous space, the result is the
    a. formation of ATP.
    b. reduction of NAD$^+$.
    c. creation of a proton gradient.
    d. restoration of the Na$^+$-K$^+$ balance across the membrane.
    e. reduction of glucose to lactic acid.
    *Answer: c*

73. The respiratory chain contains three large enzyme complexes: NADH-Q reductase complex, cytochrome *c* reductase complex, and cytochrome *c* oxidase complex. The function of these enzyme complexes is to
    a. allow electrons to be transported.
    b. ensure the production of water and oxygen.
    c. regulate the passage of water through the respiratory chain.
    d. oxidize NADH.
    e. complete oxidation of pyruvate to acetate.
    *Answer: a*

74. The chemiosmotic generation of ATP is driven by
    a. osmotic movement of water into an area of high solute concentration.
    b. the addition of protons to ADP and phosphate via enzymes.
    c. oxidative phosphorylation.
    d. a difference in hydrogen ion concentration on both sides of the mitochondrial membrane.
    e. isocitrate dehydrogenase.
    *Answer: e*

# 8 Photosynthesis: Energy from the Sun

## Fill in the Blank

1. During the light reactions of photosynthesis, the synthesis of _____ is coupled to the diffusion of protons.
*Answer: ATP*

2. Atmospheric $CO_2$ enters plant leaves through openings called _____.
*Answer: stomata*

★ 3. In the 1800s, the summarized chemical reaction for photosynthesis was incorrect because it left out _____ as a product.
*Answer: water*

4. In noncyclic photophosphorylation, the electrons for the reduction of chlorophyll in photosystem II come from

_____.
*Answer: water*

5. When _____ are exposed to light and $CO_2$, four-carbon compounds are the first carbon-containing products.
*Answer: $C_4$ plants*

6. During the process of _____, rubisco catalyzes the reaction of RuBP with oxygen.
*Answer: photorespiration*

7. A group of scientists led by _____ conducted experiments demonstrating that RuBP is the $CO_2$ acceptor in the dark reactions of photosynthesis.
*Answer: Calvin*

8. When researchers shifted isolated chloroplasts from a low pH solution to a more alkaline solution, ATP synthesis occurred, even in the absence of light. This experiment was used to support the _____ mechanism of ATP formation in chloroplasts.
*Answer: chemiosmotic*

9. During cyclic photophosphorylation, the energy of photons is converted to the chemical energy of the product, _____.
*Answer: ATP*

★ 10. In $C_3$ plants, the Calvin–Benson cycle occurs in the chloroplasts of _____ cells, whereas in $C_4$ plants the cycle occurs in the _____ cells.
*Answer: mesophyll; bundle sheath*

11. In both photosynthesis and respiration, _____ synthesis is coupled to the diffusion of protons across a membrane.
*Answer: ATP*

12. The dark reactions take place in the (dark/light) _____.
*Answer: light*

★ 13. NADP is the abbreviation for _____.
*Answer: nicotinamide adenine dinucleotide phosphate*

14. The Calvin–Benson cycle is sometimes called the _____.
*Answer: dark reactions*

★ 15. The $O_2$ found in Earth's atmosphere is generated from the photosystem _____ of noncyclic photophosphorylation.
*Answer: II*

★ 16. During cyclic photophosphorylation, _____ rather than $NADP^+$ receives the electron from ferredoxin.
*Answer: plastoquinone*

17. The most abundant enzyme in the biosphere is _____.
*Answer: rubisco (or RuBP carboxylase)*

18. The absorption spectrum of chlorophyll *a* is (different from/similar to) _____ the action spectrum of chlorophyll *a*.
*Answer: similar to*

19. The wide range of wavelengths that photons can have is shown by the _____.
*Answer: electromagnetic spectrum*

## Multiple Choice

■ 1. How do red and blue light differ from one another?
a. They differ in intensity.
b. They have a different number of photons in each quantum.
c. Their wavelengths are different.
d. They differ in duration.
e. Red is radiant, whereas blue is electromagnetic.
*Answer: c*

■ 2. The wavelength of X rays is shorter than the wavelength of infrared rays. Which of the following is true?
a. X rays have more energy per photon than infrared rays have.
b. X rays have a smaller value for Planck's constant than do infrared waves.
c. X rays have a different absorption spectrum than do infrared waves.
d. X rays and infrared waves have the same frequency.
e. Infrared waves are in the ground state, whereas X rays are in the excited state.
*Answer: a*

3. A molecule has an absorption spectrum that shows maximum absorption within the wavelengths of visible light. This molecule is
a. a reducing agent.
b. a quantum.

c. a photon.
d. electromagnetic radiation.
e. a pigment.
*Answer: e*

4. When white light strikes a blue pigment, blue light is
   a. reduced.
   b. absorbed.
   c. converted to chemical energy.
   d. scattered or transmitted.
   e. used to synthesize ATP.
   *Answer: d*

5. A graph that plots the rate at which $CO_2$ is converted to glucose versus the wavelength of light illuminating a leaf is called
   a. a Planck equation.
   b. an absorption spectrum.
   c. enzyme kinetics.
   d. an electromagnetic spectrum.
   e. an action spectrum.
   *Answer: e*

6. The photosynthetic pigment chlorophyll *a* absorbs
   a. infrared light.
   b. orange-red and blue light.
   c. X rays.
   d. gamma rays.
   e. white light.
   *Answer: b*

7. Accessory pigments
   a. play no role in photosynthesis.
   b. transfer energy from chlorophyll to the electron transport chain.
   c. absorb only in the red wavelengths.
   d. allow plants to absorb visible light of intermediate wavelengths.
   e. transfer electrons to NADP.
   *Answer: d*

■8. Why is the absorption spectrum of chlorophyll *a not* identical to the action spectrum of photosynthesis?
   a. Accessory pigments contribute energy to drive photosynthesis.
   b. Chlorophyll *a* absorbs both red and blue light.
   c. Chlorophyll *a* reflects green light.
   d. Different wavelengths of light have different energies.
   e. Chlorophyll *a* can be activated by absorbing a photon of light.
   *Answer: a*

★9. In cyclic photophosphorylation, chlorophyll is reduced by
   a. NADPH.
   b. a chemiosmotic mechanism.
   c. plastoquinone.
   d. ATP.
   e. hydrogen ions liberated by the splitting of a water molecule.
   *Answer: c*

10. The energy difference between an electron excited by a photon and its ground state is
    a. less than the energy of the photon.
    b. greater than the energy of the photon.
    c. equal to the energy of the photon.
    d. related to the wavelength of the photon.

    e. Both c and d
    *Answer: e*

★11. The precise moment when light energy is captured in chemical energy is the point at which
    a. light shines on chlorophyll.
    b. water is hydrolyzed.
    c. chlorophyll is oxidized.
    d. chlorophyll is reduced.
    e. the $CO_2$ from air is captured in a sugar.
    *Answer: c*

12. Free energy is released in cyclic photophosphorylation
    a. by the formation of ATP.
    b. during the excitation of chlorophyll.
    c. during the fluorescence of chlorophyll.
    d. during each of the redox reactions of the electron transport chain.
    e. when electrons are transferred from photosystem I to photosystem II.
    *Answer: d*

13. During cyclic photophosphorylation, the energy to produce ATP is provided by
    a. heat.
    b. NADPH.
    c. ground-state chlorophyll.
    d. the redox reactions of the electron transport chain.
    e. the Calvin–Benson cycle.
    *Answer: d*

14. In noncyclic photophosphorylation, water is used for the
    a. hydrolysis of ATP.
    b. excitation of chlorophyll.
    c. reduction of chlorophyll.
    d. oxidation of NADPH.
    e. synthesis of chlorophyll.
    *Answer: c*

15. Photophosphorylation provides the Calvin–Benson cycle with
    a. protons and electrons.
    b. $CO_2$ and glucose.
    c. water and photons.
    d. light and chlorophyll.
    e. ATP and NADPH.
    *Answer: e*

16. In noncyclic photophosphorylation, the chlorophyll in photosystem I is reduced by
    a. water.
    b. an electron from the transport chain of photosystem II.
    c. two photons of light.
    d. NADPH.
    e. ATP.
    *Answer: b*

17. The enzyme ATP synthase couples the synthesis of ATP to
    a. the diffusion of protons.
    b. the reduction of $NADP^+$.
    c. the excitation of chlorophyll.
    d. the reduction of chlorophyll.
    e. $CO_2$ fixation.
    *Answer: a*

18–19. A suspension of algae is incubated in a flask in the presence of both light and $CO_2$. When it is transferred to the dark, the reduction of 3-phosphoglycerate to glyceraldehyde 3-phosphate is blocked.

**\*■18.** Why does this reaction stop when the algae are placed in the dark?
 a. It requires $CO_2$.
 b. It is an exergonic reaction.
 c. It requires ATP and NADPH + $H^+$.
 d. It requires $O_2$.
 e. Chlorophyll is not synthesized in the dark.
 *Answer: c*

**\*■19.** When the reduction of 3-phosphoglycerate (3PG) to glyceraldehyde 3-phosphate (G3P) is blocked, why does the concentration of ribulose bisphosphate (RuBP) decline?
 a. Ribulose bisphosphate is synthesized from glyceraldehyde 3-phosphate.
 b. Glyceraldehyde 3-phosphate is converted to glucose.
 c. Ribulose bisphosphate is used to synthesize 3-phosphoglycerate.
 d. Both a and b
 e. Both a and c
 *Answer: e*

20. The enzyme rubisco is found in
 a. chloroplasts.
 b. mitochondria.
 c. the cytoplasm.
 d. the nucleus.
 e. yeast.
 *Answer: a*

21. During $CO_2$ fixation, $CO_2$ combines with
 a. NADPH.
 b. 3PG.
 c. G3P.
 d. water.
 e. 1,5-ribulose bisphosphate.
 *Answer: e*

22. How many moles of $CO_2$ must enter the Calvin–Benson cycle for the synthesis of one mole of glucose?
 a. 1
 b. 2
 c. 3
 d. 6
 e. 12
 *Answer: d*

23. The NADPH required for the reduction of 3PG to G3P comes from
 a. the dark reactions.
 b. the light reactions.
 c. the synthesis of ATP.
 d. the Calvin–Benson cycle.
 e. oxidative phosphorylation.
 *Answer: b*

24. In $C_4$ plants, the function of the four-carbon compound that is synthesized in the mesophyll cells is to
 a. reduce $NADP^+$.
 b. combine with $CO_2$ to produce glucose.
 c. carry $CO_2$ to the bundle sheath cells.
 d. drive the synthesis of ATP.
 e. close the stomata.
 *Answer: c*

25. In $C_4$ plants, starch grains are found in the chloroplasts of
 a. the thylakoids.
 b. mesophyll cells.
 c. the intracellular space.
 d. the stroma.
 e. bundle sheath cells.
 *Answer: e*

26. During photorespiration, rubisco uses _____ as a substrate.
 a. $CO_2$
 b. $O_2$
 c. glyceraldehyde 3-phosphate
 d. 3-phosphoglycerate
 e. NADPH
 *Answer: b*

27. Photorespiration starts in
 a. mitochondria.
 b. chloroplasts.
 c. $C_4$ plants only.
 d. the microbodies.
 e. the cytoplasm.
 *Answer: b*

**\*28.** In plants, the reactions of glycolysis occur
 a. in $C_3$ plants only.
 b. in the mitochondria.
 c. in the chloroplasts.
 d. only in the presence of light.
 e. in the cytosol.
 *Answer: e*

29. In both photosynthesis and respiration, protons are pumped across a membrane during
 a. electron transport.
 b. photolysis.
 c. $CO_2$ fixation.
 d. reduction of $O_2$.
 e. glycolysis.
 *Answer: a*

30. The enzyme PEP carboxylase
 a. can trap $CO_2$ even at relatively low $CO_2$ concentrations.
 b. catalyzes the synthesis of RuBP.
 c. catalyzes the synthesis of 3PG.
 d. is found in the chloroplasts of bundle sheath cells.
 e. couples the synthesis of ATP to the diffusion of protons.
 *Answer: a*

**\*31.** The function of photorespiration is
 a. $CO_2$ fixation.
 b. unknown.
 c. ATP production.
 d. the generating of a proton gradient.
 e. synthesis of glucose.
 *Answer: b*

32. The NADPH required for $CO_2$ fixation is formed
 a. by the reduction of $O_2$.
 b. by the hydrolysis of ATP.
 c. during the light reactions.
 d. in $C_4$ plants only.
 e. in the mitochondria.
 *Answer: c*

33. The $O_2$ gas produced during photosynthesis is derived from
    a. $CO_2$.
    b. glucose.
    c. water.
    d. carbon monoxide.
    e. bicarbonate ions.
    *Answer: c*

■34. Photosynthesis and respiration have which of the following in common?
    a. In eukaryotes, both processes reside in specialized organelles.
    b. ATP synthesis in both processes relies on the chemiosmotic mechanism.
    c. Both use electron transport.
    d. Both require light.
    e. a, b, and c
    *Answer: e*

35. When a photon interacts with molecules such as those within chloroplasts, the photons may
    a. bounce off the molecules, having no effect.
    b. pass through the molecules, having no effect.
    c. be absorbed by the molecules.
    d. Both a and c
    e. a, b, and c
    *Answer: e*

*36. Which of the following scientific tools "cracked" the Calvin–Benson cycle?
    a. Isotopes
    b. Paper chromatography
    c. Crystallography
    d. Centrifugation and electron microscopy
    e. Both a and b
    *Answer: e*

37. In noncyclic photophosphorylation, electrons from which source replenish chlorophyll molecules that have given up electrons?
    a. $CO_2$
    b. Water
    c. NADPH + $H^+$
    d. $O_2$ gas
    e. None of the above
    *Answer: b*

38. Which of the following biological groups is dependent on photosynthesis for its survival?
    a. Vertebrates
    b. Class Mammalia
    c. Fishes
    d. Both a and b
    e. a, b, and c
    *Answer: e*

39. Photosynthesis is divided into two main phases, the first of which is a series of reactions that requires the absorption of photons. This phase is referred to as the
    a. reduction phase.
    b. dark reactions phase.
    c. carbon fixation phase.
    d. light reactions phase, or photophosphorylation.
    e. None of the above
    *Answer: d*

*40. The energy to hydrolyze water comes from
    a. oxidized chlorophyll.
    b. reduced chlorophyll.
    c. the proton gradient.
    d. ATP.
    e. NADPH + $H^+$.
    *Answer: a*

41. Heterotrophs are dependent on autotrophs for their food supply. Autotrophs can make their own food by
    a. feeding on bacteria and converting the nutrients into usable energy.
    b. using light and simple chemicals to make reduced carbon compounds.
    c. synthesizing it from water and $CO_2$.
    d. All of the above
    e. None of the above
    *Answer: b*

42. Photosynthesis is the process that uses light energy to extract hydrogen atoms from which of the following sources?
    a. Glucose
    b. Chlorophyll
    c. $CO_2$
    d. Water
    e. None of the above
    *Answer: d*

43. Which of the following occurs during the dark reactions of photosynthesis?
    a. Water is converted into hydrogen and water.
    b. $CO_2$ is converted into sugars.
    c. Chlorophyll acts as an enzyme.
    d. Nothing occurs; the plant rests in the dark.
    e. None of the above
    *Answer: b*

44. In bright light, the pH of the thylakoid space
    a. can become more acidic.
    b. can become more alkaline.
    c. stays the same; the pH of the thylakoid space never changes.
    d. can become neutral.
    e. None of the above
    *Answer: a*

■45. When a photon is absorbed by a molecule, what happens to the photon?
    a. It loses its ability to generate any energy.
    b. It raises the molecule from a ground state of low energy to an excited state.
    c. The exact relationship of the photon to the molecule is not clearly understood.
    d. It causes a change in the velocity of the wavelengths.
    e. None of the above
    *Answer: b*

46. A range of energy that cannot be seen by human eyes, but which has slightly more energy per photon than visible light, is
    a. adaptive radiation.
    b. solar radiation.
    c. gamma radiation.
    d. ultraviolet radiation.
    e. None of the above
    *Answer: d*

47. The main photosynthetic pigments in plants are
    a. chlorophyll *s* and chlorophyll *a*.
    b. chlorophyll *x* and chlorophyll *y*.
    c. retinal pigment and accessory pigment.
    d. chlorophyll *a* and chlorophyll *b*.
    e. None of the above
    *Answer: d*

48. Compared to long wavelength photons, short-wavelength photons have
    a. an insignificant amount of energy.
    b. more energy.
    c. energy not available to plant cells.
    d. a ladder of energy.
    e. an equal amount of energy.
    *Answer: b*

49. The chemiosmotic hypothesis states that the energy for the production of ATP comes from
    a. the transfer of phosphate from intermediate compounds.
    b. the reduction of NADP.
    c. a proton gradient set up across the thylakoid membrane.
    d. the oxidation of $CO_2$.
    e. Both a and b
    *Answer: c*

* 50. When $CO_2$ is added to RuBP, the first stable product synthesized is
    a. pyruvate.
    b. glyceraldehyde 3-phosphate.
    c. phosphoglycerate.
    d. ATP.
    e. Both a and b
    *Answer: c*

51. Photosynthesis takes place in plants only in the light. Respiration takes place
    a. in the dark only.
    b. in the light only.
    c. in all organisms except for plants.
    d. both with and without light.
    e. None of the above
    *Answer: d*

52. The revised, balanced equation for the generation of sugar from sunlight, water, and $CO_2$ is
    a. $6\,CO_2 + 6\,H_2O \rightarrow C_6H_{12}O_6 + O_2$.
    b. $6\,CO_2 + 12\,H_2O \rightarrow C_6H_{12}O_6 + 6\,O_2 + 6\,H_2O$.
    c. $6\,CO_2 + 6\,H_2O \rightarrow C_6H_{12}O_6 + 6\,O_2$.
    d. $12\,CO_2 + 12\,H_2O \rightarrow 2\,C_6H_{12}O_6 + 2\,O_2$.
    e. None of the above
    *Answer: b*

53. The net energy outcome of cyclic photophosphorylation is
    a. ATP.
    b. ATP and NADH.
    c. NADPH.
    d. ATP and NADPH.
    e. sugar.
    *Answer: a*

54. The concentration of $O_2$ in the air is _____ percent.
    a. 70
    b. 21
    c. 2.1
    d. 0.21
    e. 0.02
    *Answer: b*

55. In $C_4$ plants, $CO_2$ is first fixed into a compound called
    a. pyruvate.
    b. glucose.
    c. oxaloacetate.
    d. ribulose bisphosphate.
    e. 3-phosphoglycerate.
    *Answer: c*

56. In cacti, $CO_2$ is stored for use in the Calvin–Benson cycle
    a. in the stems, roots, and leaves.
    b. during the evening.
    c. in glucose molecules.
    d. in the stroma.
    e. Both a and d
    *Answer: b*

* 57. Plants classified as CAM store $CO_2$
    a. by making oxaloacetate.
    b. by making PEP carboxylase.
    c. in malic acid.
    d. in crassulacean acid.
    e. Both a and c
    *Answer: e*

58. Which of the following statements about photosynthesis is *false*?
    a. The water for photosynthesis in land plants comes primarily from the soil.
    b. $CO_2$ is taken in, and water and $O_2$ are released through stomata.
    c. Light is absolutely necessary for the production of $O_2$ and carbohydrates.
    d. Photosynthesis is the reverse of cellular respiration.
    e. All the $O_2$ gas produced during photosynthesis comes from water.
    *Answer: d*

59. What is the difference between chlorophyll *a* and chlorophyll *b*?
    a. Chlorophyll *a* has a complex ring structure whereas chlorophyll *b* has a linear structure.
    b. Chlorophyll *a* has a magnesium atom at its center whereas chlorophyll *b* has a phosphate group at its center.
    c. Chlorophyll *a* has a methyl group whereas chlorophyll *b* has an aldehyde group.
    d. A hydrocarbon tail is found only in chlorophyll *a*.
    e. Chlorophyll *a* fluoresces whereas chlorophyll *b* passes the absorbed energy to another molecule.
    *Answer: c*

60. The energy source for the synthesis of carbohydrates in the Calvin–Benson cycle is
    a. ATP only.
    b. photons.
    c. energized chlorophyll *a*.
    d. $NADPH + H^+$.
    e. $NADPH + H^+$ and ATP only.
    *Answer: e*

61. Cyclic electron transport
    a. occurs when the ratio of NADPH + H$^+$ to NADP$^+$ in the chloroplasts of some organisms is high.
    b. is a series of redox reactions.
    c. stores its released energy as a proton gradient.
    d. is completed when the electron returns to P$_{700}$$^+$.
    e. All of the above

    *Answer: e*

62. How does rubisco "decide" whether to act as an oxygenase or a carboxylase?
    a. Rubisco has 10 times more affinity for O$_2$ than CO$_2$; therefore, it favors O$_2$ fixation.
    b. If O$_2$ is relatively abundant, rubisco acts as a carboxylase.
    c. If O$_2$ predominates, rubisco fixes it and the Calvin-Benson cycle occurs.
    d. Photorespiration is more likely at low temperatures.
    e. As the ratio of CO$_2$ to O$_2$ falls in the leaf, the reaction of rubisco with O$_2$ is favored and photorespiration proceeds.

    *Answer: e*

63. When RuBP reacts with O$_2$,
    a. it cannot react with CO$_2$.
    b. carbohydrate production increases.
    c. plant growth is stimulated.
    d. net carbon fixation increases by 25 percent.
    e. two carbon molecules combine to form the four-carbon phosphoglycolate.

    *Answer: a*

64. Photosynthesis and respiration are linked through the
    a. Calvin–Benson cycle.
    b. the citric acid cycle.
    c. glycolysis.
    d. Both a and c
    e. a, b, and c

    *Answer: e*

65. What happens when a photon is absorbed by chlorophyll?
    a. Chlorophyll becomes "excited," or energized.
    b. A greater number of light wavelengths can be absorbed.
    c. ATP is split into ADP, phosphate, and energy.
    d. Hydrogen ions are released.
    e. The chlorophyll molecules fluoresce.

    *Answer: a*

# 9 Chromosomes, the Cell Cycle, and Cell Division

## Fill in the Blank

1. When a DNA molecule doubles, a chromosome is then made up of two joined _____.
   *Answer: chromatids*

★2. Prokaryotic DNA molecules are packaged by _____ proteins, which associate with DNA.
   *Answer: basic*

3. In general, the division of the cell, called _____, follows immediately upon mitosis.
   *Answer: cytokinesis*

★4. Bacteria have a short sequence called _____, where DNA synthesis begins.
   *Answer: ori*

★5. Bacteria have a short sequence called _____, where DNA synthesis ends.
   *Answer: ter*

6. During prophase I of meiosis, a unique event occurs that results in the formation of recombinant chromosomes. This event is termed _____.
   *Answer: crossing over*

7. The structure present during mitosis that is composed of two identical DNA molecules complexed with proteins and joined at the centromere is called a _____.
   *Answer: chromosome*

8. The stage of the cell cycle during which DNA replicates is called the _____.
   *Answer: S phase*

9. _____ is the fusion of two gametes.
   *Answer: Fertilization*

10. During prometaphase, the chromatids are held together by _____.
    *Answer: cohesion*

11. Occasionally, a homologous chromosome pair fails to separate during anaphase I of meiosis. One of the resulting cells lacks a copy of this chromosome, whereas the other contains both members of the homologous pair. These cells are called _____ cells.
    *Answer: aneuploid*

12. During a process known as _____, a piece of one chromosome can break off and become joined to a different chromosome.
    *Answer: translocation*

13. The orderly distribution of genetic information occurs in prokaryotic cells by a process known as _____.
    *Answer: fission*

14. A cell with three homologous sets of chromosomes is called a _____ cell.
    *Answer: triploid*

15. The heritable information of the cell is _____.
    *Answer: DNA*

16. The process that ensures that genetic information is passed on to a cell's daughter cells is _____.
    *Answer: mitosis*

17. The process that ensures that only one of each pair of chromosomes is included in a gamete is _____.
    *Answer: meiosis*

18. The main role of nucleosomes in eukaryotic cells is to _____ the DNA.
    *Answer: package*

19. The G2 phase always follows _____ phase.
    *Answer: S*

20. The G in G1 and G2 is short for _____.
    *Answer: gap*

21. The chromatin _____ during prophase.
    *Answer: condenses*

22. The milestone event that defines entry into prometaphase is _____.
    *Answer: loss of the nuclear envelope*

23. In plants, a _____ forms at the equatorial region of the cell.
    *Answer: cell plate*

24. The cell plate is derived from the _____ of the cell.
    *Answer: Golgi apparatus*

25. The "invisible thread" that pinches cells apart during cell division is made of _____ and _____.
    *Answer: actin; myosin*

26. A zygote usually has _____ copies of each chromosome.
    *Answer: two*

27. The _____ is the number, form, and type of chromosomes found in a cell.
    *Answer: karyotype*

28. A _____ is one of a pair of chromosomes having the same overall genetic composition and sequence.
    *Answer: homolog*

29. Nondisjunction causes the production of _____ cells.
    *Answer: aneuploid*

30. Down syndrome can be caused by an extra chromosome
    _____.
    *Answer: 21*

## Multiple Choice

*1. During prokaryotic cell division, two chromosomes separate from each other and distribute into the daughter cells by
   a. attachments to microtubules.
   b. a mitotic spindle.
   c. repellent forces.
   d. attachment to separating membrane regions.
   e. All of the above
   *Answer: d*

2. Bacteria typically have _____, whereas eukaryotes have
   _____.
   a. one chromosome that is circular; many that are linear
   b. several chromosomes that are circular; many that are linear
   c. one chromosome that is linear; many that are circular
   d. two chromosomes that are circular; eight that are linear
   e. None of the above
   *Answer: a*

3. Chromosomes contain large amounts of different interacting proteins, which are known as
   a. pentanes.
   b. hexosamines.
   c. histones.
   d. protein hormones.
   e. histamines.
   *Answer: c*

4. The appropriate decisions to enter the S phase and the M phase of the cell cycle depend on a pair of biochemicals called
   a. actin and myosin.
   b. Cdk's and cyclin.
   c. ligand and receptor.
   d. MSH and MSH-receptor.
   e. ATP and ATPase.
   *Answer: b*

5. The molecules that make up a chromosome are
   a. DNA and RNA.
   b. DNA and proteins.
   c. proteins and lipids.
   d. nucleotides and nucleosides.
   e. proteins and phospholipids.
   *Answer: b*

■6. During mitosis and meiosis the chromatin compacts. Which of the following processes takes place more easily because of this compaction?
   a. The orderly distribution of genetic material to two new nuclei
   b. The replication of the DNA
   c. Exposing the genetic information on the DNA
   d. The unwinding of DNA from around the histones
   e. The disappearance of the nuclear membrane
   *Answer: a*

7. The basic structure of chromatin has sometimes been referred to as beads on a string of DNA. These beads are called

   a. chromosomes.
   b. chromatids.
   c. supercoils.
   d. interphases.
   e. nucleosomes.
   *Answer: e*

8. DNA replication occurs
   a. during both mitosis and meiosis.
   b. only during mitosis.
   c. only during meiosis.
   d. during the S phase.
   e. during G2.
   *Answer: d*

*9. Around _____ base pairs wrap around each core particle in a _____.
   a. 1000; solenoid
   b. 20000; chromosome
   c. 146; nucleosome
   d. 1000; chromosome
   e. 20,000; solenoid
   *Answer: c*

■10. When cyclin binds Cdk,
   a  the cell transitions from G2 to S.
   b. kinase activation occurs.
   c. chromosomes condense.
   d. the cell quickly enters M phase.
   e. the cell begins apoptosis.
   *Answer: b*

11. Mature nerve cells are incapable of cell division. These cells are probably in
   a. G1.
   b. the S phase.
   c. G2.
   d. mitosis.
   e. meiosis.
   *Answer: a*

12. A cell cycle consists of
   a. mitosis and meiosis.
   b. G1, the S phase, and G2.
   c. prophase, metaphase, anaphase, and telophase.
   d. interphase and mitosis.
   e. meiosis and fertilization.
   *Answer: d*

*13. Evidence in yeast suggests that the maturation-promoting factor of sea urchins is
   a. a cyclin.
   b. MFP.
   c. an S nuclease.
   d. a Cdk.
   e. a Cdk/cyclin phosphatase.
   *Answer: d*

14. The cells of the intestinal epithelium are continually dividing, replacing dead cells lost from the surface of the intestinal lining. If you examined a population of intestinal epithelial cells under the microscope, most of the cells would
   a. be in meiosis.
   b. be in mitosis.
   c. be in interphase.
   d. have condensed chromatin.
   e. Both b and d
   *Answer: c*

*15. DNA damage by UV radiation causes the synthesis of
a. p53.
b. DNA.
c. Cdk.
d. cyclin.
e. p21.
*Answer: e*

*16. The uncondensed length of human DNA found in chromosomes is _____, whereas a typical cell is 10 μm in length.
a. 5 μm
b. 2 μm
c. 2 meters
d. 20 meters
e. 2.54 inches
*Answer: c*

17. The products of mitosis are
a. one nucleus containing twice as much DNA as the parent nucleus.
b. two genetically identical cells.
c. four nuclei containing half as much DNA as the parent nucleus.
d. four genetically identical nuclei.
e. two genetically identical nuclei.
*Answer: e*

18. The mitotic spindle is composed of
a. chromosomes.
b. chromatids.
c. microtubules.
d. chromatin.
e. centrosomes.
*Answer: c*

19. Centrosomes
a. are constricted regions of phase chromosomes.
b. determine the plane of cell division.
c. are the central region of the cell.
d. are the region where the membrane constricts during cytokinesis.
e. are part of cilia.
*Answer: b*

20. When dividing cells are examined under a light microscope, chromosomes first become visible during
a. interphase.
b. the S phase.
c. prophase.
d. G1.
e. G2.
*Answer: c*

21. The structures that line up the chromatids on the equatorial plate during metaphase are called
a. asters.
b. polar and kinetochore microtubules.
c. centrosomes.
d. centrioles.
e. histones.
*Answer: b*

22. The microtubules of the mitotic spindle attach to a specialized structure in the centromere region of each chromosome called the
a. kinetochore.

b. nucleosome.
c. equatorial plate.
d. aster.
e. centrosome.
*Answer: a*

23. After the centromeres separate during mitosis, the chromatids, now called _____, move toward opposite poles of the spindle.
a. centrosomes
b. kinetochores
c. half spindles
d. asters
e. daughter chromosomes
*Answer: e*

24. In plant cells, cytokinesis is accomplished by the formation of a(n)
a. aster.
b. membrane furrow.
c. equatorial plate.
d. cell plate.
e. spindle.
*Answer: d*

*25. The distribution of mitochondria between the daughter cells during cytokinesis
a. is random.
b. is directed by the mitotic spindle.
c. is directed by the centrioles.
d. results in the mitochondria remaining in the parent cell.
e. occurs only during meiosis.
*Answer: a*

26. Genetically diverse offspring result from
a. mitosis.
b. cloning.
c. sexual reproduction.
d. cytokinesis.
e. fission.
*Answer: c*

27. During asexual reproduction, the genetic material of the parent is passed on to the offspring by
a. homologous pairing.
b. meiosis and fertilization.
c. mitosis and cytokinesis.
d. karyotyping.
e. chiasmata.
*Answer: c*

*28. Meiosis can occur
a. in all organisms.
b. only when an organism is diploid.
c. only in multicellular organisms.
d. only in haploid organisms.
e. only in single-celled organisms.
*Answer: b*

29. All zygotes are
a. multicellular.
b. diploid.
c. animals.
d. clones.
e. gametes.
*Answer: b*

30. In all sexually reproducing organisms, the diploid phase of the life cycle begins at
    a. spore formation.
    b. gamete formation.
    c. meiosis.
    d. mitosis.
    e. fertilization.
    *Answer: e*

31. The members of a homologous pair of chromosomes
    a. are identical in size and appearance.
    b. contain identical genetic information.
    c. separate to opposite poles of the cell during mitosis.
    d. are found only in haploid cells.
    e. are present only after the S phase.
    *Answer: a*

32. The diagnosis of Down syndrome is made by examining the individual's
    a. spores.
    b. karyotype.
    c. chromatin.
    d. nucleosomes.
    e. kinetochores.
    *Answer: b*

33. During meiosis, the sister chromatids separate during
    a. anaphase II.
    b. anaphase I.
    c. the S phase.
    d. synapsis.
    e. telophase II.
    *Answer: a*

34. The exchange of genetic material between chromatids on homologous chromosomes occurs during
    a. interphase.
    b. mitosis and meiosis.
    c. prophase I.
    d. anaphase I.
    e. anaphase II.
    *Answer: c*

35. At the end of the first meiotic division, each chromosome consists of
    a. chiasmata.
    b. a homologous chromosome pair.
    c. four copies of each DNA molecule.
    d. two chromatids.
    e. a pair of polar microtubules.
    *Answer: d*

36. The four haploid nuclei found at the end of meiosis differ from one another in their exact genetic composition. Some of this difference is the result of
    a. cytokinesis.
    b. replication of DNA during the S phase.
    c. separation of sister chromatids at anaphase II.
    d. spindle formation.
    e. crossing over during prophase I.
    *Answer: e*

37. Diploid cells of the fruit fly *Drosophila* have 10 chromosomes. How many chromosomes does a *Drosophila* gamete have?
    a. 1
    b. 2
    c. 5
    d. 10
    e. 20
    *Answer: c*

38. During meiosis I in humans, one of the daughter cells receives
    a. only maternal chromosomes.
    b. a mixture of maternal and paternal chromosomes.
    c. the same number of chromosomes as a diploid cell.
    d. a sister chromatid from each chromosome.
    e. one-fourth the amount of DNA in the parent nucleus.
    *Answer: b*

■39. The fact that most monosomies and trisomies are lethal to human embryos illustrates the
    a. importance of the orderly distribution of genetic material during meiosis.
    b. exchange of genetic information during crossing over.
    c. advantage of sexual reproduction to the survival of a population.
    d. fact that each chromosome contains a single molecule of DNA.
    e. formation of haploid gametes as a result of meiosis.
    *Answer: a*

*40. A triploid nucleus cannot undergo meiosis because
    a. the DNA cannot replicate.
    b. not all of the chromosomes can form homologous pairs.
    c. the sister chromatids cannot separate.
    d. cytokinesis cannot occur.
    e. a cell plate cannot form.
    *Answer: b*

*41. A bacterial cell gives rise to two genetically identical daughter cells by a process known as
    a. nondisjunction.
    b. mitosis.
    c. meiosis.
    d. fission.
    e. fertilization.
    *Answer: d*

42. Which of the following is *not* part of sexual reproduction?
    a. The segregation of homologous chromosomes during gamete formation
    b. The fusion of sister chromatids during fertilization
    c. The fusion of haploid cells from a diploid zygote
    d. The reduction in chromosome number during meiosis
    e. The production of genetically distinct gametes during meiosis
    *Answer: b*

43. Chromatin condenses to form discrete, visible chromosomes
    a. early in G1.
    b. during S.
    c. during telophase.
    d. during prophase.
    e. at the end of cytokinesis.
    *Answer: d*

44. Chromosomes "decondense" into diffuse chromatin
    a. at the end of telophase.
    b. at the beginning of prophase.
    c. at the end of interphase.
    d. at the end of metaphase.
    e. only in dying cells.
    *Answer: a*

45. Microtubules that form the mitotic spindle tend to originate from or terminate in
    a. centromeres and telomeres.
    b. euchromatin.
    c. centrioles and telomeres.
    d. the nuclear envelope.
    e. centrioles and kinetochores.
    *Answer: e*

■46. Asexual reproduction produces genetically identical individuals because
    a. chromosomes do not have to replicate.
    b. it involves chromosome replication without cytokinesis.
    c. no meiosis or fertilization takes place.
    d. cell division occurs only in meiosis.
    e. the mitotic spindle prevents nondisjunction.
    *Answer: c*

■47. One difference between mitosis and meiosis I is that
    a. homologous chromosome pairs synapse during mitosis.
    b. chromosomes do not replicate in the interphase preceding meiosis.
    c. homologous chromosome pairs synapse during meiosis but not during mitosis.
    d. spindles composed of microtubules are not required during meiosis.
    e. sister chromatids separate during meiosis but not during mitosis.
    *Answer: c*

48. Genetic recombination occurs during
    a. prophase of meiosis I.
    b. the interphase preceding meiosis II.
    c. the mitotic telophase.
    d. fertilization.
    e. the formation of somatic cells.
    *Answer: a*

49. The number of chromosomes is reduced to half during
    a. anaphase of mitosis and meiosis.
    b. meiosis II.
    c. meiosis I.
    d. fertilization.
    e. interphase.
    *Answer: c*

■50. The total DNA content of each daughter cell is reduced during meiosis because
    a. chromosomes do not replicate during the interphase preceding meiosis I.
    b. chromosomes do not replicate between meiosis I and II.
    c. half of the chromosomes from each gamete are lost during fertilization.
    d. sister chromatids separate during anaphase of meiosis I.
    e. chromosome arms are lost during crossing over.
    *Answer: b*

**■51. Many chromosome abnormalities (trisomies and monosomies) are not observed in the human population because
    a. they are lethal and cause spontaneous abortion of the embryo early in development.
    b. all trisomies and monosomies are lethal early in childhood.
    c. meiosis distributes chromosomes to daughter cells with great precision.

d. they are so difficult to count.
    e. the human meiotic spindle is self-correcting.
    *Answer: a*

*52. In a haploid organism, most mitosis occurs
    a. after fertilization and before meiosis.
    b. after meiosis and before fertilization.
    c. between meiosis I and II.
    d. during G1.
    e. in diploid cells.
    *Answer: b*

*53. The event in the cell division process that clearly involves microfilaments is
    a. chromosome separation during anaphase.
    b. movement of chromosomes to the metaphase plate.
    c. chromosome condensation during prophase.
    d. disappearance of the nuclear envelope during prophase.
    e. cytokinesis in animal cells.
    *Answer: e*

54. Trisomies and monosomies can result from accidents that occur during meiosis called
    a. nondisjunctions.
    b. inversions.
    c. reciprocal translocations.
    d. recombinations.
    e. acrocentricities.
    *Answer: a*

■55. Chromosome number is reduced during meiosis because the process consists of
    a. two cell divisions without any chromosome replication.
    b. a single cell division without any chromosome replication.
    c. two cell divisions in which half of the chromosomes are destroyed.
    d. two cell divisions and only a single round of chromosome replication.
    e. four cell divisions with no chromosome replication.
    *Answer: d*

56. Which of the following phases of the cell cycle is *not* part of interphase?
    a. M
    b. S
    c. G1
    d. G2
    e. G0
    *Answer: a*

57. During mitotic anaphase, chromatids migrate
    a. from the poles of the cell toward the metaphase plate.
    b. from the metaphase plate toward the poles.
    c. toward the nuclear envelope.
    d. along with their sister chromatids toward one pole.
    e. along with the other member of the homologous pair toward the metaphase plate.
    *Answer: b*

58. Which of the following does *not* occur during mitotic prophase?
    a. Disappearance of the nuclear envelope
    b. Chromosome condensation
    c. Migration of centrioles toward the cell poles
    d. Synapsis of homologous chromosomes
    e. Formation of the mitotic spindle
    *Answer: d*

*59. Human males have _____ different types of chromosomes.
   a. 23
   b. 24
   c. 46
   d. 48
   e. 92
   *Answer: b*

60. Which of the following is *not* true of homologous chromosome pairs?
   a. They come from only one of the individual's parents.
   b. They usually contain slightly different versions of the same genetic information.
   c. They segregate from each other during meiosis I.
   d. They synapse during meiosis I.
   e. Each contains two sister chromatids at the beginning of meiosis I.
   *Answer: a*

61. Which of the following is *not* true of sister chromatids?
   a. They arise by replication during S phase.
   b. They segregate from each other during each mitotic anaphase.
   c. They usually contain identical versions of the same genetic information.
   d. They segregate from each other during meiosis I.
   e. They are joined during prophase and metaphase at their common centromere.
   *Answer: d*

62. Chromatin consists of
   a. DNA and histones.
   b. DNA, histones, and many other nonhistone proteins.
   c. mostly RNA and DNA.
   d. RNA, DNA, and nonhistone proteins.
   e. DNA alone.
   *Answer: b*

63. The DNA of a eukaryotic cell is
   a. double-stranded.
   b. single-stranded.
   c. circular.
   d. complex inverted.
   e. conservative.
   *Answer: a*

64. Cells that do not divide are usually arrested in
   a. S.
   b. G1.
   c. G2.
   d. M.
   e. prophase.
   *Answer: b*

65. At the milestone that defines metaphase, the chromosomes
   a. separate.
   b. come together.
   c. are at opposite poles.
   d. line up.
   e. cross over.
   *Answer: d*

66. At the milestone that defines anaphase, the chromosomes
   a. separate.
   b. come together.
   c. are at opposite poles.
   d. line up.
   e. cross over.
   *Answer: a*

67. At the milestone that defines telophase, the chromosomes
   a. separate.
   b. come together.
   c. are at opposite poles.
   d. line up.
   e. cross over.
   *Answer: c*

68. The major drawback of asexual reproduction is
   a. that it takes too little time.
   b. the variation generated.
   c. that it prevents change.
   d. that it requires cytokinesis.
   e. the lack of variation among the progeny.
   *Answer: e*

69. A haploid cell is a cell
   a. in which the genes are arranged haphazardly.
   b. containing only one copy of each chromosome.
   c. that has resulted from the process of mitosis.
   d. with twice the number of chromosomes of a diploid cell.
   e. None of the above
   *Answer: b*

■70. The second meiotic division of meiosis is important because
   a. it returns the chromosome number to diploid before fertilization.
   b. it allows for crossing over and random distribution of chromosomes.
   c. it reduces cell size by dividing the cytoplasm in half.
   d. otherwise chromosome copies would double at each fertilization.
   e. fertilization requires this step.
   *Answer: d*

71. Interleukins and erythropoietin are
   a. growth factors.
   b. Cdk's.
   c. cyclins.
   d. antitumor agents.
   e. intracellular signaling molecules.
   *Answer: a*

72. Half of all human cancers have defective _____ associated with their cells.
   a. p53
   b. p21
   c. Cdk
   d. cyclin
   e. DNA polymerase
   *Answer: a*

73. Each diploid cell of a human female contains _____ of each type of chromosome.
   a. one
   b. two
   c. four
   d. a total of 23
   e. a total of 46
   *Answer: b*

74. The importance of synapsis and the formation of chiasmata is that
   a. reciprocal exchange of chromosomal sections occurs.
   b. the DNA on homologous chromosomes recombines.

c. genetic variation increases.
d. they provide evidence that an exchange of genetic material has occurred.
e. All of the above
*Answer: e*

75. A triploid plant has
   a. one extra chromosome.
   b. one extra set of chromosomes.
   c. three chromosomes.
   d. three times the chance of surviving that a monoploid has.
   e. None of the above
   *Answer: b*

76. Nucleosomes are composed of
   a. centromeres and DNA.
   b. microtubules and condensins.
   c. kinetochores and centromeres.
   d. histones and DNA.
   e. polar microtubules and kinetochore microtubules.
   *Answer: d*

*77. The energy to move chromosomes during mitosis is provided by
   a. centrioles.
   b. DNA polymerization.
   c. migration of the centrosomes.
   d. formation of the cell plate.
   e. ATP.
   *Answer: e*

78. During bacterial cell division, the two DNA molecules are separated by
   a. centrosomes.
   b. spindle fibers.
   c. nucleosomes.
   d. cell elongation.
   e. aneuploidy.
   *Answer: d*

79. The process of programmed cell death is called
   a. necrosis.
   b. lysis.
   c. apoptosis.
   d. cell displacement.
   e. cellular suicide.
   *Answer: c*

80. An indicator of programmed cell death is
   a. fragmented chromatin.
   b. swelling of the membrane.
   c. cell lysis.
   d. loss of transcription control.
   e. All of the above
   *Answer: a*

*81. In order for a prokaryote cell to divide, which of the following must occur?
   a. A reproductive signal, replication, segregation of DNA, and cytokinesis
   b. DNA replication, crossing over, and segregation of DNA
   c. DNA replication and segregation of DNA.
   d. Cell growth and cytokinesis
   e. DNA replication, segregation of DNA, and cytokinesis
   *Answer: a*

82. Which of the following is true for mitosis?
   a. The chromosome number in the resulting cells is halved.
   b. DNA replication is completed in prophase.
   c. Crossing over occurs during prophase.
   d. Two genetically identical daughter cells are formed.
   e. It consists of two nuclear divisions.
   *Answer: d*

■83. A person with Klinefelter Syndrome has 44 chromosomes and three sex chromosomes (XXY). The resulting aneuploidy is caused by
   a. nondisjunction.
   b. crossing over.
   c. a mutation.
   d. an enzyme deficiency.
   e. failure of DNA to replicate.
   *Answer: a*

**84. How does sexual reproduction increase genetic variability?
   a. By the exchange of genetic information between male and female gametes during meiosis I
   b. By random separation of the chromosome pairs
   c. By the union of male and female gametes
   d. Through crossing over, independent assortment, and random fertilization
   e. By random assortment of male and female chromosomes
   *Answer d*

85. Cyclin-dependent kinases (Cdk's) catalyze the phosphorylation of targeted proteins. Phosphorylation
   a. makes the targeted protein hydrophilic.
   b. makes the targeted protein hydrophobic.
   c. changes the shape and function of the targeted protein.
   d. gives the protein a three-dimensional shape.
   e. blocks the cell cycle from proceeding.
   *Answer: c*

86. Regulation of the cell cycle is dependent upon cyclins and cyclin-dependent kinases. The key(s) that allows a cell to progress beyond the restriction point is (are)
   a. cdk1 and cyclin B.
   b. cyclin D and p21.
   c. cyclin A and Cdk2.
   d. phosphorylation of RB by Cdk4 and Cdk2
   e. external signals from growth factors.
   *Answer: d*

87. The optimal time to observe sizes and shapes of chromosomes is
   a. prophase.
   b. metaphase.
   c. anaphase.
   d. telophase.
   e. cytokinesis.
   *Answer: b*

88. The spindle checkpoint is a process
   a. that is responsible for the breakdown of separase.
   b. in which the centromeres separate.
   c. that determines whether all kinetochores are attached to the spindle.
   d. that makes sure the polar microtubules from each pole overlap.
   e. that ensures the attachment of kinetochores to microtubules.
   *Answer: c*

89. Which of the following is true of plant cytokinesis?
    a. It begins when telophase ends.
    b. A division furrow completely separates the cytoplasm.
    c. It is under the control of $Ca^{2+}$.
    d. Vesicles fuse to form a cell plant.
    e. The spindle fibers break down to form a cell plate.
    *Answer d*

**90. How does mitotic prophase differ from prophase I of meiosis?
    a. Chromatin supercoils only in mitotic prophase.
    b. The nuclear envelope disappears only in prophase I of meiosis.
    c. Synapsis occurs in mitotic prophase and but not in meiotic prophase I.
    d. The chromatids separate in mitotic prophase, not in prophase I of meiosis.
    e. Crossing over is characteristic of prophase I of meiosis but not of mitotic prophase.
    *Answer: e*

91. How does a nucleus in G2 differ from one in the G1 phase?
    a. G2 has double the amount of DNA as G1.
    b. DNA synthesis occurs only in the G1 phase.
    c. Inactive cells are arrested only in the G2 phase.
    d. During G2, the cell prepares for S phase.
    e. All of the above
    *Answer: a*

92. Paired chromatids separate and move to opposite poles because
    a. sister chromatids attach to microtubules in opposite halves of the spindle.
    b. separase hydrolyzes cohesion, allowing chromatid separation.
    c. tubulin fibers move the chromatids away from the equatorial plate.
    d. polar microtubules push the chromatids to the poles.
    e. centromeres separate.
    *Answer: a*

# 10 Genetics: Mendel and Beyond

## Fill in the Blank

1. A _____ trait is one that can be passed from one generation to another.
   *Answer: heritable*

2. A _____ is a portion of DNA that resides at a particular locus or site on a chromosome and encodes a particular function.
   *Answer: gene*

3. To determine the overall probability of independent events, _____ the probabilities of the individual events.
   *Answer: multiply*

4. One particular allele of a gene may be defined as _____ or standard, because it is present in most individuals and gives rise to an expected trait, or phenotype.
   *Answer: wild type*

5. Geneticists make use of _____ frequencies to map chromosomes, that is, to locate genetic loci on the chromosome.
   *Answer: recombinant*

6. A cross between two parents that differs by a single trait is a _____ cross.
   *Answer: monohybrid*

7. The physical appearance of a character is the _____, whereas the genetic constitution is the _____.
   *Answer: phenotype; genotype*

8. A cross between two heterozygous parents that differs by two independent traits is a _____ cross.
   *Answer: dihybrid*

9. Genes at different loci on the same chromosome can separate and recombine to form _____.
   *Answer: recombinant chromosomes*

10. The region of the chromosome occupied by a gene is called a _____.
    *Answer: locus*

11. When a cross is made and a trait disappears in the $F_1$ generation, only to reappear in the $F_2$, the trait is probably _____.
    *Answer: recessive*

12. A female who is heterozygous for a recessive sex-linked character is a _____.
    *Answer: carrier*

13. When the expression of one gene depends on the expression of another gene, the genes demonstrate _____.
    *Answer: epistasis*

14. When many genes contribute to the phenotype, variation is said to be _____.
    *Answer: continuous*

15. A _____ is an observable feature, such as flower color; a _____ is a particular form of a character, such as a white flower.
    *Answer: character; trait*

16. The AB phenotype found in individuals with $I^A I^B$ genotype is an example of _____.
    *Answer: codominance*

17. To determine the probability of an event that can occur in two or more different ways, _____ the individual probabilities.
    *Answer: add*

18. Quantitative variation is the result of the interaction of _____ and _____.
    *Answer: genes; environment*

19. Mendel's laws of inheritance can be applied to human genetics through the study of _____.
    *Answer: pedigrees*

20. Recombination is most likely between two loci that are _____.
    *Answer: far apart*

## Multiple Choice

1. Gregor Mendel presented his genetics project orally in
   a. 1565.
   b. 1665.
   c. 1765.
   d. 1865.
   e. 1965.
   *Answer: d*

2. Mendel concluded that each pea has two units for each character, and each gamete contains one unit. Mendel's "unit" is now referred to as a(n)
   a. gene.
   b. character.
   c. allele.
   d. transcription factor.
   e. None of the above
   *Answer: a*

3. Mendel's research was rediscovered when studies by _____ were published.
   a. de Vries
   b. Correns
   c. Tschermak
   d. All of the above
   e. None of the above
   *Answer: d*

4. A particular genetic cross in which the individual in question is crossed with an individual known to be homozygous for a recessive trait is referred to as a
   a. parental cross.
   b. dihybrid cross.
   c. filial generation mating.
   d. reciprocal cross.
   e. test cross.
   *Answer: e*

5. Although the law of independent assortment is generally applicable, when two loci are on the same chromosome, the phenotypes of the progeny sometimes do not fit the phenotypes predicted. This is due to
   a. translocation.
   b. inversions.
   c. chromatid affinities.
   d. linkage.
   e. reciprocal chromosomal exchanges.
   *Answer: d*

6. When a given trait is the result of multigene action, one of the genes may mask the expression of one or all other genes. This phenomenon is termed
   a. epistasis.
   b. epigenesis.
   c. dominance.
   d. incomplete dominance.
   e. None of the above
   *Answer: a*

7. Which of the following is *not* a characteristic that makes an organism suitable for genetic studies?
   a. A small number of chromosomes
   b. A short generation time
   c. Ease of cultivation
   d. The ability to control crosses
   e. The availability of a variation for traits
   *Answer: a*

8. A key factor that allowed Mendel to interpret the results of his breeding experiments was that
   a. the varieties of peas he used were "true-breeding."
   b. peas naturally self-pollinate.
   c. peas can reproduce asexually.
   d. pollination could be controlled.
   e. Both a and d
   *Answer: e*

9. Crossing spherical-seeded pea plants with wrinkled-seeded pea plants resulted in progeny that all had spherical seeds. This indicates that the wrinkled-seed trait is
   a. codominant.
   b. dominant.
   c. recessive.
   d. Both a and b
   e. Both a and c
   *Answer: c*

10–12. Two different groups of imaginary schmoos live in geographically separated locations and rarely interbreed. On one occasion, a big-footed white schmoo does mate with a small-footed brown schmoo. Three offspring result: one big-footed brown schmoo and two small-footed brown schmoos.

■10. Which statement about the inheritance of color in schmoos is most likely to be correct?
   a. Brown is dominant to white.
   b. White is dominant to brown.
   c. White and brown are codominant.
   d. Both a and c
   e. This cannot be answered without more information.
   *Answer: a*

■■11. Which statement about the inheritance of footedness in schmoos is most likely to be correct?
   a. Big is dominant to small.
   b. Small is dominant to big.
   c. Big and small are codominant.
   d. Both a and c
   e. This cannot be answered without more information.
   *Answer: e*

■12. If big feet ($B$) in schmoos is dominant to small feet ($b$), the genotype of the big-footed white parent schmoo with respect to the foot gene can
   a. only be $bb$.
   b. only be $BB$.
   c. only be $Bb$.
   d. either be $bb$ or $BB$.
   e. either be $bb$ and $Bb$.
   *Answer: c*

13. When reciprocal crosses produce identical results, the trait is
   a. sex-linked.
   b. not sex-linked.
   c. not autosomally inherited.
   d. Both a and c
   e. Both b and c
   *Answer: b*

14. The physical appearance of a character is called
   a. the genotype.
   b. the phenotype.
   c. an allele.
   d. a trait.
   e. a gene.
   *Answer: b*

15. Different forms of a gene are called
   a. traits.
   b. phenotypes.
   c. genotypes.
   d. alleles.
   e. None of the above
   *Answer: d*

16. When genes for two different characters segregate in a cross, what type of cross is it?
   a. Monohybrid
   b. Dihybrid
   c. Trihybrid
   d. $F_1$
   e. $F_2$
   *Answer: b*

■17. In Mendel's experiments, if the allele for tall (*T*) plants was incompletely dominant over the allele for short (*t*) plants, what offspring would have resulted from crossing two *Tt* plants?
   a. ¼ tall; ½ intermediate height; ¼ short
   b. ½ tall; ¼ intermediate height; ¼ short
   c. ¼ tall; ¼ intermediate height; ½ short
   d. All the offspring would have been tall.
   e. All the offspring would have been of intermediate height.
   *Answer: a*

18. The site on the chromosome occupied by a gene is called a(n)
   a. allele.
   b. region.
   c. locus.
   d. type.
   e. phenotype.
   *Answer: c*

**■19. In the garden peas used in Mendel's experiments, the spherical seed character (*SS*) is completely dominant over the wrinkled seed character (*ss*). If the characters for height were incompletely dominant, such that *TT* was tall, *Tt* was intermediate, and *tt* was short, what would have resulted from crossing a spherical-seeded, short (*SStt*) plant to a wrinkled-seeded, tall (*ssTT*) plant?
   a. ½ would be smooth-seeded and intermediate height; ½ would be smooth-seeded and tall.
   b. All the progeny would be smooth-seeded and tall.
   c. All the progeny would be smooth-seeded and short.
   d. All the progeny would be smooth-seeded and intermediate in height.
   e. This cannot be answered without more information.
   *Answer: d*

■20. If Mendel's crosses between spherical-seeded tall plants and wrinkled-seeded short plants had produced many more than ¹⁄₁₆ wrinkled-seeded short plants in the $F_2$ generation, he might have concluded that
   a. the spherical seed and tall traits are linked.
   b. the wrinkled seed and short traits are unlinked.
   c. all traits in peas assort independently of each other.
   d. all traits in peas are linked.
   e. None of the above
   *Answer: a*

*21. It has been found that at a certain locus of the human genome, 200 different alleles exist in the population. Each person has at most _____ alleles.
   a. 1
   b. 2
   c. 100
   d. 200
   e. 400
   *Answer: b*

22. An organism that produces either male gametes or female gametes, but not both, is called
   a. monoecious.
   b. dioecious.
   c. heterozygous.
   d. homozygous.
   e. parthenogenic.
   *Answer: b*

■23. It is predictable that half of the human babies born will be male and half will be female because
   a. of the segregation of the X and Y chromosomes during male meiosis.
   b. of the segregation of the X chromosomes during female meiosis.
   c. all eggs contain an X chromosome.
   d. Both a and b
   e. Both a and c
   *Answer: a*

24. A human male carrying an allele for a trait on the X chromosome is
   a. heterozygous.
   b. homozygous.
   c. hemizygous.
   d. monozygous.
   e. holozygous.
   *Answer: c*

■25. Cleft chin is a sex-linked dominant trait. A man with a cleft chin marries a woman with a round chin. What percent of their female progeny will show the cleft chin trait?
   a. 0
   b. 25
   c. 50
   d. 75
   e. 100
   *Answer: e*

■26. What percent of their male progeny will show the cleft chin trait?
   a. 0
   b. 25
   c. 50
   d. 75
   e. 100
   *Answer: c*

27. A linkage group corresponds to
   a. a group of genes on different chromosomes.
   b. the linear order of chromomeres on a chromosome.
   c. the length of a chromosome.
   d. a group of genes on the same chromosome.
   e. None of the above
   *Answer: d*

■28. It would have been very difficult for Mendel to draw conclusions about the patterns of inheritance if he had used cattle instead of peas. Why?
   a. Cattle reproduce asexually.
   b. Cattle have small numbers of offspring.
   c. Cattle do not have observable phenotypes.
   d. Cattle do not have genotypes.
   e. Cattle do not have autosomes.
   *Answer: b*

29. Epistasis refers to
   a. a group of genes that are close together.
   b. the interaction of two genes so that a new phenotype is produced.
   c. the expression of two genes in the same individual.
   d. the linear order of genes on a chromosome.
   e. the expression of one gene masking the expression of another.
   *Answer: e*

30–32. An agouti mouse that is heterozygous at the agouti and albino loci (*AaBb*) is mated to an albino mouse that is heterozygous at the agouti locus (*aaBb*). Non-albino mice without the dominant agouti allele are black.

**30. What percent of the progeny do you expect to be albino?
   a. 0
   b. 12.5
   c. 37.5
   d. 50
   e. 100
   *Answer: d*

**31. What percent of the progeny do you expect to be agouti?
   a. 0
   b. 12.5
   c. 37.5
   d. 50
   e. 100
   *Answer: c*

**32. What percent of the progeny do you expect to be black?
   a. 0
   b. 12.5
   c. 37.5
   d. 50
   e. 100
   *Answer: b*

33. The complete phenotype of an organism is dependent on
   a. genotype.
   b. penetrance.
   c. expressivity.
   d. polygenes.
   e. All of the above
   *Answer: e*

34. When a dihybrid black, straight-winged fly is crossed to a double-recessive brown, curly-winged fly, the frequency at which black curly-winged and brown straight-winged flies are seen in the progeny is called the _____ frequency.
   a. mutation
   b. mitotic
   c. meiotic
   d. allele
   e. recombinant
   *Answer: e*

35. Alleles for genes located on mitochondrial DNA are said to be maternally inherited. What is the reason for this pattern of inheritance?
   a. The egg and sperm contribute equal numbers of cytoplasmic organelles to the zygote.
   b. The egg contributes virtually all of the cytoplasmic organelles to the zygote.
   c. Half of the nuclear chromosomes in the zygote come from the father.
   d. Half of the nuclear chromosomes in the zygote come from the mother.
   e. All of the nuclear chromosomes in the zygote come from the mother.
   *Answer: b*

36. Which of the following methods was *not* used by Mendel in his study of the genetics of the garden pea?
   a. Maintenance of true-breeding lines
   b. Cross-pollination
   c. Microscopy
   d. Production of hybrid plants
   e. Quantitative analysis of results
   *Answer: c*

37. In Kölreuter's studies, reciprocal crosses
   a. always gave identical results.
   b. only involved heterozygous individuals.
   c. supported the blending hypothesis of inheritance.
   d. could be done only with homozygous individuals.
   e. consisted of an $F_1$ and an $F_2$ generation.
   *Answer: a*

38. Which of the following statements about Mendelian genetics is *false*?
   a. Alternative forms of genes are called alleles.
   b. A locus is a gene's location on its chromosome.
   c. Only two alleles can exist for a given gene.
   d. A genotype is a description of the alleles that represent an individual's genes.
   e. Individuals with the same phenotype can have different genotypes.
   *Answer: c*

39. Segregation of alleles occurs
   a. during gamete formation.
   b. at fertilization.
   c. during mitosis.
   d. during the random combination of gametes to produce the $F_2$ generation.
   e. only in monohybrid crosses.
   *Answer: a*

•40. A pea plant with red flowers is test crossed, and one-half of the resulting progeny have red flowers, while the other half have white flowers. You know that the genotype of the test-crossed parent was
   a. *RR*.
   b. *Rr*.
   c. *rr*.
   d. either *RR* or *Rr*.
   e. This cannot be answered without more information.
   *Answer: b*

**41. Mendel performed a cross between individuals heterozygous for three different traits: yellow versus green seeds (green is dominant), red versus white flowers (red is dominant), and green versus yellow pods (green is dominant). What fraction of the offspring would have been expected to have green seeds, red flowers, and green pods?
   a. $^{27}/_{64}$
   b. $^{12}/_{64}$
   c. $^{9}/_{64}$
   d. $^{6}/_{64}$
   e. $^{3}/_{64}$
   *Answer: a*

•42. Classical albinism results from a recessive allele. Which of the following is the expected ratio for the progeny when a normally pigmented male with an albino father has children with an albino woman?
   a. $^{3}/_{4}$ normal; $^{1}/_{4}$ albino
   b. $^{3}/_{4}$ albino; $^{1}/_{4}$ normal
   c. $^{1}/_{2}$ normal; $^{1}/_{2}$ albino
   d. All normal
   e. All albino
   *Answer: c*

■43. In humans, a widow's peak is caused by a dominant allele *W*, and a continuous hairline is caused by a recessive allele *w*. Short fingers are caused by a dominant allele *S*, and long fingers are caused by a recessive allele *s*. Suppose a woman with a continuous hairline and short fingers and a man with a widow's peak and long fingers have three children. One child has short fingers and a widow's peak, one has long fingers and a widow's peak, and one has long fingers and a continuous hairline. What are the genotypes of the parents?
   a. Female *wwSS*; male *WWss*
   b. Female *wwSs*; male *Wwss*
   c. Female *wwSs*; male *WWss*
   d. Female *WwSs*; male *WwSs*
   e. None of the above
   *Answer: b*

■44. In garden peas, the allele for tall plants is dominant over the allele for short plants. A true-breeding tall plant is crossed with a short plant, and one of their offspring is test crossed. Out of 20 offspring resulting from the test cross, about _____ should be tall.
   a. 0
   b. 5
   c. 10
   d. 15
   e. 20
   *Answer: c*

**■45. Which of the following phenomena is *not* observed using only dihybrid crosses?
   a. Crossing over
   b. Segregation of alleles
   c. Independent assortment of alleles
   d. Recessive lethal alleles
   e. All of the above
   *Answer: e*

46. Separation of the alleles of a single gene into different gametes is called
   a. synapsis.
   b. segregation.
   c. independent assortment.
   d. heterozygous separation.
   e. recombination.
   *Answer: b*

■47. In mice, short hair is dominant to long hair. If a short-haired individual is crossed with a long-haired individual and both long- and short-haired offspring result, you can conclude that
   a. the short-haired individual is homozygous.
   b. the short-haired individual is heterozygous.
   c. the long-haired individual is homozygous.
   d. the long-haired individual is heterozygous.
   e. This cannot be answered without more information.
   *Answer: b*

**■48. In dogs, erect ears and barking while following a scent are due to dominant alleles; droopy ears and silence while following a scent are due to recessive alleles. A dog homozygous for both traits is mated to a droopy-eared, silent follower. If the two genes are unlinked, the expected $F_1$ phenotypic ratios should be
   a. 9:3:3:1.
   b. 1:1.

   c. 16:0.
   d. 1:2:1.
   e. None of the above
   *Answer: c*

**■49. In cocker spaniels, black color (*B*) is dominant over red (*b*), and solid color (*S*) is dominant over spotted (*s*). If the offspring between *BBss* and *bbss* individuals are mated with each other, what fraction of their offspring will be expected to be black and spotted? Assume the genes are unlinked.
   a. $^1/_{16}$
   b. $^9/_{16}$
   c. $^1/_9$
   d. $^3/_{16}$
   e. $^3/_4$
   *Answer: e*

50. If the same allele has two or more phenotypic effects, it is said to be
   a. codominant.
   b. a marker.
   c. linked.
   d. pleiotropic.
   e. hemizygous.
   *Answer: d*

51. In the ABO blood type system,
   a. A, B, and O are codominant.
   b. A, B, and O are incompletely dominant.
   c. A and B are codominant.
   d. O is incompletely dominant to A and B.
   e. A is dominant to B, and B is dominant to O.
   *Answer: c*

■52. In Netherlands dwarf rabbits, a gene showing intermediate inheritance produces three phenotypes. Rabbits that are homozygous for one allele are small rabbits; individuals homozygous for the other allele are deformed and die; heterozygous individuals are dwarf. If two dwarf rabbits are mated, what proportion of their surviving offspring should be dwarf?
   a. $^1/_4$
   b. $^1/_3$
   c. $^1/_2$
   d. $^2/_3$
   e. $^3/_4$
   *Answer: d*

**■53. In tomatoes, tall is dominant to short, and smooth fruits are dominant to hairy fruits. A plant homozygous for both dominant traits is crossed with a plant homozygous for both recessive traits. The $F_1$ progeny are tested and crossed with the following results: 78 tall, smooth fruits; 82 dwarf, hairy fruits; 22 tall, hairy fruits; and 18 dwarf, smooth fruits. These data indicate that the genes are
   a. on different chromosomes.
   b. linked, but do not cross over.
   c. linked and show 10 percent recombination.
   d. linked and show 20 percent recombination.
   e. linked and show 40 percent recombination.
   *Answer: d*

■54. White eyes is a recessive sex-linked trait in fruit flies. If a white-eyed female fruit fly is mated to a red-eyed male, their offspring should be
   a. 50 percent red-eyed, 50 percent white-eyed for both sexes.
   b. all white-eyed for both sexes.
   c. all white-eyed males, all red-eyed females.
   d. all white-eyed females, all red-eyed males.
   e. 50 percent red-eyed males, 50 percent white-eyed males, all red-eyed females.
   *Answer: c*

■55. A dominant allele *K* is necessary for normal hearing. A dominant allele *M* on a different locus results in deafness no matter what other alleles are present. If a *kkMm* individual is crossed with a *Kkmm* individual, what percentage of the offspring will be deaf?
   a. 0
   b. 25
   c. 50
   d. 75
   e. None of the above
   *Answer: d*

■56. The genetic disease blue sclera is determined by an autosomal dominant allele. The eyes of individuals with this allele have bluish sclera. These same individuals may also suffer from fragile bones and deafness. This is an example of
   a. incomplete dominance.
   b. pleiotropy.
   c. epistasis.
   d. codominance.
   e. linkage.
   *Answer: b*

★■57. The blue sclera allele has 90 percent penetrance for producing blue sclera, 60 percent penetrance for fragile bones, and 40 percent penetrance for deafness. If these probabilities of penetrance are independent, what percent of individuals with the blue sclera allele will have deafness, blue sclera, and fragile bones?
   a. 22
   b. 40
   c. 60
   d. 90
   e. None of the above
   *Answer: a*

■58. Y-linked genes include a gene that produces hairy pinnae (the external ear). A male with hairy pinnae will pass this trait
   a. usually to his sons, but rarely also to a daughter.
   b. only to his sons.
   c. only to his daughters.
   d. only to his grandsons.
   e. to all his children if the mother is a carrier.
   *Answer: b*

59. Which of the following organelles contain(s) DNA?
   a. Nucleus
   b. Chloroplast
   c. Mitochondria
   d. Ribosome
   e. a, b, and c
   *Answer: e*

60. Two strains of true-breeding plants that have different alleles for a certain character are crossed. Their progeny are called
   a. the P generation.
   b. the $F_1$ generation.
   c. the $F_2$ generation.
   d. $F_1$ crosses.
   e. $F_2$ progeny.
   *Answer: b*

61. A mutation at a single locus causes a change in many different characters. This an example of a(n) _____ effect.
   a. polygene
   b. epigenetic
   c. cytoplasmic
   d. multiple negativity
   e. pleiotropic
   *Answer: e*

★■62. In guppies, fan tail is dominant to flesh tail, and rainbow color is dominant to pink. $F_1$ female guppies are crossed with flesh-tailed, pink-colored males, and the following progeny are observed: 401 fan-tailed, pink-colored; 399 flesh-tailed, rainbow-colored; 98 flesh-tailed, pink-colored; and 102 fan-tailed, rainbow-colored guppies. The map distance between these two genes is
   a. 80 cM.
   b. 25 cM.
   c. .8 cM.
   d. 20 cM.
   e. None of the above
   *Answer: d*

63. How many linkage groups are present in a female human?
   a. 46
   b. Thousands
   c. 23
   d. 2
   e. 496
   *Answer: c*

■64. Tall pea plants are crossed to short, and the progeny are medium height. The $F_1$ plants are crossed together, but the progeny observed among the $F_2$ have nine different size classes. This character's mode of inheritance is
   a. pleiotropic.
   b. epistasis.
   c. multiallelic.
   d. polygenic.
   e. hypostatic.
   *Answer: d*

65. The approximate total number of genes in the human nuclear genome is closest to
   a. 30.
   b. 300.
   c. 3,000.
   d. 30,000.
   e. 300,000.
   *Answer: d*

66. Sex in humans is determined by
   a. a gene called *SRY* found on the Y chromosome.
   b. a gene called *SRY* found on the X chromosome.
   c. a gene found on an autosomal chromosome called *SDG*.

d. the simple presence or absence of a Y chromosome.
e. a gene called *SDG* found on the Y chromosome.
*Answer: a*

67. Humans have _____ genes in their mitochondria.
    a. 300
    b. 3,000
    c. 30,000
    d. 300,000
    e. 37
    *Answer: e*

68. Mendel has contributed significantly to the study of genetics. One of his major contributions was
    a. the use of statistics and probability to analyze data.
    b. a complete description of the process of meiosis.
    c. the understanding that phenotypes are affected by the environment.
    d. the discovery that dominance is always complete.
    e. the finding that heritable traits combine or blend together.
    *Answer: a*

■69. If a trait not expressed in the F$_1$ generation reappears in the F$_2$ generation, the inheritance of the trait in question is a(n) example of
    a. codominance.
    b. dominance and recessiveness.
    c. incomplete dominance.
    d. epistasis.
    e. a sex linked trait.
    *Answer: b*

■70. A dihybrid cross
    a. results in a genotypic ratio of 2:1.
    b. involves genes located on the sex chromosomes.
    c. results in offspring of lower quality than that of the parents.
    d. results in two different phenotypes in the F$_2$ generation.
    e. is a cross between identical double heterozygotes.
    *Answer: e*

■71. The ABO blood groups in humans are determined by a multiple allelic system in which $I^A$ and $I^B$ are codominant and are dominant to $i^O$. If an infant born to a type O mother also is type O, possible genotypes for the father are
    a. O or A.
    b. A or B.
    c. O only.
    d. O, A, or B.
    e. impossible to determine.
    *Answer: d*

# 11 DNA and Its Role in Heredity

## Fill in the Blank

1. The X-ray crystallographs of the English chemist _____ were essential for the discovery of the structure of the DNA molecule.
   *Answer: Rosalind Franklin*

2. Since the DNA molecule is continuous, nucleotide pair after nucleotide pair, their information must lie in the _____ sequence of the nitrogenous bases.
   *Answer: linear*

3. Arthur Kornberg showed that DNA could replicate in the test tube if it contained intact DNA for a template, a mixture of the four precursors (the four nucleoside triphosphates), and _____.
   *Answer: DNA polymerase*

4. The material that changed R strain pneumococcus into the virulent S strain was originally referred to as the _____.
   *Answer: transforming principle*

5. The basic units of DNA and RNA molecules are the
   _____.
   *Answer: nucleotides (or nitrogenous bases)*

6. The experiments of Meselson and Stahl established the _____ of DNA.
   *Answer: semiconservative replication*

★7. The nitrogenous bases classified as purines are _____ and _____.
   *Answer: adenine; guanine*

★8. The nitrogenous bases classified as pyrimidines are _____ and _____.
   *Answer: cytosine; thymine*

★■9. Using Meselson and Stahl's experimental system for studying the mode of replication of DNA, researchers analyze the genetic information from a life-form from Mars. After the first round of replication, they see two distinct bands in the cesium chloride gradient. This finding is consistent with _____ replication.
   *Answer: conservative*

10. The enzyme that replicates the lagging strand is _____.
    *Answer: DNA polymerase III*

11. In prokaryotes the enzyme that replicates the leading strand is called _____.
    *Answer: DNA polymerase III*

12. The region of DNA where replication begins is the

    _____.
    *Answer: origin of replication*

13. The fragments of RNA and DNA found on the lagging strand of DNA prior to RNA removal and ligation are called _____.
    *Answer: Okazaki fragments*

★14. The purines take up (more/less) _____ space in the center of a DNA molecule than the pyrimidines do.
    *Answer: more*

15. In a sequencing reaction, the shortest sequences are those that end closer to the (5'/3') _____ end than to the (5'/3') _____ end of the synthesized molecule.
    *Answer: 5'; 3'*

16. The _____ function of the DNA polymerase reduces the number of mistakes by the square of the frequency of the error rate.
    *Answer: proofreading*

17. The repetitive sequences at the end of many chromosomes are called _____.
    *Answer: telomeres.*

18. An enzyme that catalyzes the addition of any lost telomere sequences is _____.
    *Answer: telomerase*

## Multiple Choice

■1. Before the discovery of DNA, why was the hereditary material thought to be made of proteins and not nucleic acids?
   a. Nucleic acids are made up of 20 different bases, whereas proteins are made up of only 5 amino acids.
   b. Protein subunits can combine to form larger proteins.
   c. Proteins seemed to be much more chemically diverse.
   d. Proteins can be enzymes.
   e. None of the above
   *Answer: c*

■2. How can DNA, made up of only 4 different bases, encode the information necessary to specify the workings of an entire organism?
   a. DNA molecules are extremely long.
   b. DNA molecules are found in the nucleus.
   c. DNA is transcribed into RNA and then into proteins with specific functions.
   d. DNA is eventually translated into proteins, which are made up of 20 different amino acids.
   e. None of the above
   *Answer: a*

3. In Griffith's experiments, what happened when heat-killed S strain pneumococci were injected into a mouse along with live R strain pneumococci?
   a. DNA from the live R was taken up by the heat-killed S, converting it to R and killing the mouse.
   b. DNA from the heat-killed S was taken up by the live R, converting it to S and killing the mouse.
   c. Proteins released from the heat-killed S killed the mouse.
   d. RNA from the heat-killed S was translated into proteins that killed the mouse.
   e. Nothing
   *Answer: b*

4. Experiments designed to identify the transforming principle were based on
   a. purifying each of the macromolecule types from a cell-free extract.
   b. removing each of the macromolecules from a cell, then testing its type.
   c. selectively destroying the different macromolecules in a cell-free extract.
   d. Both a and b
   e. Both a and c
   *Answer: c*

5. The Hershey–Chase experiment determined that
   a. protein and DNA are the hereditary materials of viruses.
   b. protein, not DNA, is the hereditary material of viruses.
   c. viruses do not contain hereditary material.
   d. DNA, not protein, is the hereditary material of viruses.
   e. None of the above
   *Answer: d*

6. The rules formulated by Erwin Chargaff state that
   a. A = T and G = C in any molecule of DNA.
   b. A = C and G = T in any molecule of DNA.
   c. A = G and C = T in any molecule of DNA.
   d. A = U and G = C in any molecule of RNA.
   e. DNA and RNA are made up of the same four nitrogenous bases.
   *Answer: a*

7. Purines include
   a. cytosine, uracil, and thymine.
   b. adenine and cytosine.
   c. adenine and thymine.
   d. cytosine and thymine.
   e. adenine and guanine.
   *Answer: e*

8. The structure of DNA is characterized by a
   a. right- or left-handed double helix and antiparallel strands.
   b. right-handed double helix and antiparallel strands.
   c. right-handed single helix.
   d. right-handed single helix and parallel strands.
   e. All of the above
   *Answer: b*

9. The antiparallel relationship of the two strands of DNA refers to the
   a. twisted configuration of the strands.
   b. alternative branching pattern of the strands.
   c. alignment of the strands, such that one strand starts with a 3′ carbon, the other with a 5′ carbon.
   d. view at one end of the molecule–one strand has an A wherever the other has a T, and one has a G wherever the other has a C.
   e. All of the above
   *Answer: c*

10. The nitrogenous bases (and the two strands of the DNA double helix) are held together by
    a. weak van der Waals forces.
    b. covalent bonds.
    c. hydrogen bonds.
    d. Both a and b
    e. Both a and c
    *Answer: c*

11. Why is RNA incorporated into the DNA molecule during DNA replication?
    a. RNA primase adds bases that act as primers.
    b. RNA primase is able to use DNA as a template.
    c. RNA primase must be incorporated into the holo-enzyme complex.
    d. DNA polymerase I and III can only add on to an existing strand.
    e. All of the above
    *Answer: d*

12. The structure of DNA explains which three major properties of genes?
    a. They contain information, direct the synthesis of proteins, and are contained in the cell nucleus.
    b. They contain nitrogenous bases, direct the synthesis of RNA, and are contained in the cell nucleus.
    c. They replicate exactly, are contained in the cell nucleus, and direct the synthesis of cellular proteins.
    d. They encode the organism's phenotype, are passed on from one generation to the next, and contain nitrogenous bases.
    e. They contain information, replicate exactly, and can change to produce a mutation.
    *Answer: e*

13. Semiconservative replication of DNA involves
    a. each of the original strands acting as a template for a new strand.
    b. only one of the original strands acting as a template for a new strand.
    c. the complete separation of the original strands, the synthesis of new strands, and the reassembly of double-stranded molecules.
    d. the use of the original double-stranded molecule as a template.
    e. None of the above
    *Answer: a*

14. The molecules that function to replicate DNA in the cell are
    a. DNA nucleoside triphosphates.
    b. DNA polymerases.
    c. nucleoside polymerases.
    d. DNAses.
    e. ribonucleases.
    *Answer: b*

15. Which of the following is the correct order of events for synthesis of the lagging strand?

a. Primase adds RNA primer, DNA polymerase III creates a stretch, DNA polymerase I removes the primer, and ligase seals the gaps.
b. Primase adds primer, DNA polymerase I removes the primer, DNA polymerase III extends the segment, and ligase seals the gap.
c. Ligase adds bases to the primase, the primase generates polymerase I, polymerase III adds to the stretch, helicase winds the DNA.
d. Helicase unwinds the DNA, primase creates a primer, DNA polymerase I elongates the stretch, DNA polymerase III removes the primer, and ligase seals the gaps in the DNA.
e. None of the above

*Answer: a*

16. During replication, the new DNA strand is synthesized
a. in the 3′ to 5′ direction.
b. in the 5′ to 3′ direction.
c. in both the 3′ to 5′ and 5′ to 3′ directions from the replication fork.
d. from one end to the other, in the 3′ to 5′ or the 5′ to 3′ direction.
e. None of the above

*Answer: b*

▪17. Why were fragments like those now called Okazaki fragments expected before they were discovered?
a. DNA replicates in the 5′ to 3′ direction.
b. The replication fork moves forward along a double-stranded DNA molecule.
c. DNA replicates in the 3′ to 5′ direction on the lagging strand.
d. RNA primase places short RNA primer sequences along the DNA molecule.
e. DNA polymerase I can connect short segments.

*Answer: b*

▪18. DNA replication in eukaryotes differs from replication in bacteria because
a. synthesis of the new DNA strand is from 3′ to 5′ in eukaryotes and from 5′ to 3′ in bacteria.
b. synthesis of the new DNA strand is from 5′ to 3′ in eukaryotes and from 3′ to 5′ in bacteria.
c. there are many replication forks in each eukaryotic chromosome and only one in bacterial DNA.
d. synthesis of the new DNA strand is from 5′ to 3′ in eukaryotes and is random in prokaryotes.
e. Okazaki fragments are produced in eukaryotic DNA replication but not in prokaryotic DNA replication.

*Answer: c*

19. In the cells of prokaryotes, methylated guanine contributes to
a. an increased rate of DNA replication.
b. a slowed rate of DNA replication.
c. the correct separation of DNA strands.
d. proofreading of replicated strands.
e. correcting mismatched pairs of bases.

*Answer: e*

*20. In eukaryotes, Okazaki fragments are about _____ base pairs long.
a. 50
b. 150
c. 1,500
d. 150,000
e. 15,000,000

*Answer: b*

21. Which of the following molecules functions to transfer information from one generation to the next?
a. DNA
b. mRNA
c. tRNA
d. Proteins
e. Lipids

*Answer: a*

22. Mutations are
a. heritable changes in the sequence of DNA bases that produce an observable phenotype.
b. heritable changes in the sequence of DNA bases.
c. mistakes in the incorporation of amino acids into proteins.
d. heritable changes in the mRNA of an organism.
e. None of the above

*Answer: b*

23. In order to show that DNA is the "transforming principle," Avery, MacLeod, and McCarty showed that DNA could transform nonvirulent strains of pneumococcus. Their hypothesis was strengthened by their demonstration that
a. enzymes that destroyed proteins also destroyed transforming activity.
b. enzymes that destroyed nucleic acids also destroyed transforming activity.
c. enzymes that destroyed complex carbohydrates also destroyed transforming activity.
d. the transformation activity was destroyed by boiling.
e. other strains of bacteria also could be transformed successfully.

*Answer: b*

24. During infection of *E. coli* cells by bacteriophage T2,
a. proteins are the only phage components that enter the infected cell.
b. both proteins and nucleic acids enter the cell.
c. only protein from the infecting phage can also be detected in progeny phage.
d. only nucleic acids enter the cell.
e. more than one infecting phage particle is required to produce infection.

*Answer: d*

25. Bacteriophage nucleic acids were labeled by carrying out an infection of *E. coli* cells growing in
a. $^{14}C$-labeled $CO_2$.
b. $^{3}H$-labeled water.
c. $^{32}P$-labeled phosphate.
d. $^{35}S$-labeled sulfate.
e. $^{18}O$-labeled water.

*Answer: c*

26. Information sources used by Watson and Crick to determine the structure of DNA included
a. electron micrographs of individual DNA molecules.
b. light micrographs of bacteriophage particles.
c. light micrographs of individual bacteria chromosomes.
d. nuclear magnetic resonance analysis of DNA.
e. X-ray crystallography of double-stranded DNA.

*Answer: e*

27. Double-stranded DNA looks a little like a ladder that has been twisted into a helix, or spiral. The side supports of the ladder are
    a. individual nitrogenous bases.
    b. alternating bases and sugars.
    c. alternating bases and phosphate groups.
    d. alternating sugars and phosphates.
    e. alternating bases, sugars, and phosphates.
    *Answer: d*

28. The steps of the ladder are
    a. individual nitrogenous bases.
    b. pairs of bases.
    c. alternating bases and phosphate groups.
    d. alternating sugars and bases.
    e. alternating bases, sugars, and phosphates.
    *Answer: b*

29. If a double-stranded DNA molecule contains 30 percent T, it must contain _____ percent G.
    a. 20
    b. 30
    c. 40
    d. 50
    e. 60
    *Answer: a*

**•30. You have analyzed the DNA isolated from a newly discovered virus, and you have found that its base composition is 32 percent A, 17 percent C, 32 percent G, and 19 percent T. What would be a reasonable explanation of this observation?
    a. The virus must be extraterrestrial.
    b. In some viruses, double-stranded DNA is made up of base pairs containing two purines or two pyrimidines.
    c. Some of the T was converted to C during the isolation procedure.
    d. The genome of the phage is single-stranded, not double-stranded.
    e. The genome of the phage must be circular, not linear.
    *Answer: d*

•31. For the viral DNA described in the previous question, what would be the base composition of the complementary DNA?
    a. 32% A, 17% C, 32% G, 19% T
    b. 32% T, 17% G, 32% C, 19% A
    c. 32% C, 17% A, 32% G, 19% T
    d. 25% A, 25% G, 25% C, 25% T
    e. 32% A, 32% T, 18% C, 18% G
    *Answer: b*

*32. In the Meselson–Stahl experiment, the conservative model of DNA replication is ruled out by which of the following observations?
    a. No completely "heavy" DNA is observed after the first round of replication.
    b. No completely "light" DNA ever appears, even after several replications.
    c. The product that accumulates after two rounds of replication is completely "heavy."
    d. Completely "heavy" DNA is observed throughout the experiment.
    e. Three different DNA densities are observed after a single round of replication.
    *Answer: a*

33. During DNA replication
    a. one parental strand must be degraded to allow the other strand to be copied.
    b. the parental strands must separate so that both can be copied.
    c. the parental strands come back together after the passage of the replication fork.
    d. origins of replication always give rise to single replication forks.
    e. two replication forks diverge from each origin but one always lags behind the other.
    *Answer: b*

•34. The enzyme DNA ligase is required continuously during DNA replication because
    a. fragments of the leading strand must be joined together.
    b. fragments of the lagging strand must be joined together.
    c. the parental strands must be joined back together.
    d. 3′-deoxynucleoside triphosphates must be converted to 5′-deoxynucleoside triphosphates.
    e. the complex of proteins that work together at the replication fork must be kept from falling apart.
    *Answer: b*

35. In DNA replication, each newly made strand is
    a. identical in DNA sequence to the strand from which it was copied.
    b. complementary in sequence to the strand from which it was copied.
    c. oriented in the same 3′ to 5′ direction as the strand from which it was copied.
    d. an incomplete copy of one of the parental strands.
    e. a hybrid molecule consisting of both ribo- and deoxyribonucleotides.
    *Answer: b*

•36. Which feature of the Watson–Crick model of DNA structure explains its ability to function in replication and gene expression?
    a. Each strand contains all the information present in the double helix.
    b. There are structural and functional similarities between DNA and RNA.
    c. The double helix is right-handed, not left-handed.
    d. DNA replication does not require enzyme catalysts.
    e. Bases are exposed in the major groove of the double helix.
    *Answer: a*

37. The Hershey–Chase experiment persuaded most scientists that
    a. bacteria can be transformed.
    b. DNA is indeed the carrier of hereditary information.
    c. DNA replication is semiconservative.
    d. the transforming principle requires host factors.
    e. All of the above
    *Answer: b*

38. Which of the following features summarizes the molecular architecture of DNA?
    a. The two strands run in opposite directions.
    b. The molecule twists in the same direction as the threads of most screws.
    c. The molecule is a double-stranded helix.
    d. It has a uniform diameter.

e. All of the above

*Answer: e*

39. The fidelity of DNA replication is astounding. During DNA synthesis, the error rate is on the order of one wrong nucleotide per
    a. 1000.
    b. 100,000.
    c. $10^{10}$.
    d. $10^{13}$–$10^{16}$.
    e. one trillion.

    *Answer: c*

40. Chargaff's rule states that
    a. DNA must be replicated before a cell can divide.
    b. viruses enter cells without their protein coat.
    c. only protein from the infecting phage can also be detected in progeny phage.
    d. only nucleic acids enter the cell during infection.
    e. the amount of cytosine equals the amount of guanine.

    *Answer: e*

41. The first scientist(s) to suggest a mode of replication for DNA was (were)
    a. Linus and Pauling.
    b. Hershey and Chase.
    c. Rosalind Franklin.
    d. Watson and Crick.
    e. Meselson and Stahl.

    *Answer: d*

42. In eukaryotic cells, each chromosome has
    a. one origin of replication.
    b. two origins of replication.
    c. many origins of replication.
    d. only one origin of replication per nucleus.
    e. None of the above

    *Answer: c*

43. When adding the next monomer to a growing DNA strand, the monomer is added to which carbon of the deoxyribose?
    a. 1′
    b. 2′
    c. 3′
    d. 4′
    e. 5′

    *Answer: c*

44. The energy necessary for making a DNA molecule comes directly from the
    a. sugar.
    b. ATP.
    c. release of phosphates.
    d. NADPH.
    e. NADH.

    *Answer: c*

45. A deoxyribose nucleotide is a
    a. deoxyribose plus a nitrogenous base.
    b. sugar and a phosphate.
    c. deoxyribose plus a nitrogenous base and a phosphate.
    d. ribose plus a nitrogenous base.
    e. nitrogenous base bonded at the 5′ end to a sugar–phosphate backbone.

    *Answer: c*

46. The enzyme that removes the RNA primers is called
    a. DNA ligase.
    b. primase.
    c. reverse transcriptase.
    d. helicase.
    e. DNA polymerase I.

    *Answer: e*

47. The enzyme that restores the phosphodiester linkage between adjacent fragments in the lagging strand during DNA replication is
    a. DNA ligase.
    b. primase.
    c. reverse transcriptase.
    d. helicase.
    e. DNA polymerase I.

    *Answer: a*

48. A deoxyribose nucleoside is a
    a. deoxyribose plus a nitrogenous base.
    b. sugar and a phosphate.
    c. deoxyribose plus a nitrogenous base and a phosphate.
    d. ribose plus a nitrogenous base.
    e. nitrogenous base bonded at the 5′ end to a sugar–phosphate backbone.

    *Answer: a*

49. The building blocks for a new DNA molecule are deoxyribose
    a. nucleoside monophosphates.
    b. nucleoside diphosphates.
    c. nucleoside triphosphates.
    d. nucleotide diphosphates.
    e. nucleotide triphosphates.

    *Answer: c*

50. The force that holds DNA together in a double helix is
    a. the force of the twist.
    b. covalent bonds.
    c. ionic bonds.
    d. ionic interactions.
    e. hydrogen bonds.

    *Answer: e*

51. The enzyme that unwinds the DNA prior to replication is called
    a. DNA polymerase III.
    b. DNA ligase.
    c. single-stranded DNA binding protein.
    d. primase.
    e. helicase.

    *Answer: e*

52. The first repair of mistakes during DNA replication is made by
    a. the mismatch repair system.
    b. DNA polymerase.
    c. excision repair.
    d. SOS repair.
    e. postreplication repair.

    *Answer: b*

53. The error rate of changing an incorrect base with another incorrect base during proofreading is one in _____ bases.
    a. 10
    b. 100
    c. 1,000
    d. 10,000
    e. 1,000,000

    *Answer: d*

■54. If Hershey and Chase had found $^{35}$S in both the pellet and the supernatant, what would have been their likely conclusion about the nature of DNA replication?
a. A protein must be the information molecule.
b. No conclusion would have been possible from these results.
c. DNA is the genetic information molecule.
d. Phage must have stuck to the bacteria.
e. Phosphorus was in the information molecule.
*Answer: b*

■■55. An alien DNA-like molecule is isolated from the frozen remains of a life-form found beneath the Martian polar ice caps. In this sample, for every base designated Q, there is twice that amount of base R, and for every base Z, there is twice that amount of base S. If the molecule contains 12 percent R, what percentage of Z would you expect?
a. 6
b. 12
c. 24
d. 27.33
e. 54.66
*Answer: d*

■■56. Select the molecular model that would best fit the data presented in the previous question.
a. The molecule is single-stranded.
b. The molecule is antiparallel.
c. The molecule is triple-stranded.
d. The molecule is helical.
e. The molecule is helical, double-stranded, and antiparallel.
*Answer: c*

■■57. If Meselson and Stahl had observed one intermediate, slightly smeared band after growing bacteria for one generation, and then after two generations again had found one slightly smeared band, what would have been their likely conclusion about the mode of DNA replication?
a. DNA replicates semiconservatively.
b. DNA replicates conservatively.
c. DNA replicates semidiscontinuously.
d. DNA replicates dispersively.
e. None of the above
*Answer: d*

*58. Pyrophosphate is a
a. building block for DNA synthesis.
b. by-product of DNA synthesis.
c. precursor to DNA synthesis.
d. fire phosphate used in nucleic acid metabolism.
e. All of the above
*Answer: b*

59. Which one of the following is *not* found in DNA?
a. Carbon
b. Oxygen
c. Nitrogen
d. Hydrogen
e. Sulfur
*Answer: e*

■60. What was most remarkable about the Griffith experiment?
a. Smooth bacteria could survive heating.
b. DNA, not protein, was found to be the genetic molecule.

c. Materials from dead organisms could affect and change living organisms.
d. Nonliving viruses could change living cells.
e. None of the above statements apply to the Griffith experiment.
*Answer: c*

61. The Hershey–Chase experiment
a. proved that DNA replication is semiconservative.
b. used $^{32}$P to label protein.
c. used $^{35}$S to label DNA.
d. helped prove that DNA is the genetic molecule.
e. Both b and c
*Answer: d*

62. Griffith was able to distinguish the two strains of pneumococcus by means of
a. the appearance of the colonies in culture.
b. differences in their lethality in mice.
c. their sizes.
d. their odors.
e. Both a and b
*Answer: e*

63. The maximum length of a DNA sequence that can be determined using current technology is approximately _____ base pairs.
a. 50
b. 100
c. 700
d. 1,000
e. 5,000
*Answer: c*

64. In PCR, _____ creates single-stranded template molecules.
a. heat
b. high salt concentration
c. DNA polymerase
d. exonuclease
e. a primer
*Answer: a*

65. Ideally, PCR _____ increases the amount of DNA during additional cycles.
a. additively
b. gradually
c. linearly
d. systematically
e. exponentially
*Answer: e*

66. DNA polymerase lengthens a polynucleotide strand by
a. building short DNA fragments that are later linked together.
b. adding lost DNA sequences to the 3' end.
c. linking purines with pyrimidines.
d. covalently linking new nucleotides to a previously existing strand.
e. threading the existing DNA through a replication complex.
*Answer: d*

67. Fourteen human DNA polymerases have been identified. Which of the following statements about them is true?
a. They all are involved in DNA replication.
b. One of the DNA polymerases opens the replication fork, one forms the primer, another removes the primer, and the rest are involved in protein synthesis.

c. The functions of DNA polymerase, helicase, and primase are known; the functions of the others are unknown.

d. Each DNA polymerase requires a specific primer to function.

e. Only one of the fourteen is involved in DNA replication; the others are involved in primer removal and DNA repair.

*Answer: e*

68. Why don't cells last the entire lifetime of an organism?
    a. The removal of the RNA primer following DNA replication leads to a shortening of  the chromosome and eventual cell death.
    b. The enzyme telomerase is readily destroyed by the environment, resulting in cell death.
    c. DNA replication is subject to errors, causing premature cell death.
    d. Okazaki fragments disrupt protein synthesis, resulting in premature cell death.
    e. The repeating telomeric sequence of TTAGGG interferes with normal DNA replication and leads to cell death.

    *Answer: a*

■69. The strands that make up DNA are antiparallel. This means that
    a. one strand is positively charged and the other is negatively charged.
    b. the base pairings create unequal spacing between the two DNA strands.
    c. the 5′ to 3′ direction of one strand is counter to the 5′ to 3′ direction of the other strand.
    d. the twisting of the DNA molecule has shifted the two strands.
    e. purines bond with purines and pyrimidines bond with pyrimidines.

    *Answer: c*

70. What accounts for the uniform diameter of the DNA molecule?
    a. The two sides of the DNA molecule are held together by hydrogen bonds.
    b. A purine always bonds with a pyrimidine.
    c. One side of the DNA molecule has an unconnected 5′ phosphate group and the opposite end has an unconnected 3′ hydroxyl group.
    d. The 3′ carbon of one deoxyribose and the 5′ carbon of another deoxyribose bond together.
    e. The alternating sugar and phosphate backbone coils around the outside of the helix.

    *Answer: b*

# 12 From DNA to Protein: Genotype to Phenotype

## Fill in the Blank

1. Prototrophs ("original eaters") grow on minimal media, whereas _____ ("increased eaters") require specific additional nutrients.
   *Answer: auxotrophs*

2. The basic units of DNA and RNA molecules are the
   _____.
   *Answer: nucleotides*

3. RNA differs from DNA in base composition in that it contains _____ instead of thymine.
   *Answer: uracil*

4. The strand of DNA that is transcribed into RNA is the _____ strand.
   *Answer: template*

5. The part of a protein that determines whether translation will continue in the cytosol or at the endoplasmic reticulum is the _____ sequence.
   *Answer: signal*

6. The excess of codons (64) over amino acids (20) indicates that the genetic code is _____.
   *Answer: redundant*

7. An mRNA molecule with several ribosomes attached at the same time is called a(n) _____.
   *Answer: polysome*

8. Small ribosomal subunits are dispersed into smaller components when placed into a detergent solution. Upon removal of the detergent, the components interact to create new intact subunits by a process called _____.
   *Answer: self-assembly*

9. A mutation that causes a change in the nitrogenous base sequence of a DNA molecule but no change in the amino acid sequence it codes for is called a(n) _____ mutation.
   *Answer: silent*

10. The fact that some tRNA molecules do not have to pair exactly is called _____.
    *Answer: wobble*

11. The portion of the tRNA molecule that complementary base-pairs with the mRNA is called the _____.
    *Answer: anticodon*

12. The lethal toxin that is produced by the castor bean plant and that blocks protein synthesis is _____.
    *Answer: ricin*

13. The one-gene, one-enzyme hypothesis resulted from the work of _____.
    *Answer: Beadle and Tatum*

14. The synthesis of DNA from RNA is called _____.
    *Answer: reverse transcription*

15. A tRNA that has bonded to an amino acid is referred to as _____ tRNA.
    *Answer: charged*

## Multiple Choice

1. After irradiating *Neurospora*, Beadle and Tatum collected mutants that require arginine to grow. These mutants
   a. will not grow on minimal media but will grow on minimal media with arginine.
   b. will grow on minimal media and on minimal media with arginine.
   c. will not grow on minimal media and will not grow on minimal media with arginine.
   d. will grow on minimal media but will not grow on minimal media with arginine.
   e. None of the above
   *Answer: a*

■2. Within a group of mutants with the same growth requirement (i.e., the same overt phenotype), mapping studies determined that individual mutations were on different chromosomes. This indicates that
   a. the same gene governs all the steps in a particular biological pathway.
   b. different genes can govern different individual steps in the same biological pathway.
   c. different genes govern the same step in a particular biological pathway.
   d. all biological pathways are governed by different genes.
   e. genes do not govern steps in biological pathways.
   *Answer: b*

3. The study of *Neurospora* mutants grown on various supplemented media led to
   a. a determination of the steps in biological pathways.
   b. the "one-gene, one-enzyme" theory.
   c. the idea that genes are "on" chromosomes.
   d. Both a and b
   e. Both a and c
   *Answer: d*

4. The rates of DNA mutations are _____ in different organisms.
   a. the same
   b. constant
   c. different
   d. dependent on health
   e. dependent on temperature
   *Answer: c*

5. Genes code for
   a. enzymes.
   b. polypeptides.
   c. RNA.
   d. All of the above
   e. None of the above
   *Answer: d*

■6. How can DNA, which is made up of only four different bases, encode the information necessary to specify the workings of an entire organism?
   a. DNA molecules are extremely long.
   b. DNA molecules form codons of three bases that code for amino acids.
   c. The same DNA sequence can be used repeatedly.
   d. DNA can be replicated with low error rates.
   e. All of the above
   *Answer: e*

■7. How does RNA differ from DNA?
   a. RNA contains uracil instead of thymine and it is usually single-stranded.
   b. RNA contains uracil instead of thymine and it is usually double-stranded.
   c. RNA contains thymine instead of uracil and it is usually single-stranded.
   d. RNA contains uracil instead of cytosine.
   e. None of the above
   *Answer: a*

8. Which of the following statements about the flow of genetic information is true?
   a. Proteins encode information that is used to produce other proteins of the same amino acid sequence.
   b. RNA encodes information that is translated into DNA, and DNA encodes information that is translated into proteins.
   c. Proteins encode information that can be translated into RNA, and RNA encodes information that can be transcribed into DNA.
   d. DNA encodes information that is translated into RNA, and RNA encodes information that is translated into proteins.
   e. None of the above
   *Answer: d*

9. RNA polymerase uses the _____ DNA template to synthesize a _____ mRNA.
   a. 5′ to 3′; 5′ to 3′
   b. 3′ to 5′; 3′ to 5′
   c. 5′ to 3′; 3′ to 5′
   d. 3′ to 5′; 5′ to 3′
   e. Examples of all of the above have been found.
   *Answer: d*

10. Which of the following molecules transfers information from the nucleus to the cytoplasm?
    a. DNA
    b. mRNA

    c. tRNA
    d. Proteins
    e. Lipids
    *Answer: b*

★11. mRNA is synthesized in the _____ direction, which corresponds to the _____ of the protein.
    a. 5′ to 3′; N terminus to C terminus
    b. 3′ to 5′; C terminus to N terminus
    c. 5′ to 3′; C terminus to N terminus
    d. 3′ to 5′; N terminus to C terminus
    e. Examples of all of the above have been found.
    *Answer: a*

12. Which of the following molecules transfers information from one generation to the next?
    a. DNA
    b. mRNA
    c. tRNA
    d. Proteins
    e. Lipids
    *Answer: a*

13. Which of the following molecules transfers information from mRNA to protein?
    a. DNA
    b. mRNA
    c. tRNA
    d. Proteins
    e. Lipids
    *Answer: c*

14. A sequence of three RNA bases can function as a
    a. codon.
    b. anticodon.
    c. gene.
    d. Both a and b
    e. Both a and c
    *Answer: d*

■15. The difference between mRNA and tRNA is that
    a. tRNA has a more elaborate three-dimensional structure, due to extensive base pairing.
    b. tRNAs are usually much smaller than mRNAs.
    c. mRNA has a more elaborate three-dimensional structure, due to extensive base pairing.
    d. Both a and b
    e. None of the above
    *Answer: d*

16. Termination of transcription involves a
    a. stop codon.
    b. terminator sequence.
    c. termiproteator.
    d. hairline slip.
    e. series of A's.
    *Answer: b*

17. Ribosomes are a collection of _____ that are needed for
    _____.
    a. small proteins; translation.
    b. proteins and small RNAs; translation.
    c. proteins and tRNAs; transcription.
    d. proteins and mRNAs; translation.
    e. mRNAs and tRNAs; translation.
    *Answer: b*

18. In protein synthesis, the endoplasmic reticulum
    a. is the site where mRNA attaches.
    b. is the site where all ribosomes bind.

c. is the site of translation of membrane-bound and exported proteins.

d. produces tRNAs.

e. brings together mRNA and tRNA.

*Answer: c*

▪19. Retroviruses do not follow the "central dogma" of DNA→RNA→protein because they

a. contain RNA that is used to make DNA.

b. contain DNA that is used to make more RNA.

c. contain DNA that is used to make tRNA.

d. contain only RNA as the genetic material.

e. do not contain either DNA or RNA as the genetic material.

*Answer: a*

20. Mutations are

a. heritable changes in the sequence of DNA bases that produce an observable phenotype.

b. heritable changes in the sequence of DNA bases.

c. mistakes in the incorporation of amino acids into proteins.

d. heritable changes in the mRNA of an organism.

e. None of the above

*Answer: b*

21. The type of mutation that stops translation of a protein is a(n)

a. missense mutation.

b. nonsense mutation.

c. frame-shift mutation.

d. aberration.

e. None of the above

*Answer: b*

22. The type of mutation that is an insertion or a deletion of a single base is a(n)

a. missense mutation.

b. nonsense mutation.

c. frame-shift mutation.

d. aberration.

e. None of the above

*Answer: c*

23. The "central dogma" of molecular biology states that

a. information flow between DNA, RNA, and protein is reversible.

b. information flow in the cell is from protein to RNA to DNA.

c. information flow in the cell is unidirectional, from DNA to RNA to protein.

d. the DNA sequence of a gene can be predicted if we know the amino acid sequence of the protein it encodes.

e. the genetic code is ambiguous but not degenerate.

*Answer: c*

24. Transcription is the process of

a. synthesizing a DNA molecule from an RNA template.

b. assembling ribonucleoside triphosphates into an RNA molecule without a template.

c. synthesizing an RNA molecule using a DNA template.

d. synthesizing a protein using information from a messenger RNA.

e. replicating a single-stranded DNA molecule.

*Answer: c*

25. A transcription start signal is called a(n)

a. initiation codon.

b. promoter.

c. origin.

d. operator.

e. nonsense codon.

*Answer: b*

26. Initiation of transcription requires

a. a temporary stoppage of DNA replication.

b. a temporary separation of the strands in the DNA template.

c. destruction of one of the strands of the DNA template.

d. relaxation of positive supercoils in the DNA template.

e. induction of positive supercoils in the DNA template.

*Answer: b*

27. The adapters that allow translation of the four-letter nucleic acid language into the 20-letter protein language are called

a. aminoacyl tRNA synthetases.

b. transfer RNAs.

c. ribosomal RNAs.

d. messenger RNAs.

e. ribosomes.

*Answer: b*

28. The number of codons that actually specify amino acids is

a. 20.

b. 23.

c. 45.

d. 60.

e. 61.

*Answer: e*

29. The genetic code is best described as

a. redundant but not ambiguous.

b. ambiguous but not redundant.

c. both ambiguous and redundant.

d. neither ambiguous nor redundant.

e. nonsense.

*Answer: a*

30. The three codons in the genetic code that do not specify amino acids are called

a. missense codons.

b. start codons.

c. stop codons.

d. promoters.

e. initiator codons.

*Answer: c*

31. In eukaryotes, ribosomes become associated with endoplasmic reticulum membranes when

a. a signal sequence on the mRNA interacts with a receptor protein on the membrane.

b. a signal sequence on the ribosome interacts with a receptor protein on the membrane.

c. a signal sequence at the amino terminus of the protein being synthesized interacts with a receptor protein on the ribosome.

d. a signal sequence on the protein being synthesized interacts with a membrane attachment site on the ribosome.

e. the messenger RNA passes through a pore in the membrane.

*Answer: d*

32. Viruses that violate the "central dogma" through the use of an enzyme that makes DNA copies of an RNA molecule are called
    a. bacteriophage.
    b. retroviruses.
    c. RNA viruses.
    d. DNA viruses.
    e. enveloped viruses.
    *Answer: b*

33. Sickle-cell disease is caused by a _____ mutation.
    a. nonsense
    b. missense
    c. frame-shift
    d. temperature sensitive
    e. silent
    *Answer: b*

34. The classic work of Beadle and Tatum, later refined by others, provided evidence for
    a. the one-gene, one-enzyme hypothesis.
    b. the one-gene, one-polypeptide hypothesis.
    c. the mechanism by which information in genes is translated into traits.
    d. the effects of some mutations on organisms.
    e. All of the above
    *Answer: e*

35. RNA polymerase is a(n)
    a. RNA-directed DNA polymerase.
    b. RNA-directed RNA polymerase.
    c. DNA-directed RNA polymerase.
    d. typical enzyme.
    e. form of RNA.
    *Answer: c*

36. The direction of synthesis for a new mRNA molecule is _____ from a _____ template strand.
    a. 5′ to 3′; 5′ to 3′
    b. 5′ to 3′; 3′ to 5′
    c. 3′ to 5′; 5′ to 3′
    d. 3′ to 5′; 3′ to 5′
    e. 5′ to 5′; 3′ to 3′
    *Answer: b*

37. The region of DNA in prokaryotes to which RNA polymerase binds most tightly is the
    a. promoter.
    b. poly C center.
    c. enhancer.
    d. operator site.
    e. minor groove.
    *Answer: a*

38. Promoters are made of
    a. protein.
    b. carbohydrate.
    c. lipid.
    d. nucleic acids.
    e. amino acids.
    *Answer: d*

*39. The energy for transcription is derived from
    a. ATP.
    b. GTP.
    c. ATP and GTP.

d. the phosphodiester linkages in the incorporated nucleoside triphosphates.
    e. glucose.
    *Answer: d*

■40. DNA is composed of two strands, only one of which typically is used as a template for RNA synthesis. By what mechanism is the correct strand chosen?
    a. Both strands are tried and the one that works is remembered.
    b. Only one strand has the start codon.
    c. The promoter acts to aim the RNA polymerase.
    d. A start factor informs the system.
    e. It is chosen randomly.
    *Answer: c*

■41. There are differences in the amount of transcription that takes place for different genes. One reason for these differences is that
    a. some promoters are more effective at transcription initiation.
    b. longer genes take longer to transcribe.
    c. the outcome is influenced by random chance.
    d. ribosomes tend to attach to transcripts even before transcription is completed.
    e. None of the above
    *Answer: a*

42. There are _____ different RNA polymerases in eukaryotes.
    a. 1
    b. 2
    c. 3
    d. 4
    e. 5
    *Answer: c*

43. In eukaryotes, RNA polymerase _____ synthesizes mRNA.
    a. I
    b. II
    c. III
    d. IV
    e. V
    *Answer: b*

44. Proteins are synthesized from the _____, in the _____ direction along the mRNA.
    a. N terminus to C terminus; 5′ to 3′
    b. C terminus to N terminus; 5′ to 3′
    c. C terminus to N terminus; 3′ to 5′
    d. N terminus to C terminus; 3′ to 5′
    e. N terminus to N terminus; 5′ to 5′
    *Answer: a*

■45. Imagine that a novel life form is found deep within Earth's crust. Evaluation of its DNA yields no surprises. However, it is found that a codon for this life form is just two bases in length. How many different amino acids could this organism be composed of?
    a. 4
    b. 8
    c. 16
    d. 32
    e. 64
    *Answer: c*

46. The error rate for RNA polymerase is _____ that for most DNA polymerases.
    a. less than
    b. equal to
    c. greater than
    d. greater for frame shifts but less for base substitutions than
    e. greater for base substitutions but less for frame shifts than
    *Answer: c*

47. Poly uracil codes for
    a. three different amino acids.
    b. poly tryptophan.
    c. mRNA.
    d. a fatty acid.
    e. poly phenylalanine.
    *Answer: e*

■48. Fewer different tRNA molecules exist than might have been expected for the complexity of its function. This is possible because
    a. the third position of the codon does not have to pair conventionally.
    b. the second position of the codon does not have to pair conventionally.
    c. the anticodon does not have the conventional bases.
    d. there are fewer amino acids than there are possible codons.
    e. the code is degenerating.
    *Answer: a*

49. The enzyme that charges the tRNA molecules with appropriate amino acids is
    a. tRNA chargeatase.
    b. amino tRNA chargeatase.
    c. transcriptase.
    d. aminoacyl-tRNA synthetase.
    e. None of the above
    *Answer: d*

50. _____ is the addition of sugar residues to the protein after translation.
    a. Glycation
    b. Glycosylation
    c. Phosphorylation
    d. Proteolysis
    e. Exonuclease digestion
    *Answer: b*

★51. In eukaryotic cells, proteins that contain covalently attached sugar residues are translated
    a. in the nucleus.
    b. in the cytoplasm.
    c. in mitochondria.
    d. on the endoplasmic reticulum.
    e. on the Golgi apparatus.
    *Answer: d*

52. It is currently believed that the enzyme that catalyzes formation of the peptide bond during translation is composed of
    a. amino acids.
    b. protein.
    c. carbohydrate.
    d. RNA.
    e. DNA.
    *Answer: d*

53. The termination of transcription is signaled by
    a. the stop codon.
    b. a sequence of nitrogenous bases.
    c. a protein bound to a certain region of DNA.
    d. rRNA.
    e. tRNA.
    *Answer: b*

54. During translation initiation, the first site occupied by a charged tRNA is the
    a. A site.
    b. B site.
    c. large subunit.
    d. T site.
    e. P site.
    *Answer: e*

55. After translation, some proteins are processed by _____, which is cleavage of the protein to make a shortened finished protein.
    a. glycation
    b. glycosylation
    c. phosphorylation
    d. proteolysis
    e. exonuclease digestion
    *Answer: d*

56. During translation elongation, the existing polypeptide chain is transferred to
    a. the tRNA occupying the A site.
    b. the tRNA occupying the P site.
    c. the ribosomal rRNA.
    d. a signal recognition particle.
    e. None of the above
    *Answer: a*

57. The stop codons code for
    a. no amino acid.
    b. methionine.
    c. glycine.
    d. halt enzyme.
    e. DNA binding protein.
    *Answer: a*

58. Breaking and rejoining of chromosomes can lead to
    a. deletions.
    b. duplications.
    c. inversions.
    d. translocations.
    e. All of the above
    *Answer: e*

★59. Damage to DNA can be caused by _____ absorbed by thymine in DNA, causing interbase covalent bonds.
    a. X-rays
    b. cosmic radiation
    c. ultraviolet radiation
    d. smoke
    e. cigarettes
    *Answer: c*

60. The major phenotypic expression of genotype is in
    a. proteins.
    b. tRNA.
    c. mRNA.
    d. nucleic acids.
    e. a mutation.
    *Answer: a*

61. The link between mRNA and a protein is
    a. tRNA.
    b. a promoter.
    c. RNA polymerase.
    d. DNA polymerase.
    e. a start codon.
    *Answer: a*

62. How is it possible for single-stranded RNA to fold into complex shapes?
    a. Phosphodiester linkages form between the phosphate and the sugar ribose.
    b. Internal base pairings make this possible: adenine with uracil and cytosine with guanine.
    c. Uracil's methyl group binds to adenine spiraling the molecule.
    d. The single strand "twists" around itself.
    e. The RNA binds to proteins, creating conformation (three-dimensional shape).
    *Answer: b*

63. Which of the following do (does) *not* follow the "central dogma"?
    a. Yeast
    b. Onion cells
    c. Bread mold
    d. Skin cells
    e. Retroviruses
    *Answer: e*

64. What events must take place to ensure that the protein made is the one specified by mRNA?
    a. tRNA must read mRNA correctly.
    b. tRNA must carry the amino acid that is correct for its reading of the mRNA.
    c. Covalent bonding between the base pairs must occur.
    d. Both a and b
    e. All of the above
    *Answer: d*

65. Which of the following statements is true of codons and anticodons?
    a. The codon bonds covalently with the anticodon.
    b. The base sequences are the same.
    c. There are 64 codons and 61 anticodons.
    d. Activating enzymes link codons and anticodons.
    e. At contact, the codon and the anticodon are anti-parallel to each other.
    *Answer: e*

66. In the past, diphtheria was a major cause of childhood death. The bacteria that causes diphtheria
    a. breaks peptide bonds, destroying healthy proteins.
    b. inhibits protein synthesis.
    c. blocks the synthesis of bacterial cell walls.
    d. produces a lethal toxin that affects mRNA and ribosomal movement.
    e. inhibits the translocation of mRNA along the ribosome.
    *Answer: d*

# 13 *The Genetics of Viruses and Prokaryotes*

## Fill in the Blank

1. Bacteria that house bacteriophage that are not lytic are called _____.
*Answer: lysogenic*

2. _____ are nonessential genetic elements that exist as free, independently replicating, circular DNA molecules, separate from the bacterial chromosome.
*Answer: Plasmids*

3. When the synthesis of an enzyme is turned off in response to an external biochemical cue (such as an excess in tryptophan), the enzyme is said to be _____.
*Answer: repressible*

4. Bacteriophage DNA that is stably integrated into a bacterial chromosome is called a _____.
*Answer: prophage*

5. Bacteriophage that can integrate their DNA into the host bacterial chromosome are called _____.
*Answer: temperate*

6. Pieces of DNA that move from place to place in the bacterial chromosome are called _____.
*Answer: transposable elements (or transposons)*

7. Genes that produce single mRNAs containing information for more than one protein are _____.
*Answer: operons*

8. The region of the gene that binds RNA polymerase is the
_____.
*Answer: promoter*

9. The site on the operon DNA where a repressor binds is the _____ sequence.
*Answer: operator*

10. In prokaryotes, _____ genes encode repressor proteins.
*Answer: regulatory*

11. _____ is a positive control process that relies on increasing the affinity of promoters for RNA polymerase.
*Answer: Catabolic repression*

12. The first virus that was discovered was the _____ virus.
*Answer: tobacco mosaic*

13. Antibiotics are ineffective against _____ infections.
*Answer: viral*

14. A _____ is the name of an individual viral particle when it is outside its host.
*Answer: virion*

15. A _____ is a single circular RNA molecule that has no protein component but is infectious.
*Answer: viroid*

16. Antibiotic resistance in pathogenic bacteria is an example of _____.
*Answer: evolution*

17. The plasmids that encode the genes needed for conjugation are called _____.
*Answer: fertility factors (or F factors)*

18. The basic viral unit is a _____.
*Answer: virion*

19. Genes that are essential to a cell's survival can be determined by using a process called _____.
*Answer: transposon mutagenesis*

## Multiple Choice

*1. There are many advantages to working experimentally with bacteria instead of mice. Which of the following is *not* one of these advantages?
a. Bacteria have a smaller amount of DNA.
b. Bacteria are nonpathogenic.
c. Bacteria reproduce very rapidly.
d. Bacteria are easy to grow.
e. Bacteria are usually haploid.
*Answer: b*

2. The transfer of genes by a bacteriophage vector characterizes which type of gene transfer in bacteria?
a. Transformation
b. Conjugation
c. Transduction
d. Both a and b
e. All of the above
*Answer: c*

3. The term "lysogeny" refers to
a. the stable integration of bacteriophage DNA into the bacterial chromosome.
b. the excision of bacteriophage DNA from the bacterial chromosome.
c. the lysing of a bacterium by a bacteriophage.
d. mutation induced by a bacteriophage.
e. exchange of genetic material between a bacteriophage and a bacterium.
*Answer: a*

4. Viruses were first discovered in 1892 when
   a. a filtrate was found to be infectious.
   b. Albert Virusa became ill from smoking infected tobacco.
   c. Stanley found that crystallized viral preparations contained protein and DNA.
   d. they were first observed using a light microscope.
   e. they were observed as plaques for the first time.
   *Answer: a*

5. The genetic information of viruses is
   a. DNA.
   b. RNA.
   c. single-stranded.
   d. double-stranded.
   e. All of the above
   *Answer: e*

6. A strain of virus that always lyses the cells it infects is called
   a. temperate.
   b. virulent.
   c. lytic.
   d. lysogenic.
   e. deadly.
   *Answer: b*

7. The HIV virus that causes AIDS is a(n)
   a. arbovirus.
   b. double-stranded DNA virus.
   c. single-stranded DNA virus.
   d. porcine virus.
   e. retrovirus.
   *Answer: e*

8. The polio virus and the HIV virus differ in that only the polio virus can
   a. produce and use reverse transcriptase.
   b. use RNA as its genetic material.
   c. use RNA as genetic information and mRNA, without generating a DNA molecule.
   d. infect both horses and humans.
   e. All of the above
   *Answer: c*

9. Plants have a tough cell wall through which viruses cannot pass. Viruses enter plants
   a. through wounds.
   b. by means of insect vectors.
   c. by digesting the cell wall.
   d. Both a and b
   e. Both b and c
   *Answer: d*

10. Once inside a plant cell, viruses spread by
    a. diffusion.
    b. Brownian motion.
    c. plasmodesmata.
    d. attaching to a shared endoplasmic reticulum.
    e. the movement of water each time it rains.
    *Answer: c*

11. A patient is told that he has contracted a disease that is caused by an arbovirus. He most likely
    a. ate some food that carried the virus.
    b. kissed an infected person.
    c. should have used insect repellent.

d. was near an infected person who was sneezing.
e. will be contagious.
*Answer: c*

12. Viroids are composed of
    a. DNA and protein.
    b. protein.
    c. RNA.
    d. RNA and protein.
    e. DNA.
    *Answer: c*

13. Viruses replicate _____ bacteria.
    a. more slowly than
    b. at the same rate as
    c. more quickly than
    d. more quickly or more slowly, depending on the virus, than
    e. Viruses cannot replicate.
    *Answer: c*

14. Viruses are composed of
    a. nucleic acids only.
    b. proteins only.
    c. nucleic acids and proteins.
    d. nucleic acids, proteins, and organelles.
    e. nucleic acids and proteins, although a few also have organelles.
    *Answer: c*

15. A viroid differs from a virus in that it
    a. is not infectious.
    b. is composed of nucleic acids only.
    c. is man-made.
    d. has both protein and lipid components.
    e. All of the above
    *Answer: b*

*16. The scientist who first crystallized tobacco mosaic virus was
    a. Dmitri Ivanovsky.
    b. Martinus Beijerinick.
    c. Jackson Roberston.
    d. Wendell Stanley.
    e. Rosalind Franklin.
    *Answer: d*

17. A virion with a lipid and protein membrane is likely to infect
    a. animal cells.
    b. plant cells.
    c. bacteria.
    d. fungi.
    e. All of the above
    *Answer: a*

18. Lysis of the host cell is caused by
    a. the cells' bursting due to the large number of viral particles.
    b. the cells' opening up in an attempt to dump out the viruses.
    c. an attack on the cell wall by a product of the viral gene.
    d. a yet unknown mechanism.
    e. None of the above
    *Answer: c*

19. Beginning with a single bacterium, how many cells would be present after four hours of growth if they can double every 20 minutes?
    a. 12
    b. 24
    c. 64
    d. 4,096
    e. 34,217,728
    *Answer: d*

20. The term "auxotroph" refers to
    a. a mutant bacterium that requires nutrients not required by wild-type bacteria.
    b. a mutant bacterium that requires no nutrients.
    c. a mutant bacterium that can synthesize a nutrient that wild-type bacteria cannot.
    d. a mutant bacterium that can metabolize a nutrient that wild-type bacteria cannot.
    e. a bacterium that can metabolize sugars.
    *Answer: a*

▪21. When a *met⁻bio⁻* strain of bacteria is mixed with a *thr⁻leu⁻* strain, wild-type bacteria result at a rate of one in every $10^7$ cells. Why is it not possible that this is the result of mutation?
    a. The wild type would have to be a double mutation, which is extremely rare.
    b. The wild type would have to be a mutation in four genes, which is extremely rare.
    c. The wild type would have to be a deletion, which is extremely rare.
    d. Mutations can occur only in response to a mutagen.
    e. Bacterial genes do not mutate.
    *Answer: a*

22. The *met⁺leu⁺* bacteria that result from a cross between *met⁺leu⁻* and *met⁻leu⁺* bacteria are called
    a. pili.
    b. auxotrophs.
    c. parentals.
    d. recombinants.
    e. None of the above
    *Answer: d*

23. The function of the pili is to
    a. store the F plasmid.
    b. make the initial contact between an F⁺ and an F⁻ cell that precedes conjugation.
    c. uptake DNA during transformation.
    d. transfer the DNA between mating partners during conjugation.
    e. None of the above
    *Answer: b*

24. The transfer of genes by uptake of DNA from the medium characterizes which type of gene transfer in bacteria?
    a. Conjugation
    b. Sexduction
    c. Transduction
    d. Transformation
    e. None of the above
    *Answer: d*

25. Which of the following is *not* a mechanism of bacterial genetic recombination?
    a. Transformation
    b. Conjugation
    c. Transduction
    d. Catabolite repression
    e. Both a and b
    *Answer: d*

26. In transduction,
    a. only a particular part of the bacterial chromosome can be transferred.
    b. a part of the bacterial chromosome may be transferred.
    c. only the F plasmid can be transferred.
    d. only the part of the bacterial chromosome near the F plasmid can be transferred.
    e. None of the above
    *Answer: b*

27. An R factor is a(n)
    a. plasmid that carries genes for antibiotic resistance.
    b. episome that carries genes for antibiotic resistance.
    c. region of the bacterial chromosome that carries genes for antibiotic resistance.
    d. small portion of the F plasmid.
    e. measure of how well the bacterium is insulated from the cold.
    *Answer: a*

28. For a plasmid to survive within the cytoplasm of a cell as it divides, it must
    a. be integrated into the bacterial chromosome.
    b. have an R factor.
    c. have an F plasmid.
    d. have an origin of replication.
    e. All of the above
    *Answer: d*

▪29. What is the difference between a plasmid and a transposable element?
    a. Plasmids exist in many copies in the cell, whereas a transposable element exists in a single copy.
    b. A plasmid replicates independently of the cell; a transposable element does not.
    c. A plasmid has an origin of replication; a transposable element does not.
    d. A plasmid exists independently in the cell cytoplasm, whereas a transposable element is integrated into another larger DNA molecule.
    e. None of the above
    *Answer: d*

30. A population of genetically identical bacteria that arose from a single cell is known as a
    a. plaque.
    b. lysogen.
    c. clone.
    d. bacterial culture.
    e. conjugation.
    *Answer: c*

31. Which of the following is *not* true of the F plasmid?
    a. It can replicate independently of the bacterial chromosome.
    b. It can be transferred from a donor to a recipient bacterium during mating.
    c. It is involved in conjugation.
    d. It contains genes that encode the pili.
    e. It contains genes that encode antibiotic resistance.
    *Answer: e*

32. Which of the following statements is true for both F plasmids and bacteriophage?
    a. They have a protein coat enclosing the DNA.
    b. They can lyse bacteria.
    c. They can participate in conjugation.
    d. They can incorporate themselves into the bacterial chromosome.
    e. Both a and b
    *Answer: d*

33. Which of the following is *not* true of conjugation between an F⁻ bacterium and an F⁺ bacterium?
    a. A conjugation tube is formed between the two bacteria.
    b. The F⁺ cell transfers a copy of the F plasmid to the F⁻ cell.
    c. The F⁻ cell receives a copy of the F plasmid.
    d. The F⁻ cell becomes F⁺.
    e. The F⁺ cell becomes F⁻.
    *Answer: e*

■34. R factors, or resistance factors, carried by plasmids or bacteria are more widely detected now than in the past because
    a. the presence of antibiotics has stimulated the evolution of R factors.
    b. techniques for detection have improved.
    c. current bacterial culture conditions favor the selection of R factors.
    d. Both a and c
    e. None of the above
    *Answer: a*

35. When *E. coli* are grown in a medium with little lactose,
    a. all of the enzymes of the lactose operon are present in very small quantities.
    b. all of the enzymes of the lactose operon are present in large quantities.
    c. no enzymes of the lactose operon are present.
    d. β-galactosidase and permease are present in small quantities, but transacetylase is present in large quantities.
    e. the mRNA of the lactose operon is not present at all.
    *Answer: a*

36. A promoter is the region of
    a. a plasmid that binds the enzymes for replication.
    b. the mRNA that binds to a ribosome.
    c. DNA that binds RNA polymerase.
    d. the mRNA that binds tRNAs.
    e. None of the above
    *Answer: c*

37. The frequency of transcription of a particular bacterial gene is controlled by the
    a. DNA sequence of the particular promoter.
    b. availability of the promoter to RNA polymerase.
    c. number of ribosomes that are available in the cell.
    d. Both a and b
    e. a, b, and c
    *Answer: d*

38. The three basic parts of an operon are the
    a. promoter, the operator, and the structural gene(s).
    b. promoter, the structural gene(s), and the termination codons.
    c. promoter, the mRNA, and the termination codons.
    d. structural gene(s), the mRNA, and the tRNAs.

    e. None of the above
    *Answer: a*

39. An inducer
    a. combines with a repressor and prevents it from binding the promoter.
    b. combines with a repressor and prevents it from binding the operator.
    c. binds to the promoter and prevents the repressor from binding to the operator.
    d. binds to the operator and prevents the repressor from binding at this site.
    e. binds to the termination codons and allows protein synthesis to continue.
    *Answer: b*

40. The RNA transcribed from an operon is
    a. transcribed by the ribosomes to make more repressor.
    b. translated by the ribosomes to make more inducer.
    c. translated by the ribosomes to make the enzymes.
    d. translated by the ribosomes to make more RNA polymerase.
    e. usually not translated by the ribosomes at all.
    *Answer: c*

41. The genes that encode repressor proteins are
    a. repressor genes.
    b. operons.
    c. inducer genes.
    d. regulatory genes.
    e. None of the above
    *Answer: d*

42. Which operon is turned "off" in response to molecules present in the environment of the cell?
    a. Repressible
    b. Suppressible
    c. Impressible
    d. Inducible
    e. Degraded
    *Answer: a*

43. In a repressible operon, the repressor molecule
    a. must first be activated by a corepressor.
    b. can repress the transcription of the operon on its own.
    c. is a molecule made from the operon.
    d. binds to the mRNA.
    e. must first be made negative to control the operon.
    *Answer: a*

44. Catabolite repression refers to the
    a. increased transcription from many operons when glucose is present in the medium.
    b. shutdown of transcription from many operons when glucose is present in the medium.
    c. increased activity of inducers caused by glucose in the medium.
    d. Both a and b
    e. Both a and c
    *Answer: b*

45. To be activated, the CRP must first bind
    a. the repressor molecule.
    b. the repressor protein.
    c. the activator protein.
    d. the corepressor molecule.
    e. cAMP.
    *Answer: e*

*46. When the operator is unbound, the binding of the CRP–cAMP complex _____ the binding of RNA polymerase at the promoter.
   a. increases by two-fold
   b. decreases by tenfold
   c. increases by 50-fold
   d. blocks
   e. All of the above are sometimes true, depending on the concentration of lactose.
   *Answer: c*

47. When the concentration of glucose in the medium falls, the concentration of _____ rises.
   a. CRP
   b. cAMP
   c. repressors
   d. inducers
   e. None of the above
   *Answer: b*

48. The function of the promoter is to tell the RNA polymerase
   a. where to start transcribing the DNA.
   b. which strand of the DNA to read.
   c. where to stop transcribing the DNA.
   d. Both a and b
   e. All of the above
   *Answer: d*

49. The mechanism by which the inducer causes the repressor to detach from the operator is an example of
   a. catabolite repression.
   b. transcription.
   c. transposition.
   d. allosteric modification.
   e. recombination.
   *Answer: d*

50. _____ acts as a corepressor to block transcription of the tryptophan operon.
   a. cAMP
   b. Lactose
   c. Tryptophan
   d. Methionine
   e. CRP
   *Answer: c*

51. The CRP–cAMP complex binds _____ of the operon.
   a. close to the RNA polymerase binding site
   b. close to the operator
   c. inside one of the structural genes
   d. at the termination point
   e. None of the above
   *Answer: a*

*■52. It is found that a certain enzyme is synthesized whenever the solution in which the cells are growing lacks substance X. This phenomenon is most likely an example of _____ gene regulation.
   a. inducible
   b. positive
   c. negative
   d. repressible
   e. positive-negative
   *Answer: d*

*■53. It is found that a certain enzyme is synthesized whenever the solution in which the cells are growing contains substance X. This phenomenon is most likely an example of _____ gene regulation.
   a. inducible
   b. positive
   c. negative
   d. repressible
   e. positive-negative
   *Answer: a*

54. Cells control the amount of enzymes by
   a. blocking transcription.
   b. hydrolyzing the RNA.
   c. preventing translation.
   d. hydrolyzing the protein.
   e. All of the above
   *Answer: e*

55. What effect does the presence of ample glucose have on the amount of *lac* operon transcription?
   a. It increases the cAMP concentration, which in turn causes a decreased rate of transcription.
   b. It decreases the cAMP concentration, which in turn causes an increased rate of transcription.
   c. It increases the rate of transcription.
   d. It decreases the rate of transcription.
   e. None of the above
   *Answer: d*

56. The *trp* operon
   a. codes for proteins needed for tryptophan synthesis.
   b. codes for the proteins to metabolize tryptophan.
   c. is activated by the presence of tryptophan.
   d. is inducible.
   e. All of the above
   *Answer: a*

*57. The effects of Cro and cI on bacteriophage λ are
   a. competitive.
   b. cooperative.
   c. coordinate.
   d. inverted.
   e. additive.
   *Answer: a*

*58. The total minimum number of genes necessary for a life form grown in laboratory conditions is
   a. 48.
   b. 337.
   c. 470.
   d. 3,876.
   e. 10,082.
   *Answer: b*

59. The retrovirus HIV enters a host cell
   a. by fusion of its envelope with the host's plasma membrane.
   b. by endocytosis.
   c. by vectors.
   d. through cytoplasmic connections between cells (plasmodesmata).
   e. by phagocytosis.
   *Answer: a*

60. Which of the following is *not* true of *Bacillus anthracis*?
    a. It causes anthrax.
    b. It is transmitted by spores.
    c. It is a prime weapon for bioterrorism.
    d. It causes fever, vomiting, and headache.
    e. There is no treatment.
    *Answer: e*

61. Which of the following statements is true of viruses?
    a. They are acellular.
    b. They can regulate the movements of substances into and out of the cell.
    c. They can reproduce outside of living cells.
    d. They are large and therefore easy to study.
    e. They are readily destroyed by antibiotics.
    *Answer: a*

62. Why are antibiotics ineffective treatments against viruses?
    a. Viruses can remain inactive until the antibiotic disintegrates.
    b. The viruses may reproduce immediately and destroy the antibiotic.
    c. Viruses have RNA instead of DNA.
    d. Viruses do not have a cell wall and the ribosomal biochemistry of bacteria.
    e. The nucleic acid of viruses is single-stranded rather than double-stranded.
    *Answer: d*

63. Genetic diversity in a bacterial population comes about due to
    a. conjugation.
    b. transformation.
    c. transduction.
    d. acquisition of new genes.
    e. All of the above
    *Answer e*

64. How are inducible and repressible systems similar?
    a. They both control catabolic pathways.
    b. They both control biosynthetic pathways.
    c. In both systems the regulatory molecules function by binding to the operator.
    d. They both block transcription.
    e. Both systems are unique to prokaryotes.
    *Answer: c*

65. A virion consists of
    a. a capsid and nucleic acid genome.
    b. a promoter and RNA polymerase.
    c. RNA and R factors.
    d. a small bacterial chromosome.
    e. plasmids that carry genes for antibiotic resistance.
    *Answer: a*

66. Prokaryotes and viruses are useful for the study of genetics and molecular biology because
    a. they contain much less DNA than eukaryotes do.
    b. they grow and reproduce rapidly.
    c. they are haploid.
    d. Both a and b
    e. a, b, and c
    *Answer: e*

67. The occurrence of "super bugs" is expected to increase. "Super bugs" are the result of
    a. universal gene segments that have been copied many times.
    b. extensive genetic exchanges between species.
    c. transposons that have inserted themselves into a gene, causing mutation.
    d. parasites that grow and take control of normal cell activities.
    e. a gene deficiency for the electron transport chain.
    *Answer: b*

68. Commonly used antibiotics such as streptomycin, tetracycline, and erythromycin are produced by
    a. *Chlamydia trachomatis*.
    b. *Rickettsia prowazekii*.
    c. *Mycobacterium tuberculosis*.
    d. *Streptomyces coelicolor*.
    e. *E. coli*.
    *Answer: d*

# 14 The Eukaryotic Genome and Its Expression

## Fill in the Blank

1. DNA in eukaryotes is wrapped around special proteins to form structures that look like beads on a string. These structures are called _____.
   *Answer: nucleosomes*

2. Segments of a gene that are transcribed but do not encode part of the protein product are called _____.
   *Answer: introns*

3. DNA sequences called _____ can replicate and insert themselves into different parts of the genome.
   *Answer: transposable elements*

4. Members of a gene family that do not make a functional gene product are called _____.
   *Answer: pseudogenes*

5. The RNA of snRNPs are complementary to _____ sequences located at the intron–exon boundaries.
   *Answer: consensus*

■6. When you stain a preparation of epithelial cells from a female rat, you observe one highly condensed chromosome in interphase nuclei. This is the inactive X chromosome, or _____.
   *Answer: Barr body*

7. During the development of amphibian oocytes, the number of genes encoding the rRNA increases. This increase is called _____.
   *Answer: gene amplification*

8. Those DNA sequences that are not represented in the mRNA are referred to as intervening sequences, or

   _____.
   *Answer: introns*

9. Most _____ are probably duplicate genes that changed so much during evolution that they no longer function.
   *Answer: pseudogenes*

10. During RNA splicing, different snRNPs constitute a "splicing machine" called a _____.
    *Answer: spliceosome*

11. There are at least three ways in which _____ may be controlled: (1) Genes may be inactivated; (2) specific genes may be amplified; and (3) specific genes may be selectively transcribed.
    *Answer: transcription*

12. In both prokaryotes and eukaryotes, the _____ is a stretch of DNA to which RNA polymerase binds to initiate transcription.
    *Answer: promoter*

13. The gene that codes for an RNAi that binds to *Xist* RNA at the active X chromosome is _____.
    *Answer: **Tsix***

14. The process of discerning protein product and function of a known gene sequence is called _____.
    *Answer: annotation*

15. A DNA sequence that signals the end of transcription in eukaryotes is called the _____.
    *Answer: terminator*

## Multiple Choice

1. Among eukaryotes, there is _____ relationship between complexity and genome size.
   a. an inverse
   b. a direct
   c. not always an inverse
   d. not always a direct
   e. no
   *Answer: d*

2. In eukaryotic cells, transcription and translation
   a. are separated: Translation occurs in the nucleus, and transcription occurs in the cytoplasm.
   b. occur together in the cytosol.
   c. occur together in the nucleus.
   d. are separated: Translation occurs in the cytoplasm, and transcription occurs in the nucleus.
   e. are separated, except for proteins that bind to the DNA and ribosomes, which are translated in the nucleus.
   *Answer: d*

★3. A lily, which has 18 times more DNA than a human, codes for _____ proteins.
   a. more
   b. the same number of
   c. fewer
   d. 337
   e. 6,330
   *Answer: c*

4. If it were stretched out, the DNA from a human cell would be _____ in length.
   a. 2 mm
   b. 2 m
   c. 2 km
   d. 2 μm
   e. None of the above
   *Answer: b*

5. Human cells have approximately _____ times the DNA found in prokaryotic cells.
   a. 10
   b. 100
   c. 1,000
   d. 10,000
   e. 100,000
   *Answer: c*

6. Chromosomes of eukaryotic DNA must have
   a. DNA sequences that make up telomeres and centromeres.
   b. proteins that are centromeres and DNA that form telomeres.
   c. a 5′ G cap.
   d. an inactivation center.
   e. None of the above
   *Answer: c*

7. The phrase "beads on a string" describes the structural appearance of
   a. condensed chromosomes.
   b. lampbrush chromosomes.
   c. nucleosomes.
   d. the solenoid structure of DNA.
   e. the 30 nm DNA fibers.
   *Answer: c*

8. Cancer cells are often found to have abnormally high amounts of
   a. nucleosomes.
   b. spliceosomes.
   c. centromeres.
   d. topoisomerase.
   e. telomerase.
   *Answer: e*

9. Telomerase is important to eukaryotic cells because
   a. telomeres tend to get shortened with each cell division.
   b. telomeres tend to get longer with each cell division.
   c. telomerase digests telomeres to proper length.
   d. the leading strand of DNA causes the telomeres to shorten.
   e. it aids in making artificial chromosomes.
   *Answer: a*

10. Each human chromosome must have
    a. a centromeric sequence.
    b. telomeric sequences.
    c. an origin of replication.
    d. All of the above
    e. None of the above
    *Answer: d*

11. RNA synthesis in eukaryotes occurs
    a. solely in mitochondria.
    b. primarily in the Golgi apparatus.
    c. primarily in the endoplasmic reticulum.

d. primarily in the nucleus.
e. solely in the cytoplasm.
*Answer: d*

12. The three RNA polymerases of eukaryotes catalyze _____ synthesis.
    a. rRNA
    b. mRNA
    c. tRNA
    d. All of the above
    e. None of the above
    *Answer: d*

13. RNA polymerase II by itself cannot bind to the chromosome and initiate transcription. It can bind and act only after the assembly of regulatory proteins called _____ factors.
    a. translation
    b. posttranslation
    c. initiation
    d. transcription
    e. None of the above
    *Answer: d*

14. RNA translation in eukaryotes occurs
    a. solely in mitochondria.
    b. primarily in the Golgi apparatus.
    c. primarily in the endoplasmic reticulum.
    d. primarily in the nucleus.
    e. solely in the cytoplasm.
    *Answer: e*

15. The RNA polymerase that produces mRNA molecules in eukaryotic cells is
    a. RNA polymerase I.
    b. RNA polymerase II.
    c. RNA polymerase III.
    d. primase.
    e. reverse transcriptase.
    *Answer: b*

16. In bacteria, transcription
    a. is separate from translation.
    b. occurs in the nucleus.
    c. is not separate from translation.
    d. is the process of synthesizing proteins.
    e. is continuous for all genes.
    *Answer: c*

17. *Caenorhabditis elegans* is a
    a. yeast.
    b. virus.
    c. fruit fly.
    d. bacterium.
    e. roundworm.
    *Answer: e*

18. *C. elegans* has approximately _____ genes.
    a. 19,000
    b. 6,000
    c. 4,200
    d. 3,200
    e. 370
    *Answer: a*

*19. In comparison to a simple roundworm, *Drosophila melanogaster*, the fruit fly, has _____ genes.
    a. about the same number of
    b. fewer

c. many more
d. a few more
e. an unknown number of
*Answer: b*

20. The genome size of fruit flies is _____ roundworms.
    a. about the same as
    b. smaller than
    c. much larger than
    d. a bit larger than
    e. This is currently not known.
    *Answer: c*

21. _____ is the study involved with assigning functions to DNA sequences.
    a. Biology
    b. Genetics
    c. Cytogenetics
    d. Genomics
    e. All of the above
    *Answer: d*

22. *Arabidopsis thaliana*, or shale crest, is a favorite model organism for study because
    a. hundreds can grow and reproduce in a small space.
    b. it is easy to manipulate.
    c. it has a small amount of repetitive DNA.
    d. it has a small genome.
    e. All of the above
    *Answer: e*

23. Transposable elements can
    a. alter transcription of a gene.
    b. cause a mutation.
    c. cause gene duplication.
    d. All of the above
    e. None of the above
    *Answer: d*

24. SINEs and LINEs are both
    a. highly repetitive DNA sequences.
    b. transposable elements.
    c. control sequences.
    d. retrotransposons.
    e. DNA transposons.
    *Answer: b*

25. DNA transposons
    a. work by using reverse transcriptase.
    b. are transcribed but not translated.
    c. are also called retrotransposons.
    d. move to new locations from old locations.
    e. are circular DNA molecules.
    *Answer: d*

■26. Transposable elements are found in both prokaryotes and eukaryotes. One of the major differences between the copying of transposable elements in eukaryotes and in prokaryotes is that in some eukaryotes the elements
    a. must be denatured before they can be copied.
    b. all require an RNA intermediate.
    c. are not always copied.
    d. do not need an enzyme.
    e. None of the above
    *Answer: c*

27. The importance of transposons is
    a. that they provide mobile promoters.
    b. that they alter gene activities.
    c. that they provide for genetic change when required.
    d. that they inactivate unnecessary or undesirable genes.
    e. not known.
    *Answer: e*

28. The mRNA will form hybrids only with the template strand of DNA because
    a. DNA will not reanneal at high temperatures.
    b. salt concentration affects DNA reannealing.
    c. DNA will not reanneal at low temperatures.
    d. RNA–DNA hybridization follows the base-pairing rules.
    e. denatured DNA will not reanneal after it is diluted.
    *Answer: d*

29. DNA sequences found in introns provide
    a. amino acid sequence information.
    b. regulatory information.
    c. no known useful information.
    d. structure for the gene.
    e. alternative DNA splicing possibilities.
    *Answer: c*

30. A gene family is a set of genes that over time has changed slightly, extensively, or not at all. Which of the following statements is true for gene families?
    a. They must always be on the same chromosome.
    b. They must always code for the exact same proteins.
    c. One copy must retain the original function.
    d. They usually differ in their exons because these are coding segments.
    e. None of the above
    *Answer: c*

31. The different members of the β-globin gene family
    a. are expressed differently in different tissues.
    b. are expressed differently at different times of development.
    c. are expressed in the same way in different tissues.
    d. are expressed in the same way at different times of development.
    e. have no known function.
    *Answer: b*

32. Which of the following is a posttranscriptional modification of mRNA found in eukaryotes?
    a. A 5′ cap
    b. A 3′ cap
    c. A poly T tail
    d. A polyadenylation at the 5′ end
    e. None of the above
    *Answer: a*

33. Which of the following is *not* part of RNA processing in eukaryotes?
    a. Splicing of exons
    b. Reverse transcription
    c. Addition of a 5′ cap
    d. Addition of a poly A tail
    e. Intron removal
    *Answer: b*

34. Consensus sequences (short segments of DNA) appear in the transcribed regions of various genes. These sequences appear to be involved in
    a. directing the polymerases to the appropriate place on the DNA for transcription to begin.
    b. the splicing of introns out of the DNA.
    c. allowing the transcription to stop at the appropriate spot.
    d. catalyzing the synthesis of a protein.
    e. None of the above
    *Answer: b*

35. The tail added to pre-mRNA
    a. is coded for by DNA.
    b. is composed of poly T.
    c. is important for mRNA stability.
    d. is attached to its 5′ end.
    e. All of the above
    *Answer: c*

36. The expression of some genes can be regulated in part by the pattern of RNA splicing. This is an example of
    a. DNA methylation.
    b. transcriptional regulation.
    c. catalytic RNA activity.
    d. posttranscriptional control.
    e. the endosymbiotic theory.
    *Answer: d*

37. Which of the following best describes the function of the addition of a methylated guanosine cap to the 5′ end of primary mRNA?
    a. It contains all of the coding and noncoding sequences of the DNA template.
    b. It provides the mRNA molecule with a poly A tail.
    c. It facilitates the binding of mRNA to ribosomes.
    d. It forms hydrogen bonds.
    e. It helps transfer amino acids to the ribosomes.
    *Answer: c*

38. Energy for RNA splicing comes
    a. from the nucleotides in the RNA.
    b. from ATP.
    c. directly from photosynthesis.
    d. always from glucose.
    e. from the splicing enzyme.
    *Answer: b*

39. RNA processing in eukaryotes involves the
    a. addition of a G cap.
    b. addition of a poly A tail.
    c. removal of introns.
    d. splicing of exons.
    e. All of the above
    *Answer: e*

40. Exons are
    a. spliced out of the original transcript.
    b. spliced together from the original transcript.
    c. spliced to introns to form the final transcript.
    d. much larger than introns.
    e. larger than the original coding region.
    *Answer: b*

41. The binding of snRNPs to consensus sequences is necessary for
    a. gene duplication.
    b. addition of a poly A tail.
    c. capping an hnRNA.
    d. RNA splicing.
    e. transcription.
    *Answer: d*

42. What are the three processes that must be completed before transcripts can be translated in eukaryotes?
    a. Binding of snRNPs, addition of a poly A tail, splicing of introns
    b. Binding of snRNPs, transporting, synthesizing of ribose
    c. Capping, transporting, synthesizing of ribose
    d. Binding of snRNPs, capping, splicing
    e. Splicing, capping, addition of a poly A tail
    *Answer: e*

43. snRNPs are
    a. exon–intron boundary regions.
    b. small nuclear ribonucleoprotein particles.
    c. protein fragments removed from the snRNA molecules.
    d. signal ribosomal nuclear proteins.
    e. glucose conjugated trapezoids.
    *Answer: b*

44. The theoretical basis for DNA–DNA or DNA–RNA hybridization studies is
    a. base complementarity between nucleic acid strands.
    b. enzyme action.
    c. DNA and RNA looping.
    d. Both a and c
    e. None of the above
    *Answer: a*

45. When eukaryotic DNA is hybridized with mRNA, the hybrid molecules contain loops of double-stranded DNA called
    a. retroviruses.
    b. introns.
    c. exons.
    d. transcripts.
    e. puffs.
    *Answer: b*

46. The regions of DNA in a eukaryotic gene that encode a polypeptide product are called
    a. enhancers.
    b. mRNAs.
    c. hnRNAs.
    d. exons.
    e. leader sequences.
    *Answer: d*

47. Exons are
    a. translated.
    b. found in most prokaryotic genes.
    c. removed during RNA processing.
    d. Both a and b
    e. Both a and c
    *Answer: a*

48. A set of related genes that encode similar proteins is a(n)
    a. pseudogene.
    b. intron.
    c. gene family.
    d. retrovirus.
    e. oncogene.
    *Answer: c*

49. Most gene families probably arose by
    a. RNA processing.
    b. RNA splicing.
    c. DNA methylation.
    d. transcriptional regulation.
    e. gene duplication.
    *Answer: e*

50. The modified G cap on eukaryotic mRNAs is found
    a. at the 5′ end.
    b. at the 3′ end.
    c. in the consensus sequence.
    d. in the poly A tail.
    e. in snRNA.
    *Answer: a*

51. Poly A tails
    a. are added after transcription.
    b. are encoded by a sequence of thymines in the DNA.
    c. are found in all mRNAs.
    d. have no function.
    e. are removed during RNA processing.
    *Answer: a*

52. Which of the following is an enhancer?
    a. Protein
    b. RNA
    c. DNA
    d. Carbohydrate
    e. Enzyme
    *Answer: c*

53. Which of the following is a transcription factor?
    a. Protein
    b. RNA
    c. DNA
    d. Carbohydrate
    e. Enzyme
    *Answer: a*

54. In eukaryotic cells, a repressor
    a. is made of DNA.
    b. binds to the enhancer region to block transcription.
    c. is located both upstream and downstream from the promoter.
    d. binds to the operator to block RNA polymerase.
    e. binds to a silencer to reduce transcription rates.
    *Answer: e*

55. The region of a gene that binds RNA polymerase to initiate transcription is
    a. an exon.
    b. the consensus sequence.
    c. heterochromatin.
    d. the cap.
    e. the promoter.
    *Answer: e*

56. When an enhancer is bound, it
    a. increases the stability of a specific mRNA.
    b. stimulates transcription of a specific gene.
    c. stimulates transcription of all genes.
    d. stimulates splicing of a specific mRNA.
    e. stimulates splicing of all mRNAs.
    *Answer: b*

57. Which of the following is *not* a feature of TATA boxes?
    a. They bind a specific transcription factor.
    b. They are found in the region of the promoter.
    c. They are part of the intron consensus sequence.
    d. They help specify the starting point for transcription.
    e. They contain thymine–adenine base pairs.
    *Answer: c*

58. Transcription of eukaryotic genes requires
    a. binding of RNA polymerase to the promoter.
    b. binding of several transcription factors.
    c. capping of mRNA.
    d. Both a and b
    e. a, b, and c
    *Answer: d*

59. A DNA sequence, which can be distant to the gene, stimulates transcription when bound by a protein. This sequence is a(n)
    a. TATA box.
    b. operon.
    c. enhancer.
    d. promoter.
    e. consensus sequence.
    *Answer: c*

60. The TATA box is a(n)
    a. sequence common to the promoter region of many genes.
    b. square-shaped sequence.
    c. enhancer consensus sequence.
    d. activator sequence necessary for proper translation.
    e. None of the above
    *Answer: a*

61. DNA binding motifs include all of the following *except*
    a. helix-straight-helix.
    b. helix-turn-helix.
    c. helix-loop-helix.
    d. leucine zipper.
    e. zinc finger.
    *Answer: a*

62. The switching of mating type in yeast involves
    a. selection.
    b. mutation.
    c. differential gene expression.
    d. aging.
    e. transposition.
    *Answer: e*

63. A cell is found that contains three Barr bodies. This cell has
    a. three Y chromosomes.
    b. three X chromosomes.
    c. four Y chromosomes
    d. four X chromosomes.
    e. three nucleoli.
    *Answer: d*

64. DNA methylation in eukaryotic chromosomes involves adding a methyl to the
    a. 5′ position of G.
    b. 5′ position of C.
    c. proteins bound to the DNA.
    d. RNA molecules.
    e. ribose.
    *Answer: b*

65. One of the genes that is known to be transcribed from the inactive X chromosome is
    a. *Xist*.
    b. *ZIST*.
    c. inactivation controller protein.
    d. lithozist.
    e. methyl-X.
    *Answer: a*

66. Mary Lyon, Liane Russell, and Ernest Beutler discovered
    a. the basis of hormone action.
    b. X chromosome inactivation.
    c. Barr bodies.
    d. melanosomes.
    e. heterochromatin.
    *Answer: b*

67. The Barr body is evidence for
    a. X chromosome inactivation.
    b. cell death.
    c. ion pumps.
    d. posttranslational control of eukaryote gene expression.
    e. None of the above
    *Answer: a*

68. You stain a preparation of normal rat epithelial cells and examine them under the microscope. Each interphase nucleus contains a single Barr body. What can you conclude about these cells?
    a. The cells are in meiotic prophase.
    b. All of the chromatin in these cells is inactive.
    c. The DNA in these cells has replicated.
    d. The cells are not transcribing any genes.
    e. The cells came from a female rat.
    *Answer: e*

69. The interphase cells of normal female mammals have a stainable nuclear body called a Barr body. This body is
    a. an inactive X chromosome.
    b. made of fat droplets.
    c. made of fragments of mRNA.
    d. made of extra chromosomal pieces.
    e. None of the above
    *Answer: a*

70. Heterochromatin
    a. contains poly A tails.
    b. is usually not transcribed.
    c. does not contain any DNA.
    d. is not replicated during the S phase.
    e. is found only in prokaryotes.
    *Answer: b*

71. You are studying the expression of the globin gene. In neurons, this gene contains many methylated cytosines. What might this mean about the expression of the globin gene in neurons?
    a. It is expressed only in males.
    b. It is expressed only in females.
    c. It is not expressed.
    d. It is regulated by posttranscriptional control.
    e. It is regulated by posttranslational control.
    *Answer: c*

72. A chemical modification that adds methyl groups to cytosine residues in some genes acts to
    a. enhance transcription.
    b. amplify the gene.

    c. inactivate the gene.
    d. stabilize the mRNA.
    e. None of the above
    *Answer: c*

73. DNA methylation
    a. is a mechanism of gene inactivation.
    b. adds methyl groups to cytosine residues in certain genes.
    c. inhibits transcription.
    d. Both a and b
    e. a, b, and c
    *Answer: e*

74. In fish and frog eggs, _____ increase from 0.2% to 68% of the total genomic DNA through selective gene amplification.
    a. tRNA
    b. rRNA
    c. rRNA gene clusters
    d. tRNA gene clusters
    e. mRNA gene clusters
    *Answer: c*

75. The cells in *Drosophila* that make eggshell (chorion) proteins increase the number of genes encoding these proteins. This is an example of
    a. gene amplification.
    b. DNA methylation.
    c. posttranscriptional control.
    d. RNA splicing.
    e. a gene family.
    *Answer: a*

76. Ubiquitin complexes proteins and delivers them to
    a. the extracellular space.
    b. mitochondria.
    c. the molecular chamber of doom.
    d. lysosomes.
    e. the Golgi apparatus.
    *Answer: c*

77. Transferrin is a protein that transports iron into cells. The iron concentration in a cell regulates the amount of transferrin protein being made, but it has no effect on the levels of transferrin mRNA. This is an example of
    a. translational control.
    b. an enhancer.
    c. transcriptional regulation.
    d. RNA processing.
    e. a promoter.
    *Answer: a*

▪78. Some metabolic pathways are regulated in part by changes in the rate of degradation of key enzymes. This is an example of
    a. operon control.
    b. transcriptional control.
    c. liquid hybridization.
    d. feedback inhibition.
    e. posttranslational control.
    *Answer: e*

▪79. Expression of some eukaryotic genes can be regulated by translational control. What is an advantage of translational control?
    a. It provides a means for rapid change in protein concentrations.

b. It prevents synthesis of excess RNA.
c. It directs proteins to their proper subcellular location.
d. It occurs only in zygotes.
e. It degrades proteins that are no longer needed.
*Answer: a*

80. In eukaryotic cells, promoters are
   a. transcribed.
   b. transcribed and translated.
   c. neither transcribed nor translated.
   d. transcribed and then removed.
   e. sequences of RNA that are spliced out.
   *Answer: c*

81. An interesting feature of the globin genes is that
   a. different ones are expressed during different stages of prenatal development.
   b. only one copy is functional.
   c. the lengths of the mRNAs are very different.
   d. they are the result of differential posttranscriptional splicing.
   e. the transcripts are longer than the coding regions.
   *Answer: a*

82. Potential control points for regulating the amount of protein synthesized in eukaryotic cells include all of the following *except*
   a. transcription regulation.
   b. DNA amplification.
   c. transcript processing.
   d. breakdown of the synthesized protein.
   e. stabilization of the mRNA.
   *Answer: d*

83. Transcription factors are
   a. RNA sequences that bind to RNA polymerase.
   b. DNA sequences that regulate transcription.
   c. proteins that bind to DNA near the promoter sequence.
   d. polysaccharides that bind to the transcripts.
   e. factors that bind to enhancers.
   *Answer: c*

84. Coordinated gene expression in eukaryotic cells is possible because
   a. one event often leads to another similar event.
   b. a transcript codes for more than one protein.
   c. of similarities in promoters for different genes.
   d. of the universal nature of the genetic code.
   e. all promoters are different.
   *Answer: c*

85. Which of the following is a promoter?
   a. Protein
   b. RNA
   c. DNA
   d. Carbohydrate
   e. Enzyme
   *Answer: c*

*86. The drought stress response in plants is an example of
   a. a transcription factor.
   b. coordinated gene expression.
   c. a way to increase water intake.
   d. Both a and b
   e. None of the above
   *Answer: b*

**87. An interesting characteristic of the control of tubulin production is that
   a. a translational repressor protein binds to the tubulin mRNA, which prevents ribosomes from attaching.
   b. alternative splicing of pre-mRNA results in the production of several different proteins.
   c. tubulin can recognize and bind to tubulin mRNA, causing an acceleration of its breakdown.
   d. ubiquitin forms a complex with the tubulin, which causes its breakdown.
   e. the 5′ guanosine cap added to the mRNA is not modified until the mRNA and is needed.
   *Answer: c*

**88. An interesting characteristic of tropomyosin production is that
   a. a translational repressor protein binds to the tropomyosin mRNA, which prevents ribosomes from attaching.
   b. alternative splicing of pre-mRNA results in the production of several different proteins.
   c. tropomyosin can recognize and bind to tropomyosin RNA causing an acceleration of its breakdown.
   d. ubiquitin forms a complex with the tropomyosin, which causes its breakdown.
   e. the 5′ guanosine cap added to the mRNA is not modified until the mRNA and is needed.
   *Answer: b*

*89. An interesting characteristic of the control of protein synthesis in tobacco hornworms is that
   a. a translational repressor protein binds to the mRNA, which prevents ribosomes from attaching.
   b. alternative splicing of pre-mRNA results in the production of several different proteins.
   c. proteins can recognize and bind to mRNA, causing an acceleration of its breakdown.
   d. ubiquitin forms a complex with the mRNA, which causes its breakdown.
   e. the 5′ guanosine cap added to the mRNA is not modified until the mRNA and is needed.
   *Answer: e*

**90. An interesting characteristic of the control of ferritin synthesis is that
   a. a translational repressor protein binds to the ferritin mRNA, which prevents ribosomes from attaching.
   b. alternative splicing of pre-mRNA results in the production of several different proteins.
   c. ferritin can recognize and bind to ferritin RNA, causing an acceleration of its breakdown.
   d. ubiquitin forms a complex with the ferritin, which causes its breakdown.
   e. the 5′ guanosine cap added to the mRNA is not modified until the mRNA and is needed.
   *Answer: a*

91. Mature mRNA can be edited by
   a. addition of new nucleotides.
   b. methylation.
   c. alternative splicing.
   d. attachment of ubiquitin.
   e. intron removal.
   *Answer: a*

92. Gene expression can be regulated
    a. before transcription.
    b. during transcription and before translation.
    c. during translation.
    d. after translation.
    e. All of the above
    *Answer: e.*

93. Nucleosomes disaggregate to allow transcription and then reaggregate
    a. through alternative splicing.
    b. by acetylation and deacetylation.
    c. through alternation of nucleotides.
    d. by attaching ubiquitin.
    e. through insertion of nucleotides.
    *Answer: b*

94. The similarity between the human genome and the genome of the puffer fish indicates that
    a. chromosomal rearrangements account for phenotypic differences.
    b. humans and puffer fish have a large amount of repetitive DNA.
    c. the complexity of an organism is not determined by genes alone.
    d. there are more human mRNAs than human genes.
    e. alternative splicing may be a key to the levels of complexity in organisms.
    *Answer: c*

# 15 Cell Signaling and Communication

## Fill in the Blank

1. Nitric oxide and _____ are needed to relax the smooth muscle cells of the blood vessels.
   *Answer: acetylcholine*

★2. _____ could be eliminated as a needed signal to relax smooth muscle cells if a membrane-permeable cGMP existed.
   *Answer: Nitric oxide (NO)*

3. Cells of many multicellular animals communicate directly by means of _____, which allow them to link their cytoplasms.
   *Answer: gap junctions*

4. Signals that act on the same cells as those that generated them are called _____.
   *Answer: autocrine*

5. Signals that diffuse from one cell to other types of cells are called _____.
   *Answer: paracrine*

6. The molecule that binds to the receptor is called a _____.
   *Answer: ligand*

7. Phosphatidyl inositol-bisphosphate is a _____.
   *Answer: lipid*

8. In the signal pathway that includes phospholipase C, _____ opens calcium channels.
   *Answer: inositol triphosphate ($IP_3$)*

9. The signaling process from signal detection to final response is called the _____.
   *Answer: signal transduction pathway*

10. Conformation and activity of a targeted protein can be altered by _____.
   *Answer: phosphorylation*

## Multiple Choice

■1. Why might a certain signaling molecule affect one type of cell in an organism and not another?
   a. Each different type is capable of interpreting just one kind of signal to prevent confusion.
   b. Each different type is capable of interpreting signals necessary for its functions.
   c. All cells can interpret the different signaling molecules in different ways.
   d. Not all the cells are exposed to all the different signals.
   e. Cells do not interpret the signals; the signals must interpret the cells.
   *Answer: b*

2. Cells receive which of the following signals?
   a. Light
   b. Sound
   c. Odorants
   d. Hormones
   e. All of the above
   *Answer: e*

3. Signals get to target cells in multicellular organisms via
   a. circulation and diffusion.
   b. conduction and diffusion.
   c. circulation and lymph.
   d. chaperon trafficking and transmembrane transport.
   e. cytoskeletal trafficking and perfusion.
   *Answer: a*

4. A signal pathway is
   a. the path a signal takes to find its target cell.
   b. a group of signals along a concentration gradient.
   c. the coordinates created from concentration gradients that tell cells where they are located in multicellular organisms.
   d. the series of events that occur in response to a signal's being detected.
   e. a nerve propagation.
   *Answer: d*

★5. The EnvZ protein of *E. coli* changes shape in response to
   a. ion concentration.
   b. ligand binding.
   c. light.
   d. sound.
   e. $O_2$.
   *Answer: a*

★6. The EnvZ protein undergoes a conformational change to become an active
   a. channel protein.
   b. kinase.
   c. phosphorylase.
   d. environmental gene.
   e. All of the above
   *Answer: b*

*7. The OmpR protein is a(n)
  a. DNA binding protein.
  b. channel protein.
  c. channel-blocking protein.
  d. osmotic pressure-detecting protein.
  e. kinase.
  *Answer: a*

8. In general, all cell signaling causes
  a. altered gene expression.
  b. an influx of ions.
  c. protein kinase activity.
  d. G protein activation.
  e. a change in receptor conformation.
  *Answer: e*

9. The major categories of signal receptors are
  a. inside and outside.
  b. enzyme and ion channel.
  c. transmembrane and cytoplasmic.
  d. protein kinase and cAMP.
  e. sensory and molecular.
  *Answer: c*

10. Steroids bind
  a. to the outer face of transmembrane proteins.
  b. to cytoplasmic receptors.
  c. within the lipid bilayer.
  d. around the nuclear membrane.
  e. directly to DNA.
  *Answer: b*

11. Steroids typically affect
  a. gene transcription.
  b. ion channels.
  c. enzyme activity.
  d. biochemical pathways.
  e. aggression.
  *Answer: a*

12. Typically, large polar signals interact directly with
  a. cytoplasmic receptors.
  b. transmembrane receptors.
  c. G proteins.
  d. adenylyl cyclase.
  e. calmodulin.
  *Answer: b*

13. Insulin is a(n)
  a. transmembrane protein kinase.
  b. intracellular signaling molecule.
  c. form of sugar.
  d. extracellular ligand.
  e. derivative of glucose.
  *Answer: d*

14. Some signal receptors are
  a. ion channels.
  b. protein kinases.
  c. G protein-linked.
  d. DNA binding proteins.
  e. All of the above
  *Answer: e*

■15. Different types of cells that respond to a certain signal do not necessarily respond in the same way because
  a. different types of cells may have different signal pathways for the same signal.
  b. different cells have different metabolic needs.

  c. not all cells respond to all signals.
  d. the cell may already have received the signal.
  e. None of the above
  *Answer: a*

16. Protein kinase is
  a. an enzyme that makes cAMP.
  b. the enzyme that makes cGMP.
  c. the substrate molecule for kinase.
  d. an enzyme that phosphorylates.
  e. None of the above
  *Answer: d*

17. One of the substrates for protein kinase is
  a. cAMP.
  b. cGMP.
  c. G proteins.
  d. ATP.
  e. GTP.
  *Answer: d*

18. In eukaryotic cells, a substrate for phosphorylation is
  a. cAMP.
  b. cGMP.
  c. a specific tyrosine in a target protein.
  d. a specific glycine in a target protein.
  e. All of the above
  *Answer: c*

19. Insulin alters cellular metabolism
  a. by binding with two receptor subunits on the outer cell surface.
  b. by alerting cells about the availability of glucose.
  c. through kinase activity.
  d. through G protein activation.
  e. by opening glucose channels in the membrane.
  *Answer: a*

20. The insulin receptor is a
  a. G protein.
  b. cAMP molecule.
  c. kinase.
  d. phosphodiesterase.
  e. phosphatase.
  *Answer: c*

21. The receptor-associated proteins called G proteins
  a. bind GTP.
  b. can activate or inhibit an effector.
  c. interact with membrane-associated internal proteins to influence their function.
  d. Both a and b
  e. a, b, and c
  *Answer: e*

22. The Ras protein is
  a. a G protein.
  b. a protein that activates a kinase.
  c. part of a pathway called a phosphorylation cascade.
  d. the cause of a certain cancer when it is defective.
  e. All of the above
  *Answer: e*

■23. In order for a G protein to play its part in moving events forward in a signal pathway,
  a. GDP must be released and a GTP must occupy the nucleotide-binding site.
  b. GTP must be released and a GDP must occupy the nucleotide-binding site.

c. cGMP must occupy the otherwise empty nucleotide-binding site.

d. cGMP must leave the otherwise occupied nucleotide-binding site.

e. the G protein must interact with a receptor protein.

*Answer: a*

■24. If a G protein were unable to release its bound nucleotide but could hydrolyze it, signal transduction would

a. not move beyond this point.

b. be continuous beyond this point.

c. be unaffected.

d. be constantly switching on and off.

e. be unpredictable.

*Answer: a*

■25. If a G protein could release its bound nucleotide but was not able to hydrolyze it, signal transduction would

a. not move beyond this point.

b. be continuous beyond this point.

c. be unaffected.

d. be constantly switching on and off.

e. be unpredictable.

*Answer: b*

*26. In heart muscles, the G protein that associates with the epinephrine receptor

a. inhibits adenylate cyclase.

b. activates adenylate cyclase.

c. causes a rise in cGMP.

d. causes a decline in cGMP.

e. None of the above

*Answer: b*

*27. In smooth muscle cells of the blood vessels, the G protein that associates with the epinephrine receptor

a. inhibits adenylate cyclase.

b. activates adenylate cyclase.

c. causes a rise in cGMP.

d. causes a decline in cGMP.

e. None of the above

*Answer: a*

28. Cytoplasmic receptors are used for

a. epinephrine.

b. cAMP.

c. steroids.

d. insulin.

e. All of the above

*Answer: c*

29. Cytoplasmic receptors are, or are closely associated with,

a. ligands.

b. G proteins.

c. transmembrane receptors.

d. the endoplasmic reticulum.

e. DNA-binding proteins.

*Answer: e*

30. When a cell with a steroid receptor is exposed to that steroid, the one event certain to take place is that the

a. receptor will change shape upon binding the steroid.

b. cell will produce much more of a muscle cell protein, actin.

c. cell will get switched to an "on" state.

d. Both a and b

e. None of the above

*Answer: a*

31. Transducers

a. change signals from one form to another.

b. alter gene expression.

c. are simple switches.

d. pass a certain signal forward.

e. All of the above

*Answer: a*

*32. The Ras protein is part of a

a. steroid receptor.

b. receptor-linked enzyme system.

c. cytoplasmic signal receptor system.

d. protein kinase cascade.

e. random, accelerated sorting system.

*Answer: d*

*33. At what point in the signal transduction pathway is Ras is activated?

a. Early in the pathway

b. At the midpoint

c. Late in the pathway

d. All along the pathway

e. None of the above

*Answer: a*

*34. The Ras protein is a

a. G protein.

b. protein kinase.

c. cell-surface receptor.

d. a, b, and c

e. None of the above

*Answer: a*

*35. The reason that the defective Ras protein found in many cases of bladder cancer causes uncontrolled cell division is that

a. GDP fails to be released.

b. kinase activity is defective.

c. the cGMP site remains occupied.

d. Ras is bound permanently to GTP.

e. All of the above

*Answer: d*

36. The molecule cAMP is

a. coupled adenine monophosphate.

b. a signal molecule.

c. a second messenger.

d. adenylyl cyclase.

e. cyclase adenylyl-phosphate.

*Answer: c*

■37. An experimenter finds that the cell membrane is necessary for a signal to elicit its effect, but that the membrane needs to be present only after the signal and the cytoplasm have been incubated together. If the experimenter removes the signal from the cytoplasm before adding the membrane, the effect can still be detected. Which of the following conclusions would be consistent with these observations?

a. The signal interacts with a cytoplasmic receptor.

b. The signal pathway involves an interaction with a membrane-associated component.

c. The signal is unnecessary, but the membrane is necessary.

d. The signal pathway starts at the membrane and then progresses into the cytoplasm.

e. Both a and b

*Answer: e*

■38. There are analogs to cAMP that are able to pass through a membrane unimpeded. If you added this form of cAMP to liver cells, what would you expect to happen?
   a. The same events as those that would occur with the addition of epinephrine.
   b. The cells would open their $Na^+$ channels.
   c. The receptor-associated G protein would activate.
   d. The receptor-associated G protein would inactivate because this molecule normally causes the elevated cAMP.
   e. Adenylyl cyclase would activate.
   *Answer: a*

39. Adenylate cyclase
   a. is a cyclic nucleotide.
   b. catalyzes cAMP from ATP.
   c. produces a G protein.
   d. is a protein kinase.
   e. is a second messenger molecule.
   *Answer: b*

■40. There are analogs to cAMP that are able to pass through a membrane unimpeded. If you added this form of cAMP to odorant receptor nerve cells, what would you expect to happen?
   a. The same events as those that would occur with the addition of epinephrine
   b. The cell would open its $Na^+$ channels.
   c. The receptor-associated G protein would activate.
   d. The receptor-associated G protein would inactivate because this molecule normally causes the elevated cAMP.
   e. The cAMP is odorless, so nothing would happen.
   *Answer: b*

41. The concentration of cAMP in a cell is increased by
   a. phosphodiesterase.
   b. cGMP.
   c. a protein kinase.
   d. an ion channel.
   e. a G protein.
   *Answer: e*

42. Phospholipase C is activated by
   a. cAMP.
   b. an elevated $Ca^{2+}$ concentration.
   c. a specific G protein.
   d. cGMP.
   e. GTP.
   *Answer: c*

43. Activated phospholipase C
   a. is a second messenger molecule.
   b. opens $Ca^{2+}$ channels.
   c. produces DAG and $IP_3$.
   d. activates G proteins.
   e. cleaves cAMP's phosphodiester linkage.
   *Answer: c*

44. The second messenger $IP_3$
   a. opens $Ca^{2+}$ channels.
   b. activates protein kinase C.
   c. hydrolyzes phosphatidyl inositol-bisphosphate.
   d. activates G proteins.
   e. is released from the cell as a paracrine-signaling molecule.
   *Answer: a*

45. Diacylglycerol
   a. opens $Ca^{2+}$ channels.
   b. activates protein kinase C.
   c. hydrolyzes phosphatidyl inositol-bisphosphate.
   d. activates G proteins.
   e. is released from the cell as a paracrine-signaling molecule.
   *Answer: b*

46. To be activated, protein kinase C must
   a. interact with $IP_3$.
   b. bind $Ca^{2+}$.
   c. bind DAG.
   d. Both a and b
   e. Both b and c
   *Answer: e*

47. The point of signal convergence in the phospholipase C signal pathway is the point at which
   a. the G protein activates the enzyme.
   b. the receptor binds the ligand.
   c. activation of the enzyme hydrolyzes the lipid.
   d. the $Ca^{2+}$ and DAG bind protein kinase C.
   e. calcium binds to it.
   *Answer: d*

48. Which of the following is an example of the action of $Ca^{2+}$ as a second messenger?
   a. G protein activation
   b. The opening of calcium channels by $IP_3$
   c. The activation of the enzyme that hydrolyzes the lipid
   d. The binding of $Ca^{2+}$ to protein kinase C
   e. Both b and d
   *Answer: e*

49. Calmodulin is activated when it binds
   a. a cAMP molecule.
   b. $IP_3$.
   c. three $K^+$.
   d. four $Ca^{2+}$.
   e. GTP.
   *Answer: d*

50. Gap junctions
   a. allow for communication among plant cells.
   b. permit metabolic cooperation among linked cells.
   c. are linked by fused membranes.
   d. allow proteins and carbohydrates to pass through them.
   e. have a desmotubule that fills up most of the opening of the gap junction.
   *Answer: b*

51. Transduction involves a cascade of events. This process is beneficial to the organism because it
   a. opens several ion channels simultaneously.
   b. releases a large amount of energy through phosphorylation.
   c. amplifies and distributes weak signals.
   d. allows for communication between intracellular and extracellular signals.
   e. allows a large variety of signals to be processed.
   *Answer: c*

52. The concentration of protein kinase is regulated by
    a. protein phosphatases.
    b. GTPases.
    c. phosphodiesterase.
    d. membrane pumps and ion channels.
    e. drugs such as nitroglycerin.
    *Answer: a*

53. Drugs that promote penile erection
    a. open ion channels.
    b. catalyze the formation of cAMP from ATP.
    c. activate a series of proteins.
    d. are NO synthesis activators.
    e. catalyze phosphorylation.
    *Answer: d*

# 16 Recombinant DNA and Biotechnology

## Fill in the Blank

1. A _____ animal has recombinant DNA integrated into its own genetic material.
   *Answer: transgenic*

2. Enzymes that cleave double-stranded DNA at specific sites are _____.
   *Answer: restriction enzymes*

3. A short, single-stranded region at the end of a DNA fragment is a _____.
   *Answer: sticky end*

4. An enzyme that can covalently link two DNA fragments is _____.
   *Answer: DNA ligase*

5. A _____ is a virus or plasmid that can replicate its DNA and foreign DNA within a cell without being degraded.
   *Answer: cloning vector*

6. A circular cloning vector that replicates autonomously within a cell is a _____.
   *Answer: plasmid*

7. A cloning vector that replicates and destroys a bacterial cell is a _____.
   *Answer: bacteriophage*

8. _____ is obtained by reverse transcription of RNA.
   *Answer: Complementary DNA*

9. A collection of DNA molecules that represents a population of RNAs is a _____.
   *Answer: cDNA library*

★■10. DNA of the bacterial host is not cleaved by its own restriction enzymes because of the activity of specific _____, which add methyl groups to certain bases within the recognition sites.
   *Answer: methylases*

11. Restriction enzymes recognize sites by a _____ sequence of bases.
   *Answer: specific*

12. An organism's DNA is isolated, cut by restriction endonucleases, and inserted into vectors where it is cloned. The result is a collection of clones called a _____.
   *Answer: genomic library*

13. In order to synthesize cDNA from mRNA, an important retroviral enzyme called _____ is essential.
   *Answer: reverse transcriptase*

14. VNTR stands for "_____."
   *Answer: variable number of tandem repeats*

15. An undifferentiated cell that divides and differentiates into specialized cells is a _____.
   *Answer: stem cell*

16. A natural way of inhibiting mRNA translation catalyzes the breakdown of the targeted mRNA. This technique uses _____.
   *Answer: interference RNA (or RNAi)*

17. A system that tests for protein interactions in a living cell is the _____.
   *Answer: two-hybrid system*

18. The characterization of an individual by his or her DNA base sequence is known as _____.
   *Answer: DNA fingerprinting*

## Multiple Choice

1. A fragment of DNA with "sticky" ends
   a. can form hydrogen bonds with another fragment with complementary "sticky" ends.
   b. will be readily degraded in the test tube.
   c. is the starting point for RNA polymerase.
   d. is the recognition sequence for restriction endonucleases.
   e. None of the above
   *Answer: a*

2. The enzyme that can join pieces of DNA together is
   a. RNA polymerase.
   b. DNA polymerase.
   c. DNA ligase.
   d. β-galactosidase.
   e. None of the above
   *Answer: c*

★3. Restriction enzymes cleave DNA at specific sequences by hydrolyzing
   a. the 3′ hydroxyl of one nucleotide and the 5′ phosphate of the next one.
   b. at the 1′ carbons to cleave the nitrogenous bases.
   c. at the 2′ carbons to cleave hydroxyl groups.
   d. two phosphodiester linkages on the same strand.
   e. four phosphodiester linkages, two on each strand.
   *Answer: a*

4. Restriction enzymes are used naturally in a bacterial cell to
   a. defend against foreign DNA.
   b. cleave large sections of bacterial DNA.
   c. digest extra copies of plasmid DNA.
   d. replicate the bacterial chromosomal DNA.
   e. produce RNA from DNA.
   *Answer: a*

**5. Why is the DNA of the host cell not cleaved by the restriction endonucleases it produces?
   a. The restriction enzymes are contained within lysosomes.
   b. The restriction enzymes can only cleave RNA.
   c. The restriction enzymes are made on the rough endoplasmic reticulum and exported out of the cell.
   d. The bacterial DNA is altered by methylation and is not a substrate for restriction endonucleases.
   e. None of the above
   *Answer: d*

6. Restriction enzymes were first identified in _____ cells.
   a. plant
   b. eukaryotic
   c. prokaryotic
   d. protist
   e. animal
   *Answer: c*

7. Which of the following processes makes use of the nucleic acid base-pairing rules?
   a. DNA replication
   b. Transcription
   c. Translation
   d. Sequencing of genes
   e. All of the above
   *Answer: e*

8. Which of the following sequences is a DNA palindrome?
   a. GCTATCG
   b. AAAAAA
   c. GCATGC
   d. ACGTAC
   e. All of the above
   *Answer: c*

9. In order to join a fragment of human DNA to bacterial or yeast DNA, both the human DNA and the bacterial (or yeast) DNA must first be treated with the same
   a. DNA ligase.
   b. DNA polymerase.
   c. restriction enzymes.
   d. DNA gyrase.
   e. None of the above
   *Answer: c*

*10. EcoRI makes staggered cuts when it cleaves DNA, creating single-stranded tails called "sticky ends." These ends will form a complementary base pair. Which of the following conditions must exist for this to happen?
   a. The presence of specific helicases
   b. High enough temperatures
   c. Methyl groups at each end
   d. Low enough temperatures
   e. None of the above
   *Answer: d*

11. Genetic engineering is important to botanists. The goal of this technology in plants is to
   a. cause plant disease.
   b. isolate beneficial genes from one species and introduce them into another plant species for a superior crop.
   c. create new species of plants to replace the extinct ones.
   d. Both a and c
   e. None of the above
   *Answer: b*

12. The specific purpose for cloning DNA fragments out of plants or animals is to
   a. splice together other DNA fragments from dissimilar genomes.
   b. locate a specific gene and prepare DNA probes.
   c. collect appropriate vectors for genetic engineering.
   d. prepare synthetic oligonucleotides.
   e. None of the above
   *Answer: b*

13. When a DNA solution is added to a gel, the DNA will migrate
   a. toward the positive pole.
   b. toward the negative pole.
   c. perpendicular to the positive and negative poles.
   d. due to diffusion.
   e. away from the positive pole.
   *Answer: a*

14. Electrophoresis separates DNA fragments of different sizes, but this technique does not indicate which of the fragments contains the DNA piece of interest. This problem is solved by
   a. measuring the sizes of the bands on the gel.
   b. removing the bands from the gel and hybridizing them with a known strand of DNA complementary to the gene of interest.
   c. knowing the isoelectric points of the piece in question.
   d. identifying the molecular weights of the fragments in question.
   e. None of the above
   *Answer: b*

15. When DNA migrates in a gel, the longer DNA
   a. generally fails to move.
   b. migrates the most quickly because it has a greater charge.
   c. moves the most slowly.
   d. is found closest to the positive pole.
   e. moves at the same pace as shorter DNA.
   *Answer: c*

**16. If the bacteriophage T7 DNA does not contain any EcoRI sites, how can an E. coli cell protect itself from T7 infection?
   a. The cell sequesters the phage DNA in a vesicle.
   b. The cell makes a number of restriction enzymes, each with a different recognition site.
   c. The cell destroys its ribosomes and protein synthetic machinery.
   d. The cell destroys all of the cellular DNA replication enzymes.
   e. The cell cannot possibly protect itself.
   *Answer: b*

17. Which of the following is *not* a usual characteristic of a cloning vector?
   a. It contains at least one restriction endonuclease recognition sequence.
   b. It must integrate into the host chromosome.

c. It must be a relatively small piece of DNA.

d. It must be able to replicate within the host cell.

e. It carries a gene for antibiotic resistance.

*Answer: b*

■18. Transfection of plant cells is more difficult than transfection of prokaryotic cells because

a. plant cells are larger than prokaryotic cells.

b. plant cells contain more DNA than prokaryotic cells do.

c. DNA must get through plant cell walls.

d. Both a and b

e. Both a and c

*Answer: c*

19. A genomic library

a. is a collection of foreign DNA molecules inserted into cloning vectors.

b. ideally represents the entire genome of an organism.

c. is a collection of RNA molecules inserted into cloning vectors.

d. Both a and b

e. Both b and c

*Answer: d*

20. The expression of a cloned gene in a bacterium is easily regulated if it is

a. inserted anywhere in the bacterial chromosome.

b. incorporated into a phage genome.

c. spliced into a self-replicating plasmid.

d. spliced onto the *lac* operon promoter.

e. left free in the bacterial cell.

*Answer: d*

21. Which of the following is *not* usually a source of DNA for cloning?

a. Pieces of genomic, chromosomal DNA

b. DNA made by the reverse transcription of mRNA

c. DNA made by the reverse transcription of tRNA

d. DNA synthesized in the laboratory

e. All of the above

*Answer: c*

22. DNA migrates in an electric field because it

a. is positively charged.

b. is not charged.

c. is negatively charged.

d. combines with the gel molecules.

e. is hydrophobic.

*Answer: c*

*23. The disease called crown gall is caused by

a. the insertion of a transposable element carried on the Ti plasmid.

b. the transcription of the Ti plasmid in the plant cells.

c. the transfer of bacterial genomes into the plant cell genome.

d. the rampant multiplication of *A. tumefaciens* bacteria within the plant.

e. None of the above

*Answer: a*

*24. If a cloned DNA molecule is inserted into either of the antibiotic sites in the plasmid pBR322 (*amp$^r$* or *tet$^r$*), the antibiotic gene then becomes

a. transferred to a gene upstream of the site on the chromosome.

b. immediately transcribed into mRNA.

c. inactivated.

d. complemented with a corresponding sequence.

e. None of the above

*Answer: c*

25. Which of the following could be used for detecting genetic disorders?

a. cDNA probes that hybridize with DNA regions that are missing or have mutant nucleotide sequences near or within the DNA sequence responsible for the genetic disorder

b. PCR amplification of the region responsible for the genetic disorder

c. Restriction mapping of sequences that have different restriction sites and are associated with the disease

d. Both a and c

e. a, b, and c

*Answer: e*

26. Complementary DNA (cDNA) is made using

a. synthetic oligonucleotides.

b. amino acid sequences from structural genes.

c. DNA as the template.

d. mRNA as a template.

e. None of the above

*Answer: d*

27. When DNA is introduced into eukaryotic cells and is integrated, the cell is

a. transformed.

b. transduced

c. conjugated.

d. expressed.

e. transfected.

*Answer: e*

28. The production of double-stranded cDNA utilizes

a. hybridization between the poly A tails of mRNA and oligo dT.

b. reverse transcriptase.

c. DNA polymerase.

d. Both a and b

e. a, b, and c

*Answer: e*

29. In the production of double-stranded cDNA, mRNA is used as

a. a template.

b. a tail.

c. one strand of the completed cDNA.

d. Both a and b

e. Both a and c

*Answer: a*

30. An organism that is modified by introduction of a DNA sequence from another organism is called

a. transformed.

b. transgenic.

c. transfected.

d. transduced.

e. a multi-species.

*Answer: b*

31. A YAC is a(n)

a. animal of interest because of its unusual mutations.

b. yeast artificial chromosome.

c. yellow activation center.

d. Y chromosome.

e. None of the above

*Answer: b*

32. YACs have
    a. a centromere.
    b. telomeres.
    c. origins of replication.
    d. more than 10,000 base pairs.
    e. All of the above
    *Answer: e*

■33. A researcher inserts a DNA segment at the *Bam*HI recognition site within a plasmid; this site is located within the tetracycline resistance gene. This plasmid also has a gene for ampicillin resistance. Following DNA transformation, the researcher must differentiate the bacteria that have taken up the recombinant DNA from those that have taken up either the foreign DNA only, or intact plasmids. In doing so, the researcher should select the bacteria that
    a. will grow on ampicillin, but that are sensitive to tetracycline.
    b. are sensitive to both antibiotics.
    c. are resistant to both antibiotics.
    d. will grow on tetracycline, but that are sensitive to ampicillin.
    e. grow only on an enriched medium.
    *Answer: a*

■34. Sticky ends are "sticky" because they are
    a. single-stranded.
    b. from one to four bases long.
    c. complementary to other sticky ends.
    d. poly A tails.
    e. part RNA and part DNA.
    *Answer: c*

■35. A second screening step is often necessary to find bacteria that have taken up plasmids with foreign DNA inserts because
    a. there are many cells without any plasmids at all.
    b. only the plasmids with foreign DNA are taken up into cells.
    c. very often a plasmid without foreign DNA is taken up by a cell.
    d. Both a and b
    e. Both a and c
    *Answer: c*

■■36. From the list below, choose a reasonable sequence of steps for cloning a piece of foreign DNA into a plasmid vector, introducing the plasmid into bacteria, and verifying that the plasmid and insert are present.
    1. Transform competent cells
    2. Select for the lack of antibiotic resistance gene #1 function
    3. Select for the plasmid antibiotic resistance gene #2 function
    4. Digest vector and foreign DNA with *Eco*RI, which inactivates antibiotic resistance gene #1
    5. Ligate the digested DNA together
    a. 4, 5, 1, 3, 2
    b. 4, 5, 1, 2, 3
    c. 1, 3, 4, 2, 5
    d. 3, 2, 1, 4, 5
    e. None of the above
    *Answer: a*

37. A knockout experiment involves
    a. homologous recombination.
    b. transfected embryonic cells.
    c. introduction of cells into a developing embryo.
    d. use of a genetic marker.
    e. All of the above
    *Answer: e*

38. DNA chips have _____ attached.
    a. bacteriophage λ DNA libraries
    b. DNA sequences up to 20 nucleotides long
    c. cDNA
    d. radionucleotides
    e. All of the above
    *Answer: b*

39. Antisense RNA prevents the expression of specific genes by
    a. binding to the complementary RNA.
    b. inhibiting mRNA translation.
    c. causing a mutation that interrupts the gene.
    d. Both a and b
    e. Both b and c
    *Answer: d*

40. A cell or an organism that contains foreign DNA inserted into its own genetic material is termed
    a. transgenic.
    b. polygenic.
    c. engineered.
    d. foreign.
    e. xenophobic.
    *Answer: a*

41. Principal sources of genes or DNA fragments used in recombinant DNA work include
    a. genomic libraries.
    b. cDNA samples.
    c. artificially prepared oligonucleotides.
    d. Both a and b
    e. a, b, and c
    *Answer: e*

★42. An important vector for manipulation of plant genes comes from the bacterium *A. tumefaciens* and is called
    a. an *Eco*RI plasmid.
    b. a raze bacteriophage.
    c. a pangene-site vector.
    d. a Ti plasmid.
    e. None of the above
    *Answer: d*

43. The advantage of tissue plasminogen activator (TPA) over streptokinase is that streptokinase
    a. is too effective.
    b. might trigger an immune response, whereas TPA will not.
    c. costs much more than TPA.
    d. Both a and b
    e. Both b and c
    *Answer: b*

44. The advantage of a viral vector over a plasmid vector is that
    a. viral vectors can often carry much larger DNA inserts.
    b. plasmid vectors are unreliable.
    c. viral vectors require less medium.
    d. Both a and b
    e. a, b, and c
    *Answer: a*

45. A plasmid is isolated, digested with a restriction enzyme, and run on a gel using electrophoresis. The fragments closest to the well where the DNA was loaded are _____ than the fragments farthest away.
    a. shorter
    b. longer
    c. duller
    d. less original
    e. Both a and c
    *Answer: b*

46. Antisense RNA is
    a. employed to prevent the translation of a specific mRNA.
    b. RNA that is complementary to a specific mRNA.
    c. the noncoding strand of the DNA molecule.
    d. Both a and b
    e. None of the above
    *Answer: d*

•47. A single hair is found at the scene of a crime. Which technology would you use to determine if the hair could have come from a certain suspect?
    a. PCR
    b. DNA sequencing
    c. Fragment cloning
    d. Probing
    e. Antisense RNA
    *Answer: a*

48. Plant cells can be used as host cells to develop a transgenic plant. Plant cells are good hosts because
    a. their cells divide rapidly.
    b. they have a small genome size.
    c. of their totipotency.
    d. they are easy to grow and manipulate.
    e. they have numerous genetic markers.
    *Answer: c*

49. Initially bacteria were used in recombinant DNA experiments. However, the use of bacteria is not ideal because they
    a. grow and divide too quickly.
    b. have numerous genetic markers.
    c. have a large genome.
    d. cannot excise introns from the initial eukaryotic RNA transcript.
    e. are difficult to culture in the laboratory.
    *Answer: d*

50. Foreign DNA can enter a host cell as part of a vector. A vector has all of the following characteristics *except*
    a. genes that confer antibiotic resistance.
    b. the ability to form a circular plasmid containing the new DNA.
    c. an appropriate recognition sequence.
    d. a reporter gene.
    e. a smaller size than that of the host chromosomes.
    *Answer: b*

51. The roles of genes during development have been studied using
    a. the knockout technique.
    b. DNA chips.
    c. RT-PCR.
    d. antisense RNA.
    e. the two-hybrid system.
    *Answer: a*

52. Expression vectors have extra sequences needed for a foreign gene to be expressed in a host cell. An extra sequence unique to eukaryotes is the
    a. sequence for termination.
    b. sequence for promotion.
    c. sequence for ribosome binding.
    d. poly A addition sequence.
    e. sequence for translation.
    *Answer: d*

53. *B. thuringiensis* is a biodegradable insecticide. It is not widely used because
    a. it has a toxic effect on other organisms.
    b. it kills pests and beneficial insects.
    c. its toxicity magnifies with use.
    d. the toxin genes cannot be isolated and cloned.
    e. it must be applied repeatedly.
    *Answer: e*

54. One type of emphysema can be treated with a protein called α-1-antitrypsin (α-1AT). A large supply of α-1AT can be obtained from
    a. human serum.
    b. transgenic sheep.
    c. salt-tolerant tomato plants.
    d. transgenic rice.
    e. plasminogen.
    *Answer: b*

55. Salt-tolerant tomato plants are an example of
    a. a transgenic crop that is adapted to its environment.
    b. tailoring the environment to the needs of crop plants.
    c. a chloroplast enzyme system that has been inhibited.
    d. plants that are able to make β-carotene.
    e. a medically useful product of biotechnology.
    *Answer: a*

56. DNA fingerprinting can be used to
    a. identify an agricultural product that came from a specific crop.
    b. identify thoroughbred racehorses.
    c. diagnose infections.
    d. analyze historical events.
    e. All of the above
    *Answer: e*

57. DNA is negatively charged at neutral pH due to
    a. purines binding with pyrimidines.
    b. an OH⁻ group at the 3′ end.
    c. hydrogen bonding.
    d. its antiparallel structure.
    e. its phosphate groups.
    *Answer: e*

# 17 Molecular Biology and Medicine

## Fill in the Blank

1. In addition to defective alleles, a major source of genetic ill health is _____ aberrations.
   *Answer: chromosomal*

2. Sex-linked recessive diseases affect _____ more than _____.
   *Answer: men; women*

★3. Sex-linked dominant diseases affect _____ more than _____.
   *Answer: women; men*

4. The name of an X-linked recessive disorder resulting in muscular deterioration is _____.
   *Answer: Duchenne muscular dystrophy*

5. _____ is an X-linked recessive trait affecting the clotting of blood.
   *Answer: Hemophilia*

6. _____ is the process by which an abnormal gene is replaced by a normal one.
   *Answer: Gene therapy*

7. In the treatment of a person with defective adenosine deaminase genes, success was short-lived when it involved the transfer of genes to her white blood cells. Treatment might have been more effective if the genes had been transferred to bone marrow _____ cells.
   *Answer: stem*

8. In testing a fetus for harmful alleles, DNA obtained from amniocentesis can be amplified by an artificial cycling process called _____.
   *Answer: PCR*

9. The most thoroughly studied example of a disease that arises via somatic mutation is _____.
   *Answer: cancer*

10. An estimated one-half of spontaneous abortions that occur during the first trimester of pregnancy are attributed to _____.
    *Answer: chromosomal abnormalities*

11. An _____ arises by mutation of a normal gene that controls some aspect of the normal development of the cell.
    *Answer: oncogene*

12. All of the proteins produced by an organism are known as its _____.
    *Answer: proteome*

## Multiple Choice

1. Which two diseases are caused by defects in phenylalanine metabolism?
   a. Alkaptonuria and cystic fibrosis
   b. Sickle-cell disease and cystic fibrosis
   c. Alkaptonuria and sickle-cell disease
   d. Fragile-X syndrome and prion disease
   e. Phenylketonuria and alkaptonuria
   *Answer: e*

2. Sickle-cell disease is caused by a _____ mutation.
   a. frameshift
   b. base substitution
   c. deletion
   d. insertion
   e. tandem repeat
   *Answer: b*

3. A common form of inherited mental retardation is associated with
   a. Duchenne muscular dystrophy.
   b. cystic fibrosis.
   c. familial hypercholesterolemia.
   d. fragile-X syndrome.
   e. hemophilia.
   *Answer: d*

4. Triplet repeats
   a. are due to inadequate amounts of FMR1 protein.
   b. are limited to disease-causing genes.
   c. expand readily in human genes but not in nonhuman genes.
   d. occur in certain genetic disorders.
   e. lead to decreased methylation of cytosines.
   *Answer: d*

5. In addition to the existence of inherited genetic diseases, there are also non-inherited genetic diseases caused by mutations in _____ cells.
   a. somatic
   b. germ
   c. white blood
   d. connective tissue
   e. All of the above
   *Answer: e*

*6. Metabolic diseases such as PKU and alkaptonuria are caused by a(n)
a. nonfunctional or missing enzyme.
b. abnormal number of chromosomes.
c. mutation in the mitochondrial DNA.
d. virus.
e. abnormal structural protein.
*Answer: a*

7. Although the exact cause of the mental retardation associated with PKU is not known, medical science does know that
a. it occurs when the X chromosome breaks into pieces.
b. high levels of phenylalanine are involved.
c. it is currently untreatable.
d. Both b and c
e. All of the above
*Answer: b*

8. The first genetic disease for which an amino acid abnormality was tracked down was
a. PKU.
b. alkaptonuria.
c. fragile-X syndrome.
d. sickle-cell disease.
e. hemophilia.
*Answer: d*

9. Sickle-cell disease is the result of a
a. deletion.
b. nonsense mutation.
c. frameshift mutation.
d. base pair substitution.
e. chromosomal deletion.
*Answer: d*

*10. The probability of a person's carrying a defective allele for a certain disease depends on
a. environment.
b. ancestry.
c. genetic predisposition.
d. unknown factors.
e. None of the above
*Answer: b*

11. What is the frequency of cystic fibrosis in the human population?
a. 1 in 2.5
b. 1 in 25
c. 1 in 250
d. 1 in 2,500
e. 1 in 25,000
*Answer: d*

12. Tissues affected by cystic fibrosis include the
a. lungs.
b. liver.
c. pancreas.
d. gut.
e. All of the above
*Answer: e*

13. People with cystic fibrosis often die in their
a. teens and twenties.
b. twenties and thirties.
c. thirties and forties.
d. forties and fifties.
e. None of the above
*Answer: b*

14. The gene that is affected by cystic fibrosis normally encodes a protein that
a. stimulates mitochondria.
b. regulates gene expression.
c. synthesizes mucus.
d. encodes a chloride channel in the cell membrane.
e. None of the above
*Answer: d*

15. People with phenylketonuria cannot metabolize
a. alanine.
b. phenylalanine.
c. glutamic acid.
d. tryptophan.
e. tyrosine.
*Answer: b*

16. The principle consequence of phenylketonuria is
a. muscle atrophy.
b. kidney failure.
c. mental retardation.
d. skeletal problems.
e. Both a and b
*Answer: c*

17. Following a diagnosis of phenylketonuria, infants are restricted to a diet low in
a. alanine.
b. phenylalanine.
c. glutamic acid.
d. tryptophan.
e. tyrosine.
*Answer: b*

*18. Among the African-American population, the frequency of sickle-cell disease is about
a. 1 in 2,000.
b. 1 in 655.
c. 1 in 60.
d. 1 in 20.
e. None of the above
*Answer: b*

19. Individuals homozygous for the sickle-cell trait produce abnormal
a. keratin.
b. hemoglobin.
c. myosin.
d. tyrosinase.
e. None of the above
*Answer: b*

*20. The gene associated with Duchenne muscular dystrophy
a. codes for a protein called dystrophin.
b. is recessive.
c. encodes components of plasma membranes of skeletal muscle cells.
d. is X-linked.
e. All of the above
*Answer: e*

*21. Familial hypercholesterolemia is caused by an allele that is classified as
a. a deletion.
b. a hot spot.
c. dominant.
d. recessive.
e. X-linked.
*Answer: c*

■22. The reason most serious genetic diseases are rare is that
 a. each individual is unlikely to carry a genetic disease.
 b. a person is unlikely to mate with a carrier of the same mutant allele.
 c. genetic diseases are usually dominant.
 d. genetic diseases are usually not serious.
 e. None of the above
 *Answer: b*

23. Familial hypercholesterolemia is a genetic disease
 a. that causes elevated cholesterol levels in the blood.
 b. in which a liver-cell membrane receptor is defective.
 c. that leads to a higher likelihood of strokes.
 d. All of the above
 e. None of the above
 *Answer: d*

24. Transmissible spongiform encephalopathies are the causative agents for
 a. scrapie.
 b. "mad cow disease."
 c. kuru.
 d. Both a and c
 e. All of the above
 *Answer: e*

25. The infectious prion is a
 a. virus.
 b. viron.
 c. yeast.
 d. bacteria.
 e. protein.
 *Answer: e*

26. Estimates are that _____ percent of all people have diseases that are genetically influenced.
 a. 0.06
 b. 1
 c. 10
 d. 6
 e. 60
 *Answer: e*

*27. Normal brain cells have the membrane protein _____, whereas abnormal TSE-affected brains have _____.
 a. $PrP^C$; $PrP^{SC}$
 b. $PrP^{SC}$; $PrP^C$
 c. $PrP^r$; $PcP^C$
 d. $PcP^C$; $PrP^r$
 e. $PscP^{SC}$; $PcP^C$
 *Answer: a*

28. Genomic imprinting causes
 a. male and female genetic contributions to be equal.
 b. different expression for DNA contributed from males as compared to DNA from females.
 c. methylation of DNA.
 d. cells to differentiate.
 e. Both c and d
 *Answer: b*

**29. The advantage of genetic screening for Tay-Sachs disease is that it
 a. makes treatment possible.
 b. makes possible the reduction of negative effects through dietary control.

 c. helps prevent the disease by informing heterozygotic mates of their condition and the possible dangers to a child.
 d. makes early treatment with gene therapy possible.
 e. All of the above
 *Answer: c*

30. "Hot spots" are locations where
 a. there are cytosine residues.
 b. there are 5-methylcytosine residues.
 c. there are adenine residues.
 d. there is guanidine.
 e. all bases are equally prone to mutation.
 *Answer: b*

■31. Scientists who study genetic disease are involved in
 a. characterizing symptoms of the disease.
 b. developing epidemiological data.
 c. defining the pattern of inheritance.
 d. moving from the Mendelian to the molecular level of analysis.
 e. All of the above
 *Answer: e*

■32. Differences in RFLP banding patterns indicate that
 a. the two different DNAs being tested possess different base pairs.
 b. mRNA is not transcribed.
 c. the genes map to different chromosomes.
 d. Both a and c
 e. None of the above
 *Answer: a*

33. A RFLP
 a. is a restriction fragment length polymorphism.
 b. is inherited in a Mendelian fashion.
 c. can be used as a genetic marker.
 d. can be useful to help define a discrete gene.
 e. All of the above
 *Answer: e*

34. When a person with defective adenosine deaminase (ADA) receives gene therapy, which of the following steps is performed?
 a. Leukocytes are obtained from the person undergoing treatment.
 b. Leukocytes from the person are transferred to culture dishes.
 c. Leukocytes from the person are transfected with normal ADA genes.
 d. Transfected cells are returned to the patient.
 e. All of the above
 *Answer: e*

35. Human gene therapy requires
 a. gene isolation.
 b. introduction of DNA into target cells.
 c. inclusion of a promoter sequence.
 d. Both a and b
 e. a, b, and c
 *Answer: e*

■36. Which of the following is an ethical issue that arises from the genetic screening of fetuses?
 a. The recommendation of abortion for cases in which a defective allele has been detected
 b. The question of privacy: Who has rightful access to the results of screening?

c. Whether humans should be "playing God" with our genetic makeup
d. How we determine which genetic conditions warrant screening and which diseases "merit" gene therapy
e. All of the above
*Answer: e*

37. An example of a genetic disease that causes a defect not in an enzyme, but in a structural protein, is
a. sickle-cell disease.
b. hemophilia.
c. Duchenne muscular dystrophy.
d. PKU.
e. Both a and b
*Answer: c*

38. Cancer is caused by
a. mutagens.
b. carcinogens.
c. viruses.
d. substances found in natural foods.
e. All of the above
*Answer: e*

■39. How do all forms of cancer differ from other diseases?
a. Control of cell division is lost in cancerous cells. They divide rapidly and continuously.
b. Cancers arise in several different tissues.
c. Cancer cells metastasize.
d. Both a and c
e. a, b, and c
*Answer: e*

40. What is the medical term for noncancerous tumors that remain in place?
a. Malignant
b. Benign
c. Harmless
d. Innocuous
e. None of the above
*Answer: b*

41. The spread of tumor cells throughout the body is termed
a. cell focusing.
b. metastasis.
c. benign neglect.
d. orientation.
e. None of the above
*Answer: b*

42. Metastasis proceeds via the cancer cells'
a. extension into surrounding tissues.
b. differentiation into normal cells.
c. entrance into the bloodstream or the lymphatic system.
d. Both a and c
e. a, b, and c
*Answer: d*

43. Carcinomas are defined as cancers that arise in
a. muscle.
b. bone.
c. the liver.
d. the lung, breast, colon, or liver.
e. None of the above
*Answer: d*

44. Sarcomas are cancers of
a. the brain.
b. the skin.
c. bone, blood vessels, or muscle tissue.
d. the lung, breast, colon, or liver.
e. None of the above
*Answer: c*

45. Lymphomas and leukemias affect the cells
a. that give rise to the white and red blood cells.
b. of the brain.
c. of blood vessels.
d. of connective tissue.
e. of muscle and connective tissue.
*Answer: a*

46. The first cancer-causing virus to be identified was
a. Epstein-Barr virus.
b. T cell leukemia virus.
c. Rous sarcoma virus in chickens.
d. porcine sarc virus.
e. benzene 1 virus.
*Answer: c*

47. Which of the following is thought to be involved in the development of cancer cells?
a. Viruses
b. Chemicals
c. Ultraviolet radiation and X rays
d. Excessive exposure to sunlight
e. All of the above
*Answer: e*

48. The development of cancer cells results from changes in genes required for
a. collagen synthesis.
b. normal growth.
c. the production of keratin sulfate.
d. the production of myosin and actin.
e. None of the above
*Answer: b*

49. Eighty-five percent of all human cancers fall into which of the following categories?
a. Carcinomas
b. Lymphomas
c. Leukemias
d. Sarcomas
e. Both b and c
*Answer: a*

50. Agents that can cause mutations in the DNA of host cells include
a. chemical carcinogens.
b. radiation.
c. tumor-induced viruses.
d. enzyme kinetics.
e. a, b, and c
*Answer: e*

51. Cell division is regulated in part by a group of proteins called _____, which circulate in the blood and trigger the normal division of cells.
a. follicle-stimulating hormones
b. erythropoietins
c. anabolic steroids
d. growth factors
e. None of the above
*Answer: d*

52. A mutation that causes an amino acid change in one of the hemoglobin subunits
    a. always causes disease in an individual who is homozygous for the mutant allele.
    b. always causes disease, even in an individual who is heterozygous.
    c. sometimes causes disease in an individual who is homozygous.
    d. is always a dominant mutation.
    e. changes the charge of the protein.
    *Answer: c*

53. The basic concept of the "two-hit" hypothesis is that for a person to develop cancer,
    a. both copies of a tumor suppressor gene must mutate.
    b. one tumor suppressor and one oncogene must mutate.
    c. both copies of an oncogene must mutate.
    d. All of the above
    e. None of the above
    *Answer: a*

54. The tumor suppressor gene *p53* codes for a product that
    a. kills cancerous cells.
    b. causes apoptosis.
    c. stops cell division during G1.
    d. triggers an immune response.
    e. All of the above
    *Answer: c*

55. Today, people with hemophilia A are treated with
    a. screened human-derived blood products.
    b. porcine blood products.
    c. a yeast clotting factor.
    d. a product generated from recombinant DNA technology.
    e. None of the above
    *Answer: d*

56. The entire human genome
    a. was completely sequenced in 2003.
    b. was completely sequenced in 1986.
    c. will require another two decades to be completely sequenced.
    d. project is 10 percent complete.
    e. cannot ever be completely sequenced.
    *Answer: a*

*57. The project to sequence the entire 180 million base pairs of the fruit fly involved the _____ method.
    a. hierarchical sequencing
    b. Maxwell–Gilbert
    c. random-plasmid-select
    d. shotgun sequencing
    e. All of the above
    *Answer: d*

58. A disease caused by an abnormal autosomal dominant allele is
    a. PKU.
    b. sickle-cell disease.
    c. cystic fibrosis.
    d. familial hypercholesterolemia.
    e. fragile X syndrome.
    *Answer: d*

59. Which of the following is an example of an inherited cancer?
    a. Kaposi's sarcoma
    b. T cell leukemia
    c. Anogenital cancer
    d. Lymphoma
    e. Some breast cancers
    *Answer: e*

60. Gleevec, which is used to treat chronic myelogenous leukemia, is considered a revolutionary drug because it
    a. is a competitive inhibitor that inactivates abnormal kinase.
    b. inhibits cell division in all cells.
    c. has broad applications to other forms of leukemia.
    d. increases phosphorylation.
    e. prevents mitotic spindles from forming.
    *Answer: a*

61. Prions
    a. contain DNA.
    b. contain RNA.
    c. affect protein conformation.
    d. are mutated viruses.
    e. are mutated genes.
    *Answer: c*

62. The most common pattern of inheritance of a sex-linked recessive condition is from
    a. carrier mother to son.
    b. father to son.
    c. carrier mother to daughter.
    d. father to daughter.
    e. carrier mother and carrier father to child.
    *Answer: a*

63. Sequencing of the human genome revealed that fewer than 2 percent of the 3.2 billion base pairs are coding regions. This finding indicates that
    a. there is great variation in gene size.
    b. genes are not evenly distributed over the genome.
    c. the functions of many genes are unclear.
    d. the diversity of proteins develops posttranscriptionally.
    e. almost all of the genome is the same in all people.
    *Answer: d*

64. Humans can make more proteins than their number of genes might suggest because of
    a. variations in posttranscriptional and posttranslational regulation.
    b. genetic recombinations during meiosis.
    c. mutations.
    d. environmental factors.
    e. a wide variety of activating enzymes.
    *Answer: a*

65. All of the following have been applications of the human genome sequencing project *except* the
    a. identification of disease-related genes.
    b. development of medications matched to the genetic and functional individuality of patients.
    c. search for polymorphisms in specific human populations.
    d. development of a cancer vaccine.
    e. discovery of an increased number of genetic markers.
    *Answer: d*

# 18 Natural Defenses against Disease

## Fill in the Blank

1. The final large lymph duct connects to a major _____ near the _____.
   *Answer: vein; heart*

2. The _____ immune response is a specific defense system against pathogens that is carried out by antibodies in the blood.
   *Answer: humoral*

3. The blood fluid that carries leukocytes but not red blood cells is _____.
   *Answer: lymph*

4. _____ are specific molecules on the surface of T cells that react to antigenic determinants.
   *Answer: T cell receptors*

5. The fusion of B cells and tumor cells in culture produces _____, which make monoclonal antibodies.
   *Answer: hybridomas*

6. _____ are soluble signal proteins released by T cells.
   *Answer: Cytokines*

7. When the immune recognition of self fails, a(n) _____ disease results.
   *Answer: autoimmune*

8. HIV, the retrovirus that causes AIDS, uses the enzyme _____ to make a DNA copy of the viral genome.
   *Answer: reverse transcriptase*

9. The _____ regions of the heavy chains of the antibody molecule determine whether the antibody remains part of the plasma membrane of the cell or is secreted into the bloodstream.
   *Answer: constant*

10. Of the phagocytes neutrophils and macrophages, _____ live longer.
    *Answer: macrophages*

11. The concept that antigenic determinants stimulate clones of B cells that were already making specific antibodies against those antigens is called the _____ theory.
    *Answer: clonal selection*

12. The ability of the human body to remember a specific antigen explains why _____ has eliminated diseases such as smallpox, diphtheria, and polio.
    *Answer: immunization*

13. Highly specified protein molecules called _____ carry out the humoral immune response against invaders in the fluids.
    *Answer: immunoglobulins*

14. T cells are educated, and those that have receptors for "self" are eliminated in the _____.
    *Answer: thymus*

15. Two broad groups of cells are the important cells of the immune system: the _____, which sometime differentiate to produce antibodies, and the _____, some of which are involved with elimination of virus-infected cells.
    *Answer: B cells; T cells*

16. Tears, nasal drips, and saliva possess an enzyme called _____ that degrades the cell walls of many bacteria.
    *Answer: lysozyme*

17. Types of defenses that provide general protection against a wide variety of pathogens are classified as _____ defenses.
    *Answer: nonspecific*

18. A cell signaling pathway stimulates defense processes, such as the production of complement proteins, cytokines, and interferons. The receptor in this pathway is _____.
    *Answer: toll*

## Multiple Choice

1. Which of the following is *not* one of the first lines of defense against invading pathogens?
   a. Skin
   b. Mucus secretion
   c. Lysozyme in tears
   d. T cell receptors
   e. Low pH of the stomach
   *Answer: d*

2. Blood is a fluid tissue with a noncellular fluid called
   a. lymph.
   b. leukocyte.
   c. plasma.
   d. lymphocyte.
   e. immunoglobulin.
   *Answer: c*

3. Nonspecific responses include all of the following components *except*
   a. macrophages.
   b. antibodies.
   c. eosinophils.
   d. neutrophils.
   e. interferons.
   *Answer: b*

4. Which of the following is *not* an adaptation for preventing a pathogen from penetrating the body surface?
   a. Presence of a normal flora
   b. Sneeze reflex
   c. Mucus-covered body surfaces
   d. Low pH
   e. Immunological tolerance
   *Answer: e*

5. Allergies involve
   a. IgE.
   b. mast cells.
   c. basophils.
   d. histamine release.
   e. All of the above
   *Answer: e*

*6. Which of the following is *not* true of interferons?
   a. They bind to receptors on cell surfaces.
   b. All are glycoproteins.
   c. They prevent viral replication.
   d. They are found only in mammals.
   e. They confer a generalized resistance to viral diseases.
   *Answer: d*

7. _____ is the generalized bodily response to infections and is accompanied by redness, swelling, heat (increased temperature), and pain.
   a. Shock
   b. Inflammation
   c. DNA repair
   d. AIDS
   e. None of the above
   *Answer: b*

8. Which of the following activities is *not* normally involved in preventing pathogens from infecting the mucous membranes of vertebrate animals?
   a. Beating cilia
   b. Lysozyme production
   c. Secretion of HCl
   d. Production of bile salts in the small intestine
   e. Secretion of interferon
   *Answer: e*

9. *E. coli* bacteria live in our large intestines and do not normally cause disease. These microorganisms are called
   a. pathogens.
   b. antibodies.
   c. normal flora.
   d. phagocytes.
   e. the complement system.
   *Answer: c*

*10. Each milliliter of blood normally contains about _____ red blood cells and _____ white blood cells.
   a. 5 billion; 7 million
   b. 5 million; 7 billion
   c. 5 thousand; 7 thousand

   d. 5 million; 7 thousand
   e. 5 million; 7 million
   *Answer: a*

11. Lysozyme is an enzyme that
   a. digests proteins.
   b. causes viral-infected cells to lyse.
   c. attacks bacterial cell walls.
   d. is produced by bacteria.
   e. Both a and b
   *Answer: c*

12. Pathogens that reach the digestive tract are often destroyed by
   a. bile salts.
   b. B cells.
   c. stomach acids.
   d. T cells.
   e. Both a and c
   *Answer: e*

13. Immunoglobulins are composed of _____, which are composed of _____ chain(s).
   a. octomers; two heavy and two light
   b. tetramers; one large and one small
   c. dimers; one heavy and one light
   d. tetramers; two heavy and two light
   e. dimers; one large and one small
   *Answer: d*

14. Phagocytes kill pathogenic bacteria by
   a. endocytosis.
   b. producing antibodies.
   c. complement fixation.
   d. stimulating T cells.
   e. causing inflammation.
   *Answer: a*

15. Which of the following is a nonspecific defense mechanism that protects animals against pathogenic microorganisms?
   a. Sealing off the damaged tissue
   b. Production of phytoalexins
   c. Production of antibodies
   d. Humoral immune response
   e. Inflammation
   *Answer: e*

16. An individual with influenza is unlikely to develop a second viral infection because the infected cells produce a glycoprotein called
   a. phytoalexin.
   b. interferon.
   c. immunoglobulin.
   d. antigen.
   e. IgG.
   *Answer: b*

17. The humoral immune system acts primarily against
   a. intracellular viruses.
   b. circulating bacteria.
   c. tissue transplants.
   d. Both a and b
   e. Both a and c
   *Answer: b*

18. B cells will react
   a. nonspecifically with any foreign matter they encounter.

b. only with tissue transplants.

c. with all of the antigenic determinants on a specific antigen.

d. with only one specific antigenic determinant on an antigen.

e. only with specific antibody molecules.

*Answer: d*

19. When an individual is first exposed to the smallpox virus, there is a delay of several days before significant numbers of specific antibody molecules and T cells are produced. However, a second exposure to the virus causes a large and rapid production of antibodies and T cells. This response is an example of

a. antigenic determinants.

b. phytoalexins.

c. phagocytosis.

d. interferon production.

e. immunological memory.

*Answer: e*

20. When an animal encounters an antigen for the second time, it is capable of producing a massive and rapid immune response to the antigen. The cells responsible for this rapid response are called _____ cells.

a. memory

b. effector

c. humoral

d. immunization

e. antigenic

*Answer: a*

21. According to the clonal selection theory,

a. antibodies are not produced until the animal encounters a specific antigen.

b. antigens determine the three-dimensional structure of antibodies.

c. all B cells have identical genotypes.

d. an antigen stimulates the proliferation of a specific group of B cells.

e. B cells give rise to specific T cells.

*Answer: d*

22. One explanation for the absence of antiself B cells in the bloodstream is

a. the presence of memory cells.

b. immunological memory.

c. clonal deletion.

d. interferon production.

e. their destruction by natural killer cells.

*Answer: c*

23. Immunological tolerance occurs

a. after an exposure to an antigen early in development.

b. when an antigen has no antigenic determinants.

c. during the clonal growth of B cells.

d. as a result of class switching.

e. when interleukins activate T cells.

*Answer: a*

24. Failure to distinguish "self" from "nonself" can result in

a. clonal deletion.

b. the production of suppressor T cells.

c. the development of AIDS.

d. an autoimmune disease.

e. a deficiency in complement proteins.

*Answer: d*

25. When a T cell is activated by an antigen, it

a. secretes antibodies.

b. proliferates.

c. dies.

d. becomes a hybridoma.

e. becomes a plasma cell.

*Answer: b*

26. Which of the following is *not* a characteristic of plasma cells?

a. They arise from B cells.

b. They are effector cells.

c. They secrete antibodies.

d. They survive in the animal for many years.

e. They have large amounts of endoplasmic reticulum.

*Answer: d*

27. Hay fever is an allergic response to pollen. Which type of antibody molecule is being produced?

a. IgG

b. IgM

c. IgD

d. IgA

e. IgE

*Answer: e*

28. Monoclonal antibodies

a. recognize a single antigenic determinant.

b. are produced in the spleen.

c. are produced by animals injected with a single antigen.

d. are memory cells.

e. are a complex mixture of different antibody classes.

*Answer: a*

29. The region of the antibody that binds to the antigen is the

a. constant region of the heavy chain.

b. constant region of the light chain.

c. variable region.

d. Both a and b

e. Both b and c

*Answer: c*

■30. Since the joining and random deletion mechanisms account only for a part of antibody diversity, what else is involved in producing vast numbers of unique immunoglobulins?

a. Mutation

b. Inversion

c. Cell fusion

d. Cell surface proteins

e. All of the above

*Answer: a*

31. The genes for immunoglobulin heavy chains are _____ the genes for the light chains.

a. located on the same chromosomes as

b. spliced together with

c. expressed when the cells are exposed to antigens, as are

d. located on different chromosomes from

e. rearranged during development, as are

*Answer: d*

32. A plasma cell is producing IgM molecules that recognize an antigenic determinant on an influenza virus. After several days, the cell begins to produce IgG molecules that recognize the same antigenic determinant. This process is called
    a. activation.
    b. RNA splicing.
    c. gene mutation.
    d. class switching.
    e. an autoimmune disease.
    *Answer: d*

33. T cell receptors recognize and bind
    a. $T_C$ cells.
    b. B cells.
    c. processed antigens.
    d. $T_H$ cells.
    e. IgM antibodies.
    *Answer: c*

34. The class I MHC proteins are _____ in the animal body.
    a. secreted by B cells
    b. found only on T cells
    c. on the surface of all nucleated cell types
    d. only produced early in development
    e. the same in all individuals of the same strain
    *Answer: c*

35. The retrovirus HIV specifically destroys $T_H$ cells and thus disrupts
    a. the humoral immune response.
    b. the cellular immune response.
    c. both the humoral and cellular immune responses.
    d. the inflammatory response.
    e. the complement cascade.
    *Answer: c*

36. Which of the following cells is *not* normally involved in the functioning of the immune system?
    a. Phagocytes
    b. Red blood cells
    c. Lymphocytes
    d. B cells
    e. T cells
    *Answer: b*

37. Which of the following is *not* associated with a non-specific defense mechanism?
    a. Inflammation
    b. Mucous membranes
    c. Cytokines
    d. Phagocytes
    e. Interferons
    *Answer: c*

38. Which of the following is *not* true of both B cells and T cells?
    a. They are types of lymphocytes.
    b. They are found in the lymph.
    c. They secrete antibodies.
    d. They give rise to both effector cells and memory cells.
    e. They originate in bone marrow.
    *Answer: c*

39. Select the feature that is characteristic of both the humoral and cellular immune responses.
    a. Specificity
    b. Cell–cell communication via interleukin
    c. Component of immunological memory

d. Recognition of diverse antigenic determinants
e. Recognition of self from nonself
*Answer: b*

40. Which of the following cellular immune components is the functional equivalent of an immunoglobulin?
    a. T cell receptor
    b. An antigenic determinant
    c. An antibody
    d. Processed antigen bound to class II MHC proteins
    e. A complement cascade
    *Answer: a*

41. A fundamental postulate of the clonal selection theory is that
    a. the antigen specifies the structure of the antibody directed against it.
    b. an antigen can lead to the selection of only a single line of B cells.
    c. a B cell makes only one specific antibody.
    d. the production of diverse antibodies is not genetically based.
    e. a mechanism must exist to prevent the production of antiself lymphocytes.
    *Answer: c*

42. Which of the following is *not* a feature of studies that attempt to explain the body's recognition of self?
    a. Monoclonal antibodies
    b. The clonal deletion theory
    c. Self-identifying cell surface proteins
    d. Immunological tolerance
    e. Nonidentical twin cattle with blood of mixed types
    *Answer: a*

43. Which of the following is *not* true of an immunoglobulin such as IgG?
    a. It is a protein molecule with a quaternary structure.
    b. It consists of subunits held together by hydrogen bonding.
    c. It is a tetramer with two heavy chains and two light chains.
    d. It has variable and constant regions in each subunit.
    e. It has two antigen-binding sites.
    *Answer: b*

44. Which of the following is *not* characteristic of an immunoglobulin?
    a. The variable regions determine the type of antigen that will bind to it.
    b. The constant regions determine the class of the immunoglobulin.
    c. Disulfide bonds occur within and between the polypeptide chains.
    d. The antigen-binding sites are formed by the variable portions of the light chains only.
    e. The two halves are identical.
    *Answer: d*

45. Which of the following features is characteristic of the immunoglobulin class IgG?
    a. It is found in blood immediately after first exposure to the antigen.
    b. It is involved in inflammation and allergic reactions.
    c. It forms multi-immunoglobulin complexes.
    d. It is produced in the greatest amount after the second exposure to the antigen.
    e. It is found in saliva, tears, milk, and gastric secretions.
    *Answer: d*

46. Which of the following features is *not* characteristic of the complement system?
    a. It consists of 20 different proteins.
    b. It is involved in a cascade of reactions.
    c. It lyses invading cells.
    d. It consists of highly folded membranes that capture pathogens.
    e. It interacts with phagocytes.
    *Answer: d*

47. Which of the following features is *not* characteristic of T cell receptors?
    a. They consist of two polypeptide chains.
    b. They are glycoproteins.
    c. They are able to bind to intact antigen.
    d. They are able to bind MHC proteins.
    e. They have constant and variable regions.
    *Answer: c*

48. Which of the following is *not* a normal activity of helper T cells?
    a. Release of lytic signals when bound to processed antigen on the surface of a virus-infected cell
    b. Binding to class II MHC and processed antigen on the surface of macrophages
    c. Binding to class II MHC and processed antigen on the surface of B cells
    d. Release of chemical signals
    e. Proliferation and differentiation into memory and effector cells
    *Answer: a*

49. Which of the following statements is true about AIDS?
    a. The chances of getting the disease are far greater if one of the partners already has a sexually transmitted disease.
    b. AIDS stands for "autoimmune deficiency syndrome."
    c. The HIV virus of AIDS does not kill people directly.
    d. Both a and c
    e. a, b, and c
    *Answer: d*

50. To ensure survival, pathogenic organisms must
    a. enter a host.
    b. multiply in the host.
    c. prepare to infect the next host.
    d. Both a and b
    e. a, b, and c
    *Answer: e*

51. Which of the following are two general types of responses mounted by the immune system against invaders?
    a. The humoral immune response and the cellular immune response
    b. The humoral immune response and the antihumoral immune response
    c. Complementation and clonal deletion
    d. The cellular immune response and the antihumoral immune response
    e. None of the above
    *Answer: a*

52. Antibodies are produced by
    a. fibroblasts.
    b. mesenchyme cells.
    c. adrenal cortex cells.
    d. cells of the anterior pituitary.
    e. plasma cells.
    *Answer: e*

53. A person can produce _____ of distinct antibodies directed against antigenic determinants, even when these have never been encountered.
    a. dozens
    b. hundreds
    c. thousands
    d. millions
    e. trillions
    *Answer: d*

54. There appear to be two mechanisms of self-tolerance in the immune system,
    a. clonal selection and clonal deletion.
    b. clonal deletion and clonal anergy.
    c. clonal proliferation and suppressor T cell action.
    d. suppressor T cell action and clonal selection.
    e. None of the above
    *Answer: b*

55. In order to synthesize antibodies, a plasma cell must
    a. be activated by the binding of specific antigens to the antibodies carried on the B cell surface.
    b. interact with a helper T cell.
    c. develop an extensive endoplasmic reticulum and Golgi complex.
    d. Both a and b
    e. a, b, and c
    *Answer: e*

56. The immunoglobulin class IgA is found
    a. circulating in the blood.
    b. in mucus secretions.
    c. associated with mast cells near the surface of the skin
    d. shortly after first exposure to an antigen.
    e. All of the above
    *Answer: b*

57. All of the following are necessary for a humoral response *except*
    a. $T_C$ cells.
    b. $T_H$ cells.
    c. B cells.
    d. cytokines.
    e. antigenic determinant.
    *Answer: a*

58. Which of the following is *not* a characteristic of an inflammatory reaction?
    a. Release of histamine
    b. Invasion of the region by phagocytes
    c. Binding of IgG
    d. Dilation of capillaries
    e. Escape of blood plasma
    *Answer: c*

59. To make a hybridoma, a plasma cell is fused to a _____ cell.
    a. carcinoma
    b. B
    c. macrophage
    d. tumor
    e. liver
    *Answer: d*

60. An important organ that helps to protect against auto-immune disease is the
    a. kidney.
    b. spleen.
    c. thymus.
    d. adrenal gland.
    e. tonsils.
    *Answer: c*

61. The core of HIV contains
    a. a protease.
    b. reverse transcriptase.
    c. integrase.
    d. two identical RNA molecules.
    e. All of the above
    *Answer: e*

*62. Viral membrane proteins of HIV are synthesized
    a. in the core particle.
    b. in the infected cell's nucleus.
    c. on the surface of the ER.
    d. in the cytoplasm.
    e. in the Golgi apparatus.
    *Answer: c*

63. Cells that get infected by HIV generally have
    a. CD4 membrane proteins.
    b. gp120.
    c. gp41.
    d. antigen-bound T cell receptors.
    e. All of the above
    *Answer: a*

*64. HAART is
    a. a hypersensitive AIDS patient.
    b. highly active antiretroviral therapy.
    c. HIV and AIDS advanced retroviral treatment.
    d. highly advanced AIDS RNA treatment.
    e. hereditary AIDS acquired and retransmitted.
    *Answer: b*

65. An immune response that is characterized by a short lag time and high production of antibodies is the _____ response.
    a. primary immune
    b. secondary immune
    c. attenuation
    d. effector cell
    e. polyclonal
    *Answer: b*

66. A medical "smart bomb" consists of the practical application of monoclonal antibodies, also known as
    a. immunoassays.
    b. passive immunization.
    c. immunotherapy.
    d. DNA vaccination.
    e. attenuation.
    *Answer: c*

67. The primary immune response is characterized by
    a. short lag time.
    b. a high rate of antibody production.
    c. a large production of T cells.
    d. a large production of antibodies.
    e. the addition of memory cells to the immune system.
    *Answer: e*

68. Antigen-presenting cells include
    a. macrophages.
    b. dendritic cells.
    c. B cells.
    d. Both a and b
    e. a, b and c
    *Answer: e*

69. T cells
    a. have immunoglobulin receptors.
    b. have antibodies as effector molecules.
    c. develop from activated B cells.
    d. are secreted by plasma cells.
    e. release perforin.
    *Answer: e*

*70. Binding of a molecule from a pathogen to a receptor initiates a signal transduction pathway. Which is the correct order of events in this pathway?
    a. Phosphorylation of NF-κB, binding of NF-κB to the promoter, activation of toll receptor, transcription of defense genes
    b. Binding of NF-κB to the promoter, phosphorylation of NF-κB, activation of toll receptor, transcription of defense genes
    c. Activation of toll receptor, phosphorylation of NF-κB, binding of NF-κB to the promoter, transcription of defense genes
    d. Activation of toll receptor, Binding of NF-κB to the promoter, phosphorylation of NF-κB, transcription of defense genes
    e. Binding of NF-κB to the promoter, activation of toll receptor, transcription of defense genes, phosphorylation of NF-κB
    *Answer: c*

71. All of the following are disorders of the immune system *except*
    a. AIDS.
    b. multiple sclerosis.
    c. insulin-dependent diabetes mellitus.
    d. HMC transplant rejection.
    e. rheumatoid arthritis.
    *Answer: d*

72. The five classes of antibodies differ from each other in terms of
    a. the composition of their light and heavy chains.
    b. their function.
    c. the number of antigen binding sites.
    d. whether they are immunoglobulins or proteins.
    e. the site of B cell differentiation.
    *Answer: b*

73. A person suffering from AIDS would likely have which one of the following diseases?
    a. Rheumatoid arthritis.
    b. Lupus erythematosis.
    c. Multiple sclerosis.
    d. Kaposi's sarcoma.
    e. Insulin-dependent diabetes mellitus.
    *Answer: d*

# 19 Differential Gene Expression in Development

## Fill in the Blank

1. Although we often stress the embryo in discussing animal development, development is a process that continues throughout all stages of life, ceasing only with _____.
   *Answer: death*

2. _____ occurs through cell division and cell expansion.
   *Answer: Growth*

3. A _____ cell has the ability to give rise to every type of cell in the adult body.
   *Answer: totipotent*

4. An important sequence of 180 base pairs of DNA called the _____ is involved in mutant homeotic mutations. The sequence, which has been found in both animals and plants, encodes a portion of some proteins called a homeodomain.
   *Answer: homeobox*

5. Bicoid protein (a morphogen) affects _____.
   *Answer: transcription*

6. The series of events that is caused by the expression of certain genes and that programs cell death is called
   _____.
   *Answer: apoptosis*

7. Cells of adult tissues that need frequent cell replacement, such as the skin and the blood system, are called _____.
   *Answer: stem cells*

8. A sea urchin egg is said to have _____ because the distribution of cytoplasm components at one end is different from that at the other end.
   *Answer: polarity*

9. The number and polarity of segments formed during the development of insect larvae is determined by three classes of _____ genes.
   *Answer: segmentation*

10. The _____ is a small sequence of DNA that is found in some of the segmentation genes of *Drosophila* as well as in some genes of all animals with segmented body plans.
    *Answer: homeobox*

11. One factor that regulates pattern formation is _____, signals that indicate where one group of cells lies in relation to other cells.
    *Answer: positional information*

12. The process by which cells become functionally distinct is called cellular _____.
    *Answer: differentiation*

13. _____ produce cytoplasmic determinants and exert their effects on the embryo regardless of the genotype of the father.
    *Answer: Maternal effect genes*

14. Substances that are produced in one place, diffuse to another, and cause pattern formation are called _____.
    *Answer: morphogens*

15. The _____ cell controls the fate of six other cells involved in the formation of the vulva in *Caenorhabditis elegans*.
    *Answer: anchor*

## Multiple Choice

1. Three major processes that reveal a great deal about animal development include
   a. determination, differentiation, and pattern formation.
   b. blastulation, gastrulation, and physiology.
   c. immunoregulation, neurulation, and histogenesis.
   d. oncogenesis, differentiation, and histogenesis.
   e. None of the above
   *Answer: a*

2. In its earliest stage of development, an animal or plant is called a(n)
   a. fetus.
   b. embryo.
   c. germ cell.
   d gamete.
   e. All of the above
   *Answer: b*

3. Development occurs
   a. only during growth of an organism.
   b. throughout the life of an organism.
   c. only in nondividing cells.
   d. in ectoderm and endoderm but not in mesoderm.
   e. only in animals.
   *Answer: b*

4. The ability of scientists to clone an entire carrot plant from a differentiated root cell indicates that the cell
   a. contains the entire carrot genome and can express the appropriate genes at the right time.
   b. contains an incomplete carrot genome but can still express the appropriate genes at the right time.
   c. is not totipotent.
   d. is a zygote.
   e. is unable to undergo any further development.
   *Answer: a*

5. Molecules found in the egg cytoplasm that play a role in directing embryonic development are called
   a. instars.
   b. imaginal discs.
   c. mesenchyme proteins.
   d. polar proteins.
   e. cytoplasmic determinants.
   *Answer: e*

6. During initial development, plant cells tend to _____ and animal cells tend to _____.
   a. grow; grow
   b. divide; expand
   c. expand; divide
   d. differentiate; determine
   e. determine; differentiate
   *Answer: c*

7. The human body has approximately _____ functionally distinct kinds of cells.
   a. 12
   b. 24
   c. 100
   d. 200
   e. 1,000
   *Answer: d*

■8. The use of nuclear transplantation allowed developmental biologists to address the fundamental question of
   a. why a frog embryo develops into a frog and not into some other organism.
   b. whether or not cell differentiation is due to loss of DNA from cells as they divide.
   c. the effect of mitochondria on growth rates.
   d. Both a and c
   e. a, b, and c
   *Answer: b*

9. The process by which cells organize to create the form of a multicellular organism is called
   a. morphogenesis.
   b. differentiation.
   c. determination.
   d. restriction.
   e. metamorphosis.
   *Answer: a*

10. Which of the following is *not* true of animal development?
    a. Genes regulate development.
    b. Development occurs by progressive loss of DNA.
    c. Blastomeres are early embryonic cells.
    d. The sea urchin blastopore forms the archenteron.
    e. Cells actively migrate during development.
    *Answer: b*

11. Cells must become _____ before they _____.
    a. aged; divide
    b. large; divide
    c. elongated; divide
    d. determined; differentiate
    e. developed; shape
    *Answer: d*

*12. Cells of different types
    a. express different genes.
    b. express the same genes.
    c. express some of the same genes.
    d. have different DNA sequences.
    e. None of the above
    *Answer: d*

13. Differentiation is caused by
    a. loss of DNA.
    b. determination.
    c. morphogenesis.
    d. differential gene expression.
    e. nuclear transplantation.
    *Answer: d*

14. Experiments by Briggs and King and by Gurdon provided graphic evidence that
    a. neurulation is fixed.
    b. differentiation is not irreversible.
    c. amphibian embryos can cease development for long periods of time.
    d. Both a and c
    e. a, b, and c
    *Answer: b*

15. Pattern formation is necessary for
    a. morphogenesis.
    b. differentiation.
    c. determination.
    d. restriction.
    e. metamorphosis.
    *Answer: a*

16. As cells become specialized, they must first
    a. be differentiated.
    b. be determined.
    c. lose broad developmental potential.
    d. gain a fate.
    e. become functionally distinct.
    *Answer: b*

17. Once cells become fixed in a final functional and physical state, they are
    a. determined.
    b. committed.
    c. differentiated.
    d. totipotent.
    e. morphogized.
    *Answer: c*

18. The fertilized egg is
    a. fully competent.
    b. totipotent.
    c. pluripotent.
    d. determined.
    e. haploid.
    *Answer: b*

19. The transgenic sheep Dolly produces _____ in her milk.
    a. human growth hormone
    b. IGF
    c. insulin
    d. interferon
    e. α-1-antitrypsin
    *Answer: e*

20. Transplantation has been accomplished
    a. with nuclei from cells of an adult ewe's udder.
    b. with very small amounts of cytoplasm as the graft.
    c. in mammals.
    d. in frogs.
    e. All of the above
    *Answer: e*

21. The Dolly experiment addressed the fundamental question of
    a. whether nuclei of mammals differentiate irreversibly.
    b. the effect of cytozymes on cellular differentiation.
    c. the effect of mitochondria on growth rates.
    d. Both b and c
    e. a, b, and c
    *Answer: a*

22. The current theory of why the nucleus of an udder cell "dedifferentiated" to generate Dolly is that the
    a. cells were cultured.
    b. cells were starved and stalled at G1.
    c. cells were starved and stalled at G2.
    d. cells were fed and dividing.
    e. egg failed to have its genetic material removed properly.
    *Answer: b*

*23. Therapeutic cloning is a proposed procedure that may combine nuclear transplantation and stem cell technologies. Which of the following steps is *not* required in therapeutic cloning?
    a. Removal of eggs from a female donor
    b. Enucleation of an egg
    c. Fertilization of the egg with a donor sperm
    d. Fusion of a donor cell with an enucleated egg
    e. Stimulation of the egg to cause cell division
    *Answer: c*

24. Embryonic induction
    a. cannot explain the formation of the vertebrate eye.
    b. was first described by Briggs and King.
    c. initiates a sequence of differential gene expression.
    d. does not occur in the adult.
    e. is an example of a tissue's inducing itself.
    *Answer: c*

*25. Maternal effect genes
    a. affect embryos regardless of the genotype of the father.
    b. establish the dorsal–ventral axis of the embryo.
    c. establish the anterior–posterior axis of the embryo.
    d. lead to the production of specific cytoplasmic determinants.
    e. All of the above
    *Answer: e*

*26. The DNA binding domain for *MyoD1* is
    a. helix-loop-helix.
    b. zinc finger.
    c. leucine zipper.
    d. helix-turn-helix.
    e. hydrogen bonding.
    *Answer: a*

27. When the protein from the *MyoD1* gene is injected into a fat cell, the cell
    a. becomes a muscle cell.
    b. becomes a muscle cell only until the protein breaks down.
    c. becomes a hybrid fat–muscle cell called a factual cell.
    d. remains a fat cell because differentiation has already occurred.
    e. becomes a selector cell.
    *Answer: a*

*28. Cutting sea urchin embryos into two parts (upper and lower halves) at the eight-cell stage results in
    a. two normal larvae.
    b. one abnormal larva.
    c. two dwarfed larvae.
    d. no larvae.
    e. two abnormal larvae.
    *Answer: e*

*29. Female *Drosophila* homozygous for the *bicoid* maternal effect gene mutation produce larvae with no head or thorax. When cytoplasm from the anterior end of wild-type eggs is inoculated into the anterior end of *bicoid* mutant eggs, the treated eggs develop normally. The normal development in the mutant eggs is cause by
    a. disruption of mutant *bicoid* gene expression after their inoculation with wild-type egg cytoplasm.
    b. normal bicoid gene expression.
    c. expression of the *nanos* gene.
    d. induction of normal development by adjacent embryo cells.
    e. the inoculation of normal egg cytoplasm.
    *Answer: e*

30. Which of the following statements about eye development is *false*?
    a. The optic vesicle forms before the lens placode does.
    b. The lens placode is lateral to the optic vesicle.
    c. The lens placode undergoes invagination.
    d. The optic cup connects with the optic nerve.
    e. The eye develops because of cytoplasmic determinants.
    *Answer: e*

**31. If a permeable barrier allowing free movement of molecules is placed between optic vesicles and the surface cells in the future eye region,
    a. no lenses will form.
    b. lenses will form.
    c. lens placodes, but no lenses, will form.
    d. no lens placodes will form, but lenses will.
    e. no lens placodes or lenses will form.
    *Answer: b*

32. Induction of surface tissue to form a lens placode requires
    a. intercellular biochemical communication.
    b. an underlying optic vesicle.
    c. a signal from the optic vesicle.
    d. Both a and b
    e. a, b, and c
    *Answer: e*

33. The developmental fate of cells in the chick wing bud is determined, at least in part, by the proximity of cells from the posterior base of the bud. This is the result of
    a. homeotic mutation.
    b. positional information.
    c. cytoplasmic determinants.
    d. totipotency.
    e. irreversible differentiation.
    *Answer: b*

34. *Caenorhabditis elegans*
    a. has a fixed number of cells in the adult form.
    b. has no mutations to help in its analysis.
    c. lacks a zygote.
    d. is a nematode roundworm of significant research value.
    e. Both a and d
    *Answer: e*

35. The anchor cell influences the differentiation and morphogenesis of several surrounding cells by means of
    a. cytoplasmic determinants.
    b. a random events generator.
    c. induction.
    d. P granules.
    e. None of the above
    *Answer: c*

36. Which of the following statements about *Caenorhabditis elegans* is *false*?
    a. It is a roundworm.
    b. Genetic analysis of it has been productive.
    c. It has a fixed number of cells in the nervous system.
    d. It can reproduce asexually.
    e. Its life cycle lasts less than one week.
    *Answer: d*

37. If the anchor cell is destroyed in *Caenorhabditis elegans*,
    a. no vulva will form.
    b. no anchorettes will form.
    c. secondary vulval precursors will become primary vulval cells.
    d. another cell will differentiate into an anchor cell.
    e. signals from the roundworm's surface will direct readjustments.
    *Answer: a*

■38. The roundworm *Caenorhabditis elegans* has been used for the detailed analysis of animal development. Which of the following is *not* a characteristic that makes this organism useful for such studies?
    a. The adult contains fewer than 1,000 cells.
    b. Development from the zygote to the adult takes only a few days.
    c. The body is relatively transparent.
    d. Symmetry in the adult body is the result of symmetrical cell divisions.
    e. Mutations in genes that control development have been identified.
    *Answer: d*

39. Large-cell lymphoma is caused by a mutation in _____. This gene is analogous to *ced-9* in *Caenorhabditis elegans*.
    a. *ced-3*
    b. *ced-4*

    c. *bcl-2*
    d. *lin-s*
    e. *MyoD1*
    *Answer: c*

40. The genes that regulate the differentiation of whorls in *Arabidopsis thaliana* are called
    a. organ identity genes.
    b. dimer genes.
    c. whorl locator genes.
    d. sepals, petals, stamens, and carpels.
    e. central axis control genes.
    *Answer: a*

*■41. In a type 3 whorl, genes of classes B and C are expressed. As a result of this expression, it would be expected that _____ would be found in those whorl cells.
    a. AA dimers
    b. AB dimers
    c. BB dimers
    d. All of the above
    e. None of the above
    *Answer: d*

42. If a second pair of wings develops in *Drosophila*, the insect probably has a _____ mutation.
    a. chronogene
    b. gap gene
    c. homeotic
    d. segmentation gene
    e. cytoplasm
    *Answer: c*

43. The _____ involves a cluster of genes that controls the development of the abdomen and posterior thorax of *Drosophila* sp.
    a. *bithorax* mutation
    b. segment polarity
    c. *Antennapedia* complex
    d. pair rule
    e. imaginal disc
    *Answer: a*

44. Body segmentation in *Drosophila* sp. is controlled by which sequence of gene action?
    a. Segment polarity, pair rule, gap
    b. Gap rule, pair true, segment polarity
    c. Gap, pair rule, segment polarity
    d. *Bithorax*, pair rule, segment polarity
    e. None of the above
    *Answer: c*

45. The homeobox gene complex
    a. codes for a protein that is part of many different transcription factors.
    b. is found in segmented organisms.
    c. plays a role in development similar to that of the MADS box genes in plants.
    d. encodes a 60-amino acid sequence called the homeodomain.
    e. All of the above
    *Answer: e*

46. Homeotic mutations
    a. do not create developmental abnormalities.
    b. affect the number of body segments.

c. alter eye color in *Drosophila* sp.

d. are easily studied in the adult.

e. produce changes in segment identity.

*Answer: e*

47. The _____ genes determine the number and polarity of the segments in an insect larva.

a. homeobox

b. determinant

c. segmentation

d. positional

e. involution

*Answer: c*

48. Homeobox DNA is found

a. in all organisms except humans.

b. only in tomatoes and sea urchins.

c. only in organisms with segmented body plans.

d. in both plants and animals.

e. only in *Drosophila*.

*Answer: d*

49. The proteins made from genes containing a homeobox

a. bind to DNA.

b. are found only in the cytoplasm.

c. regulate the transcription of other genes.

d. Both a and c

e. Both b and c

*Answer: d*

50. The mutant *Drosophila* called *Antennapedia*

a. has legs growing in place of antennae.

b. grows wings in place of eyes.

c. is a homeotic mutant.

d. Both a and c

e. Both b and c

*Answer: d*

*51. A syncytium in *Drosophila* is expected

a. early during development.

b. near the head region.

c. in larvae.

d. in the unfertilized egg.

e. during the time when segmentation genes are expressed.

*Answer: a*

52. In a general sense, mutations in developmental biology

a. can be explained by changes in DNA.

b. can alter body segmentation.

c. help explain congenital malformations in humans.

d. can be studied experimentally with the techniques of molecular biology.

e. All of the above

*Answer: e*

53. Programmed cell death is crucial to normal development. The scientific term for this kind of cell death is

a. apoptosis.

b. program X.

c. terminal differentiation.

d. death by default.

e. sonic hedgehog.

*Answer: a*

54. The genes *ced-3*, *ced-4*, and *ced-9* are all involved with regulating

a. muscle cell differentiation.

b. positional information.

c. morphogenesis.

d. egg polarization.

e. apoptosis.

*Answer: e*

55. If the gene *ced-9* becomes nonfunctional, the result is

a. webbed feet and hands.

b. cell death.

c. a fetus with no muscle cells.

d. loss of symmetry.

e. loss of nerve function.

*Answer: b*

# 20 Animal Development: From Genes to Organism

## Fill in the Blank

1. The time from conception to birth, or the period of pregnancy, is called _____.
   *Answer: gestation*

★2. The primitive groove of bird and mammalian embryos is analogous to the _____ of amphibians and sea urchins.
   *Answer: blastopore*

3. _____ are central to the process of anterior–posterior determination and differentiation.
   *Answer: Homeobox genes*

4. The proteinaceous shell that surrounds the eggs of many mammalian species is called the _____.
   *Answer: zona pellucida*

5. Presumed mechanisms of determination include determination by cytoplasm segregation and determination by _____, in which certain tissues cause other tissues to develop in a certain manner.
   *Answer: induction*

6. When the developmental fate of a cell does not change, even when the cell's surroundings are altered, the cell is said to be _____.
   *Answer: determined*

★7. The region of the frog egg that is darkly pigmented is known as the _____ pole.
   *Answer: animal*

8. _____ is a birth defect caused by the failure of the closure of the posterior region of the neural tube.
   *Answer: Spina bifida*

9. The process in development by which germ layers form and take specific positions relative to each other is

   _____.
   *Answer: gastrulation*

10. The rod that forms from the chordomesoderm and gives structural support to the developing embryo is called the

    _____.
    *Answer: notochord*

★11. In chicks, the first extraembryonic membrane to form is the _____.
    *Answer: yolk sac*

★12. In mammals, the first extraembryonic membrane to form is the _____.
    *Answer: trophoblast*

## Multiple Choice

■1. Development occurs
   a. only during growth of the organism.
   b. throughout the life of the organism.
   c. only in nondividing cells.
   d. in ectoderm and endoderm, but not in mesoderm.
   e. only in animals.
   *Answer: b*

2. The large amount of yolk in birds' eggs results in
   a. gradual metamorphosis.
   b. complete metamorphosis.
   c. incomplete cleavage.
   d. incomplete mitosis during cleavage.
   e. bicoid larvae.
   *Answer: c*

3. If one of the blastomeres is removed from a developing mouse embryo, the remaining cells will go on to develop into a normal mouse. This is an example of
   a. regulative development.
   b. the zone of polarizing activity.
   c. cleavage.
   d. gastrulation.
   e. ingression.
   *Answer: a*

4. During cleavage, the number of cells in a developing frog embryo increases. The cytoplasm in these new cells
   a. comes from the egg cytoplasm.
   b. is synthesized by the blastomeres.
   c. does not contain any yolk.
   d. is the vegetal pole.
   e. undergoes mitosis.
   *Answer: a*

5. The mesoderm
   a. is located on the outside of the embryo.
   b. lies between the endoderm and the ectoderm.
   c. is found in blastula-stage embryos.
   d. gives rise to the linings of the gut.
   e. is formed during cleavage.
   *Answer: b*

6. The formation of the endoderm during gastrulation in frogs results from
   a. movement of cells from the surface layer to the interior.
   b. migration of cells within the blastocoel.
   c. formation of columnar cells at the vegetal pole.
   d. rapid cell division.
   e. migration of secondary mesenchyme cells.
   *Answer: a*

7. During the second trimester of pregnancy in humans,
   a. the blastocyst implants in the uterine lining.
   b. the mother goes through noticeable hormonal responses.
   c. the fetal digestive system starts to function.
   d. the fetal brain undergoes cycles of sleep and waking.
   e. limbs of the fetus elongate, and fingers and toes become well formed.
   *Answer: e*

8. In bird eggs, cells migrate into the interior of the embryo through the primitive streak. In nonmammalian embryos with smaller amounts of yolk, this involution occurs at the
   a. archenteron.
   b. blastopore.
   c. notochord.
   d. mesenchyme.
   e. endoderm.
   *Answer: b*

9. The earliest stage of development is called the _____ stage.
   a. fetal
   b. embryonic
   c. germinal
   d gametic
   e. Both a and c
   *Answer: b*

**10. If cells from the neural tube of a frog embryo are transplanted onto the ventral surface of a second embryo, the transplanted tissue will still go on to develop into tissues of the nervous system. The transplanted cells are
   a. differentiated.
   b. totipotent.
   c. discontinuous.
   d. determined.
   e. endodermal.
   *Answer: d*

▪11. When the blastopore dorsal lip is grafted from one frog embryo onto a second embryo, the second dorsal lip will
   a. change the polarity of the adjacent segments.
   b. block gastrulation.
   c. change the developmental fate of the surrounding cells.
   d. change the prospective potency of the surrounding cells.
   e. cause rapid cell division.
   *Answer: c*

12. Which of the following statements about neurulation is *false*?
   a. Neural crest cells give rise to peripheral nerves.
   b. Body segmentation develops during neurulation.
   c. Hox genes control differentiation along the anterior–posterior body axis.
   d. Neural crest cells are migratory.
   e. None of the above
   *Answer: e*

13. Invagination
   a. requires unique cell movement.
   b. creates unique positions for new cell interactions.
   c. occurs in sea urchins.
   d. forms the archenteron in echinoderms.
   e. All of the above
   *Answer: e*

14. Gastrulation is the stage of development
   a. when neural tube formation begins.
   b. when sea urchins begin to form primary mesenchyme.
   c. when new embryonic tissue begins to form in the frog embryo.
   d. that precedes cleavage.
   e. Both b and c
   *Answer: e*

15. The gray crescent
   a. is observable in the zygote and the two-celled frog embryo.
   b. is a homeobox gene.
   c. controls cellular affinities.
   d. induces the optic cup to form.
   e. can be mimicked by retinoic acid.
   *Answer: a*

16. An organism with extensive yolk, such as the chick,
   a. has complete cleavage.
   b. has incomplete cleavage.
   c. forms a blastoderm but no blastocoel.
   d. has yolk evenly distributed in the egg.
   e. fails to synthesize DNA during cleavage.
   *Answer: b*

▪17. Because the human embryo is able to split at the 64-cell level of organization to produce two viable progeny, it is said to exhibit _____ development.
   a. mosaic
   b. determinative
   c. definitive
   d. classical vertebrate
   e. regulated
   *Answer: e*

18. Proteins in the egg cytoplasm that play a role in directing embryonic development are called
   a. instars.
   b. imaginal discs.
   c. mesenchyme proteins.
   d. polar proteins.
   e. cytoplasmic determinants.
   *Answer: e*

19. Place the following developmental events in their proper chronological sequence: (1) formation of the neural tube, (2) movement of neural folds, and (3) thickening of neural ectoderm.
   a. 1, 2, and 3
   b. 2, 1, and 3
   c. 3, 1, and 2
   d. 3, 2, and 1
   e. None of the above
   *Answer: d*

20. The gray crescent is the region of the egg
   a. that is opposite the site of sperm penetration.
   b. where gastrulation will begin.
   c. that was pigmented before the cytoplasm rearranged itself.
   d. where the dorsal lip of the blastopore will be.
   e. All of the above
   *Answer: e*

21. The sea urchin differs from insects in early cleavage. The sea urchin undergoes _____ cleavage, whereas insects undergo _____ cleavage.
    a. complete; superficial
    b. complete; incomplete
    c. superficial; incomplete
    d. incomplete; complete
    e. superficial; complete
    *Answer: a*

22. Which of the following is *not* true of animal development?
    a. Genes regulate development.
    b. Development occurs by progressive loss of DNA.
    c. Blastomeres are early embryonic cells.
    d. The sea urchin blastopore forms the archenteron.
    e. Cells actively migrate.
    *Answer: b*

23. The embryos of complex multicellular animals must establish spatial coordinates in order for development to progress. An important reference coordinate for developing frogs is the
    a. location of the vegetal pole.
    b. location of the animal pole.
    c. point of sperm penetration.
    d. location where gastrulation begins.
    e. direction of the sun.
    *Answer: c*

*24. Which of the cells below develop into a human embryo?
    a. Trophoblast
    b. Extraembryonic
    c. Inner cell mass
    d. Cumulus
    e. All of the above
    *Answer: c*

25. The fertilized egg is
    a. fully competent.
    b. totipotent.
    c. pluripotent.
    d. determined.
    e. haploid.
    *Answer: b*

26. The structure in birds and mammals that is most analogous to the dorsal lip of the blastopore found in frogs is
    a. the primitive streak.
    b. the archenteron.
    c. the yolk plug.
    d. Hensen's node.
    e. the notochord.
    *Answer: d*

27. The trophoblast cells in frogs
    a. are important to placenta formation.
    b. protect the egg from sunlight.
    c. are used for gas exchange.
    d. help with implantation.
    e. are absent.
    *Answer: e*

28. Gastrulation in birds differs in many ways from gastrulation in frogs. The main feature that both have in common is that
    a. three germ layers are established.
    b. cells from the surface of the blastocyst migrate to form the gut.
    c. the neural tube forms.

d. somites are created.
e. All of the above
    *Answer: a*

29. After gastrulation, the mesodermal cells contribute to the developing
    a. brain and nervous system.
    b. skeletal system and muscles.
    c. inner lining of the gut and respiratory tract.
    d. sweat glands.
    e. liver and pancreas.
    *Answer: b*

30. After gastrulation, the ectodermal cells contribute predominantly to the developing
    a. brain, nervous system, and sweat glands.
    b. skeletal system and muscles.
    c. inner lining of the gut and respiratory tract.
    d. liver and pancreas.
    e. None of the above
    *Answer: a*

31. After gastrulation, the endodermal cells contribute predominantly to the developing
    a. brain, nervous system, and nails.
    b. skeletal system and muscles.
    c. lining of the digestive and respiratory tracts.
    d. sweat glands and milk secretory glands.
    e. None of the above
    *Answer: c*

32. Which of the following is *not* true of the umbilical cord?
    a. It contains blood vessels from the embryo.
    b. It contains blood vessels from the mother.
    c. It is derived from the allantois.
    d. It joins the placenta and the embryo.
    e. It carries nutrients and wastes.
    *Answer: b*

33. The formation of the zygote is synonymous with
    a. copulation.
    b. fertilization.
    c. intercourse.
    d. recombination.
    e. capacitation.
    *Answer: b*

34. Which of the following is *not* true of the first trimester of pregnancy?
    a. There is rapid cell division.
    b. It is the period when the embryo is the most sensitive to radiation and drugs.
    c. The fetus grows rapidly in size.
    d. The mother experiences nausea and mood swings.
    e. There are high levels of estrogen and progesterone circulating in the mother's blood.
    *Answer: c*

35. Bottle cells are
    a. cells of the trophoblast.
    b. cells of the inner cell mass.
    c. fluid-filled cells used for storage.
    d. bottle-shaped cells found around the blastopore.
    e. another name for granulosa cells.
    *Answer: d*

36. The movement of cells toward the blastopore is called
    a. mass cellular migration.
    b. embryonic cellular integration.
    c. cellular destiny.
    d. furrowing.
    e. epiboly.
    *Answer: e*

37. One of Hans Spemann's important experiments involved
    a. dividing human embryos into equal halves.
    b. killing a single cell in a sea urchin embryo to study the effects.
    c. dividing a frog embryo in two using a human hair.
    d. removing cytoplasm from muscle cells and transplanting it into a fertilized egg.
    e. All of the above
    *Answer: c*

38. The epiblast and the hypoblast are structures found during _____ development.
    a. human
    b. frog
    c. human and frog
    d. chicken
    e. human and chicken
    *Answer: e*

39. In humans, the amnion forms from the
    a. hypoblast.
    b. epiblast.
    c. chorion.
    d. trophoblast.
    e. yolk sac.
    *Answer: b*

40. In mammals and birds, the _____ is the structure from which the embryo is derived.
    a. trophoblast
    b. cumulus cells
    c. blastocyst
    d. epiblast
    e. hypoblast
    *Answer: d*

41. The cells that form the neural tube come from the
    a. notochord.
    b. mesoderm.
    c. endoderm.
    d. ectoderm.
    e. neuroderm.
    *Answer: a*

42. The _____ is an important structure for waste storage in birds and some mammals, but not in humans.
    a. allantois
    b. yolk sac
    c. placenta
    d. umbilical cord
    e. amnion
    *Answer: a*

43. The sperm contributes _____ to the zygote.
    a. a nucleus
    b. half of the mitochondria and a nucleus
    c. a centriole and a nucleus
    d. cilium and a nucleus
    e. All of the above
    *Answer: c*

44. In frogs, the sperm penetrates
    a. in the region of the vegetal pole.
    b. in the region of the animal pole.
    c. in the region of the gray crescent.
    d. anywhere.
    e. somewhere on the border of the animal and vegetal poles.
    *Answer: b*

45. If a cell is removed from an 8-cell embryo, and a particular portion of the organism fails to form, the development is termed
    a. regulative.
    b. controlled.
    c. banished.
    d. irreversible.
    e. mosaic.
    *Answer: e*

46. _____ are used by mesenchyme cells to move along extracellular matrix molecules.
    a. Amoeboids
    b. "Walking feet"
    c. Filopodia
    d. Sliding cell receptors
    e. None of the above
    *Answer: c*

47. The segmented characteristic of human embryonic development is evident from bricklike structures that form along the notochord, called
    a. neural tubes.
    b. mesoderm.
    c. ectoderm.
    d. somites.
    e. somatomeres.
    *Answer: d*

48. Which of the following does *not* occur during cleavage?
    a. Rapid DNA synthesis
    b. Cell growth
    c. Differential distribution of nutrients and information molecules
    d. A rapid series of cell divisions
    e. Formation of blastula
    *Answer: b*

49. Which of the following statements about neurulation is *false*?
    a. Neurulation initiates the nervous system.
    b. The notochord gives structural support to the embryo.
    c. Bulges at the posterior end of the neural tube become the brain.
    d. The incidence of neural tube defects can be lowered if pregnant women take folic acid.
    e. Both a and b
    *Answer: c*

50. Which of the following statements about gestation is *false*?
    a. Small mammals have shorter gestation periods than do large mammals.
    b. Amniocentesis is performed during the first trimester of pregnancy.
    c. Chorionic villus sampling can be used to detect genetic diseases as early as 8 weeks of gestation.
    d. Human pregnancy has a duration of about 9 months.
    e. None of the above
    *Answer: b*

# 21 Development and Evolutionary Change

## Fill in the Blank

1. The genome has often been believed to provide a _____ for an organism's development.
   *Answer: blueprint*

2. The sciences of genetics and embryology have come together to form a new discipline called _____.
   *Answer: evolutionary developmental biology*

3. Early discoveries in evolutionary developmental biology have shown that genes regulating development are highly _____ throughout the course of evolution.
   *Answer: conserved*

4. The genetic instructions for forming embryos are provided by _____ in vertebrates and invertebrates.
   *Answer: homologous genes*

5. _____ provide positional information to cells along the anterior–posterior axis of the body of organisms.
   *Answer: Homeobox genes*

6. _____ is the shift in relative timing of two different developmental processes.
   *Answer: Heterochrony*

7. The ability of an organism to change its development in response to environmental conditions is called _____.
   *Answer: developmental plasticity*

8. Throughout their lives, plants produce clusters of undifferentiated, actively dividing cells called _____.
   *Answer: meristems*

9. _____ allows morphological changes to occur without disruption of the entire organism.
   *Answer: Modularity*

## Multiple Choice

1. The genome encodes instructions for making
   a. enzymes.
   b. receptors.
   c. signal molecules.
   d. structural molecules.
   e. All of the above
   *Answer: e*

2. Darwin determined that barnacles were crustaceans by comparing _____ of different crustaceans.
   a. adult forms
   b. early developmental stages
   c. reproductive behavior

   d. the feeding behavior
   e. differences in the eggs
   *Answer: b*

■3. In *Drosophila*, the formation of legs where the antennae should be results from
   a. the *Antennapedia* mutation.
   b. injection of *Pax6* cDNA from a mouse.
   c. the *bithorax* mutation.
   d. Both a and b
   e. Both b and c
   *Answer: a*

4. The *bithorax* mutation in *Drosophila* causes
   a. development of one set of wings.
   b. growth of legs where antennae should be.
   c. development of two sets of wings.
   d. growth of eyes in the legs.
   e. changes in the anterior–posterior axis.
   *Answer: c*

■5. Changes in the regulation of development can lead to evolutionarily important morphological changes. These include
   a. mutations in developmental genes.
   b. changes in the time or place of expression of developmental genes.
   c. changes in the behavior of an organism.
   d. Both a and b
   e. None of the above
   *Answer: d*

■6. Insects do not have abdominal legs because
   a. the mutated *Ubx* gene represses expression of the *dll* gene.
   b. the *dll* gene is expressed.
   c. the mutated *Ubx* gene is not expressed.
   d. only the *dll* gene is expressed.
   e. the mutated *Ubx* and *dll* genes are both expressed.
   *Answer: a*

■7. Except for insects, most arthropods have abdominal legs because
   a. the mutated *Ubx* gene represses expression of the *dll* gene.
   b. the normal *Ubx* gene does not repress expression of the *dll* gene.
   c. the normal *Ubx* gene represses expression of the *dll* gene.
   d. the *dll* gene is not expressed.
   e. the normal *Ubx* gene is not expressed.
   *Answer: b*

8. The process of programmed cell death is called
   a. apoptosis.
   b. heterochrony.
   c. gene expression.
   d. mutation.
   e. learning.
   *Answer: a*

9. In a chicken, the absence of expression of the *gremlin* gene leads to
   a. the separation of toes in the adult.
   b. webbed toes in the adult.
   c. the development of four wings in the adult.
   d. no developmental changes.
   e. differences in the anterior–posterior axis.
   *Answer: a*

■10. If the Gremlin protein is applied to the hindlimbs of a chicken,
   a. the feet will develop normally.
   b. the hindlimbs will not develop at all.
   c. the toes will be separated in the feet.
   d. the feet will develop webbing.
   e. Both a and c
   *Answer: d*

■11. In the arthropods, what is the event that likely led to the two different lineages: one that has legs growing from the abdomen and the other that does not?
   a. A mutation in the *Ubx* gene
   b. A mutation in the *dll* gene
   c. A mutation in the *gremlin* gene
   d. Normal divergence
   e. The *bithorax* mutation
   *Answer: a*

12. The process by which the timing of two independent developmental processes shifts is called
   a. mutation.
   b. modularity.
   c. heterochrony.
   d. gene duplication.
   e. plasticity.
   *Answer: c*

■13. Heterochrony leads to the evolution of new species
   a. through single gene mutations that lead to the morphological changes in an organism.
   b. by changing the timing of gene expression, leading to morphological changes.
   c. through mutations in homeobox genes.
   d. through environmental signals that cause mutation.
   e. by synchronizing changes in the modules of developing organisms.
   *Answer: b*

■14. How does modularity allow gene duplication to cause structural changes?
   a. Gene duplication affects all aspects of an organism's development.
   b. Gene duplications that occur in one module of an organism do not disrupt development in other modules.
   c. Gene duplications that occur in one module of an organism disrupt development in other modules.

d. Gene duplications cause development to cease.
   e. None of the above
   *Answer: b*

■15. Which of the following statements about plant and animal development is *false*?
   a. Plant cells do not move relative to one another.
   b. Plants and animals show equal developmental plasticity.
   c. Animals show little developmental plasticity.
   d. Only plants possess meristems, clusters of undifferentiated cells.
   e. Animal embryo cells exhibit complex patterns of movement.
   *Answer: b*

16. When a herbivore eats a plant, the plant may respond by growing new leaves or by producing toxins. This ability to respond to environmental conditions is called
   a. heterochrony.
   b. developmental plasticity.
   c. mutation.
   d. gene duplication.
   e. modularity.
   *Answer: b*

17. _____ allow(s) a plant to develop and form new organs as long as it grows.
   a. Meristems
   b. Mutations
   c. Gene duplication
   d. Modularity
   e. Developmental plasticity
   *Answer: a*

18. Which two sets of genes regulate important developmental processes in both plants and animals?
   a. MADS box and homeobox
   b. MADS box and *Ubx*
   c. *Ubx* and *dll*
   d. Homeobox and *BMP4*
   e. *Antennapedia* and *gremlin*
   *Answer: a*

■19. Why do plants exhibit greater developmental plasticity than animals?
   a. Only in plants does the modular construction allow some parts to change independently of others.
   b. Plants are more sensitive to the environment than animals are.
   c. Plants have evolved this mechanism to compensate for a lack of mobility.
   d. Both a and c
   e. Both b and c
   *Answer: c*

20. Which of the following signals does *not* accurately predict future environmental conditions?
   a. Increasing day length
   b. Sunrise
   c. Decreasing day length
   d. Temperature change
   e. Precipitation
   *Answer: b*

■21. What is the effect of environmental signals on the development of organisms?
   a. Development stays the same regardless of the environment because genes have an autonomous existence.
   b. In many species, adults develop into particular forms depending on environmental signals.
   c. Some organisms develop differently depending on their interactions with other organisms.
   d. Both b and c
   e. a, b, and c
   *Answer: d*

■22. Larvae of the moth *Nemoria arizonaria* develop into different forms depending on the season in which they hatch. What environmental signal causes this difference?
   a. Temperature at the time in which they hatch
   b. The part of the oak on which they feed
   c. Day length at the time in which they hatch
   d. The level of rain that has recently fallen
   e. The amount and intensity of sunlight present as they develop
   *Answer: b*

■23. What is the mechanism explaining the development of the two forms of larvae of the moth *Nemoria arizonaria*?
   a. The different food sources provide different nutrients.
   b. Summer caterpillars do not need camouflage because they are hidden by the leaves of the tree.
   c. A chemical in oak leaves induces the twiglike form in summer caterpillars.
   d. The timing of development is different depending on the food source.
   e. Homeotic mutations show differing activity based on different food sources.
   *Answer: c*

■24. House mice that are raised in a relatively microbe-free environment
   a. fail to develop normal-functioning intestines.
   b. are much healthier than those raised under normal conditions.
   c. lack gene induction that promotes gut capillary development.
   d. lack intestinal gene mutations present in mice raised under normal conditions.
   e. Both a and c
   *Answer: e*

★25. A male anemonefish can change sex and become female when the leader female is removed from the social group. How is this adaptation beneficial?
   a. This mechanism assures that the robust genes from large, formerly male fish are propagated extensively.
   b. This adaptation allows the social group to continue to reproduce.
   c. This adaptation reduces the genetic variation in the social group.
   d. This adaptation allows for a harmonious, quasi-hermaphroditic social group.
   e. Both a and c
   *Answer: b*

★26. The developmental changes associated with predator-induced developmental plasticity in a given species (such as *Daphnia*) may not always occur because
   a. predators are sometimes beneficial to the species as a whole.

   b. predators may not be active in an organism's environment.
   c. developmental genes must be mutated in order for changes to occur.
   d. secondary environmental factors must also be present in order for development to be affected.
   e. predators usually eat their prey before any developmental changes can take place.
   *Answer: b*

★27. The large "helmet" sometimes developed by the water flea *Daphnia* may prevent it from being eaten. However, this characteristic is not always present in *Daphnia* populations. If this structure is adaptive against predators, why isn't it maintained in greater frequency in the population?
   a. The structure develops only in the offspring of adult female *Daphnia* that have come into contact with fly larvae prior to laying their eggs.
   b. The helmet structure disrupts other developmental modules.
   c. *Daphnia* with helmets reproduce at a lower rate than those without helmets.
   d. Development of the helmet is associated with other environmental influences that may or may not be present.
   e. Predator flies have been coevolving with *Daphnia* and therefore are often able to ingest the larger helmets as well as the smaller ones.
   *Answer: c*

28. Plants respond to environmental conditions by varying all of the following *except* for
   a. their size.
   b. their shape.
   c. the number of flowers.
   d. the number of seeds.
   e. the size of the seeds.
   *Answer: e*

29. Which of the following environmental influences are not likely to have caused the evolution of developmental changes in organisms?
   a. Chemicals released by human pollution
   b. Temperature
   c. Presence of predators
   d. Rainfall
   e. All of the above
   *Answer: a*

30. Learning is a modification of development in the sense that it
   a. takes place in the larvae of an organism.
   b. takes a great deal of effort and time, during which the organism is prevented from doing other things (i.e., there are costs involved).
   c. allows an organism to adjust its behavior to changing environmental conditions.
   d. can take place throughout the life of an organism.
   e. b, c, and d
   *Answer: e*

# 22 The History of Life on Earth

## Fill in the Blank

1. A massive extinction occurred at the end of the _____ period. It may have been caused over a long period of time (ten million years) by the coalescing of the continents into the supercontinent, Pangaea.
   **Answer: Permian**

2. The period within the Mesozoic era in which frogs, salamanders, and lizards first appeared, and in which one lineage of dinosaurs gave rise to birds, is termed the _____ period.
   **Answer: Jurassic**

3. Earth's crust consists of solid _____ approximately 40 kilometers thick that float on a liquid mantle.
   **Answer: plates**

4. Many patterns in the fossil record suggest that there are long periods during which rates of morphological evolutionary change are extremely slow. These periods are called _____.
   **Answer: stasis**

★5. There have been _____ mass extinctions in the history of life, and _____ periods of rapid diversification of organisms.
   **Answer: six; three**

6. _____ is a change in a species that happens during its lifetime.
   **Answer: Biological evolution**

7. Some ancient insects have been perfectly preserved in _____ formed by tree resin.
   **Answer: amber**

## Multiple Choice

1. The movements of continents were important causes of extinction during the history of life on Earth. Which of the following results of continental movement were important in these extinctions?
   a. Changes in sea level
   b. Separation of biotas
   c. Mixing of biotas
   d. Changes in climate
   e. All of the above
   **Answer: d**

2. Over much of its history, the climate of Earth was
   a. about the same as it is today.
   b. considerably cooler than it is today.
   c. considerably warmer than it is today.
   d. unknown; we have no information about climates in the past.
   e. much more variable annually.
   **Answer: c**

3. Conditions for the preservation of deceased organisms (fossilization) are best in environments
   a. lacking oxygen.
   b. with high levels of oxygen.
   c. that are warm and moist.
   d. that are cold and dry.
   e. with constant temperature.
   **Answer: a**

■4. When a new form of organism appears in the fossil record in a particular area, we are able to conclude that
   a. it evolved there rapidly.
   b. it migrated there from another location.
   c. there is a gap in the fossil record and it evolved there slowly.
   d. All of the above
   e. None of the above
   **Answer: d**

■5. Radioisotopes are often used to determine the time of death of fossilized remains. Tritium has a half-life of 12.3 years. That means that 24.6 years after an organism dies it will have _____ of the original radioactive tritium.
   a. all
   b. ½
   c. none
   d. ¼
   e. ⅛
   **Answer: d**

6. The coal beds we now mine for energy are the remains of trees of the
   a. Precambrian period (600 mya).
   b. Cambrian period (600–500 mya).
   c. Ordovician period (500–440 mya).
   d. Silurian period (440–400 mya).
   e. Carboniferous period (about 300 mya).
   **Answer: e**

7–10. The Mesozoic era (about 245–66 mya) had three periods: the Triassic, the Jurassic, and the Cretaceous. Match the following events with the correct time frame from the list below. Each choice may be used once, more than once, or not at all.
   a. Mesozoic era (about 245–66 mya)
   b. Triassic period (about 245–195 mya)
   c. Jurassic period (about 195–138 mya)
   d. Cretaceous period (about 138–66 mya)

7. The first mammals evolved from reptiles.
   *Answer: c*

8. Individual continents acquired distinctive terrestrial floras and faunas.
   *Answer: a*

9. Salamanders, and lizards first appeared.
   *Answer: c*

10. A great radiation of reptiles began.
    *Answer: b*

11. The Cenozoic era (66 mya–present) is often called the age of
    a. bacteria, because they cause so many diseases.
    b. fishes, because they are an important food resource.
    c. mammals, because of the extensive radiation of this group.
    d. plants, because they convert radiant energy from the sun into chemical energy.
    e. viruses, because viruses such as AIDS have evolved recently.
    *Answer: c*

*12. The last glaciers retreated from temperate latitudes about _____ years ago. As a result, many temperate ecological communities have occupied their current locations for no more than _____ years.
    a. 10,000,000; a million
    b. 1,000,000; a hundred thousand
    c. 100,000; tens of thousands of
    d. 15,000; a few thousand
    e. 1,000; a few hundred
    *Answer: d*

13. Which of the following statements about the Cenozoic era is *false*?
    a. Australia and Antarctica were still attached.
    b. Flowering plants dominated world forests.
    c. The climate got hotter and wetter.
    d. There was an extensive radiation of mammals.
    e. *Homo sapiens* arrived in North and South America.
    *Answer: c*

14. Which of the following is *not* considered to be a plausible hypothesis for the causes of mass extinctions?
    a. Extraterrestrial causes, such as meteorite or asteroid collisions
    b. Glaciations
    c. Massive volcanic activity
    d. Competition among organisms
    e. All of the above
    *Answer: d*

*15. How many species of organisms have been identified in the fossil record?
    a. 300
    b. 3,000
    c. 30,000
    d. 300,000
    e. 3,000,000
    *Answer: d*

*16. There are thought to be _____ mass extinctions in the history of life on Earth and _____ periods during which many new evolutionary lineages originated.
    a. six; no
    b. three; three
    c. six; six
    d. three; six
    e. six; three
    *Answer: e*

*17. The major difference between the earliest explosion in evolutionary lineages (the Cambrian, about 500 mya), and the Triassic, which was more recent, is that
    a. new phyla originated during the Cambrian.
    b. new phyla originated during the Triassic.
    c. the Cambrian explosions were probably caused by an asteroid colliding with Earth.
    d. there were more species during the Cambrian, and thus more competition.
    e. Both b and c
    *Answer: a*

18. During the Mesozoic era (245–66 mya), Pangaea separated into individual continents. As a result
    a. many species became extinct.
    b. many new phyla evolved.
    c. individual continents acquired distinctive terrestrial floras and faunas.
    d. flight evolved to allow organisms to migrate among continents.
    e. All of the above
    *Answer: c*

19. Which of the following continents is oldest?
    a. Laurasia
    b. Pangaea
    c. Antarctica
    d. Australia
    e. Both a and c
    *Answer: b*

20. During Earth's history, the climate has been
    a. about the same as it is now.
    b. uniform and considerably colder than it is now.
    c. considerably colder than it is now, but with numerous short warm periods.
    d. considerably warmer than it is now.
    e. considerably warmer than it is now, but with numerous short cold periods.
    *Answer: e*

21. What can be said about a typical environment that will become the site of a rich fossil bed, and of the origin of organisms that become fossils there?
    a. The environment is anaerobic, and most organisms originated elsewhere.
    b. The environment is anaerobic, and most organisms originated locally.
    c. The environment is aerobic, and most organisms originated elsewhere.
    d. The environment is aerobic, and most organisms originated locally.
    e. The environment is anaerobic, and about equal numbers of organisms originated elsewhere and locally.
    *Answer: a*

**⁎■22.** The half-life of a particular radioactive isotope of element X is 100 years. If you know that 400 years ago a fossil contained 10 mg of this isotope, about how much would you expect to find in that fossil today?
a. 5 mg
b. 2.5 mg
c. 0.6 mg
d. 0.3 mg
e. 0.15 mg
*Answer: c*

23. Modern biota evolved during the _____ era.
a. Proterozoic
b. Cenozoic
c. Mesozoic
d. Paleozoic
e. Both a and b
*Answer: b*

24. Select the time division during which the world biota became provincialized due to continental drift.
a. Precambrian
b. Paleozoic
c. Mesozoic
d. Cenozoic
e. Both b and c
*Answer: c*

25. Select the time division during which the major diversification of angiosperms, birds, and mammals occurred.
a. Precambrian
b. Paleozoic
c. Mesozoic
d. Cenozoic
e. Both c and d
*Answer: d*

26. Select the period during which the first insects and amphibians appeared on land.
a. Cambrian
b. Devonian
c. Permian
d. Triassic
e. Tertiary
*Answer: b*

27. What types of organisms would you *not* expect to see during a walk in a Permian forest?
a. Dragonflies
b. Amphibians
c. Club mosses and horsetails
d. Flowering plants
e. Gymnosperms
*Answer: d*

28. During which period did the "age of reptiles" begin?
a. Permian
b. Triassic
c. Jurassic
d. Cretaceous
e. Tertiary
*Answer: b*

29. During the Cretaceous period of the Mesozoic era,
a. the continents Laurasia and Gondwana first formed.
b. horses first appeared.
c. birds and mammals diverged from a common reptilian stock.

d. a second major extinction eliminated most land vertebrates.
e. bony fishes dominated the seas.
*Answer: d*

30. During the Tertiary period of the Cenozoic era,
a. the dinosaurs become extinct.
b. the "age of fishes" began.
c. many major "ice ages" occurred.
d. the genus *Homo* first appeared.
e. the main radiation of birds took place.
*Answer: e*

31. Which of the following statements about the fauna of the Cambrian, Paleozoic, or modern era is true?
a. The diversity of body plans is greatest in the modern fauna.
b. Most of the phyla that were present in the Paleozoic fauna are now extinct.
c. The number of families has steadily increased in the modern fauna.
d. Most of the species that were part of the Cambrian fauna are alive today.
e. None of the above
*Answer: c*

32. Which of the following statements about the diversity of the modern fauna is true?
a. As the diversity of insects declined, the diversity of flowering plants increased.
b. The evolution of different ecological communities was important in creating the changes in diversity in the modern fauna.
c. Provinciality due to continental drift was an unimportant factor after the Cambrian era.
d. As diversity increased, organisms became more specialized and less interdependent.
e. The numbers of species during the Cenozoic has been rather stable.
*Answer: b*

33. Of the six mass extinctions on Earth, the extinction at the end of the _____ eliminated the most phyla.
a. Cambrian
b. Devonian
c. Permian
d. Triassic
e. Cretaceous
*Answer: a*

34. Spine reduction in sticklebacks is usually related to
a. genetic drift.
b. a reduction in predation in fresh water.
c. a dominant allele.
d. movement into a more stable environment.
e. differences in the type of predators encountered in freshwater and marine habitats.
*Answer: e*

**■35.** Which of the following statements about evolution would be supported by most biologists?
a. Evolution is understandable, but not predictable.
b. Evolution results in greater complexity.
c. Given specific initial conditions, evolution is predetermined.
d. As evolution progresses, extinction rates decrease.
e. Mutation is more important than environment in directing the course of evolution.
*Answer: a*

36. Evolution of species on Earth
    a. has stopped.
    b. occurred only in the distant past.
    c. occurred after the Cambrian explosion.
    d. has occurred throughout Earth's history and is still under way.
    e. None of the above
    *Answer: d*

*37. Earth's crust consists of solid tectonic plates approximately _____ km thick that "float" on a liquid mantle.
    a. 4
    b. 40
    c. 400
    d. 4,000
    e. None of the above
    *Answer: b*

38. The possible collision of a large meteorite with Earth 65 million years ago may explain
    a. the theory of Gondwana.
    b. Alfred Wegner's concept of continental drift.
    c. the mass extinction that occurred at the end of the Cretaceous period.
    d. contemporary atmospheric conditions on Earth.
    e. None of the above
    *Answer: c*

39. More than _____ percent of the species that have ever lived on Earth since the beginning of time are now extinct.
    a. 10
    b. 50
    c. 75
    d. 90
    e. 99
    *Answer: e*

40. Three times during the history of life, many new evolutionary lineages originated. Select the correct choice below that represents these periods.
    a. The Cambrian explosion, the Paleozoic explosion, and the Triassic explosion
    b. The Jurassic revolution, the Cambrian explosion, and the Hadean period
    c. The Cambrian explosion, the Archean extinction, and the Jurassic explosion
    d. The Paleozoic explosion, the Mesozoic explosion, and the Cenozoic explosion
    e. The Archean extinction, the Jurassic explosion, and the Triassic explosion
    *Answer: a*

41. The most probable cause of the mass extinction at the end of the Permian period was
    a. a volcano.
    b. a giant meteorite hitting Earth.
    c. major predation among species.
    d. aggregation of the continents into Pangaea.
    e. Both b and d
    *Answer: e*

42. Scientists proposed that a meteorite was the cause of the mass extinction 66 million years ago. Their hypothesis was based on the discovery of
    a. high concentrations of iridium in the rock layer separating the Cretaceous and Tertiary periods.
    b. fossils in the rock layer separating the Cretaceous and Tertiary periods.
    c. a crater site.
    d. high concentrations of argon 39 in the rock layer separating the Cretaceous and Tertiary periods.
    e. evidence suggesting unusual volcanic activity during that period.
    *Answer: a*

43. Most kingdoms evolved prior to the Cambrian period (about 600 mya). Which kingdom evolved after the Cambrian; that is, which was the last kingdom to evolve?
    a. Prokaryotes
    b. Protists
    c. Fungi
    d. Plants
    e. Animals
    *Answer: d*

44. The appearance of multicellular organisms coincided with increased levels of _____ in Earth's atmosphere.
    a. sulfur
    b. hydrogen
    c. nitrogen
    d. carbon
    e. oxygen
    *Answer: e*

*45. The Ediacaran fauna consisted mostly of
    a. prokaryotic organisms.
    b. soft-bodied invertebrates similar to present-day forms.
    c. soft-bodied invertebrates unlike present-day forms.
    d. trilobites and other hard-shelled forms.
    e. primitive horses.
    *Answer: c*

46. Which of the following is believed to be a cause of the mass extinction at the end of the Devonian period?
    a. Two continents colliding
    b. An asteroid colliding with Earth
    c. A large volcano
    d. The breaking apart of Pangaea and subsequent rise in the ocean
    e. None of the above
    *Answer: b*

# 23 The Mechanisms of Evolution

## Fill in the Blank

1. Evolution is the accumulation of _____ changes within populations over time.
   *Answer: heritable*

2. The frequencies of the variants at each locus in a population are called _____.
   *Answer: allele frequencies*

3. The genetic expression of traits (e.g., "homozygous recessive") is called the organism's _____.
   *Answer: genotype*

4. The physical expression of a trait (e.g., height or eye color) describes an organism's _____.
   *Answer: phenotype*

5. Short-term changes in a population's gene pool are often referred to as _____.
   *Answer: microevolution*

6. The _____ is the sum total of genetic information in a population at any given moment. It includes every allele at every locus in every organism.
   *Answer: gene pool*

7. The relative reproductive contribution of an individual to subsequent generations is termed the individual's

   _____.
   *Answer: fitness*

8. When individuals with intermediate values of traits have the highest fitness, _____ is said to be operating.
   *Answer: stabilizing selection*

9. A population that is not changing (i.e., it has constant genotype and allele frequencies from generation to generation) is said to be in _____.
   *Answer: equilibrium*

10. _____ involves movement of individuals to a new location, followed by breeding.
    *Answer: Gene flow*

11. The differential contribution of offspring resulting from different heritable traits is called _____.
    *Answer: natural selection*

12. The idea of natural selection is most closely associated with _____, who proposed it in 1859 in his book *The Origin of Species*.
    *Answer: Charles Darwin*

13. In a small population, a change in the allele frequencies that results from chance is called _____.
    *Answer: genetic drift*

14. Genetic drift can be brought about either by a severe reduction in population size, known as a population _____, or by a small number of individuals' establishing a new population, which results in a _____.
    *Answer: bottleneck; founder effect*

15. Populations of a European clover, *Trifolium repens*, produce cyanide, which increases their resistance to herbivores such as mice and slugs; however, it also increases their susceptibility to frost. Populations from northeast Europe produce relatively little cyanide compared to populations in southwest Europe. Sub-populations vary genetically in this way because they are subjected to different _____.
    *Answer: selective pressures*

16. A _____ is a change in DNA sequence.
    *Answer: mutation*

## Multiple Choice

■1. Assume that a population is in Hardy–Weinberg equilibrium for a trait controlled by one locus and two alleles. If the frequency of the recessive allele is 0.90, what is the frequency of the dominant allele?
   a. 0.10
   b. 0.19
   c. 0.81
   d. None of the above
   e. This cannot be answered without more information.
   *Answer: a*

■2. There is a gene that causes people to have crumbly earwax. This gene is expressed as a complete dominant: Individuals who are homozygous dominants (*CC*) or heterozygous (*Cc*) have crumbly earwax. Homozygous recessives (*cc*) have gooey earwax. On Paradise Island there are 100 people, 75 of whom have crumbly earwax. Assuming Hardy–Weinberg conditions, what is the frequency of the *c* allele on Paradise Island?
   a. 0.25
   b. 0.50
   c. 0.87
   d. None of the above
   e. This cannot be answered without more information.
   *Answer: b*

■3. In a population at Hardy–Weinberg equilibrium, the frequency of heterozygotes is 0.64. What is the frequency of the homozygous dominants?
a. 0.08
b. 0.64
c. 0.80
d. None of the above
e. This cannot be answered without more information.
*Answer: e*

■4. In a population at Hardy–Weinberg equilibrium, the frequency of the *a* allele is 0.60. What is the frequency of individuals heterozygous for the *A* gene?
a. 0.16
b. 0.24
c. 0.48
d. None of the above
e. This cannot be answered without more information.
*Answer: c*

5. There are five conditions that must be met for a population to be in Hardy–Weinberg equilibrium. Which of the following is *not* one of those conditions?
a. Nonrandom mating
b. Large population size
c. No migration
d. No natural selection
e. No mutations
*Answer: a*

6. Which of the following populations would demonstrate a population bottleneck?
a. The population of El Paso, Texas, after it has moved in its entirety to Patagonia
b. Eight male and eight female elephant seals that have survived the wreck of the Exxon *Valdez*
c. A million male orangutans
d. Six male orangutans collected from a natural population in Sumatra and moved to the San Diego Zoo
e. None of the above
*Answer: b*

7. _____ is the effect produced when a bee carries pollen from one population to another.
a. Gene flow
b. A population bottleneck
c. A founder event
d. Genetic equilibrium
e. Assortative mating
*Answer: a*

8. In a large population, mutation pressure in the absence of selection
a. probably has little effect on the gene pool.
b. produces major evolutionary changes.
c. is what has made us different from the dinosaurs.
d. is usually beneficial.
e. never occurs.
*Answer: a*

★9. Which of the following is *not* true of mutation?
a. Mutation creates the raw material that makes evolution possible.
b. Most mutations are harmful or neutral.
c. Mutation rates are very low for most loci.
d. Mutations are a likely cause of deviations from Hardy–Weinberg proportions in a population.

e. Mutations probably have little effect on the gene pool of a large population.
*Answer: d*

■10. Over the long run, mutations are important to evolution because
a. they are the original source of genetic variation.
b. once an allele is lost through mutation, another mutation to that same allele cannot occur.
c. most mutation rates are one in a thousand.
d. whether good or bad, mutations increase the fitness of an individual.
e. mutations are usually beneficial to the progeny.
*Answer: a*

11. _____ selection occurs when the extremes of a population contribute relatively few offspring to the next generation as compared to average members of the population.
a. Corrective
b. Directional
c. Stabilizing
d. Disruptive
e. Natural
*Answer: c*

12. _____ selection occurs when one extreme of a population contributes more offspring to the next generation than average members of the population do.
a. Corrective
b. Directional
c. Stabilizing
d. Disruptive
e. Natural
*Answer: b*

13. Which of the following agents of evolution adapts a population to its environment?
a. Nonrandom mating
b. Natural selection
c. Migration
d. Genetic drift
e. Mutation
*Answer: b*

14. The raw material for evolutionary change is _____ variation.
a. phenotypic
b. genetic
c. geographical
d. environmentally-induced
e. behavioral
*Answer: b*

15. A population evolves when
a. environmentally-induced variation is constant between generations.
b. individuals with different genotypes survive or reproduce at different rates.
c. the environment changes on a seasonal basis.
d. members reproduce by cloning.
e. juvenile and adult stages require different environments.
*Answer: b*

★16. Selection acts on _____ variation; however, evolution depends on _____ variation.
a. phenotypic; genetic
b. genetic; phenotypic

c. genetic; environmentally-induced
d. environmentally-induced; phenotypic
e. environmentally-induced; genetic
*Answer: a*

*17. Limpets growing high in the intertidal zone, where they experience heavy wave action, are more conical than individuals of the same species growing in the subtidal zone, where they are protected from waves. Individuals transplanted from the high intertidal zone to the subtidal zone add new growth, which produces a flatter, subtidal shape. This experiment suggests that the difference in
a. phenotypes is environmentally-induced.
b. genotypes is environmentally-induced.
c. phenotypes is genetically based.
d. genotypes is due to natural selection.
e. This cannot be answered without more information.
*Answer: a*

*18. Which of the following is an example of environmentally-induced variation?
a. Water fleas grown in cool or calm water develop round heads. If they are moved to warm or turbulent water, they develop pointed "helmets" on their heads.
b. Plants collected along an altitudinal cline vary in size when grown in their native habitat, yet when grown in a greenhouse they are all the same size.
c. Limpets growing high in the intertidal zone, where they experience heavy wave action, are more conical than individuals of the same species growing in the subtidal zone, where they are protected from waves. Individuals transplanted from the high intertidal zone to the subtidal zone add new growth, which produces a flatter, subtidal shape.
d. Leaves on the same tree or shrub often differ in shape and size. In oaks, leaves closer to the top receive more wind and sunlight and are more deeply lobed than leaves lower down.
e. All of the above
*Answer: e*

19–23. Suppose you have a population of flour beetles with 1,000 individuals. Normally the beetles are a red color; however, this population is polymorphic for a mutant autosomal body color, black, designated by $b/b$. Red is dominant to black, so $B/B$ and $B/b$ genotypes are red. Assume the population is in Hardy–Weinberg equilibrium, with $f(B) = p = .5$ and $f(b) = q = .5$.

*19. What would be the expected frequencies of the red and black phenotypes?
a. .5 red, .5 black
b. .75 red, .25 black
c. .25 red, .75 black
d. None of the above
e. This cannot be answered without more information.
*Answer: b*

*20. What would be the expected frequencies of the homozygous dominant, heterozygous, and homozygous recessive after 100 generations if the population is under the conditions of Hardy–Weinberg?
a. .75, .20, .05
b. .25, .5, .25

c. All red, because it is the natural color
d. All black, because all red alleles would mutate to black
e. .5, .2, .3
*Answer: b*

*21. What would be the expected red and black allele frequencies if all Hardy–Weinberg conditions were met, except that 1,000 black individuals migrated into the population?
a. .75, .25
b. .25, .75
c. .5, .5
d. They would not change because the population would still be in Hardy–Weinberg equilibrium.
e. None of the above
*Answer: b*

*22. What would be the allele frequencies if a population bottleneck occurred and only four individuals survived: one female red heterozygote and three black males?
a. .875, .125
b. .125, .875
c. .25, .75
d. .75, .25
e. .5, .5
*Answer: b*

*23. If the population in the previous question mated randomly, what would be the allele frequencies of their offspring?
a. .875, .125
b. .125, .875
c. .25, .75
d. .75, .25
e. .5, .5
*Answer: b*

24. A gene pool consists of all the alleles
a. of an individual's genotype.
b. present in a specific population.
c. that occur in a species throughout its evolutionary existence.
d. that contribute to the next generation of a population.
e. of a biome.
*Answer: b*

25. An important feature of sexual reproduction is that it
a. is unique to animals.
b. produces variation through genetic recombination.
c. uses mitosis in producing gametes.
d. is more efficient than asexual reproduction.
e. is always associated with selective mating.
*Answer: b*

26. In a West African finch species, birds with large or small bills survive better than birds with intermediate-sized bills. The type of natural selection operating on these bird populations is _____ selection.
a. directional
b. disruptive
c. stabilizing
d. nonrandom
e. deme
*Answer: b*

27–28. Hardy–Weinberg frequencies for the three genotypes *AA*, *Aa*, and *aa*, at all values of *p* and *q*, are shown below.

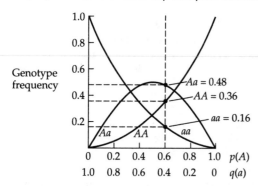

27. As the frequency of the homozygous recessive genotype increases, the frequency of the
a. heterozygous genotype decreases continuously.
b. homozygous dominant increases.
c. heterozygous genotype increases continuously.
d. dominant allele approaches 1.
e. recessive allele approaches 1.
*Answer: e*

28. Which of the following is supported by the graph above?
a. When *aa* = 0.45, *Aa* = 0.45.
b. When *aa* = 0.16, *q* = 0.60.
c. When *p* = 0.33, *q* = 0.77.
d. When *Aa* = 0.49, *p* = 0.70.
e. When *AA* = 0.50, *aa* = 0.
*Answer: a*

29. In a population of organisms, the frequency of the heterozygous genotype is always
a. greater than the frequency of the homozygous recessive genotype.
b. equal to one minus the sum of the homozygous genotypes.
c. equal to two times the frequencies of the homozygous genotypes.
d. greater than either homozygous genotype.
e. equal to the frequency of the homozygous dominant genotype minus the frequency of the homozygous recessive genotype.
*Answer: b*

*30. Cheetahs are a very homogeneous species. The lack of genetic variability among cheetahs may be attributed to
a. gene flow.
b. sexual selection.
c. population bottleneck.
d. high mutation rate.
e. high immigration rates.
*Answer: c*

31. Genetic equilibrium in a population refers to
a. equal numbers of dominant and recessive alleles.
b. equal numbers of females and males.
c. unchanging allele frequencies in successive generations.
d. lack of mutations affecting the observed phenotypes.
e. proportional numbers of each genotype.
*Answer: c*

32. Genetic drift as an evolutionary factor is
a. greater in a population with small numbers than in a population with large numbers.
b. greater in a population with much genetic variation than in a population with little genetic variation.
c. responsible for the selection of mutations.
d. connected to the movements of alleles between populations of a single species.
e. proportional to the size of a population: the larger the population, the greater the force.
*Answer: a*

33. In the Hardy–Weinberg equation, the homozygous dominant individuals in a population are represented by
a. $p^2$.
b. $2pq$.
c. $q^2$.
d. $p$.
e. $q$.
*Answer: a*

34–37. The two graphs below represent phenotypic distribution of a character in a population of organisms over time. The first curve represents the distribution at an initial sampling time, while the second represents a sampling of a later generation.

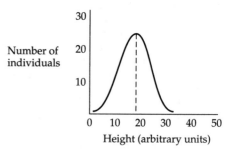

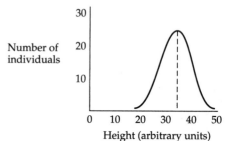

34. The average size of an individual at the initial sampling is _____ units.
a. 5
b. 10
c. 15
d. 20
e. 25
*Answer: d*

35. At the initial sampling, there are as many individuals 15 units tall as there are individuals _____ units tall.
a. 5
b. 10
c. 15
d. 20
e. 25
*Answer: e*

*36. The process illustrated by these graphs is called
   a. stabilizing selection.
   b. directional selection.
   c. disruptive selection.
   d. convergent evolution.
   e. divergent evolution.
   *Answer: b*

37. The graphs suggest that
   a. taller individuals migrated away from the original population.
   b. shorter individuals migrated into the original population.
   c. taller individuals were more successful at reproducing than short individuals were.
   d. in the initial population, there were more tall individuals than short ones.
   e. the individuals in the population continued to grow between sample times.
   *Answer: c*

38. A mutation occurs in one of your lung cells. Which of the following is true?
   a. You have evolved to be better adapted to your environment.
   b. You will soon die because most mutations are lethal.
   c. You will be sterile and no longer be able to have children.
   d. The human species will have evolved because this mutation will be passed on to your children.
   e. This mutation does not affect human evolution because it will not be passed on to your offspring.
   *Answer: e*

39. Natural selection acts directly on the
   a. genotype to produce new mutations.
   b. phenotype to produce new mutations.
   c. genotype to favor existing mutations.
   d. phenotype to favor traits due to existing mutations.
   e. genotype to inhibit new mutations.
   *Answer: d*

40–43. Biologists use the term "fitness" when speaking of evolution. Below are descriptions of four male cats. Answer the following questions based on this information.

| Name | Tabby | Chessy | Tony | Tiger |
|---|---|---|---|---|
| Size | 12 lbs | 10 lbs | 6 lbs | 8 lbs |
| Number of kittens fathered | 19 | 25 | 20 | 20 |
| Kittens surviving to adulthood | 15 | 14 | 14 | 19 |
| Age at death | 13 yrs | 16 yrs | 12 yrs | 9 yrs |

Comments: Tabby is the largest and strongest cat. Chessy has mated with the most females. Tony was lost on a family vacation but adapted to street life and lived two more years. Tiger died from an infection after a cat fight.

*40. Which cat contributed the most genes to the next generation?
   a. Tabby
   b. Chessy
   c. Tony
   d. Tiger
   e. This cannot be answered without more information.
   *Answer: b*

*41. Which cat contributed the most genes to the gene pool?
   a. Tabby
   b. Chessy
   c. Tony
   d. Tiger
   e. This cannot be answered without more information.
   *Answer: d*

*42. Which cat contributed the most to the ongoing evolution of the species?
   a. Tabby
   b. Chessy
   c. Tony
   d. Tiger
   e. This cannot be answered without more information.
   *Answer: d*

*43. Which cat is the "fittest"?
   a. Tabby
   b. Chessy
   c. Tony
   d. Tiger
   e. This cannot be answered without more information.
   *Answer: d*

44–46. In a population of 200 individuals, 72 are homozygous recessive for the character of eye color ($cc$). One hundred individuals from this population die due to a fatal disease. Thirty-six of the survivors are homozygous recessive. Answer the following questions.

▪44. In the original population, the frequency of the dominant allele is
   a. 0.16.
   b. 0.36.
   c. 0.40.
   d. 0.48.
   e. 0.60.
   *Answer: c*

▪45. In the new population, the frequency of the dominant allele is
   a. 0.16.
   b. 0.36.
   c. 0.40.
   d. 0.48.
   e. 0.60.
   *Answer: c*

▪46. How many heterozygous individuals are expected in the new population?
   a. 16
   b. 36
   c. 40
   d. 48
   e. 60
   *Answer: d*

47. Which of the following is *not* an effect of sexual recombination on alleles?
   a. Greater evolutionary potential
   b. Change in the frequency of specific alleles
   c. Greater genotypic variety
   d. New combinations of genetic material
   e. Greater phenotypic variety
   *Answer: b*

48. An allele that does *not* affect the fitness of an organism is called a _____ allele.
    a. neutral
    b. directional
    c. distributed
    d. positive
    e. natural
    *Answer: a*

# 24 Species and Their Formation

## Fill in the Blank

1. The process by which one evolutionary unit splits into two units which thereafter evolve as distinct lineages is known as _____.
*Answer: speciation*

2. _____ occurs when one species evolves into two daughter species after a physical barrier to movement develops within its range.
*Answer: Geographic speciation*

3. Two populations that are _____ are said to belong to different species.
*Answer: reproductively isolated*

4. The offspring of genetically dissimilar parents, such as adults from two geographically isolated populations, are called _____.
*Answer: hybrids*

5. _____ are factors that reduce the possibility of interbreeding between geographically isolated populations that somehow have come into contact.
*Answer: Isolating mechanisms*

6. _____ isolating mechanisms prevent the formation of a zygote, thus reducing the probability that hybrids will be formed.
*Answer: Prezygotic*

7. Once fertilization between isolated populations has occurred, _____ isolating mechanisms may prevent the hybrids from surviving.
*Answer: postzygotic*

8. A common means of sympatric speciation is _____, in which the number of chromosomes is multiplied.
*Answer: polyploidy*

9. The fossil record and current distributions of organisms reveal that some species or groups have given rise to a large number of daughter species, a phenomenon called _____.
*Answer: evolutionary radiation*

10. A factor promoting cohesion of species is _____, the movement of individuals so that they reproduce in a population other than their original population.
*Answer: gene flow*

11. Two hybrid species of sunflowers, *Tragopogon mirus* and *T. miscellus*, have four sets of chromosomes; this condition is called _____.
*Answer: tetraploidy*

12. Subdivision of a gene pool when members of the daughter species are not geographically separated is called
_____.
*Answer: sympatric speciation*

13. Organisms that are found only in one specific area are considered _____ to that area.
*Answer: endemic*

## Multiple Choice

1. Evolutionary biologists believe that all the species present today, along with all the species that lived in the past, are descended from
a. a single ancestral species.
b. thousands of different origins.
c. one ancestral species per kingdom.
d. one ancestral species per phylum.
e. two or three ancestral species per kingdom.
*Answer: a*

2. Ernst Mayr's definition of species states that species are groups of _____ interbreeding natural populations that are _____ isolated from each other.
a. actually; reproductively
b. potentially; reproductively
c. actually or potentially; reproductively
d. actually or potentially; geographically
e. potentially; geographically
*Answer: c*

3. Ernst Mayr's definition of species does *not* include
a. groups that actually or potentially interbreed.
b. evolutionary units evolving separately from other units.
c. gene exchange.
d. members in a single geographic location.
e. groups that are reproductively separated from other groups.
*Answer: d*

■4. The phrase "natural population" is important to the definition of species because
  a. if two populations interbreed only in captivity, they are members of the same species.
  b. if two populations co-occur but do not interbreed in nature, they are separate species.
  c. if two populations do not co-occur, they must be different species because they cannot interbreed.
  d. if two populations can potentially interbreed, they are different species.
  e. if two populations interbreed for the first time, their offspring form a new species.
  *Answer: b*

5–8. Match the following descriptions of speciation with the appropriate terms from the list below. Terms may be used once, more than once, or not at all.
  a. geographic speciation
  b. sympatric speciation
  c. polyploidy

5. Two species of Japanese ladybird beetles occur in the same area. *Epilachnea niponica* feeds on thistles, and *Epilachnea yasutomii* feeds on other plants that grow among thistles. Adults of both species feed and mate on their own host plant. The two species hybridize in the laboratory, but not in nature. Although we cannot be sure which process actually produced speciation in this case, it is likely the two separate species are a result of _____.
  *Answer: b*

6. If the number of chromosomes of a species of lizard were to double (perhaps through an error in meiosis), resulting in female offspring that could produce young from unfertilized eggs, we would say speciation had occurred through _____.
  *Answer: c*

7. The red tubular-flowered gilias of western North America are a group of species living in the Mojave Desert. Originally considered to be a single species, the group now contains five species: three diploids and two tetraploids (having four sets of chromosomes). These five species, which are similar in appearance and sterile in all interspecific mating combinations, are an example of _____.
  *Answer: c*

8. Different populations of platyfish live in the rivers of eastern Mexico. The subpopulations living in different streams have been diverging since an ancestral platyfish colonized all the streams. Some subpopulations have diverged so much that they cannot interbreed and produce viable offspring with individuals from other subpopulations. Thus, these populations of platyfish exist at various stages in the process of _____.
  *Answer: a*

9. Which of the following factors probably was *not* important in the formation of the 14 species of Darwin's finches on the Galápagos Islands?
  a. Geographical isolation from the mainland
  b. Polyploidy
  c. Different habitats on the different islands
  d. Geographical isolation of one island from another
  e. Different food supplies on the different islands
  *Answer: b*

10. Which of the following would *not* result in reproductive isolation between two populations that are reunited following geographic isolation?
  a. The two populations produce successful hybrids.
  b. The two populations have different breeding seasons.
  c. There are physiological differences between the two populations so that they cannot produce viable offspring.
  d. Members of one population do not find members of the other population attractive as mating partners.
  e. The two populations have different courtship behaviors.
  *Answer: a*

11. Which of the following is a postzygotic isolation mechanism?
  a. Temporal isolation
  b. Behavioral isolation
  c. Reduced viability of hybrids
  d. Variation in mating pheromones
  e. Differences in courtship behavior
  *Answer: c*

12. Which of the following is a prezygotic isolation mechanism?
  a. Abnormal meiosis following fertilization
  b. Infertile hybrids
  c. Reduced viability of hybrids
  d. Abnormal mitosis following fertilization
  e. Variation in mating pheromones
  *Answer: e*

13. _____ involves the subdivision of a gene pool even though members of the daughter species overlap in their range during the speciation process. Often this is accomplished by a multiplication of the number of chromosomes.
  a. Polyploidy
  b. Hybrid zonation
  c. Sympatric speciation
  d. Geographic speciation
  e. None of the above
  *Answer: c*

14. Which of the following is *not* true of polyploidy?
  a. Polyploidy is more common in animals than in plants.
  b. Many species of flowering plants arose by polyploidy.
  c. Animals that speciate by polyploidy are often parthenogenetic.
  d. Polyploidy can create new species quickly if the polyploid individuals can self-fertilize.
  e. Polyploid siblings are capable of reproducing with one another.
  *Answer: a*

15. Which mode of speciation is thought to be most important in speciation of most organisms?
  a. Polyploidy
  b. Hybrid zonation
  c. Sympatric speciation
  d. Geographic speciation
  e. All of the above
  *Answer: d*

*16. Which of the following is *not* true of the genetics of speciation?
   a. Sympatric species need not have diverged very much from each other genetically.
   b. Speciation in *Drosophila* has not involved major reorganization of the genome.
   c. Species that show a great deal of morphological variation may be relatively similar genetically.
   d. Gene flow is important in the creation of new species.
   e. Differences between closely related species occur due to the same mechanisms that operate within species.
   *Answer: d*

■17. Rapid speciation is thought to occur among animals with complex behavior patterns because
   a. the timing of reproduction is behaviorally mediated.
   b. behavioral differences usually reflect physiological differences.
   c. those organisms have difficulty identifying members of their own species and often mate with members of other species.
   d. those organisms make sophisticated discriminations among potential mates.
   e. those organisms find mates more quickly and efficiently.
   *Answer: d*

18. The deer mouse, *Peromyscus maniculatus*, is the most widely distributed small mammal in North America. It varies greatly according to its geographical location, especially in coat color, tail length, and foot length. Where would you expect deer mice to be relatively uniform?
   a. In mountainous areas, where environmental conditions change dramatically from place to place
   b. On islands, where populations are isolated
   c. Over large areas where there is little topographical or vegetational change
   d. In regions between forests and deserts, where there is significant vegetational change
   e. In regions between forests and prairies, where there is significant vegetational change
   *Answer: c*

19. Antelope in Africa went through a burst of speciation 2.5–2.9 million years ago. What was the likely trigger of the speciation event?
   a. Continental drift
   b. Genetic drift
   c. A change in the gene pool
   d. A change in climate
   e. Reproductive incompatibility
   *Answer: d*

**20. Which of the following is a reason that evolutionary radiation is likely to occur on islands?
   a. Islands may have relatively few plant and animal groups and thus present more ecological opportunities than the continents do.
   b. There are many more species on islands than on the continents.
   c. Many organisms disperse easily over oceans.
   d. Islands have more diverse habitats than continents do.
   e. Islands tend to have more favorable habitats than continents, due to the moderating effects of the oceans.
   *Answer: a*

21. The Hawaiian Islands are useful for studying evolutionary radiation because they are very isolated. Which of the following is *not* true of the biota of these islands?
   a. More than 90 percent of the plant species are endemic.
   b. Several groups of flowering plants are more diverse on these islands than their counterparts are on the mainland.
   c. There are no endemic amphibians on these islands.
   d. There are no endemic reptiles on these islands.
   e. There are no endemic mammals on these islands.
   *Answer: e*

22. Which of the following is the hypothesized sequence of events in geographic speciation?
   a. Geographic barrier, reproductive isolation, genetic divergence
   b. Geographic barrier, genetic divergence, reproductive isolation
   c. Genetic divergence, geographic barrier, reproductive isolation
   d. Genetic divergence, reproductive isolation, geographic barrier
   e. Reproductive isolation, genetic divergence, geographic barrier
   *Answer: b*

23. Mules are the offspring of parents of two different species, a horse and a donkey. These hybrids exhibit
   a. polyploidy.
   b. a shortened lifespan.
   c. behavioral isolation.
   d. reduced viability.
   e. sterility.
   *Answer: e*

24. The modern polar bear species evolved from ancestral bear populations in southern Alaska that became separated by glaciers from bear populations in the rest of North America. This type of event is called
   a. allopatric speciation.
   b. temporal isolation.
   c. mechanical isolation.
   d. sympatric speciation.
   e. None of the above
   *Answer: a*

*25. A situation in which two very similar animal species have overlapping distributions is most likely the result of
   a. allopatric speciation followed by range expansion.
   b. sympatric speciation due to polyploidy.
   c. convergent evolution of unrelated species.
   d. mechanical isolation between the two species.
   e. None of the above
   *Answer: a*

*26. Which of the following is *not* true of a hybrid zone?
   a. It occurs where two different populations come into contact.
   b. It may shift in location due to environmental changes.
   c. Its habitat may differ from that favored by the parent populations.
   d. It may disappear if isolating mechanisms develop.
   e. It may occur among animals, but it does not occur among plants.
   *Answer: e*

**▪▪27.** American and European sycamores have been isolated from each other for at least 20 million years, but they are morphologically very similar and can form fertile hybrids. According to Ernst Mayr's definition of species, these sycamores should belong to _____ species because they _____.
   a. different; are geographically isolated
   b. different; lack the opportunity to interbreed in nature
   c. different; they are evolving separately
   d. the same; are morphologically similar
   e. the same; are capable of forming fertile offspring
   *Answer: e*

**▪28.** Ginkgo trees occur in Asia and North America. Despite their geographic separation by the Pacific Ocean, biologists consider them the same species. What aspect of Ernst Mayr's definition of species accounts for this?
   a. They are reproductively isolated.
   b. They are potentially capable of exchanging genes.
   c. They are exchanging genes across the ocean.
   d. They have different evolutionary ancestries.
   e. They have formed a large hybrid zone.
   *Answer: b*

**▪29.** Why are biological species not always equivalent to taxonomic species?
   a. Taxonomic species are based on appearance, not reproductive behavior.
   b. Taxonomic species are based on reproductive behavior, not appearance.
   c. Biological species are based on appearance, not reproductive behavior.
   d. Biological species are based on genetic information; taxonomic species are based on ecological information.
   e. None of the above
   *Answer: a*

**30.** Which mode of speciation involves a geographic barrier that prevents exchange of genes?
   a. Allopatric
   b. Sympatric
   c. Polyploidy
   d. Behavioral isolation
   e. None of the above
   *Answer: a*

**▪31.** Six platyfish with different tail spotting patterns live in eastern Mexico. Five of them can interbreed and produce fertile offspring. One of them cannot interbreed with the other five. What can you conclude about these platyfish?
   a. They all belong to the same biological species.
   b. They each are a different biological species.
   c. There are two biological species in the example above.
   d. There is not enough data given to draw a conclusion.
   e. There are six biological species in the example above.
   *Answer: c*

**32.** The progeny of genetically dissimilar parents are
   a. infertile.
   b. fertile.
   c. hybrids.
   d. mutants.
   e. hermaphrodites.
   *Answer: c*

**▪▪33.** If two genetically differentiated populations reestablish contact and the resulting hybrid progeny are successful and reproduce with other members of the two populations, what will probably happen?
   a. Speciation will occur.
   b. Genetic differences will increase between the populations.
   c. Reproductive isolation will occur.
   d. The two populations will amalgamate; no new species will form.
   e. Two new species will form.
   *Answer: d*

**▪34.** If two genetically differentiated populations reestablish contact and the resulting hybrid progeny are *not* as successful in reproduction as the members of the two parental populations, what will probably happen?
   a. Reproductive isolation will occur.
   b. A stable hybrid zone will form.
   c. The two populations will amalgamate; no new species will form.
   d. The populations will become genetically identical.
   e. Two new species will form.
   *Answer: a*

**35.** Which form of reproductive isolation occurs between a horse and donkey when they mate and produce a mule?
   a. Prezygotic isolation
   b. Abnormal zygote formation
   c. Hybrid infertility
   d. Hybrid vigor
   e. Hybrid fertility
   *Answer: c*

**36.** Which of the following modes of speciation can occur most quickly?
   a. Allopatric
   b. Prezygotic isolation
   c. Sympatric
   d. Dyspatric
   e. Adaptive radiation
   *Answer: c*

**37.** Because there are no physical barriers separating the species, many biologists believe that speciation of host-specific insects occurs by
   a. sympatric speciation.
   b. hybrid infertility.
   c. allopatric speciation.
   d. hybridization.
   e. prezygotic isolation.
   *Answer: a*

**38.** What is the most important factor promoting the cohesion of a species?
   a. Mutation
   b. Natural selection
   c. Gene flow
   d. Genetic drift
   e. Reproduction
   *Answer: c*

**39.** Behavioral complexity and short generation time will tend to _____ the process of speciation.
   a. accelerate
   b. decelerate
   c. have no effect on

d. obscure

e. threaten

*Answer: a*

40. Some groups or species have given rise to large numbers of daughter species. This phenomenon is known as
   a. sympatric speciation.
   b. hybridization.
   c. evolutionary radiation.
   d. gene flow.
   e. evolution.

   *Answer: c*

41. Which group of organisms has the highest prevalence of sympatric speciation?
   a. Mammals
   b. Insects
   c. Plants
   d. Fungi
   e. Bacteria

   *Answer: c*

42. When individuals from one population disperse to a new, geographically isolated location and mate, the eventual result may be the formation of a new species. Such an outcome is called a _____ event.
   a. foundation
   b. founder
   c. radiative
   d. gene flow
   e. hybridization

   *Answer: b*

43. A species that is found only in a certain area of the planet and nowhere else is called
   a. endangered.
   b. endemic.
   c. extinct.
   d. emetic.
   e. exotic.

   *Answer: b*

44. Which of the following is *not* true of ecology and the rate of speciation?
   a. Speciation rates are correlated with the diets of mammals.
   b. Speciation rates are not related to birth rates.
   c. Speciation rates differ markedly in the fossil record.
   d. Speciation rates are consistent across all organisms that have been studied so far.
   e. Speciation rates are higher among animal species with more complex behavior patterns than they are among animals with simple behavior patterns.

   *Answer: d*

■45. Islands are often called "natural laboratories for evolutionary studies" because they
   a. are isolated from other land masses.
   b. are geologically very young.
   c. have very low speciation rates.
   d. are all ecologically similar.
   e. always are inhabited by small numbers of species.

   *Answer: a*

46. The adaptive radiation of Hawaiian silverswords has demonstrated what fact about speciation?
   a. Major morphological changes can be produced by small genetic changes.

   b. Major morphological changes are produced only by large-scale genetic changes.
   c. Many different silversword ancestors colonized Hawaii.
   d. The Hawaiian silverswords are not distinct species.
   e. Plant speciation always requires polyploidy.

   *Answer: a*

*47. Of the following reproductive isolating mechanisms, which one is the most efficient at preventing waste of reproductive effort?
   a. Hybrid inviability
   b. Hybrid sterility
   c. Gamete incompatibility
   d. Prezygotic isolation
   e. Hybrid cross-reproduction

   *Answer: d*

48. Which mode of speciation is most prevalent among larger animals?
   a. Geographic
   b. Parapatric
   c. Sympatric
   d. Polyploidy
   e. Genetic drift

   *Answer: a*

49. Which of the following is *not* true of speciation?
   a. A new species can be formed in one breeding season.
   b. Animals with complex behavior and mating patterns speciate more rapidly than animals with simpler behaviors.
   c. Members of the same species can live on separate continents.
   d. Speciation is more rapid on island archipelagos such as Hawaii and the Galápagos than it is on the continents.
   e. Sympatric speciation is the most common type among animals.

   *Answer: e*

50. The family Asteraceae contains several genera found on the Hawaiian islands. There is very little genetic variation among them, and they are thought to have evolved from a single ancestral species. What is responsible for the large morphological differences?
   a. Each island has a unique environment to which the plants have adapted.
   b. Predation has eliminated all but the remaining genera on each island.
   c. Birds traveling between islands carried different types of seeds that started the plant growth.
   d. Polyploidy during reproduction created the variation in features.
   e. Both c and d

   *Answer: a*

51. Populations (such as blue and snow geese) that occasionally interbreed and form hybrid zones
   a. cannot be different species.
   b. often result in evolutionary radiation.
   c. can maintain species differences.
   d. have little geographic variation.
   e. will produce polyploid offspring.

   *Answer: c*

**⋆■52.** Why is it so difficult to obtain evidence supporting sympatric speciation?
a. Sympatric speciation always takes a very long time to occur.
b. Sympatric speciation always involves polyploidy.
c. It is hard to distinguish true sympatric speciation from allopatric speciation that occurred in the recent past.
d. It is impossible to show genetic differences in sympatric speciation.
e. Sympatric speciation is a rare event.
*Answer: c*

53. Fertile species produced from asexual reproduction of tetraploid individuals are considered
a. diploid.
b. polyploid.
c. paraploid.
d. allopolyploid.
e. mutants.
*Answer: d*

# 25 Reconstructing and Using Phylogenies

## Fill in the Blank

1. _____ is the theory and practice of classifying organisms.
   *Answer: Taxonomy*

2. _____ is the scientific study of the diversity of organisms.
   *Answer: Systematics*

3. _____, an eighteenth-century Swedish biologist, developed the classification system used today.
   *Answer: Carolus Linnaeus*

4. A two-name classification system, referred to as _____, is used today throughout biology.
   *Answer: binomial nomenclature*

5. A trait such as the modern horse's single toe, which differs from the trait in an organism's ancestors, is called a _____.
   *Answer: derived trait*

6. Any two traits derived from a common ancestral form are called _____.
   *Answer: homologous traits*

7. Traits such as fins in aquatic mammals and fins in fishes exhibit _____; that is, they evolved in different lineages but are not found in their most recent common ancestor.
   *Answer: homoplasy*

8. Traits that evolved by _____ were formerly very different but now resemble one another because they have undergone selection to perform similar functions.
   *Answer: convergent evolution*

9. _____ classification shows evolutionary relationships and expresses them in treelike diagrams.
   *Answer: Cladistic*

10. The entire portion of a phylogeny that is descended from a common ancestor is called a _____.
    *Answer: clade*

11. The term "Rosaceae" is an example of the taxonomic category of _____.
    *Answer: family*

12. In the Linnaean system, classes are divided into _____, which are then divided into families.
    *Answer: orders*

13. The operating rule of cladogram design, that it is wiser to postulate a minimal number of changes in traits, is called _____.

    *Answer: parsimony*

## Multiple Choice

1. In modern systematics, each family name is based on
   a. the name of the order to which it belongs.
   b. a characteristic common to all members.
   c. the name of a member genus.
   d. the name of the largest member species.
   e. the Latin name for the organisms.
   *Answer: c*

2. In the Linnaean system, the suffix "-aceae" refers to a(n)
   a. genus of plants.
   b. genus of animals.
   c. family of plants.
   d. family of animals.
   e. order of plants.
   *Answer: c*

■3. North America and Great Britain both have birds called robins. These birds all have brown backs and red breasts, but they are incapable of interbreeding, are different sizes, and have different diets and habitats. Based on this information, one would expect these birds to belong to
   a. the same species.
   b. the same genus, but different species.
   c. the same family, but different genera.
   d. different species, but more information is needed for further classification.
   e. different genera, but more information is needed for further classification.
   *Answer: d*

4. Within a family, the number of species placed into each genus is determined by
   a. the evolutionary uniqueness among organisms.
   b. an even distribution of the total number of species.
   c. an even distribution of the species, based on their sizes.
   d. grouping species by their geographic ranges.
   e. clustering species based on their habitats.
   *Answer: a*

5. The wing of a bat and the wing of a bird are an example of
   a. divergent evolution.
   b. vertical evolution.
   c. convergent evolution.
   d. a derived trait.
   e. evolutionary reversal.
   *Answer: c*

6–9. Use the cladogram below to answer the following questions.

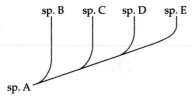

sp. B    sp. C    sp. D    sp. E

sp. A

6. Assuming the cladogram includes modern species, which of these species is (are) alive today?
   a. A, B, C, D, E
   b. B, C, D, E
   c. C, D, E
   d. D, E
   e. E
   *Answer: b*

*7. Species D and E share _____ homologous and _____ homoplastic traits.
   a. many; many
   b. many; few
   c. few; many
   d. few; few
   e. no; no
   *Answer: b*

*8. Which species shares the most recent common ancestor with species E?
   a. Species A
   b. Species B
   c. Species C
   d. Species D
   e. All of the above
   *Answer: d*

9. The position of species B in the cladogram indicates that it
   a. became extinct before species C, D, or E.
   b. has fewer derived traits than species C, D, or E.
   c. has a shorter evolutionary history than species C, D, or E.
   d. shares more common ancestors with species C than it does with species D or E.
   e. is less fit for its environment than species C, D, or E.
   *Answer: b*

10. When compared to other taxa, a taxon that shares general homologous traits but lacks special homologous traits is called a(n)
    a. clade.
    b. genus.
    c. homoplasy.
    d. outgroup.
    e. population.
    *Answer: d*

**11. In a comparison of humans and chimps with dogs and cats, a general homologous trait would be
    a. hands specialized for grasping.
    b. presence of body hair.
    c. standing on two legs.
    d. lack of a tail.
    e. poor sense of smell.
    *Answer: b*

**12. When comparing humans and chimps with dogs and cats, a special homologous trait would be
    a. bony skeleton.
    b. presence of body hair.
    c. hands specialized for grasping.
    d. mouth containing teeth.
    e. presence of circulatory system.
    *Answer: c*

13. Which of the statements below is often assumed to be true in the construction of cladograms?
    a. Derived traits appear only once in a lineage.
    b. Branching points are determined by the number of homologous traits.
    c. Derived traits are never lost.
    d. Both a and b
    e. Both a and c
    *Answer: e*

14–17. Use the table below to answer the following questions.

**Ancestral and Derived Traits Among Five Species**

|         | Trait |   |   |   |   |
|---------|-------|---|---|---|---|
| Species | 1     | 2 | 3 | 4 | 5 |
| A       | 1     | 1 | 0 | 0 | 1 |
| B       | 0     | 0 | 1 | 0 | 1 |
| C       | 1     | 1 | 0 | 0 | 1 |
| D       | 0     | 1 | 0 | 0 | 1 |

The ancestral form of each trait is coded 0, and the derived trait is coded 1.

*14. In constructing a cladogram, the oldest divergence would separate
    a. A from B, C, and D.
    b. B from A, C, and D.
    c. C from A, B, and D.
    d. D from A, B, and C.
    e. A and D from B and C.
    *Answer: b*

*15. The two species that diverged most recently from one another are
    a. A and B.
    b. A and C.
    c. A and D.
    d. B and C.
    e. B and D.
    *Answer: b*

*16. Which trait is most recently derived?
    a. 1
    b. 2
    c. 3
    d. 4
    e. 5
    *Answer: a*

*17. Which trait is most ancestral?
    a. 1
    b. 2
    c. 3
    d. 4
    e. 5
    *Answer: d*

18. Which statement is true about the use of behavioral traits in the reconstruction of phylogenies?
    a. Behavioral traits are never relevant.
    b. Behavioral traits are more important than other traits among closely related species.
    c. Knowledge of behavioral traits supports knowledge of other traits among closely related species.
    d. Knowledge of behavioral traits supports knowledge of other traits among distantly related species.
    e. Behavioral traits are always relevant.
    *Answer: c*

19. Systematists usually employ parsimony in reconstructing a phylogeny. Parsimony involves arranging the organisms such that _____ in determining the lineage.
    a. the minimal number of changes in traits is postulated
    b. the maximum number of traits is used
    c. molecular information is given priority over other traits
    d. larval traits are included
    e. ancestral traits are given priority over derived traits
    *Answer: a*

20. _____ is the study of biological diversity and its evolution.
    a. Phylogeny
    b. Systematics
    c. Taxonomy
    d. Classification
    e. Nomenclature
    *Answer: b*

21. _____ is the science of biological classification.
    a. Phylogeny
    b. Systematics
    c. Taxonomy
    d. Classification
    e. Nomenclature
    *Answer: c*

22. Classification systems have many uses. Which of the following is *not* a goal of biological classification?
    a. To depict convergent evolution
    b. To clarify relationships among organisms
    c. To help us remember organisms and their traits
    d. To clearly identify organisms being studied
    e. To provide predictive powers
    *Answer: a*

23. Taxonomic systems used by biologists are hierarchical; that is,
    a. taxonomic groups reflect shared characters, not evolutionary relationships.
    b. each higher taxonomic group contains all the groups below it.
    c. taxonomic groups reflect common habitats.
    d. a hierarchy of traits is used to establish classifications.
    e. phylogenetic relationships do not help us understand evolution.
    *Answer: b*

24. Classification systems serve four important roles. Which of the following is *not* one of those roles?
    a. To help us remember characteristics of a large number of different things
    b. To help us identify shared traits, such as hair, mammary glands, and constant high body temperature in mammals

c. To reveal the harmony of nature
    d. To provide stable, unique, unequivocal names for organisms
    e. To help reconstruct evolutionary pathways
    *Answer: c*

25. The biological classification system used today is based on the work of
    a. Charles Darwin.
    b. Barbara McClintock.
    c. Gregor Mendel.
    d. Lynn Margulis.
    e. Carolus Linnaeus.
    *Answer: e*

26. The biological classification system used today is referred to as
    a. dichotomous taxonomy.
    b. dichotomous nomenclature.
    c. dichotomous keys.
    d. binomial taxonomy.
    e. binomial nomenclature.
    *Answer: e*

27. Which of the following is the correct hierarchy of categories from most inclusive to least inclusive in the classification system used today?
    a. Division or phylum, kingdom, order, family, class, genus, species
    b. Division or phylum, kingdom, order, class, family, genus, species
    c. Division or phylum, kingdom, class, order, family, genus, species
    d. Kingdom, division or phylum, order, class, family, genus, species
    e. Kingdom, division or phylum, class, order, family, genus, species
    *Answer: e*

28. In referring to the species of an organism in writing (e.g., in a newspaper, textbook, or lab report), which of these rules should be followed?
    a. Underline or italicize genus and species
    b. The first letter of genus should be uppercase.
    c. The first letter of species should be uppercase.
    d. Both a and b
    e. a, b, and c
    *Answer: d*

29. If you saw the name of a family of organisms and it ended with "-idae" (e.g., Formicidae), you would know that it was a family of
    a. bacteria.
    b. fungi.
    c. plants.
    d. animals.
    e. protists.
    *Answer: d*

30. A consensus tree is produced by
    a. merging multiple phylogenetic trees.
    b. aligning two phylogenetic trees.
    c. subtracting two phylogenetic trees.
    d. using only molecular data.
    e. using only morphological data.
    *Answer: a*

31. When a research team employs the maximum likelihood method to construct a phylogeny, what information are they basing it on?
    a. Morphological
    b. Environmental
    c. Molecular
    d. Parsimony
    e. Fossil
    *Answer: c*

32. In comparing an adult sea squirt to vertebrate animals using morphological data, sea squirts and vertebrates would seem quite unrelated; what other information could be useful to verify this conclusion?
    a. Fossil
    b. Developmental
    c. Environmental
    d. Asystematic
    e. None
    *Answer: b*

33. The molecular structures most often considered in the construction of phylogenies are those of
    a. proteins and nucleic acids.
    b. lipids and nucleic acids.
    c. carbohydrates and lipids.
    d. proteins and lipids.
    e. carbohydrates and nucleic acids.
    *Answer: a*

34. In a cladistic classification, each taxon
    a. includes several lineages from a single common ancestor.
    b. is a single lineage and includes all—and only—the descendants of a single ancestor.
    c. shares common morphologies that are homoplasic.
    d. shares common morphologies that are derived characters.
    e. includes several lineages that share common morphologies.
    *Answer: b*

35. The excellent fossil record of horses shows that modern horses, which have one toe on each foot, evolved from ancestors that had multiple toes. A trait that differs from the ancestral trait in the lineage is called a(n) _____ trait.
    a. derived
    b. ancestral
    c. morphological
    d. biochemical
    e. fundamental
    *Answer: a*

36. Any two structures derived from a common ancestral trait are said to be
    a. analogous.
    b. morphological traits.
    c. biochemical traits.
    d. homologous.
    e. homoplasic.
    *Answer: d*

*37. Homoplasy can result from convergent evolution, in which
    a. structures that formerly were very different come to resemble one another because they have undergone selection to perform similar functions.

b. the same character evolves in different lineages, often from a common basis.
    c. a structure evolves in a lineage that is not found in their common ancestor.
    d. a character from a particular ancestral trait evolves in different lineages because of similar selection pressures.
    e. an ancestral trait evolves into different characters in different lineages.
    *Answer: a*

38. Genetic similarities among some vertebrates have been estimated by DNA sequencing. According to these data, humans are most closely related to
    a. gibbons.
    b. chimpanzees.
    c. baboons.
    d. galagos.
    e. rhesus monkeys.
    *Answer: b*

*39. In groups such as birds, _____ make(s) it difficult to resolve phylogenies using only morphological data, because the use of different traits produces different phylogenies.
    a. reverse evolution
    b. parallel evolution
    c. convergent evolution
    d. genetic drift
    e. founder events
    *Answer: c*

40. A taxon is
    a. an archaic concept not used in modern classification systems.
    b. any group of organisms treated as a unit.
    c. a single species.
    d. a group of organisms that are reproductively isolated from the organisms of other such groups.
    e. the smallest grouping in the Linnaean classification system.
    *Answer: b*

41. Which of the following choices is the conventional representation for the name of a sea star common to the rocky intertidal zone?
    a. Pisaster Ochraceous
    b. *Pisaster Ochraceous*
    c. *Pisaster ochraceous*
    d. *pisaster ochraceous*
    e. pisaster ochraceous
    *Answer: c*

42. In a scientific paper, which of the following represents, by convention, the correct binomial name of the English bluebell after the name has already been cited once (i.e., earlier in the paper)?
    a. *Edymion nonscriptus*
    b. Edymion n.
    c. *E. nonscriptus*
    d. E. n.
    e. E. nonscriptus
    *Answer: c*

43. In the Linnaean classification system, which one of the following taxa usually ends in "-idae" when used with animals?
    a. Genus
    b. Order

c. Division
d. Class
e. Family
*Answer: e*

44. Which of the following statements about classification is *false*?
    a. Members of a family are less similar than members of an included genus.
    b. An order has more members than a genus has.
    c. Families have more members than phyla have.
    d. The common ancestor shared by members of a family is a more distant ancestor than the one shared by members of an included genus.
    e. The number of species in a taxon depends on the relative similarity of the different species.
    *Answer: c*

45. All of the following can result in homoplasy *except*
    a. convergent evolution.
    b. parallel evolution.
    c. reverse evolution.
    d. descent from a common ancestor.
    e. similar selection pressures.
    *Answer: d*

46. A taxon consisting of members that do *not* share the same common ancestor is
    a. monophyletic.
    b. polyphyletic.
    c. paraphyletic.
    d. unrelated.
    e. a clade.
    *Answer: b*

47. A monophyletic group containing all the descendants of a particular ancestor and no other organisms is also know as a
    a. phylogeny.
    b. clade.
    c. composite.
    d. breeding group.
    e. species.
    *Answer: b*

48. Which of the following techniques for studying the biochemical traits of organisms could be used to understand the evolution of cytochrome *c* (a protein)?
    a. Immunological distance determination
    b. Amino acid sequencing
    c. DNA hybridization
    d. Nucleic acid base sequencing
    e. cpDNA and mtDNA comparisons
    *Answer: b*

49. Most taxonomists today believe that classification systems should be
    a. phylogenetic.
    b. monophyletic.
    c. paraphyletic.
    d. polyphyletic.
    e. changed.
    *Answer: b*

■50. In systematics and phylogeny, the fossil record is especially important because
    a. most groups are well represented.
    b. it provides the absolute timing of evolutionary events.
    c. random mutations make most biochemical methods unreliable.
    d. DNA can be extracted from the fossils and analyzed.
    e. it is the only type of data useful in reconstructing the past.
    *Answer: b*

51. In reconstructing phylogenies, an outgroup is used to
    a. distinguish between ancestral traits and derived traits.
    b. exclude a taxon from the phylogenetic group.
    c. distinguish homoplasy from convergent traits.
    d. gain knowledge of reverse evolution.
    e. Both b and c
    *Answer: a*

■52. Larval stages can help determine relationships among organisms, but care must be taken because
    a. not all organisms have a larval stage.
    b. the larval form is morphologically very different from the adult form.
    c. the larval form may closely resemble an organism that the adult stage does not.
    d. larval forms all have a notochord and all adult forms do not.
    e. Both b and d
    *Answer: c*

53. Reconstructed phylogenies can be useful for all of the following *except*
    a. studying the evolution of human language.
    b. studying the migration of human populations.
    c. determining if divergent traits occurred from reverse evolution.
    d. predicting future trends in evolution.
    e. evaluating how a protein has changed in different species.
    *Answer: d*

# 26 Molecular and Genomic Evolution

## Fill in the Blank

1. The mechanisms of morphological change were not understood until discoveries were made in _____.
**Answer: biochemistry**

2. Molecular evolution studies the patterns of _____ in molecules and the process by which this is achieved.
**Answer: evolutionary change**

3. A _____ mutation is the partial or complete replacement of a nucleotide base or longer sequence by another throughout an entire population or species.
**Answer: substitution**

4. According to the hypothesis of _____, most of the variability in molecules measured by evolutionists does not affect their functioning.
**Answer: neutral evolution**

5. _____ is a molecule similar to hemoglobin that has a greater affinity for oxygen.
**Answer: Myoglobin**

6. Lungfishes and lilies have more _____ DNA than humans but less of it is _____ DNA.
**Answer: total; coding**

7. If a protein has consistent changes over time, then its _____ ticks at a constant rate.
**Answer: molecular clock**

8. DNA that has changed little during evolution probably has the same _____ in all species.
**Answer: function**

9. A visual representation of the evolutionary relationship between species is called a _____.
**Answer: gene tree**

10. Morphological traits are not useful for comparisons between bacteria and humans because these organisms lack _____.
**Answer: comparable structures**

11. In studying lysozyme, researchers found that five amino acid substitutions on its surface affect its biological

_____.
**Answer: activity**

12. Evidence suggests that the three families of globin genes arose as gene _____.
**Answer: duplications**

13. Langurs and cows share a very similar _____ molecule.
**Answer: lysozyme**

14. To infer times of the most ancient splits in lineages, scientists must study _____ that are found in all organisms and that evolve slowly.
**Answer: molecules**

15. A _____ is a group of homologous genes with related functions that have been produced by successive rounds of duplication and mutation.
**Answer: gene family**

## Multiple Choice

1. Molecular evolution
a. studies the patterns of evolutionary change in molecules.
b. studies the processes of molecular change.
c. compares the molecules of living organisms.
d. can determine the linkages of various species based on molecular changes.
e. All of the above
*Answer: e*

■2. Why is the protein cytochrome *c* important in the study of molecular evolution?
a. It is found in all eukaryotes.
b. Its mutation rate is very low.
c. Its mutation rate fluctuates to a great extent.
d. It is found in all animals.
e. All of its mutations are adaptive changes.
*Answer: a*

3. Using the polymerase chain reaction, DNA has been amplified from human fossils more than _____ years old.
a. 40,000
b. 135,000
c. 13.5 million
d. 135 million
e. 135 billion
*Answer: a*

4. Which of the following is *not* an example of a gene that arose from an ancient gene duplication?
a. rRNA
b. tRNA
c. Hemoglobin
d. Pyruvate kinase
e. Lactate dehydrogenase
*Answer: c*

5. Which of the following is *not* true of lysozyme?
   a. It is produced in tears and saliva.
   b. It can digest bacterial cell walls.
   c. It can digest plant matter.
   d. Langur lysozyme must function at a low pH.
   e. It is found in the whites of bird eggs.
   *Answer: c*

■6. What changes did scientists discover when they studied the amino acid sequence of lysozyme in langurs and cows?
   a. The changes were on the active site of the molecule.
   b. Arginine was changed to lysine, making it more resistant to trypsin.
   c. Lysine was changed to arginine, making it less resistant to trypsin.
   d. The changes were all neutral and did not affect the function of the molecule.
   e. The lysozyme in langurs and cows can break down plant matter.
   *Answer: b*

■7. Cows and langurs have nearly identical lysozyme because
   a. lysozyme is nearly identical in all animals, not just cows and langurs.
   b. both ferment leafy food in their stomachs.
   c. both share a common ancestor.
   d. the bacteria in their foregut caused mutations in their lysozyme.
   e. None of the above
   *Answer: b*

8. Select the group that shares the most similar lysozyme molecule.
   a. Horse, cow, pig
   b. Langur, baboon, human
   c. Deer, antelope, horse
   d. Neotropical cuckoo, sparrow, cow
   e. Neotropical cuckoo, cow, langur
   *Answer: e*

9. Select the correct order, from smallest to largest, of total DNA in these organisms.
   a. Bacterium, lungfish, human
   b. Fruit fly, bacterium, human
   c. Lungfish, human, fruit fly
   d. Bacterium, human, lungfish
   e. Yeast, bacterium, fruit fly
   *Answer: d*

10. Which of the following about the difference in genome size between lungfish and humans is true?
    a. Lungfish have less total DNA than humans but more coding DNA.
    b. Lungfish have more total DNA than humans but less coding DNA.
    c. Lungfish have more total DNA than humans because they have more genes.
    d. Lungfish have less total DNA than humans because they are simpler organisms.
    e. Lungfish have the same amount of total DNA as humans but less coding DNA.
    *Answer: b*

11. Most changes in nucleic acids or amino acids
    a. change the function of a molecule.
    b. occur at constant locations.
    c. can be predicted.
    d. do not change the function of a molecule.
    e. are influenced by natural selection.
    *Answer: d*

12. A molecular clock is useful for
    a. estimating lineage divergence during evolution.
    b. figuring the rates of change of a molecule.
    c. predicting where a molecule will mutate.
    d. timing when a new gene duplication will arise.
    e. timing the aging of an organism.
    *Answer: a*

13. The consistency of molecular clocks is checked by
    a. finding the rate of mutation of a molecule.
    b. using well-dated fossil records as a comparison.
    c. artificially creating mutations of molecules in the lab and timing them.
    d. comparing similar molecules.
    e. using cytochrome *c* as a comparison.
    *Answer: b*

14. Molecules that evolve slowly are used to study
    a. very similar species.
    b. relationships of all organisms.
    c. extinct organisms.
    d. duplicate genes.
    e. organisms that diverged long ago.
    *Answer: e*

15. Molecules that have a rapid mutation rate are used to study
    a. very similar species.
    b. relationships of all organisms.
    c. extinct organisms.
    d. duplicate genes.
    e. organisms that diverged long ago.
    *Answer: a*

■16. Ribosomal RNA is a useful molecule for the study of evolution because it
    a. mutates frequently.
    b. mutates very slowly.
    c. is found only in a few organisms.
    d. varies greatly from organism to organism.
    e. does not have functional constraints.
    *Answer: b*

17. The times of lineage splits between the major branches of life—Bacteria, Archaea, and Eukarya—have been estimated using
    a. lysozyme.
    b. cytochrome *c*.
    c. small-subunit rRNA.
    d. hemoglobin.
    e. genome size.
    *Answer: c*

18. The polymerase chain reaction allows biologists to
    a. amplify DNA.
    b. sequence DNA.
    c. observe DNA.
    d. amplify RNA.
    e. sequence RNA.
    *Answer: a*

19. If there are about 100 substitutions in the myoglobin molecule per 500 million years, then two organisms differing by 75 amino acids split about _____ million years ago.
    a. 500
    b. 125
    c. 380
    d. 485
    e. 65
    *Answer: c*

20. Gene duplication may involve
    a. part of a gene.
    b. a whole gene.
    c. whole chromosomes.
    d. a, b, and c
    e. None of the above
    *Answer: d*

21. Which of the following is *not* true of gene duplication?
    a. All duplications produce functional genes.
    b. It can result in evolution of novel functions in proteins.
    c. It created much of the diversity in our genome.
    d. rRNA and tRNA arose from duplications.
    e. It can result in increased complexity.
    *Answer: a*

22. To achieve more accurate estimates of lineage divisions than data from one molecule can provide,
    a. molecular data is combined with morphological data.
    b. molecular data is combined with fossil data.
    c. the data from two or more molecules are combined.
    d. Both a and b
    e. a, b, and c
    *Answer: e*

23–26. The following questions are based on the hypothetical similarity matrix below:

| Human | Horse | Cow | Langur | Baboon | |
|---|---|---|---|---|---|
| | 9 | 11 | 1 | 0 | **Human** |
| 1 | | 1 | 12 | 13 | **Horse** |
| 1 | 18 | | 14 | 12 | **Cow** |
| 14 | 0 | 0 | | 1 | **Langur** |
| 16 | 0 | 0 | 17 | | **Baboon** |

23. What do the numbers on the upper right stand for?
    a. Similarities in amino acid sequence
    b. Differences in amino acid sequence
    c. Number of total amino acids
    d. Number of deletions in amino acids
    e. Number of additions in amino acids
    *Answer: b*

24. What do the numbers on the bottom left stand for?
    a. Similarities in amino acid sequence
    b. Differences in amino acid sequence
    c. Number of total amino acids
    d. Number of deletions in amino acids
    e. Number of additions in amino acids
    *Answer: a*

*25. Which of the following organisms are the most similar?
    a. Human and horse
    b. Horse and cow
    c. Human and baboon
    d. Langur and cow
    e. Human and langur
    *Answer: b*

*26. Which of the following organisms are the most different?
    a. Human and horse
    b. Horse and cow
    c. Human and baboon
    d. Langur and cow
    e. Human and langur
    *Answer: d*

27. The three families of human globin genes arose from
    a. spontaneous mutations.
    b. gene duplication.
    c. ancestral DNA.
    d. differential splicing of DNA.
    e. transposable elements.
    *Answer: b*

28. Select the statement about mitochondrial DNA that is *false*.
    a. It is maternally inherited.
    b. It was used to give validation to the "out of Africa" hypothesis.
    c. It mutates very slowly.
    d. It is useful for studying recent evolution of closely related species.
    e. None of the above
    *Answer: c*

*29. Ribosomal RNA has evolved very slowly because even minor changes in its base sequence will
    a. activate ribosomes.
    b. never occur.
    c. duplicate genes.
    d. inactivate ribosomes.
    e. be advantageous.
    *Answer: d*

30. The hypothesis of neutral evolution asserts that
    a. the rate of molecular mutation is influenced by natural selection.
    b. the variability in structures of molecules does not affect their functioning.
    c. closely related species have more similar molecular structures than distantly related species.
    d. organisms evolved through neutral changes in their molecules.
    e. mutations neither add nor subtract amino acids from molecules.
    *Answer: b*

31. The earliest organisms known to have both α- and β-globin genes lived about _____ years ago.
    a. 500 million
    b. 1 billion
    c. 100 million
    d. 50 million
    e. 1 million
    *Answer: a*

32. Homeobox elements have been identified in
    a. flies only.
    b. animals.
    c. plants.
    d. bacteria.
    e. Both b and c
    *Answer: e*

33. Sometimes molecular data can lead scientists to conclude that two species are more closely related than other evidence suggests. One explanation for the misleading data is
    a. convergent evolution.
    b. divergent evolution.
    c. neutral evolution.
    d. gene duplication.
    e. None of the above
    *Answer: a*

34. Ancient lineage data supports the division of living organisms into three major branches:
    a. Bacteria, Archaea, and Eukarya.
    b. Bacteria, Prokaryotes, and Eukarya.
    c. Bacteria, Eukarya, and Plants.
    d. Bacteria, Humans, and Plants.
    e. Archaea, Eukarya, and Animals.
    *Answer: a*

35. The invariant positions 14, 17, 18, and 80 in cytochrome *c* interact with
    a. the positions on the molecule that change rapidly.
    b. the other invariant positions.
    c. the heme group.
    d. hemoglobin.
    e. myoglobin.
    *Answer: c*

36. If there were mutations in the invariant positions of a cytochrome *c* molecule, what would be the result?
    a. The molecule would have a new function.
    b. The molecule would no longer be functional.
    c. The molecule would still bind the iron-containing group.
    d. The changes would not affect the function of the molecule.
    e. The molecule would bind to another metal-containing group.
    *Answer: b*

37. Genes that are nonfunctional and that probably resulted from gene duplication are called
    a. invariant genes.
    b. homeobox genes.
    c. homologous genes.
    d. pseudogenes.
    e. exons.
    *Answer: d*

38. Which of the following is *not* true of hemoglobin?
    a. It is a simpler molecule than myoglobin.
    b. It binds four molecules of oxygen.
    c. It is a more refined molecule than myoglobin.
    d. It has a lower affinity for oxygen than myoglobin has.
    e. It is a tetramer of two $\alpha$ and two $\beta$ chains.
    *Answer: a*

39. By studying the similarities of molecules in different species, scientists can determine
    a. common ancestry.
    b. the position of branches in a cladogram.
    c. which functions are similar in widely divergent species.
    d. Both a and b
    e. a, b, and c
    *Answer: e*

40. Which of the following must be true in order to use molecular clocks to date events?
    a. There must have been relatively few adaptive changes in the molecule being studied.
    b. Scientists must know what proportion of molecular clocks tick at a constant rate.
    c. The molecular clock of the molecule being studied must tick very slowly.
    d. Both a and b
    e. a, b, and c
    *Answer: d*

41. Ancient lineage divisions can be estimated by means of _____ data.
    a. molecular
    b. morphological
    c. fossil
    d. molecular and fossil
    e. molecular, morphological, and fossil
    *Answer: e*

# 27 Bacteria and Archaea: The Prokaryotic Domains

## Fill in the Blank

1. The most abundant organisms on Earth are prokaryotes, which fall into the kingdoms Archaea and _____.
   *Answer: Bacteria*

2. Bacteria unable to survive for extended periods in the absence of oxygen are termed _____.
   *Answer: obligate aerobes*

3. Photoautotrophs use light as their source of energy and _____ as their source of carbon.
   *Answer: $CO_2$*

4. Of the four nutritional categories of archaea and bacteria, most are _____, as are all animals, fungi, and many protists.
   *Answer: chemoheterotrophs*

5. In 1884 _____ developed a staining process for bacteria that is still the single most common tool in the study of bacteria.
   *Answer: Hans Gram*

6. _____ is the ability of a bacterial pathogen to enter into and multiply within the body of a host.
   *Answer: Invasiveness*

7. _____ is the ability of a bacterial pathogen to produce chemical substances injurious to the tissues of the host.
   *Answer: Toxigenicity*

8. Eukarya are evolutionarily closer to _____ than they are to Bacteria.
   *Answer: Archaea*

9. *Treponema pallidum*, the causative agent of syphilis, belongs to the group of bacteria known as the _____.
   *Answer: spirochetes*

10. *Agrobacterium tumefaciens* is an important bacterium that is especially useful for inserting genes into _____ hosts.
    *Answer: plant*

11. One of the smallest of the bacteria, _____, has a complex intracellular reproductive cycle.
    *Answer: chlamydia*

*12. The important bacterium *Escherichia coli* has a _____ shape and a _____ Gram stain.
    *Answer: rod; negative*

13. _____ have a branched system of filaments and were once thought to be fungi.
    *Answer: Actinomycetes*

14. Many Gram-negative bacteria have _____ composed of lipopolysaccharides that can cause fever and vomiting.
    *Answer: endotoxins*

15. Antibiotics such as penicillin have little, if any, effect on the cells of eukaryotes because they interfere with the synthesis of cell walls containing _____.
    *Answer: peptidoglycan*

16. Gram-positive bacteria stain _____, whereas Gram-negative bacteria stain _____.
    *Answer: violet; pink-red*

## Multiple Choice

1. The size of the smallest bacterium is _____ μm.
   a. 0.2
   b. 2
   c. 7.5
   d. 75
   e. 750
   *Answer: a*

2. Which of the following is *not* a basic unit of a prokaryotic cell?
   a. DNA
   b. RNA
   c. Enzymes for transcription and translation
   d. A system for generating ATP
   e. An immune system
   *Answer: e*

3. Which of the following is found in both eukaryotes and prokaryotes?
   a. Organelles
   b. A system for generating ATP
   c. A nucleus
   d. Chromatin
   e. Histones
   *Answer: b*

4. Which of the following is *not* a characteristic of prokaryotic cells?
   a. Mesosome
   b. A photosynthetic membrane system
   c. Peptidoglycan
   d. Circular DNA
   e. Organelles
   *Answer: e*

5. Bacteria move by means of
   a. flagella, gas vesicles, and rolling.
   b. flagella, cilia, and axial filaments.
   c. axial filaments, rolling, and pseudopods.
   d. cilia, pseudopods, and axial filaments.
   e. pseudopods, flagella, and cilia.
   *Answer: a*

6. Some cyanobacteria that form filamentous colonies possess heterocysts, which are
   a. a means of locomotion.
   b. cells specialized for nitrogen fixation.
   c. reproductive cells.
   d. endospores.
   e. None of the above
   *Answer: b*

7. Bacteria participate in
   a. digestion in animals.
   b. processing nitrogen and sulfur in soils.
   c. decomposition in all ecosystems.
   d. many industrial and commercial processes.
   e. All of the above
   *Answer: e*

8. The earliest prokaryotic fossils date back at least _____ years.
   a. 35,000
   b. 350,000
   c. 3.5 million
   d. 3.5 billion
   e. 3.5 trillion
   *Answer: d*

9. Which of the following is a broad nutritional category of bacteria that is recognized by biologists?
   a. Physioautotrophs
   b. Heteroautotrophs
   c. Chemolithotrophs
   d. Both a and b
   e. Both a and c
   *Answer: c*

*10. Bacteria may differ from one another
   a. structurally.
   b. metabolically.
   c. reproductively.
   d. Both a and c
   e. a, b, and c
   *Answer: e*

11. The highest categorization of life consists of
   a. bacteria, archaea, and eukarya.
   b. bacteria, fungi, plants, and animals.
   c. bacteria, protists, fungi, plants, and animals.
   d. plants and animals.
   e. None of the above
   *Answer: a*

12. Bacteria reproduce
   a. only asexually.
   b. only sexually.
   c. asexually and sexually by transformation, conjugation, and transduction.
   d. following mitosis.
   e. Both a and d
   *Answer: c*

*13. Prokaryotic flagella
   a. are structurally related to those of eukaryotes.
   b. are similar to cilia.
   c. are structurally unrelated to those of eukaryotes.
   d. operate the same way as those of spermatozoa.
   e. a, b, and d
   *Answer: c*

*14. The space between the outer membrane and the cell wall of a bacteriaum is called the _____ space.
   a. periplasmic
   b. negative
   c. perimembrane
   d. enzymatic
   e. resistance
   *Answer: a*

15. The chlorophyll of cyanobacteria is
   a. like that of plants.
   b. distinct from the chlorophyll of plants.
   c. bacteriochlorophyll.
   d. bacteriorhodopsin.
   e. None of the above
   *Answer: a*

16. The range of time between cell divisions for different bacteria in a vegetative state is
   a. 10 minutes to 100 years.
   b. 1 minute to 60 minutes.
   c. 10 minutes to 60 minutes.
   d. 20 minutes to 120 minutes.
   e. 1 minute to 1000 years.
   *Answer: a*

**17. The sequencing of rRNA allowed scientists to understand
   a. the clear relationships of different prokaryotic species.
   b. a signature sequence unique to archaea and eukarya.
   c. that archaea are similar to bacteria.
   d. that all DNA is relatively the same.
   e. that half of the sequences of archaea were previously unknown.
   *Answer: b*

18. Synthesis of peptidoglycan-containing cell walls is affected by
   a. archaea.
   b. antibiotics.
   c. bacteria.
   d. high toxigenicity.
   e. Gram stains.
   *Answer: b*

19. One factor that determines the consequences of a bacterial infection for the host is the ability of the bacterium to produce chemical substances injurious to the tissues of the host. The anthrax-causing bacterium, *Bacillus anthracis*, produces few toxins however, it can multiply readily and ultimately invades the entire bloodstream. We say that such a bacterium has the qualities of _____ invasiveness and _____ toxigenicity.
   a. low; high
   b. high; high
   c. low; low
   d. high; low
   e. moderate; moderate
   *Answer: d*

20. Bacteria are known for the many roles they play in biological communities. Which of the following roles is the *rarest* for this group of organisms?
    a. Pathogens
    b. Digestive aids
    c. Nitrogen and sulfur processing in soils
    d. Decomposers
    e. Uses in industry and agriculture
    *Answer: a*

21. Which one of the following is *not* a motivation for the classification of biological organisms?
    a. Convenience
    b. Showing evolutionary affinity
    c. Facilitating identification
    d. Displaying diversity
    e. All are motivations
    *Answer: e*

■22. An ecosystem based on chemolithotrophs exists 2,500 meters below the ocean surface near the Galápagos Islands. These archaea
    a. use light as energy and use carbon dioxide for carbon.
    b. use light as energy and get organic compounds from other organisms.
    c. oxidize inorganic substances for energy and use carbon dioxide for carbon.
    d. get both energy and carbon from organic compounds.
    e. oxidize organic compounds for energy and use carbon dioxide for carbon.
    *Answer: c*

23. Which of the following statements about Archaea is *false*?
    a. They include some species that live in environments with extreme salinity and low oxygen.
    b. They have peptidoglycan in their cell walls.
    c. Their rRNA differs as much from Bacteria's DNA as it does from that of Eukarya.
    d. They include some species that are obligate anaerobes, which produce all the methane in the atmosphere.
    e. They include some species that love heat and acid and may die of "cold" at 55°C (131°F).
    *Answer: b*

24. Which of the following statements about bacteria is true?
    a. Gram-positive bacteria have a lot of peptidoglycan in their cell walls and stain blue to purple.
    b. Gram-positive bacteria have relatively little peptidoglycan in their cell walls and stain pink to red.
    c. Gram-positive bacteria weigh more than a gram.
    d. Gram-positive bacteria weigh more than a milligram.
    e. Gram-negative bacteria have no peptidoglycan in their cell walls.
    *Answer: a*

25–30. Match the following descriptions of groups of bacteria with their names from the list below.
    a. Cyanobacteria
    b. Spirochetes
    c. Chlamydia
    d. Gram-negative rods
    e. Actinomycetes
    f. Mycoplasmas

25. These Gram-positive bacteria, which form myceliaum, were once classified as fungi. They include the bacteria that cause tuberculosis and produce streptomycin. Most antibiotics come from bacteria in this group.
    *Answer: e*

26. These bacteria photosynthesize using chlorophyll *a*, are a homogeneous grouping with similar rRNA sequences, and contain photosynthetic lamellae or thylakoids.
    *Answer: a*

27. These bacteria use axial filaments to move and include the bacterium that causes syphilis.
    *Answer: b*

28. These bacteria are not a monophyletic group.
    *Answer: d*

29. These bacteria are small intracellular parasites that have a unique, complex reproductive cycle and include strains that cause eye infections and venereal disease.
    *Answer: c*

30. These bacteria lack cell walls and are among the smallest cellular creatures. They are mostly plant and animal parasites.
    *Answer: f*

31. The Archaea are the closest evolutionary relatives (sister group) to the
    a. Bacteria.
    b. Bacteria and Eukarya.
    c. Eukarya.
    d. Prokaryota.
    e. None of the above
    *Answer: c*

32. Peptidoglycan is a unique feature (synapomorphy) of the
    a. Bacteria.
    b. Bacteria and Eukarya.
    c. Eukarya.
    d. Prokaryota.
    e. None of the above
    *Answer: a*

33. The purple sulfur bacteria use $H_2S$ as an electron donor and release pure sulfur as a waste product. They are examples of
    a. photoautotrophs.
    b. photoheterotrophs.
    c. chemolithotrophs.
    d. chemoheterotrophs.
    e. deep-sea, volcanic vent bacteria.
    *Answer: a*

34. Which of the following nutritional categories of bacteria can exist independently of other organisms?
    a. Photoautotrophs
    b. Photoheterotrophs
    c. Photochemotrophs
    d. Chemoheterotrophs
    e. None of the above
    *Answer: a*

35. The majority of bacteria are
    a. photoautotrophs.
    b. photoheterotrophs.
    c. chemolithotrophs.
    d. chemoheterotrophs.
    e. disease-causing.
    *Answer: d*

36. The Gram method is useful for classifying all bacteria that
    a. have two plasma membranes.
    b. have cell walls.
    c. have cell walls with at least some peptidoglycan.
    d. are prokaryotic.
    e. form endospores.
    *Answer: c*

*37. Which of the following is *not* a characteristic of endospores?
    a. Parent cells can produce more than one.
    b. They can survive harsh environmental conditions.
    c. They contain some cytoplasm and replicated nucleic acid.
    d. They are enclosed within a tough cell wall.
    e. They are a resting structure, not a reproductive structure.
    *Answer: a*

38. Which of the following is *not* a characteristic of archaea?
    a. They live in harsh environments.
    b. Their cell walls lack peptidoglycan.
    c. They are unlike most bacteria.
    d. There are similarities in their base sequences of ribosomal RNA's.
    e. They are a recently evolved group.
    *Answer: e*

39. Which of the following areas or conditions would be favored by thermoacidophiles?
    a. The stomachs of many herbivores
    b. Hot, alkaline springs
    c. Anaerobic conditions
    d. Hot, sulfur springs
    e. Deep-sea volcanic vents
    *Answer: d*

*40. Which of the following is *not* a characteristic of the methanogens?
    a. Methane is their preferred carbon source.
    b. They are associated with mammalian flatulence.
    c. They prefer anaerobic conditions.
    d. Some are thermophilic.
    e. Some live in volcanic vents on the ocean floor.
    *Answer: a*

41. If you were looking for a new heat-tolerant DNA polymerase enzyme, you would investigate
    a. thermoacidophiles.
    b. methanogens.
    c. strict halophiles.
    d. Both a and b
    e. a, b, and c
    *Answer: d*

42. Which one of the following bacterial groups includes the greatest number of species?
    a. Chlamydias
    b. Proteobacteria
    c. Firmicutes
    d. Mycoplasmas
    e. Cyanobacteria
    *Answer: b*

43. Prokaryotes, along with fungi, return tremendous quantities of organic carbon to the atmosphere as
    a. ATP.
    b. $CO_2$.
    c. photosynthesis.
    d. evolution.
    e. nitrogen gas.
    *Answer: b*

44. Spirochetes
    a. are all free-living.
    b. are the only spiral-shaped bacteria.
    c. can form chains of cells.
    d. all possess structures called axial filaments.
    e. are all parasites of other bacteria.
    *Answer: d*

45. Gram-negative rods
    a. are all closely related.
    b. include *Escherichia coli* as well as many human pathogens.
    c. can reproduce only within the cells of other organisms.
    d. all possess structures called axial filaments.
    e. include the important genera *Bacillus* and *Clostridium*.
    *Answer: b*

46. Chlamydias
    a. can form heterocysts.
    b. form a branched, filamentous mycelium.
    c. contain less DNA than any other organism.
    d. can live only within the cells of other organisms.
    e. were once classified as fungi.
    *Answer: d*

47. Cyanobacteria
    a. are a diverse, unrelated group of bacteria.
    b. use chlorophyll *a* and release oxygen during photosynthesis.
    c. are the only group of photosynthetic bacteria.
    d. can reproduce sexually.
    e. are obligate aerobes.
    *Answer: b*

48. The closest monophyletic group to the domain Bacteria is
    a. Eukarya.
    b. Eukarya and Archaea.
    c. Archaea.
    d. Bacteria and Eukarya.
    e. Prokaryotes.
    *Answer: b*

49. Actinomycetes
    a. are photoheterotrophs.
    b. are the source of many important antibiotics.
    c. are Gram-negative bacteria.
    d. were once classified as protists.
    e. are the only bacteria that divide by mitosis.
    *Answer: b*

*50. Mycoplasmas
    a. contain less DNA than most other prokaryotes do.
    b. have peptidoglycan in their cell walls.
    c. form a branched, filamentous mycelium.
    d. possess elaborate internal membrane systems.
    e. can be controlled by penicillin.
    *Answer: a*

51. Which of the following is *not* one of Koch's postulates?
    a. The microorganism is always found in the diseased individual.
    b. The microorganism taken from the diseased host can be grown in pure culture.
    c. A sample of the pure culture of the microorganism produces the disease when injected into an uninfected host.
    d. A host infected by injection from the cultured microorganism yields a new culture of the microorganism identical to the original culture.
    e. The microorganism can be transmitted by insect vectors.
    *Answer: e*

52. An extremely important set of rules for the determination of bacterial disease transmission is
    a. Ehrlich's optimal law.
    b. Koch's postulates.
    c. Occam's razor.
    d. Zeno's paradox.
    e. Darwin's law of evolution by natural selection.
    *Answer: b*

53. Which of the following is true?
    a. Bacteria are more closely related to Archaea than they are to Eukarya.
    b. Bacteria are more closely related to Eukarya than they are to Archaea.
    c. Archaea are more closely related to Bacteria than they are to Eukarya.
    d. Eukarya are more closely related to Archaea than they are to Bacteria.
    e. Eukarya are more closely related to Bacteria than they are to Archaea.
    *Answer: d*

*54. Lateral gene transfer is the process by which
    a. scientists make transgenic organisms.
    b. bacteria within a species exchange genes.
    c. bacteria acquire DNA from a different species.
    d. genes are passed to daughter cells.
    e. None of the above
    *Answer: c*

*55. Which of the following characteristics or components is (are) distinctive to archaea?
    a. Many of their genes
    b. Branched, long-chain hydrocarbons
    c. A lack of peptidoglycan
    d. Lipids with glycerol-ether linkages
    e. All of the above
    *Answer: e*

# 28 Protists and the Dawn of the Eukarya

## Fill in the Blank

1. All eukaryotes that are not plants, fungi, or animals are called _____.
   *Answer: protists*

2. The most basal group of protists includes the _____ and the _____.
   *Answer: diplomonads; parabasalids*

3. The condition of different organisms living together, one inside the other, is called _____.
   *Answer: endosymbiosis*

4. The process by which a diploid generation that produces spores alternates with a haploid generation that produces gametes is called _____.
   *Answer: alternation of generations*

5. _____ are marine protists that secrete shells of calcium carbonate. These protists are valuable in the geological dating of sedimentary rocks and in oil prospecting.
   *Answer: Foraminiferans*

6. _____ are specialized vesicles that excrete the excess water taken in by some protists.
   *Answer: Contractile vacuoles*

7. *Giardia* is a unicellular eukaryote that lacks _____.
   *Answer: mitochondria*

8. Ciliates, such as *Paramecium*, often contain two kinds of nuclei, a single _____ with DNA that is translated and transcribed, and several _____, which are typical eukaryotic nuclei.
   *Answer: macronucleus; micronuclei*

9. Paramecia have an elaborate sexual behavior called _____, in which sexual recombination but not sexual reproduction occurs.
   *Answer: conjugation*

10. The form taken by acellular slime molds under adverse environmental conditions is called a _____. This structure rapidly becomes a _____ again upon restoration of favorable conditions.
    *Answer: sclerotium; plasmodium*

11. The red tides that can kill tons of fish are caused by toxic species of _____.
    *Answer: dinoflagellates*

12. Some chlorophytes exhibit a variation of the heteromorphic life cycle called the _____ life cycle.
    *Answer: haplontic*

13. Many _____ deposit silicon in their cell walls.
    *Answer: diatoms*

14. The protist lineage leading to "green algae" also leads to _____.
    *Answer: plants*

15. The causative agent of African sleeping sickness is _____, a parasitic kinetoplastid euglenozoan.
    *Answer: **Trypanosoma***

16. *Paramecium* uses a precise form of locomotion; by beating its _____, it can move forward or backward.
    *Answer: cilia*

## Multiple Choice

1. Which of the following was an essential step in the evolution of eukaryotic cells?
   a. The development of a flexible cell surface
   b. The development of a cytoskeleton
   c. The development of a nuclear envelope
   d. The endosymbiotic acquisition of certain organelles
   e. All of the above
   *Answer: e*

2. Which of the following is *not* true of conjugation?
   a. Micronuclei disintegrate.
   b. Meiosis takes place.
   c. It is a reproductive process.
   d. Mitosis takes place.
   e. It is a sexual process of genetic recombination.
   *Answer: c*

3. The overall size that unicellular protists can achieve is limited by their
   a. energy-producing potential.
   b. metabolism.
   c. mitochondria.
   d. surface area-to-volume ratio.
   e. Both a and b
   *Answer: d*

4. Which of the following is a sexual reproductive process common to organisms in the Protista?
   a. Binary fission
   b. Multiple fission
   c. Budding and spore formation
   d. Union of gametes
   e. All of the above
   *Answer: e*

5. Perhaps the best-known _____ are the malarial parasites of the genus *Plasmodium*.
   a. turbellarians
   b. mastigophorans
   c. dinoflagellates
   d. apicomplexans
   e. poriferans
   *Answer: d*

6. The chemical signal responsible for the aggregation of myxamoebas is
   a. 3′,5′-cyclic adenosine monophosphate (cAMP).
   b. alginic acid.
   c. phycoerythrin.
   d. actin.
   e. None of the above
   *Answer: a*

7. In terms of their nutritional mode, protists can be
   a. autotrophs.
   b. absorptive heterotrophs.
   c. ingestive heterotrophs.
   d. Both a and b
   e. a, b, and c
   *Answer: e*

8. Protists are found in which of the following habitats?
   a. Marine habitats
   b. Freshwater aquatic habitats
   c. The body fluids of other organisms
   d. Damp soil
   e. All of the above
   *Answer: e*

9. Which of the following are *not* opisthokonts?
   a. Fungi
   b. Algae
   c. Animals
   d. Choanoflagellates
   e. Fishes
   *Answer: b*

10. The _____ are beautiful marine protists that secrete a glassy endoskeleton.
    a. algae
    b. protozoa
    c. radiolarians
    d. flagellates
    e. dinoflagellates
    *Answer: c*

*11. Some algae demonstrate the phenomenon of alternation of generations, in which a multicellular, diploid, spore-producing organism gives rise to a multicellular, haploid, gamete-producing organism. Which of the following statements about alternation of generations is *false*?
    a. The haploid and diploid organisms may or may not differ morphologically.
    b. The haploid and diploid organisms differ genetically.

   c. Only the haploid organism may also reproduce asexually.
   d. Haploid gametes can produce new organisms only by fusing with other gametes.
   e. Diploid sporophytes may undergo meiosis to produce haploid spores.
   *Answer: c*

12. Paramecia contain two types of nuclei: a large macronucleus and as many as 80 micronuclei. The micronuclei are typical eukaryotic nuclei, essential for genetic recombination. The macronucleus
    a. is important in sexual recombination (conjugation).
    b. contains several micronuclei.
    c. contains many copies of the genetic information.
    d. contains DNA that is not transcribed.
    e. contains DNA that is transcribed but not translated.
    *Answer: c*

13. Paramecia have an elaborate sexual behavior in which they line up against each other and fuse. This is followed by an extensive reorganization and exchange of nuclear material. The entire process is called
    a. isogamous reproduction.
    b. alternation of generations.
    c. sexual reproduction.
    d. conjugation.
    e. None of the above
    *Answer: d*

14. Which of the following is *not* a characteristic of the brown algae?
    a. Leaflike growths called thalli
    b. Storage of the products of photosynthesis as floridean starch
    c. A specialized holdfast that aids in attachment to a surface
    d. The presence of the carotenoid fucoxanthin in their chloroplasts
    e. Multicellularity
    *Answer: b*

15. _____ have created vast limestone deposits throughout the world.
    a. Radiolarians
    b. Dinoflagellates
    c. Foraminiferans
    d. Heliozoans
    e. None of the above
    *Answer: c*

16. Which of the following diseases is *not* caused by the trypanosomes?
    a. African sleeping sickness
    b. Malaria
    c. Leishmaniasis
    d. Chagas' disease
    e. East Coast fever
    *Answer: b*

17. The photosynthetic stramenopiles obtained their chloroplasts, which are surrounded by three membranes, through
    a. primary endosymbiosis.
    b. secondary endosymbiosis, retaining the chloroplast from a red alga.
    c. secondary endosymbiosis, retaining the chloroplast from a chlorophyte.

d. tertiary endosymbiosis.
e. None of the above
*Answer: b*

18. The _____ include the water molds and their terrestrial relatives, such as the downy mildews.
    a. parabasalids
    b. apicomplexans
    c. red algae
    d. euglenoids
    e. oomycetes
    *Answer: e*

19. The difference between acellular slime molds and cellular slime molds is that
    a. acellular slime molds are not motile, whereas cellular slime molds are.
    b. they have different numbers of nuclei contained within one plasma membrane.
    c. acellular slime molds ingest food by endocytosis, whereas cellular slime molds do not.
    d. acellular slime molds do not ingest food by endocytosis, whereas cellular slime molds do.
    e. acellular slime molds prefer cool, moist habitats, whereas cellular slime molds prefer dry, hot conditions.
    *Answer: b*

20. Which of the following is *false*?
    a. Diplomonads are monophyletic.
    b. Protists are monophyletic.
    c. Animals are monophyletic.
    d. Red algae are monophyletic.
    e. Eukarya are monophyletic.
    *Answer: b*

21. Which group below is the sister taxon of the diatoms?
    a. Red algae
    b. Brown algae
    c. Green algae
    d. Opisthokonts
    e. Oomycetes
    *Answer: b*

22. The _____ include the only protists that contain the full complement of photosynthetic pigments characteristic of plants.
    a. choanoflagellates
    b. parabasalids
    c. cilians
    d. red algae
    e. chlorophytes
    *Answer: e*

23. Which of the following groups contains multicellular brown algae and giant kelps?
    a. Diplomonads
    b. Chlorophytes
    c. Stramenopiles
    d. Alveolates
    e. Euglenozoans
    *Answer: c*

24. When some autotrophic *Euglena* are placed in the dark, they
    a. stop producing their photosynthetic pigment.
    b. produce an excess of photosynthetic pigment.

c. begin feeding on organic material floating in the surrounding water.
    d. die.
    e. Both a and c
    *Answer: e*

25. When organic material is digested in a food vacuole, the pH in the vacuole
    a. decreases to aid in digestion, then increases as digestion is completed.
    b. increases to aid in digestion, then decreases as digestion is completed.
    c. stays the same during and after digestion.
    d. stays the same during digestion, then increases as digestion is completed.
    e. None of the above
    *Answer: a*

26. Which of the following organisms lack mitochondria?
    a. Diplomonads
    b. Choanoflagellates
    c. Red algae
    d. Euglenozoans
    e. Both b and c
    *Answer: a*

27. Dinoflagellates are common endosymbionts of
    a. coral.
    b. fungi.
    c. other dinoflagellates.
    d. tertiary endosymbionts.
    e. None of the above
    *Answer: a*

28. Algal life cycles show extreme variation. Members of only one group, the _____, do *not* have flagellated motile cells in at least one stage of the life cycle.
    a. alveolates
    b. red algae
    c. euglenozoans
    d. diplomonads
    e. ciliates
    *Answer: b*

29. Protists are believed to be ecologically and evolutionarily important for many reasons. Which of the following is *not* true of this group?
    a. Multicellular groups evolved from protists.
    b. Photosynthetic protists play a major role in the energy balance of the living world.
    c. None of the protists are parasites.
    d. Saprobic protists are among the important decomposers and thus play a major role in the nutrient cycles of the living world.
    e. Many protists have highly differentiated bodies even though they consist of only one cell.
    *Answer: c*

30. Select the group of organisms that is *not* considered a protist.
    a. Protozoa
    b. Algae
    c. Slime molds
    d. Sponges
    e. Giant kelp
    *Answer: d*

*31. You place two different species of protists in a solution with an unknown osmotic potential. Protist A has a contractile vacuole firing rate of 5 per minute; protist B has a contractile vacuole firing rate of 12 per minute. Select a reasonable conclusion based on these observations.
a. Protist A has a more negative osmotic potential than protist B has.
b. Protist B has a more negative osmotic potential than protist A has.
c. The solution has a more negative osmotic potential than protist A has.
d. The solution has a more negative osmotic potential than protist B has.
e. No conclusions can be made unless we know the osmotic potential of the solution.
*Answer: b*

32. In the alternation of generations, the gametophyte generation is _____ and produces _____.
a. haploid; spores.
b. haploid; gametes.
c. diploid; spores.
d. diploid; gametes.
e. haploid or diploid; spores.
*Answer: b*

33. What do the protists that are responsible for sleeping sickness and the protists that are responsible for malaria have in common?
a. They are both alveolates.
b. They both have insect vectors for transmission to humans.
c. They cause the same symptoms.
d. They both have gametocyte life stages.
e. All of the above
*Answer: b*

34. *Plasmodium*, the organism that causes malaria, is a member of the
a. alveolates.
b. stramenopiles.
c. apicomplexans.
d. choanoflagellates.
e. None of the above
*Answer: c*

*35. Most ameobas have all but one of the following characteristics; only a very few amoeba species are
a. autotrophic.
b. free-living.
c. predators.
d. parasitic.
e. shelled.
*Answer: a*

*36. In the protist *Plasmodium*, which causes malaria, the gametocytes
a. develop into merozoites.
b. inhabit the salivary glands of *Anopheles* mosquitoes.
c. are the infective stage obtained from the insect vector.
d. are found inside red blood cells.
e. give rise to zygotes within the mammalian circulatory system.
*Answer: d*

37. Which of the following statements about the micronucleus and macronucleus is *false*?
a. The macronucleus is involved in genetic recombination.
b. The micronucleus is a typical eukaryotic nucleus.
c. Multiple copies of macronuclear genes are common.
d. Transcription and translation involve mostly genes found in the macronucleus.
e. The micronucleus and macronucleus are unique to ciliates.
*Answer: a*

38. Of the following structures, select the one that is *not* associated with movement or feeding in *Paramecium*.
a. Food vacuole
b. Oral groove
c. Cilium
d. Pseudopod
e. Both b and c
*Answer: d*

39. Which of the following processes is *not* part of conjugation in *Paramecium*?
a. Meiosis
b. Mitosis
c. Cytokinesis
d. Fusion of haploid nuclei
e. Breakdown of some micronuclei
*Answer: c*

40. Select the event that is *not* normally associated with the life cycle of an acellular slime mold.
a. Development of spores into myxamoebas
b. Formation of a sclerotium when conditions are adverse
c. Active feeding by the plasmodium as it engulfs food particles
d. Meiosis to form sporangiophores
e. Fusion of swarm cells to form a diploid zygote
*Answer: a*

41. Which of the following features is *not* a characteristic of the plasmodium of an acellular slime mold?
a. Many nuclei enclosed in a single plasma membrane (coenocyte)
b. Haploid nuclei
c. Cytoplasmic streaming
d. Mitosis without cytokinesis
e. Formation of a sclerotium when conditions are unfavorable
*Answer: b*

42. In the cellular slime molds, cAMP causes
a. aggregation of swarm cells to form a plasmodium.
b. formation of sporangia.
c. release of myxamoebas from fruiting bodies.
d. the onset of cytoplasmic streaming.
e. aggregation of myxamoebas to form a slug or pseudoplasmodium.
*Answer: e*

*43. The algal protist groups differ from each other in terms of their
a. principal photosynthetic storage product.
b. body plan (unicellular or multicellular).
c. lack of or possession of flagella.
d. principal photosynthetic pigments.
e. All of the above
*Answer: e*

44. Which group includes the dinoflagellates?
    a. Alveolates
    b. Green algae
    c. Red algae
    d. Stramenopiles
    e. Euglenozoans
    *Answer: a*

45. Members of which group are thought to be the closest relatives of the animals?
    a. Euglenozoans
    b. Choanoflagellates
    c. Stramenopiles
    d. Alveolates
    e. Chlorophytes
    *Answer: b*

*46. Which of the following statements does *not* apply to diatoms.
    a. Diatom cell walls are often impregnated with silicon.
    b. During mitosis, the top and bottom of the cell become the tops of the two new cells.
    c. Chrysolaminarin is made as a photosynthetic storage product.
    d. Zygotes (auxospores) are formed by gametes that lack cell walls.
    e. They can show bilateral or radial symmetry.
    *Answer: c*

47. The presence of _____ in the chloroplasts of red algae results in their characteristic color.
    a. phycoerythrin
    b. fucoxanthin
    c. β-carotene
    d. chrysolaminarin
    e. chlorophyll *b*
    *Answer: a*

48. Which of the following statements about the life cycle of the green alga *Ulva* is *false*?
    a. *Ulva* has an isomorphic life cycle.
    b. The sporophyte and gametophyte can be differentiated only microscopically.
    c. All species of *Ulva* are isogamous.
    d. Individual gametophytes produce only sperm or only eggs—never both.
    e. The diploid sporophyte produces flagellated spores.
    *Answer: c*

*49. An ecologically and evolutionarily important alga, which is a common endosymbiont of coral, is a
    a. dinoflagellate.
    b. red alga.
    c. parabasalid.
    d. euglenozoan.
    e. pseudoplasmodium.
    *Answer: a*

# 29 Plants without Seeds: From Sea to Land

## Fill in the Blank

1. In plants (and some algae such as *Ulva*), the sporophyte and gametophyte exhibit different levels of ploidy. Sporophytes are _____, whereas gametophytes are _____.
*Answer: diploid; haploid*

2. Plants, sometimes referred to as _____, may be defined as multicellular, photosynthetic eukaryotes that develop from embryos protected by tissues of the parent.
*Answer: embryophytes*

3. The vascular system can be said to have been launched by a single evolutionary event. Sometime during the Paleozoic era (before 440 mya), the sporophyte generation of a now long-extinct plant produced a new cell type, the _____.
*Answer: tracheid*

4. In the strictest sense, a _____ is a flattened photosynthetic structure emerging laterally from a main axis or stem and possessing true vascular tissue.
*Answer: leaf*

5. The sister taxon to the mosses contains the phyla known collectively as the _____.
*Answer: tracheophytes*

6. _____ are plants with large leaves and no seeds, and they require water as a medium for the transfer of male gametes. The sporophytes and gametophytes grow independently of each other.
*Answer: Ferns*

7. A tissue called _____ conducts the products of photosynthesis from sites where they are produced or released to sites where they are used or stored.
*Answer: phloem*

8. The _____ generation extends from the zygote through the adult, multicellular, diploid plant.
*Answer: sporophyte*

9. Plants that have an internal transport system are called _____. They once were called vascular plants.
*Answer: tracheophytes*

10. The vascular system in plants consists of specialized tissues. The _____ transports water and minerals from soil to aerial parts and contains the structural substance _____.
*Answer: xylem; lignin*

11. _____ are water-absorbing filaments present on the lower surfaces of the simplest liverwort gametophytes.
*Answer: Rhizoids*

12. In the nontracheophytes, the conspicuous green structure that can be seen by the naked eye is the _____.
*Answer: gametophyte*

13. The sporophytes of the mosses and tracheophytes grow by _____ cell division (a synapomorphy of these two groups), whereby a region at the growing tip provides an organized pattern of cell division, elongation, and differentiation.
*Answer: apical*

14. Some liverworts, such as *Marchantia*, reproduce sexually and vegetatively. Vegetative reproduction takes place by means of lens-shaped clumps of cells called _____.
*Answer: gemmae*

15. Although ferns as a whole are not monophyletic, most ferns do belong to a single clade, the _____ ferns.
*Answer: leptosporangiate*

## Multiple Choice

1. Much evidence indicates that _____ are a sister taxon to plants.
a. mosses
b. liverworts
c. red algae
d. charophytes
e. chlorophytes
*Answer: d*

★2. Which of the characteristics below links the green algae with plants?
a. The use of chlorophylls *a* and *b*
b. Active stomata
c. Starch as a major storage compound
d. Cellulose in cell walls
e. a, c, and d
*Answer: e*

3. Nonvascular plants have never evolved to the size of vascular plants, most likely because they lack
a. a photosynthetic mechanism.
b. an efficient mode of respiration.
c. an efficient system for conducting water and minerals.
d. nutrient and water absorption mechanisms.
e. All of the above
*Answer: c*

4. In some liverworts, there are structures that "throw" _____ from the capsule.
   a. moisture particles
   b. spores
   c. sperm
   d. ova
   e. rhizoids
   *Answer: b*

5. The evolutionary importance of plant tissue composed of tracheids is that it provided
   a. a plant vascular system and structural support.
   b. structural support and increased growth.
   c. enhanced photosynthesis and structural support.
   d. enhanced photosynthesis and a plant vascular system.
   e. None of the above
   *Answer: a*

6–8. Match the correct phylum from the list below with the following descriptions.
   a. Hepatophyta (liverworts)
   b. Anthocerophyta (hornworts)
   c. Bryophyta (mosses)

6. Gametophytes are green, leaflike layers that lie flat on the ground.
   *Answer: a*

7. The gametophyte is a branched, filamentous structure called a protonema; many types contain hydroid cells.
   *Answer: c*

8. Sporophytes exhibit indeterminate growth and develop no stalk; a single large chloroplast is present in each cell.
   *Answer: b*

9. There are approximately _____ species of ferns.
   a. 12,000
   b. 120
   c. 1,200
   d. 120,000
   e. 1.2 million
   *Answer: a*

10. Although Earth is estimated to be 5 billion years old, and although life first appeared about 4 billion years ago, plants did not appear until about _____ years ago.
    a. 3–4 billion
    b. 400–500 million
    c. 40–50 million
    d. 3–4 million
    e. 400,000–500,000
    *Answer: b*

11. Which of the following is *not* a characteristic of plants?
    a. Development from embryos protected by tissues of the parent plant
    b. Cell walls containing cellulose
    c. Chloroplasts containing chlorophylls *a* and *b*
    d. Starch as the storage carbohydrate
    e. Anaerobic respiration
    *Answer: e*

12. A universal feature of the life cycles of plants is
    a. morphologically identical gametophyte and sporophyte stages.
    b. genetically identical gametophyte and sporophyte stages.
    c. alteration of generations between haploid gametophytes and diploid sporophytes.

d. All of the above
e. None of the above
*Answer: c*

13. Plants invaded the land sometime during the Paleozoic era. In order to evolve and thrive on land, plants had to
    a. develop photosynthetic pigments and mechanisms for transporting water and minerals to aerial parts.
    b. develop starch for carbohydrate storage and mechanisms for transporting water and minerals to aerial parts.
    c. develop physical support structures and mechanisms for gamete dispersal.
    d. develop photosynthetic pigments that are not dependent on an aqueous environment and develop starch for carbohydrate storage.
    e. alternate generations and develop physical support structures.
    *Answer: c*

14. Vascular plants are thought to be the result of a single evolutionary event: the evolution of a wholly new cell type, the tracheid. This cell type
    a. provides a mechanism for the storage of a new type of carbohydrate, starch.
    b. is the first cell type to contain chloroplasts.
    c. permits fertilization in the absence of water, thus permitting plants to invade dry habitats.
    d. forms the seed.
    e. is the principal water-conducting element of the xylem in all vascular plants except the angiosperms.
    *Answer: e*

15. Several important adaptations evolved in the common ancestor of plants to allow the successful colonization of land. Which of the following is *not* one of those changes?
    a. Evolution of a water-impermeable cuticle
    b. Evolution of a carbohydrate energy-storage molecule
    c. Evolution of gametangia
    d. Initial absence of herbivores
    e. A mutualistic association with a fungus that promotes nutrient uptake
    *Answer: b*

16. Hornworts are the sister phylum to the
    a. mosses and tracheophytes.
    b. mosses.
    c. club mosses.
    d. mosses, club mosses, ferns and allies, and seed plants.
    e. nontracheophytes.
    *Answer: a*

17. The megaphyll found in ferns and seed plants is thought to have evolved
    a. from sterile sporangia.
    b. by the process of meiosis.
    c. by the process of heterospory.
    d. by the process of overtopping.
    e. spontaneously.
    *Answer: d*

18. Which of the following is *not* one of the evolutionary adaptations to land shared by all plants?
    a. Waxy protective coverings
    b. Support against gravity
    c. A means of taking up water from the soil
    d. Protective structures for the new sporophyte
    e. Water transport by xylem
    *Answer: e*

19. You find a green "leafy" bryophyte growing on your neighbor's front lawn. It is a
    a. liverwort.
    b. hornwort.
    c. moss.
    d. whisk fern.
    e. fern.
    *Answer: c*

20. Fossil vascular plants can be recognized by the presence of
    a. tracheids.
    b. antheridia.
    c. archegonia.
    d. protonema.
    e. vessels.
    *Answer: a*

21. One important consequence of the evolution of xylem was the development of a mechanism for
    a. sugar transport.
    b. sperm transport.
    c. preventing of water evaporation.
    d. rigid structural support.
    e. None of the above
    *Answer: d*

22. Sporangia can be identified as part of the fossil *Rhynia* because meiosis produces
    a. two haploid cells.
    b. four haploid cells.
    c. two haploid cells, but only one is functional.
    d. four haploid cells, but only one is functional.
    e. two diploid cells, but they remain attached.
    *Answer: b*

23. In heterosporous plants, _____ produce _____.
    a. microgametophytes; eggs
    b. microgametophytes; sperm
    c. microgametophytes; eggs and sperm
    d. megagametophytes; sperm
    e. megagametophytes; eggs and sperm
    *Answer: b*

*24. The evidence suggesting that heterospory probably affords selective advantages is that it
    a. evolved a number of times.
    b. is found only in the most advanced plants.
    c. is simpler than homospory.
    d. is found in the most primitive plants.
    e. evolved along with swimming sperm.
    *Answer: a*

25. Which of the following types of plants possess the microphyll leaf?
    a. Mosses
    b. Horsetails
    c. Club mosses
    d. Ferns
    e. Liverworts
    *Answer: c*

26. The horsetails, whisk ferns, and ferns are members of a single phylum, the
    a. Lycophyta.
    b. Anthocerophyta.
    c. Hepatophyta.
    d. Pteridophyta.
    e. Bryophyta.
    *Answer: d*

∎27. The bryophytes are dependent on water for reproduction because
    a. sperm are passively transported to eggs by water.
    b. gametogenesis occurs only when the plants are moist.
    c. eggs and sperm are released into water and then unite.
    d. sperm must swim through water to reach and fertilize eggs.
    e. None of the above
    *Answer: d*

28. All plants produce _____ by mitosis and _____ by meiosis.
    a. spores; gametes
    b. gametes; gametes
    c. gametes; spores
    d. spores; spores
    e. spores; gametes and spores
    *Answer: c*

29. Which of the following represents the most ancient surviving plant lineage?
    a. Liverworts
    b. Hornworts
    c. Mosses
    d. Ferns
    e. None of the above
    *Answer: a*

30. Which of the following is *not* a structure or characteristic that enables the nontracheophytes to obtain water and minerals in the absence of a vascular system?
    a. Growth in dense masses through which water can move by capillary action
    b. Leaflike structures that catch and hold water that splashes onto them
    c. Small size, which allows minerals to be distributed evenly by diffusion
    d. An extensive root system
    e. Both c and d
    *Answer: d*

31. The first tracheophytes belonged to the extinct phylum
    a. Rhyniophyta.
    b. Lycophyta.
    c. Bryophyta.
    d. Pteridophyta.
    e. Pinophyta.
    *Answer: a*

32. Which living tracheophyte phylum diverged earlier than all other living tracheophyte phyla?
    a. Bryophyta
    b. Pteridophyta
    c. Lycophyta
    d. Gymnospermae
    e. Angiospermae
    *Answer: c*

33. The first ferns appeared during which period?
    a. Permian
    b. Carboniferous
    c. Silurian
    d. Devonian
    e. Triassic
    *Answer: d*

34. Which of the following is true for the phylum Anthocerophyta (the hornworts)?
    a. They possess stomata.
    b. Their cells contain many chloroplasts.
    c. Their sporophytes exhibit determinate growth.
    d. They have tracheids.
    e. They evolved the precursor to flowers.
    *Answer: a*

35. Within the tracheophytes the large, prominent plant is the _____; in the nontracheophytes it is the _____.
    a. gametophyte; sporophyte
    b. sporophyte; gametophyte
    c. gametophyte; gametophyte
    d. sporophyte; sporophyte
    e. None of the above
    *Answer: b*

36. The evolutionary origin of microphyll is thought by some biologists to be
    a. sterile sporangia.
    b. rhizoids.
    c. specialized xylem.
    d. photosynthetic tissue developed between complex branching patterns.
    e. None of the above
    *Answer: a*

*37. Which of the following would *not* have been a feature of the tracheophyte plants whose fossilized remains were found by Kidston and Lang?
    a. Roots
    b. Xylem
    c. Spores in groups of two
    d. Dichotomous branching
    e. Both a and c
    *Answer: e*

38. In a heterosporous life cycle, the microspore develops into the _____ gametophyte, whereas the megaspore develops into the _____ gametophyte.
    a. female; male
    b. male; female
    c. diploid; haploid
    d. haploid; diploid
    e. None of the above
    *Answer: b*

39–42. Match the correct taxon from the list below with the following descriptions.
    a. Club mosses
    b. Horsetails
    c. Whisk ferns
    d. Ferns

39. Exhibit basal growth; sometimes called the scouring rushes because the silica deposits in their cell walls made them useful for cleaning; simple leaves form distinct whorls around the stem
    *Answer: b*

40. Originally thought to be evolutionarily ancient descendants of anatomically simple ancestors; gametophytes live below the surface of the ground and lack chlorophyll; now considered to be highly specialized and to have evolved fairly recently
    *Answer: c*

41. Diverged earlier than all other living tracheophytes; they bear simple leaves arranged spirally on a stem and exhibit apical growth; sporangia are contained within conelike structures called strobili
    *Answer: a*

42. Typically have large leaves with branching vascular strands; some can reach heights of up to 20 meters; sporangia are usually clustered in groups called sori
    *Answer: d*

*43. Stomata appear to have first arisen in which phylum?
    a. Hornworts
    b. Mosses
    c. Liverworts
    d. Club mosses
    e. Angiosperms
    *Answer: a*

44. Which of the following characteristics is *not* shared by the club mosses and the horsetails?
    a. Large, independent sporophytes
    b. Specialized vascular tissue
    c. Apical growth
    d. Simple leaves
    e. Both b and c
    *Answer: c*

45. The cyanobacteria present in the internal, mucilage-filled cavities of the hornworts serve to
    a. provide structural support for the plant.
    b. produce carbohydrates for the plant.
    c. convert atmospheric nitrogen gas into a nutrient form usable by the plant.
    d. convert atmospheric carbon gas into a nutrient form usable by the plant.
    e. Both c and d
    *Answer: c*

46. An abundant type of coal, called cannel coal, is formed almost entirely from the fossilized spores of a species in which taxon?
    a. Club mosses (Lycophyta)
    b. Mosses (Bryophyta)
    c. Horsetails (Pteridophyta)
    d. Flowering plants (Angiospermae)
    e. Hornworts (Anthocerophyta)
    *Answer: a*

47. The moss gametophyte that develops after spore germination is a branched, filamentous structure called a(n)
    a. antheridium.
    b. capsule.
    c. protonema.
    d. hydroid.
    e. None of the above
    *Answer: c*

48. The sporophyte of a tracheophyte is _____; the sporophyte of a nontracheophyte is _____.
    a. dependent; independent
    b. independent; independent
    c. dependent; dependent
    d. independent; dependent
    e. haploid; diploid
    *Answer: d*

49. Which of the following observations led to the conclusion
    that the fossilized plants found by Kidston and Lang
    were sporophytes?
    a. Spores in groups of four in the fossil sporangia
    b. The presences of rhizomes
    c. Dichotomous branching
    d. The presence of xylem
    e. Both a and d
    *Answer: a*

# 30 The Evolution of Seed Plants

## Fill in the Blank

1. Cycadophyta, Ginkgophyta, Gnetophyta, and Pinophyta are all phyla within the more inclusive clade _____, of which there are only about 750 living species.
   *Answer: gymnosperms*

2. Most gymnosperms are pollinated by the wind, and most angiosperms are pollinated by _____.
   *Answer: animals*

3. All living _____ and many angiosperms show secondary growth.
   *Answer: gymnosperms*

4. The phylum _____ is composed of the flowering plants. These plants produce seeds and have vessel elements and double fertilization.
   *Answer: Angiospermae*

5. The reproductive organ of angiosperms is the _____.
   *Answer: flower*

6. The ovary of a flowering plant develops into a _____, which consists of the mature ovary and its seeds and may also include other parts of the flower.
   *Answer: fruit*

7. Also known as seed leaves, _____ can digest endosperm, become photosynthetic, or do both.
   *Answer: cotyledons*

8. In a flower, the male organs are contained in the _____, and the female organs are contained in the _____.
   *Answer: stamen; pistil*

9. The modified leaves on a flower, called _____, can be showy to attract animals.
   *Answer: petals*

10. In the seed plants, male gametophytes are called _____.
    *Answer: pollen grains*

11. A seed may contain tissue from (how many?) _____ different generations.
    *Answer: three*

12. The sterile sporophyte structures that surround the megasporangium are called the _____.
    *Answer: integument*

13. Flowers with both megasporangia and microsporangia are said to be _____.
    *Answer: perfect*

14. Most angiosperm species are included in two large lineages: the _____ and the _____.
    *Answer: monocots; eudicots*

15. The _____ is triploid tissue that nourishes the embryonic sporophyte during its early development.
    *Answer: endosperm*

16. A stamen is composed of a _____ bearing an _____ that contains pollen-producing microsporangia.
    *Answer: filament; anther*

17. The structures that bear the megasporangia in the seed plants are called _____.
    *Answer: carpels*

18. Flowers have specialized sterile leaves. The inner specialized leaves are called _____ and the outer specialized leaves are called _____.
    *Answer: petals; sepals*

## Multiple Choice

1. Two hundred million years ago, when dinosaurs inhabited Earth, _____ were the dominant vegetation.
   a. early whisk ferns
   b. horsetails and tree ferns
   c. lycopods and ferns
   d. gymnosperms
   e. angiosperms
   *Answer: d*

★2. There has been a change in the dominant vegetation since plants first invaded the terrestrial environment about 400–500 million years ago. What was the order in which these vegetation types were dominant, from earliest to most current?
   a. Gymnosperms; horsetails, lycopods, and ferns; angiosperms
   b. Angiosperms; horsetails, lycopods, and ferns; gymnosperms
   c. Horsetails, lycopods, and ferns; gymnosperms; angiosperms
   d. Angiosperms; gymnosperms; horsetails, lycopods, and ferns
   e. Horsetails, lycopods, and ferns; angiosperms; gymnosperms
   *Answer: c*

3. A seed of a flowering plant or gymnosperm may contain tissues from _____ generation(s).
   a. one
   b. two
   c. three
   d. four
   e. five
   *Answer: c*

4. One reason for the enormous evolutionary success of seed plants is their possession of
   a. complex leaves that can photosynthesize at a faster rate than non-seed-producing plants can.
   b. seeds with food reserves for the young sporophyte.
   c. seeds with a resting stage that can remain viable for many years, germinating when conditions are favorable for growth of the sporophyte.
   d. Both b and c
   e. a, b, and c
   *Answer: d*

5. The most abundant gymnosperm phylum today contains the cone-bearing plants, such as pines. These plants belong to which phylum?
   a. Cycadophyta
   b. Ginkgophyta
   c. Gnetophyta
   d. Pinophyta
   e. None of the above
   *Answer: d*

6. A universal feature of plant life cycles is
   a. the seed stage.
   b. the archegonium.
   c. the phloem.
   d. the tracheid.
   e. alternation of generations.
   *Answer: e*

*7. In some plants, the gametophyte (the multicellular haploid plant that produces haploid gametes) is free-living and photosynthetic. Which group does *not* have a free-living gametophyte generation?
   a. Ferns
   b. Gymnosperms
   c. Angiosperms
   d. Both a and b
   e. Both b and c
   *Answer: e*

8. All plant life cycles have generations that alternate between the gametophyte generation and the _____ generation.
   a. heteromorphic
   b. sporophyte
   c. vascular
   d. archegonium
   e. antheridium
   *Answer: b*

9. Coniferous gymnosperms, such as pines, depend primarily on _____ for pollination; thus the plants produce large quantities of pollen that disperse over large areas during the spring.
   a. insects
   b. birds
   c. water
   d. wind
   e. mammals
   *Answer: d*

10. An evolutionary trend that runs throughout the plant kingdom is for the sporophyte generation to become _____ and more independent of the gametophyte, and the gametophyte generation to become _____ and more dependent upon the sporophyte.
    a. smaller; smaller
    b. larger; smaller
    c. smaller; larger
    d. larger; larger
    e. larger or smaller; larger
    *Answer: b*

11. In angiosperms, two male gametes contained within a single male gametophyte participate in fertilization. One sperm nucleus combines with the egg to produce a diploid zygote, and the other sperm nucleus combines with two other haploid nuclei of the female gametophyte. This process is called
    a. biparental inheritance.
    b. multiple paternity.
    c. double fertilization.
    d. biparental fertilization.
    e. multiple fertilization.
    *Answer: c*

12. The reproductive organ of angiosperms is the
    a. sporangium.
    b. flower.
    c. cone.
    d. archegonium.
    e. sporophyte.
    *Answer: b*

13. Plant species in which individual specimens are exclusively male or exclusively female are
    a. eudicots.
    b. heterozygous.
    c. perfect.
    d. monoecious.
    e. dioecious.
    *Answer: e*

14. A fruit that develops from several carpels of a single flower, such as a raspberry, is a(n) _____ fruit.
    a. aggregate
    b. simple
    c. multiple
    d. accessory
    e. perfect
    *Answer: a*

15. The two major clades of angiosperms are called monocots and eudicots. These plants differ in the number of
    a. sperm involved in fertilization.
    b. sexes per plant; monocots have one sex per plant, eudicots have both.
    c. sexes per plant; eudicots have male and female plants, monocots have both sexes in one plant.
    d. embryonic cotyledons.
    e. tracheid types.
    *Answer: d*

16. Which of the following clades has the greatest number of species?
    a. Angiosperms
    b. Pinophyta
    c. Gnetophyta
    d. Ginkgophyta
    e. Cycadophyta
    *Answer: a*

17. From an evolutionary standpoint, pollen is a
    a. microsporophyll.

b. megasporophyll.
c. microgametophyte.
d. megagametophyte.
e. microspore.
*Answer: c*

18. In most regions of Earth today, land flora consists predominantly of
   a. angiosperms.
   b. gymnosperms.
   c. ferns.
   d. bryophytes.
   e. club mosses.
   *Answer: a*

19–21. Refer to the diagram below of a pine seed to answer the following questions.

19. Which structure(s) represent(s) the embryo?
   a. A
   b. B
   c. C
   d. A and C
   e. B and C
   *Answer: a*

20. Which structure(s) represent(s) the female gametophyte?
   a. A
   b. B
   c. C
   d. A and C
   e. B and C
   *Answer: b*

21. The integument portion of the ovule developed into part(s) _____.
   a. A
   b. B
   c. C
   d. A and C
   e. B and C
   *Answer: c*

22. In angiosperms, double fertilization results in the development of
   a. two embryos.
   b. two embryos, but only one survives.
   c. the embryo and the endosperm.
   d. the embryo and the seed coat.
   e. the embryo and the megagametophyte.
   *Answer: c*

23. The seeds in angiosperms are located
   a. on the upper surface of the sporophylls.
   b. on the lower surfaces of the sporophylls.
   c. buried within the sporophylls.
   d. enclosed in the ovule.
   e. None of the above
   *Answer: d*

24. Flowering species that produce both megasporangia and microsporangia in each flower have _____ flowers and are _____.
   a. perfect; monoecious
   b. perfect; dioecious
   c. imperfect; monoecious
   d. imperfect; dioecious
   e. imperfect; monoecious or dioecious
   *Answer: a*

25. Which of the following seed plants have swimming (motile) sperm?
   a. Angiosperms
   b. Early angiosperms
   c. Early gymnosperms and a few early angiosperms
   d. Early gymnosperms
   e. None of the above
   *Answer: d*

26. In angiosperms, pollen is transferred from the _____ to the _____.
   a. anther; style
   b. filament; ovary
   c. anther; stigma
   d. filament; ovary
   e. anther; ovule
   *Answer: c*

*27. The diploid zygote in angiosperms develops into the
   a. embryonic axis.
   b. embryonic axis and cotyledons.
   c. embryonic axis and endosperm.
   d. embryonic axis, cotyledons, and endosperm.
   e. embryonic axis, cotyledons, endosperm, and seed coat.
   *Answer: b*

28. The pistil consists of
   a. anthers, filaments, and stamen.
   b. ovary, archegonium, and embryo.
   c. stigma, style, and ovary.
   d. sepals and petals.
   e. embryo, endosperm, and cotyledons.
   *Answer: c*

29. Seed plants are all
   a. heterosporous.
   b. dioecious.
   c. monoecious.
   d. eudicots.
   e. dependent on animals for fertilization.
   *Answer: a*

■30. One difference between gymnosperms and angiosperms is that gymnosperms
   a. do not form seeds.
   b. do not form flowers.
   c. do not have tracheid cells.
   d. rely on animals for fertilization.
   e. None of the above
   *Answer: b*

31. The function of the pollen tube is to
   a. eject pollen from the microsporangium.
   b. direct pollen to the megasporangium.
   c. digest the sporophyte tissue as it elongates toward the female gametophyte.
   d. produce pollen.
   e. attract animals to the plant to spread the pollen.
   *Answer: c*

32–36. Match the correct phylum below with the following descriptions.
   a. Cycadophyta
   b. Ginkgophyta
   c. Pinophyta
   d. Gnetophyta
   e. Angiospermae

32. Seeds are in cones; plants have needlelike or scalelike leaves; includes plants such as pines and redwoods
   *Answer: c*

33. Represented today by a single species; common during the Mesozoic era; plants have fan-shaped leaves
   *Answer: b*

34. The palmlike plants of the tropics; the least changed group of the present-day gymnosperms
   *Answer: a*

35. Consists of three very different genera that share characteristics with the angiosperms; plants have vessels in vascular tissue; two single leaves split as they grow
   *Answer: d*

36. Seeds are in fruit; plants form an endosperm; gametophytes are much reduced; plants produce flowers
   *Answer: c*

37. Which of the following characteristics is *not* unique (i.e., not a synapomorphy) to the angiosperms?
   a. The production of triploid endosperm
   b. The production of fruit
   c. Xylem that contains vessel elements and fibers
   d. A reduced gametophyte generation
   e. The production of flowers
   *Answer: d*

38. Bird-pollinated flowers
   a. are often red and odorless.
   b. have characteristic odors.
   c. have conspicuous markings that are evident only in the ultraviolet region of the spectrum.
   d. are always grouped in an inflorescence.
   e. None of the above
   *Answer: a*

*39. The following events in the angiosperm life cycle occur in which order?
   a. Division of diploid zygote, pollen grain reaches a sporophyte, fertilization, production of a pollen tube
   b. Pollen grain reaches a sporophyte, production of a pollen tube, fertilization, division of diploid zygote
   c. Pollen grain reaches a sporophyte, division of diploid zygote, production of a pollen tube, fertilization
   d. Production of a pollen tube, fertilization, division of diploid zygote, pollen grain reaches a sporophyte
   e. Fertilization, division of diploid zygote, production of a pollen tube, pollen grain reaches a sporophyte
   *Answer: b*

40. You are given two flowers of the same species from two separate plants. One flower has only a pistil, while the other has only a stamen. Based on your observations, you conclude that the flowers are _____, and the species is _____.
   a. perfect; imperfect
   b. imperfect; dioecious

   c. monoecious; imperfect
   d. dioecious; monoecious
   e. None of the above
   *Answer: b*

*41. The angiosperms are the sister phylum to which gymnosperm phylum?
   a. Cycadophyta
   b. Pinophyta
   C. Gnetophyta
   d. Ginkgophyta
   e. The specific sister taxon is unknown.
   *Answer: e*

*42. Angiosperms and their animal pollinators have coevolved in the terrestrial environment for
   a. 130 million years.
   b. 1.3 million years.
   c. 130,000 years.
   d. 13 million years.
   e. 500 million years.
   *Answer: a*

43. Of the following phyla, which is *not* classified as a gymnosperm?
   a. Cycadophyta
   b. Pinophyta
   c. Anthocerophyta
   d. Gnetophyta
   e. Ginkgophyta
   *Answer: c*

44. A structure composed of one carpel or two or more fused carpels is called a(n)
   a. stamen.
   b. anther.
   c. pistil.
   d. receptacle.
   e. filament.
   *Answer: c*

45. The corolla and the calyx often play roles in
   a. attracting animal and insect pollinators to the flower.
   b. protecting the immature flower in a bud.
   c. photosynthesis.
   d. spore dispersal.
   e. Both a and b
   *Answer: e*

46. Which of the following is true of most plant–pollinator interactions?
   a. They are highly specific.
   b. They are not highly specific.
   c. Flowers may have markings or odors to attract certain pollinators.
   d. Pollinators include bees, birds, and bats.
   e. b, c, and d
   *Answer: e*

47. The angiosperm carpel serves to
   a. protect the ovules and seeds.
   b. attract pollinators.
   c. produce sugars via photosynthesis.
   d. prevent self-pollination.
   e. Both a and d
   *Answer: e*

# 31 Fungi: Recyclers, Pathogens, Parasites, and Plant Partners

## Fill in the Blank

1. Organisms living in mutually beneficial symbiosis with other organisms are called _____.
*Answer: mutualists*

2. Sexual reproduction in the fungi is accomplished when two different mating types _____.
*Answer: fuse*

3. The cell walls of all fungi consist of the polysaccharide _____, which is also found in animals.
*Answer: chitin*

4. The body of a multicellular fungus is called a _____.
*Answer: mycelium*

5. The kingdom Fungi encompasses _____ organisms with absorptive nutrition and with chitin in their cell walls.
*Answer: heterotrophic*

6. The body cells of a multicellular fungus are organized into rapidly growing individual tubular filaments called

_____.
*Answer: hyphae*

7. Individual filaments that anchor chytrids and some other fungi to their substrate are called _____.
*Answer: rhizoids*

8. There are two types of parasitic fungi: _____, which can grow parasitically but can also grow by themselves, and _____, which can grow only on their specific hosts.
*Answer: facultative; obligate*

9. Sexual reproduction in fungi occurs between genetically distinct _____.
*Answer: mating types*

10. The number of phyla in the kingdom Fungi is _____.
*Answer: four*

11. Fungi often have a life stage that is _____, which is the stage at which a hypha has two nuclei.
*Answer: dikaryotic*

12. Hyphae of zygomycetes grow toward each other following the release of chemicals called _____.
*Answer: pheromones*

13. Lichens can _____ by producing a thallus or a soredia.
*Answer: reproduce*

14. The most basal clade of fungi, and one that was until recently classified as a protist, is the _____.
*Answer: Chytridiomycota*

15. After a parasitic fungus invades leaf tissue, the hyphae form _____, branching projections that push into the living plant cells and absorb their nutrients.
*Answer: haustoria*

## Multiple Choice

1. The absence of flagellated gametes is a synapomorphy of
a. Chytridiomycota.
b. Ascomycota.
c. Basidiomycota.
d. Ascomycota and Basidiomycota.
e. Ascomycota, Basidiomycota, and Zygomycota.
*Answer: d*

2. The names of fungal classes are based on important and characteristic structures associated with
a. sexual reproduction.
b. nutrition.
c. ecology.
d. vegetative growth.
e. cell division.
*Answer: a*

3. _____ are organisms that live on dead matter.
a. Parasites
b. Saprobes
c. Anaerobes
d. Aerobes
e. Autotrophs
*Answer: b*

4. The cell walls of all fungi consist of the polysaccharide
a. chitin.
b. cellulose.
c. starch.
d. silica.
e. pectin.
*Answer: a*

5. The body of a multicellular fungus is called a
a. dikaryon.
b. hypha.
c. rhizoid.
d. mycelium.
e. None of the above
*Answer: d*

**187**

6. Individual filaments that anchor chytrids to their substrate are called
   a. dikaryons.
   b. hyphae.
   c. rhizoids.
   d. mycelia.
   e. None of the above
   *Answer: c*

7. The cells of the body of a multicellular fungus are organized into rapidly growing individual tubular filaments called
   a. dikaryons.
   b. hyphae.
   c. rhizoids.
   d. mycelia.
   e. None of the above
   *Answer: b*

8. Which of the following statements is sufficient to identify an unknown organism as belonging to the kingdom Fungi?
   a. It is multicellular and nonphotosynthetic.
   b. It has cell walls and reproduces by spores.
   c. It has filamentous growth and obtains its food by absorption.
   d. It has prokaryotic cells and cell walls made of chitin.
   e. It is unicellular and eukaryotic.
   *Answer: c*

9. One adaptation that fungi have for absorptive nutrition, in which nutrients are absorbed across the cell surfaces, is
   a. lack of a cell wall.
   b. a low surface area-to-volume ratio.
   c. a high surface area-to-volume ratio.
   d. tolerance of low temperatures.
   e. tolerance of high temperatures.
   *Answer: c*

10. A major role of saprobic fungi in terrestrial ecosystems is to
    a. trap atmospheric $CO_2$.
    b. break down carbon compounds.
    c. parasitize animals.
    d. parasitize plants.
    e. form symbiotic mutualist relationships with protists.
    *Answer: b*

■11. Fungi have a larger surface area-to-volume ratio than do most other multicellular organisms because
   a. most hyphae are in close contact with their food.
   b. an individual mycelium can grow very large.
   c. hyphae grow together to form a mycelium.
   d. most fungi are microscopic organisms.
   e. chitinous cell walls are more permeable than cellulose cell walls are.
   *Answer: a*

12. Which of the following is *not* an economically useful aspect of fungi?
    a. Some species are used commercially to flavor foods.
    b. Some species are edible.
    c. Some species produce alcohol via fermentation.
    d. Some species produce oxygen via fermentation.
    e. Some species produce antibiotics.
    *Answer: d*

13. Many fungi are _____, associating with photosynthetic organisms to form mycorrhizae or lichens.
    a. symbiotic
    b. parasitic
    c. saprobic
    d. photosynthetic
    e. predatory
    *Answer: a*

14. Fungi can be parasitic on
    a. animals.
    b. plants.
    c. protists.
    d. other fungi.
    e. All of the above
    *Answer: e*

15–20. Match the terms from the list below with the following descriptions of fungal interactions. Each term may be used once, more than once, or not at all.
    a. Saprobic
    b. Competitive
    c. Predatory
    d. Parasitic
    e. Symbiotic

15. Fungi decay a fallen tree.
    *Answer: a*

16. Black stem rust draws nutrition from wheat. The rust damages the wheat plant.
    *Answer: d*

17. Fungi grow in association with the roots of soybeans, providing the plants with more minerals.
    *Answer: e*

18. A constricting ring formed by *Arthrobotrys* traps a nematode. Fungal hyphae invade and digest the nematode.
    *Answer: c*

19. Seed germination in most orchid species depends on the presence of a specific fungus species. The fungus derives nutrients from the seed and seedling.
    *Answer: e*

20. Some leaf-cutting ants farm fungi; they feed the fungi and later harvest and eat them. The ants may even "weed" the fungal gardens by removing other fungal species.
    *Answer: e*

■21. In a jar of jelly in your refrigerator, fungi will be more common than bacteria because fungi have a _____ tolerance for highly _____ environments.
    a. lower; hypotonic
    b. lower; hypertonic
    c. higher; hypotonic
    d. higher; hypertonic
    e. higher; cold
    *Answer: d*

22. Predatory fungi may trap prey by means of
    a. a constricting ring that traps the prey.
    b. sticky substances secreted by hyphae.
    c. mycorrhizae.
    d. Both a and b
    e. a, b, and c
    *Answer: d*

23. The fusion of two different mating types forms a dikaryon that is a heterokaryon. The term "heterokaryon" refers to the fact that

a. the hypha is haploid.

b. two nuclei fused in the course of its formation.

c. there are two different nuclei in a common hypha.

d. the two nuclei are different.

e. None of the above

*Answer: c*

24. The chytrids (phylum Chytridiomycota) are different from all other fungi in that

a. they reproduce only asexually.

b. their haploid gametes have flagella.

c. they are the only parasitic fungi.

d. they contain a fruiting body.

e. they contain a thallus.

*Answer: b*

25. Fungi that appear to reproduce only asexually are

a. chytridiomycetes.

b. zygomycetes.

c. ascomycetes.

d. basidiomycetes.

e. deuteromycetes.

*Answer: e*

26. Conidia are

a. spores produced within sporangia.

b. meiotic products.

c. mating structures found at the tips of specialized hyphae.

d. encased diploid spores from a basidiomycete.

e. a type of basidium which forms on a specialized stalk.

*Answer: c*

27. Motile gametes are found in

a. Zygomycota.

b. Ascomycota.

c. Basidiomycota.

d. deuteromycetes.

e. Chytridiomycota.

*Answer: e*

28. Baker's yeast is a

a. hemiascomycete.

b. euascomycete.

c. species of *Allomyces*.

d. basidiomycete.

e. deuteromycete.

*Answer: a*

29. Common morels are classified as

a. basidiomycetes.

b. ascomycetes.

c. zygomycetes.

d. mycorrhizae.

e. lichens.

*Answer: b*

*30. Which of the following fungi share the same phylum?

a. Cup fungi and bracket fungi

b. Truffles and morels

c. *Amanita* and powdery mildew

d. Black bread mold and pink bread mold

e. Dutch elm disease fungi and smut fungi

*Answer: b*

31. Fungi and plants probably invaded the land together in the

a. Paleozoic era.

b. Mesozoic era.

c. Cenozoic era.

d. time of the dinosaurs.

e. age of the insects.

*Answer: a*

32. Mycorrhizae are mutualistic symbiotic associations of a fungus with

a. an alga or a bacterium.

b. plant roots.

c. a lichen.

d. an animal.

e. another fungus.

*Answer: b*

33. The fruiting structure of a fungus

a. attracts predators away from the essential underground parts.

b. is an important organ for gas exchange with the atmosphere.

c. is an organ of reproduction.

d. acts as a hallucinogen for rodents and mammals.

e. serves as a landing pad for fungal pollinators.

*Answer: c*

34. Which of the following is a basidiomycete?

a. Dutch elm disease fungus

b. Chestnut blight fungus

c. Powdery mildew

d. Green fruit mold

e. Smut fungus

*Answer: e*

35. Which of the following is a synapomorphy for the kingdom Fungi?

a. Rhizoids

b. Spores

c. Heterotrophism

d. Chitin in the cell walls

e. Presence of DNA

*Answer: d*

36. Plants with active mycorrhizae

a. benefit nutritionally from this arrangement.

b. display enhanced absorption of water and minerals (especially phosphorus).

c. are heavily parasitized and die.

d. Both a and b

e. None of the above

*Answer: d*

37–40. Match the groups of Fungi in the list below with the following descriptions.

a. Zygomycetes

b. Ascomycetes, subgroup euascomycetes

c. Ascomycetes, subgroup hemiascomycetes

d. Basidiomycetes

e. Deuteromycetes

*Answer: a*

37. Perforated cross-walls; no specialized fruiting structures; includes baker's and brewer's yeast.

*Answer: c*

38. Common name is club fungi; complete cross-walls; includes puffballs, mushrooms, wheat rust, smut fungi, mycorrhizae.

*Answer: d*

39. Asci are contained within a specialized fruiting structure (ascocarp); includes molds, parasites such as the Dutch elm disease fungus, and epicurean delights such as morels and truffles.
*Answer: b*

40. No known sexual stages—sexual stages were lost in evolution or have not been found.
*Answer: e*

41. Virtually all oak trees and pine trees depend on mycorrhizal fungi to absorb nutrients. The relationship between the tree and the fungus is an example of
    a. saprobism.
    b. mutualism.
    c. symbiotism.
    d. heterotropism.
    e. Both b and c
    *Answer: e*

*42. Rusts and smuts are pathogens of cereal grains which are classified as
    a. zygomycetes.
    b. ascomycetes.
    c. basidiomycetes.
    d. deuteromycetes.
    e. lichens.
    *Answer: c*

43. Which of the following is the sister taxon to the Ascomycota?
    a. Basidiomycota
    b. Zygomycota
    c. Chytridiomycota
    d. The lichens
    e. None of the above
    *Answer: a*

44. Which of the groups below is *not* monophyletic?
    a. Ascomycetes
    b. Deuteromycetes
    c. Basidiomycetes
    d. Zygomycetes
    e. Chytrids
    *Answer: b*

45. Dikaryotic cells
    a. have two hyphae per fruiting body.
    b. contain pairs of homologous chromosomes.
    c. produce two spores per hypha.
    d. contain two nuclei per cell.
    e. contain diploid nuclei.
    *Answer: d*

46. A sexually produced spore that buds from the surface of a basidium is a
    a. zygospore.
    b. ascospore.
    c. conidiospore.
    d. basidiospore.
    e. uredospore.
    *Answer: d*

47. Euascomycetes produce conidia
    a. as a means of asexual reproduction.
    b. as a means of sexual reproduction.
    c. at the tips of specialized hyphae.
    d. in response to harsh environmental conditions.
    e. Both a and c
    *Answer: e*

48. The gills of a mushroom are specialized for
    a. respiration.
    b. food production.
    c. defense.
    d. reproduction.
    e. water storage.
    *Answer: d*

49. The algal partner in a lichen symbiosis is responsible primarily for
    a. respiration.
    b. food production.
    c. defense.
    d. reproduction.
    e. water storage.
    *Answer: b*

50. Lichens obtain nutrients by
    a. means of photosynthesis.
    b. engulfing other organisms.
    c. absorbing nutrients from the environment.
    d. means of decaying organic material.
    e. parasitizing flowering plants.
    *Answer: a*

51. In a lichen, the portion of the fungus that is involved directly in the symbiosis is the
    a. fruiting body.
    b. mycelium.
    c. spores.
    d. spore cases.
    e. blue-green bacteria.
    *Answer: b*

52. Which component would one expect to find as part of the fungal partner in a lichen?
    a. Chitin
    b. Chlorophyll
    c. Reverse transcriptase
    d. Silica
    e. Cellulose
    *Answer: a*

53. Lichens acquire energy from
    a. decaying matter.
    b. parasitism.
    c. the sun.
    d. minerals in the air and precipitation.
    e. minerals on rocks.
    *Answer: c*

54. Lichens are _____ associations of a fungus with _____.
    a. symbiotic; an alga or a bacterium
    b. saprobic; an alga or a bacterium
    c. parasitic; an alga or a bacterium
    d. symbiotic; plant roots
    e. parasitic; plant roots
    *Answer: a*

55. Soredia are characteristic of some
    a. zygomycetes.
    b. ascomycetes.
    c. basidiomycetes.
    d. deuteromycetes.
    e. lichens.
    *Answer: e*

# 32 Animal Origins and the Evolution of Body Plans

## Fill in the Blank

1. _____ symmetry is strongly correlated with cephaliza-tion: the development of sense organs and central nervous tissues at the anterior end of an animal.
*Answer: Bilateral*

2. Animals probably arose from colonial flagellated protists as a result of division of labor among their aggregated cells. Among the extant animals, the _____ bear the strongest similarity to the most recent common ancestor of animals.
*Answer: sponges*

3. Cnidaria possess _____ symmetry, having a cylindrical form with one main axis around which body parts are arranged.
*Answer: radial*

4. Platyhelminthes, nematodes, rotifers, annelids, and arthropods possess a body plan that shows _____ symmetry.
*Answer: bilateral*

5. Among the _____, the mouth arises from the blasto-pore. This embryonic feature is a synapomorphy of the group.
*Answer: protostomes*

6. Animals that lack an enclosed body cavity, such as Platyhelminthes (flatworms), are called _____.
*Answer: acoelomates*

7. A distinguishing feature of mollusks is the _____, a sheet of specialized tissue that covers the internal organs like a body wall. This structure secretes the shell.
*Answer: mantle*

8. Squids and octopuses are members of the phylum _____.
*Answer: Mollusca*

9. Although _____ superficially resemble mollusks, they possess a lophophore and are typically attached to the substrate via a short flexible stalk.
*Answer: brachiopods*

10. The oligochaetes are _____; each organism has both male and female sex organs.
*Answer: hermaphrodites*

11. In lophophorate animals, the circular or U-shaped ridge around the mouth bearing tentacles is called the _____.
*Answer: lophophore*

12. The embryos of _____ animals have two cell layers: an outer _____ and an inner _____.
*Answer: diploblastic; ectoderm; endoderm*

13. The existence of three embryonic cell layers, including an outer _____, an inner _____, and a middle layer, the _____, is a synapomorphy for protostomes and deuterostomes.
*Answer: ectoderm; endoderm; mesoderm*

14. The fluid-filled body cavities of animals function as _____ skeletons.
*Answer: hydrostatic*

15. The simplest type of symmetry, in which body parts radiate outward from a central point, is called _____ symmetry. It is widespread among the protists.
*Answer: spherical*

16. The organisms that make up the sister group of all bilaterally symmetrical animals (protostomes and deuterostomes) are the _____.
*Answer: ctenophores*

17. Derived cnidarian classes have life cycles involving two stages: the _____ and the _____.
*Answer: polyp; medusa*

18. Evidence from developmental, structural, and molecular biology indicates that the protostomes split into two major monophyletic lineages that have been evolving separately since ancient times. These two lineages are the _____ and the _____.
*Answer: lophotrochozoans; ecdysozoans*

19. Many lophotrochozoans have a type of free-living larva called a _____.
*Answer: trochophore*

20. Members of the class _____ are used medically to reduce fluid pressure and prevent blood clotting.
*Answer: Hirudinea*

## Multiple Choice

1. Which of the following is a phylum that includes animals with subdivided coeloms?
a. Ctenophora
b. Porifera
c. Annelida
d. Platyhelminthes
e. Cnidaria
*Answer: c*

■2. Bilaterally symmetrical animals can be divided into two major groups based on major evolutionary lineages that separated in the Cambrian. Those two lineages differ fundamentally in their
   a. modes of reproduction.
   b. early embryological development.
   c. mode of obtaining and storing energy.
   d. environmental requirements.
   e. metabolism.
   *Answer: b*

3–6. Refer to the diagram below of a cross section of a sponge to answer the following questions.

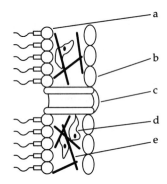

3. Which structure is a choanocyte?
   *Answer: a*

4. Which structure is an osculum?
   *Answer: c*

5. Which structure is an epidermal cell?
   *Answer: b*

6. Which structure is not a cell?
   *Answer: e*

7. Which of the following statements concerning nemato-cysts is true?
   a. They are organelles used in the capture of prey by cnidarians.
   b. They are excretory organs in Platyhelminthes.
   c. They are reproductive cells in cnidarians.
   d. They are ciliated cells in Porifera.
   e. They are excretory cells in Porifera.
   *Answer: a*

8. Which class within the Cnidaria is important geologically because the growth of its members can result in the formation of islands and atolls in tropical oceans?
   a. Hydrozoa
   b. Scyphozoa
   c. Anthozoa
   d. Turbellaria
   e. Ctenophora
   *Answer: c*

■9. Animals in the phylum Platyhelminthes are often parasitic. Which of the following is an adaptation found in Platy-helminthes that is associated with a parasitic lifestyle?
   a. Large size
   b. A highly branched gastrovascular cavity
   c. An oxygen transport system
   d. Absorption of nutrients through the body surface
   e. Both b and c
   *Answer: d*

**10. In what way does the phylum Platyhelminthes differ from the phylum Cnidaria?
   a. Platyhelminthes are radially symmetrical, whereas cnidarians are bilaterally symmetrical.
   b. Platyhelminthes have more complex internal organs than cnidarians do.
   c. Platyhelminthes are diploblastic, whereas cnidarians are triploblastic.
   d. Platyhelminthes have two openings to the gastrovas-cular cavity, whereas cnidarians have only one.
   e. Platyhelminthes have an excretory system whereas cnidarians do not.
   *Answer: b*

■11. Which of the taxa below contains the largest number of species?
   a. Ctenophora
   b. Mollusca
   c. Annelida
   d. Phoronida
   e. Polychaeta
   *Answer: c*

*12. Which of the features below is a synapomorphy for Animalia?
   a. Chitin
   b. Vascular systems
   c. Similarities in their Hox genes
   d. Bilateral symmetry
   e. Spiral cleavage
   *Answer: c*

*13. Which of the taxa below includes more than 10,000 species?
   a. Ctenophora
   b. Porifera
   c. Platyhelminthes
   d. Rotifers
   e. Monoplacophorans
   *Answer: c*

*14. Which of the features below does *not* describe the Phoronids?
   a. Fewer than 100 species
   b. Active forager and carnivorous
   c. Secretion of chitinous tubes
   d. Possession of a lophophore
   e. Length ranging from 5–25 cm
   *Answer: b*

*15. Brachiopods reached their peak abundance and diversity in the
   a. Paleozoic.
   b. Mesozoic.
   c. Cenozoic.
   d. Precambrian.
   e. Both a and b
   *Answer: e*

*16. Which phylum or phyla contain animals with a complete digestive tract?
   a. Flukes
   b. Tapeworms
   c. Cnidarians
   d. Poriferans
   e. Annelida, Rotifera, and Ctenophora
   *Answer: e*

■17. What is the main difference between a coelom and a pseudocoelom?
   a. A coelom is enclosed by muscles on the outside only; a pseudocoelom is enclosed by muscles on the inside only.
   b. A coelom does not involve an enclosed body cavity; a pseudocoelom is a cavity that is completely surrounded by muscle.
   c. A coelom is enclosed by muscles on the inside only; a pseudocoelom is a solid structure.
   d. A coelom is enclosed by muscles on the inside and the outside; a pseudocoelom has only one layer of muscle, and it rims the outside of the body cavity.
   e. None of the above
   *Answer: d*

18. The body plan of mollusks includes three unique, shared, derived characteristics that support the monophyly of the group: the _____, the _____, and the _____.
   a. foot; radula; mantle
   b. foot; mantle; shell
   c. visceral mass; radula; mantle
   d. visceral mass; mantle; shell
   e. foot; visceral mass; mantle
   *Answer: e*

19. Which of the following statements concerning the mantle cavity of mollusks is *false*?
   a. It is no longer present in octopuses.
   b. It has been modified into internal support in slugs and squids.
   c. It is used as a filtering device by bivalves.
   d. It is used as the basis of jet propulsion by cephalopods.
   e. It secretes the shell that provides external protection in most molluscan groups.
   *Answer: c*

20. Cephalization is most commonly associated with
   a. spherical symmetry.
   b. radial symmetry.
   c. sessile animals.
   d. bilateral symmetry.
   e. lophophorate animals.
   *Answer: d*

■21. Deuterostomes and protostomes differ in a number of characteristics. From the following list, select one characteristic in which they do *not* differ.
   a. Cleavage type
   b. Ability to form a blastopore
   c. Embryological origin of the mouth
   d. Derivation of mesoderm
   e. Position of the nervous system (dorsal/ventral)
   *Answer: b*

22. Spiral cleavage is a synapomorphy for a subset of lophotrochozoans; these include the
   a. Nemertea, Annelida, and Phoronida.
   b. Nemertea, Annelida, and Mollusca.
   c. Nemertea, Annelida, and Bryozoa.
   d. Nemertea, Annelida, and Rotifera.
   e. Nemertea, Annelida, and Platyhelminthes.
   *Answer: b*

★23. Which of the taxa below may gain additional nutrition from photosynthetic endosymbionts?
   a. Mollusks
   b. Ctenophores
   c. Cnidarians
   d. Flatworms
   e. Chitons
   *Answer: c*

24. Which of the following is *not* characteristic of the body plan of members of the phylum Porifera?
   a. Gastrovascular cavity
   b. No distinct tissue layers or organs
   c. No separation between the different cell layers
   d. Organization around a water canal system
   e. Asymmetry
   *Answer: a*

25. Which of the following is *not* associated with sponges?
   a. Choanocytes
   b. Porocytes
   c. Spicules
   d. Mesoderm
   e. Eggs
   *Answer: d*

26. Which of the following characteristics is *not* associated with members of the phylum Cnidaria?
   a. Alternation between polyp and medusa
   b. Three distinct body layers
   c. Nematocysts
   d. Gastrovascular cavity
   e. Planula larva
   *Answer: b*

27. Which of the following is associated with both the medusa and polyp stage in a typical hydrozoan cnidarian?
   a. Asexual reproduction
   b. Sessile lifestyle
   c. Gastrovascular cavity
   d. Production of planula larva
   e. Thick mesoglea
   *Answer: c*

28. Which of the following statements about Anthozoan cnidarians is true?
   a. They reproduce asexually only.
   b. They are commonly known as jellyfish.
   c. Only the polyp form exists.
   d. Gametes are produced directly by the medusa.
   e. They possess nematocysts.
   *Answer: c*

★29. Which of the below is the most basal lineage of cnidarians?
   a. Anthozoans
   b. Hydrozoans
   c. Scyphozoans
   d. Cephalopoda
   e. Ctenophora
   *Answer: a*

30. In free-living flatworms, which of the following functions occurs by simple diffusion?
   a. Respiration
   b. Absorption of nutrients
   c. Excretion
   d. Distribution of nutrients within the body
   e. All of the above
   *Answer: e*

31. Which of the following is *not* associated with a parasitic lifestyle as seen in tapeworms and flukes?
    a. Complex digestive systems
    b. Flattened body
    c. Life cycles with several larval stages
    d. Life cycles with multiple hosts
    e. Reduced gastrovascular cavity
    *Answer: a*

32. Of the taxa below, which one includes the fewest species?
    a. Mollusca
    b. Ctenophora
    c. Annelida
    d. Cnidaria
    e. Lophotrochozoa
    *Answer: b*

33. Which body part in a mollusk secretes the shell?
    a. Mantle
    b. Foot
    c. Visceral mass
    d. Radula
    e. Spicules
    *Answer: a*

34. Which phylum does *not* contain wormlike organisms?
    a. Platyhelminthes
    b. Ctenophora
    c. Annelida
    d. Porifera
    e. Both b and d
    *Answer: e*

35. Which phylum has an open circulatory system, a complete digestive tract, and a reduction of the coelom?
    a. Rotifera
    b. Annelida
    c. Mollusca
    d. Ctenophora
    e. Platyhelminthes
    *Answer: c*

36. You discover a segmented marine animal that develops through spiral cleavage and has parapodia. Select the most inclusive group to which this animal belongs.
    a. Ctenophora
    b. Annelida
    c. Lophotrochozoa
    d. Mollusca
    e. Platyhelminthes
    *Answer: c*

37. Clues to the evolutionary relationships of animals can be found in
    a. the fossil record.
    b. patterns of embryonic development.
    c. comparative morphology and physiology.
    d. nucleotide sequence patterns.
    e. All of the above
    *Answer: e*

38. The terms "acoelomate," "pseudocoelomate," and "coelomate" are used to describe
    a. cephalization.
    b. origin of the blastopore.
    c. ectoderm, mesoderm, and endoderm.
    d. the body cavity of animals.
    e. the vertebrate body plan.
    *Answer: d*

39. Which of the following is *not* a synapomorphy for the Deuterostomes?
    a. Ventral nervous system
    b. Larva (when present) with a food-collecting system consisting of cells with a single cilium
    c. Anterior brain that surrounds the entrance to the digestive tract
    d. Radial cleavage
    e. Both a and c
    *Answer: e*

40. Which of the following is *not* true of members of the phylum Platyhelminthes?
    a. Some are parasitic, such as those of the classes Cestoda and Trematoda.
    b. They possess a mouth but no anus.
    c. They are diploblastic.
    d. They possess some cephalization.
    e. Some possess complex life cycles.
    *Answer: c*

41. The body cavity of coelomate animals develops within the
    a. endoderm.
    b. ectoderm.
    c. mesoderm.
    d. pseudocoel.
    e. mesoglea.
    *Answer: c*

*42. Porifera differ from most other animals in that they
    a. do not form true organs.
    b. are sessile.
    c. are triploblastic.
    d. have trochophore larvae.
    e. Both a and b
    *Answer: a*

43. The Portuguese man-of-war is an example of a deadly
    a. poriferan.
    b. cnidarian.
    c. ctenophore.
    d. nematode.
    e. mollusk.
    *Answer: b*

44. Which phylum below is one of the descendants of the most recent common ancestor of lophotrochozoans?
    a. Porifera
    b. Vertebrata
    c. Mollusca
    d. Ctenophora
    e. Cnidaria
    *Answer: c*

45. Which of the following is *not* true regarding ectoprocts?
    a. They are colonial.
    b. They are wormlike in structure.
    c. They are mainly marine.
    d. They have a lophophore.
    e. Individual colony members may have specialized functions.
    *Answer: b*

46. Corals are members of which class?
    a. Anthozoa
    b. Cnidaria
    c. Porifera
    d. Ctenophora
    e. Annelida
    *Answer: a*

47. Which of the following is *not* associated with the phylum Rotifera?
    a. Pseudocoelom
    b. Complete gut
    c. Conspicuous feeding organs
    d. Radial symmetry
    e. Movement by beating cilia
    *Answer: d*

*48. Which of the following classes does *not* belong to the phylum Annelida?
    a. Scyphozoa
    b. Polychaeta
    c. Oligochaeta
    d. Hirudinea
    e. Both a and d
    *Answer: a*

49. Which of the following is *not* true concerning oligochaetes?
    a. They are hermaphroditic.
    b. They have separate nerve centers called ganglia.
    c. They have a segmented coelom.
    d. They have serially repeated organs.
    e. Their embryos undergo radial cleavage.
    *Answer: e*

50. Which of the following organisms are hermaphroditic?
    a. Leeches (Hirudinea)
    b. Earthworms (Oligochaeta)
    c. Squids
    d. Both a and b
    e. None of the above
    *Answer: d*

51. Which of the following traits is *not* shared by all animals?
    a. Special types of cell–cell junctions
    b. A common set of extracellular matrix molecules
    c. A complete gut
    d. Similarities in their small-subunit ribosomal RNAs
    e. Similarities in their Hox genes
    *Answer: c*

52. Members of the class Anthozoa (the corals) are able to survive in nutrient-poor tropical waters by forming symbiotic relationships with photosynthetic
    a. green algae.
    b. red algae.
    c. brown algae.
    d. dinoflagellates.
    e. None of the above
    *Answer: d*

53. The divergence of the common ancestor of animals into the protostome and deuterostome lineages occurred in the
    a. Precambrian.
    b. Silurian.
    c. Jurassic.
    d. Cambrian.
    e. Mesozoic.
    *Answer: d*

54. Which of the following shared, derived traits are *not* among those that unite the protostomes?
    a. Dorsal nervous system
    b. An anterior brain
    c. Free-floating larva
    d. Development of a mouth from a blastopore
    e. Spiral cleavage (in some species)
    *Answer: a*

55. Many lineages of lophotrochozoans have a type of free-living larva known as a
    a. polyp.
    b. planula.
    c. trochophore.
    d. nauplius.
    e. None of the above
    *Answer: c*

*56. The largest and most remarkable vestimentiferan annelids live near deep-ocean hydrothermal vents. Their tissues harbor endosymbiotic prokaryotes that fix carbon by oxidizing
    a. $CH_4$.
    b. $H_2S$.
    c. $C_6H_{12}O_6$.
    d. $H_2O$.
    e. Both a and b
    *Answer: b*

57. Mollusks have a rasping feeding structure known as the
    a. proboscis.
    b. rhynchocoel.
    c. radula.
    d. corona.
    e. mastax.
    *Answer: c*

58. Some cephalopods have modified their excurrent siphon and mantle to allow them to
    a. ingest large prey.
    b. alter their buoyancy.
    c. move rapidly through the water.
    d. attach to a substrate.
    e. become parasitic.
    *Answer: c*

# 33 Ecdysozoans: The Molting Animals

## Fill in the Blank

1. The distinguishing feature of ecdysozoans is the _____, a nonliving covering that provides the animal with protection and support.
*Answer: exoskeleton*

2. Growth in ecdysozoans is accomplished by _____, a periodic shedding of the exoskeleton followed by the rapid hardening of a new and larger exoskeleton that has formed under the old one.
*Answer: molting*

3. A class of arthropods that has few members living in the oceans is the _____; however, in fresh water and on land they are a dominant group.
*Answer: insects*

4. The thickened, strong, flexible, waterproof polysaccharide body covering called _____ is characteristic of the Precambrian ecdysozoan lineage leading to arthropods.
*Answer: chitin*

5. The immature stages of insects between molts are called _____.
*Answer: instars*

6. Insects and most crustaceans have tripartite body plans which include a _____, a _____, and an _____.
*Answer: head; thorax; abdomen*

7. Centipedes and millipedes, members of the phylum _____, have a head and trunk but no abdomen.
*Answer: Myriapoda*

8. An insect that exhibits gradual changes between its instars is said to undergo _____.
*Answer: incomplete metamorphosis*

9. A caterpillar changing into a butterfly is a example of complete metamorphosis. During this process, a larva transforms itself into an adult during a specialized phase in which it is a _____.
*Answer: pupa*

10. Some wormlike ecdysozoans have a relatively thin, flexible exoskeleton called a _____.
*Answer: cuticle*

11. *Trichinella* is a parasitic roundworm of the phylum _____.
*Answer: Nematoda*

12. Because a rigid exoskeleton prevents wormlike movement, animals with stiff exoskeletons require _____ that can be manipulated by muscles.
*Answer: appendages*

13. Some familiar marine arthropods are crabs, lobsters, and shrimps, all of which belong to the phylum _____.
*Answer: Crustacea*

14. Insects exchange gases by means of air sacs and tubular channels called _____.
*Answer: tracheae*

15. The lineage leading to the insects separated from the one leading to crustaceans around _____ million years ago.
*Answer: 450*

★16. Spiders are important terrestrial predators belonging to the phylum _____.
*Answer: Chelicerata*

17. Many species of crustaceans have a fold of exoskeleton that extends dorsally and laterally from the head to cover and protect some body segments. This structure is called the _____.
*Answer: carapace*

## Multiple Choice

★1. Which of the following describes locomotion in nematodes (roundworms)?
a. Rhythmically beating cilia move the animal forward.
b. Circular and longitudinal muscles work against each other and the pseudocoelom to change the shape of the animal, moving it forward.
c. Longitudinal muscles contract.
d. A series of hairs that project backward engage the substrate; back-and-forth movements propel the animal forward.
e. Water expelled through special ducts moves the animal forward.
*Answer: c*

2. Which of the following statements concerning nematodes is *false*?
   a. Nematode parasites infect many members of the animal kingdom, including many domestic animals.
   b. Nematode parasites infect many members of the plant kingdom, including many crop plants.
   c. Trichinosis is a disease caused by roundworms that is confined to pigs.
   d. Free-living nematodes are often extremely abundant.
   e. The diets of nematodes are varied.
   *Answer: c*

3. Which of the following are characteristics of exoskeletons?
   a. They are a highly efficient means of anchoring muscles, thereby providing more efficient movement.
   b. They provide protection from predators.
   c. They must be shed for the animal to grow and thus they make the animal vulnerable to predators.
   d. They provide support for walking on dry land.
   e. All of the above
   *Answer: c*

4. The _____ have bodies that are divided into two major regions, a head region that bears two pairs of appendages modified to form mouthparts, and a trunk region with four pairs of walking legs.
   a. Chelicerata
   b. Crustacea
   c. Hexapoda
   d. Nematoda
   e. Chaetognatha
   *Answer: a*

5. Which of the following statements concerning the tracheal system in insects is *false*?
   a. Tracheae are air sacs and tubular channels.
   b. Tracheae penetrate virtually every part of an insect's body.
   c. Tracheae expand and contract like lungs to move oxygen to the blood.
   d. Tracheae provide oxygen to the tissues of the insects.
   e. Tracheae extend from external openings inward to tissues throughout the body.
   *Answer: c*

6. *Trichinella spiralis*, the causative agent of the disease trichinosis, is a member of the phylum
   a. Nemertea.
   b. Platyhelminthes.
   c. Nematoda.
   d. Rotifera.
   e. Annelida.
   *Answer: c*

7. You discover an animal with bilateral symmetry, a pseudocoelom, a tubular digestive system, and a thick multilayer cuticle. Select the phylum to which this animal most likely belongs.
   a. Nematomorpha
   b. Platyhelminthes
   c. Nematoda
   d. Rotifera
   e. Annelida
   *Answer: c*

8. Which one of the following groups of arthropods includes the dragonflies?
   a. Onychophora
   b. Trilobita
   c. Chelicerata
   d. Crustacea
   e. Hexapoda
   *Answer: e*

9. Which of the following is *false*?
   a. The insect wing may have evolved from an ancestral appendage similar to that of modern crustaceans.
   b. Development of the insect wing is governed by the expression of the same gene that governs the crayfish appendage.
   c. Development of the insect wing and development of the crayfish appendage are controlled by the *pdm* gene.
   d. The *pdm* gene is expressed in both the wings and legs of insects.
   e. The wings of ancestral crustaceans and ancestral insects expressed the same gene (*pdm*).
   *Answer: e*

**10. The evolution of a chitinous exoskeleton affected many aspects of arthropod evolution. From the following list, select the aspect that was *least* affected.
   a. Division of labor among the body parts
   b. Mode of locomotion
   c. Pattern of growth
   d. Type of gas exchange system
   e. Mode of digestion
   *Answer: e*

11. How many times have wings evolved in protostomes?
   a. Once
   b. Twice
   c. Three times
   d. Four times
   e. Many times, independently
   *Answer: a*

12. From the following list of characteristics, select one that has *not* been a major theme in protostome evolution.
   a. The evolution of filter feeding structures
   b. Segmentation of the body cavity
   c. Evolution of muscles (versus cilia) for body movement
   d. Evolution of mechanisms for ingesting large prey
   e. Evolution of hard, external body parts
   *Answer: d*

13. You discover an animal that has a hemocoel, a chitinous exoskeleton, and a complete digestive tract. Select the phylum to which this animal most likely belongs.
   a. Nematoda
   b. Annelida
   c. Hexapoda
   d. Mollusca
   e. Chaetognatha
   *Answer: c*

▪14. Why are animals with thin cuticles generally restricted to moist habitats?
   a. The thin cuticle allows water to be lost across the body surface.
   b. There are fewer predators in these habitats.
   c. These animals rely on flowing water to bring them food.
   d. Both a and c
   e. None of the above
   *Answer: a*

15. The sister taxon of the kinorhychs is the
    a. Nematoda.
    b. Crustacea.
    c. Priapulida.
    d. Chelicerata.
    e. Tardigrada.
    *Answer: c*

*16. A typical crustacean larva is called a
    a. trochophore.
    b. nauplius.
    c. planula.
    d. pupa.
    e. caterpillar.
    *Answer: b*

17. Which of the following is *not* a characteristic of insects?
    a. Some parasitic species
    b. Three basic body segments
    c. A complex respiratory system, including lungs
    d. An excretory organs
    e. A digestive system
    *Answer: c*

18. Most of protostome evolution took place in
    a. oceans.
    b. terrestrial environments.
    c. freshwater environments.
    d. arboreal environments.
    e. rivers.
    *Answer: a*

*19. The centipedes and millipedes belong to which group?
    a. Insecta
    b. Arachnida
    c. Myriapoda
    d. Merostomata
    e. None of the above
    *Answer: c*

20. Flowing water is an efficient transport system for nutrients; this has led to the repeated evolution of _____ lifestyles during protostome evolution.
    a. predatory
    b. parasitic
    c. scavenging
    d. sessile
    e. None of the above
    *Answer: d*

21. Individuals of which group are so numerous that they are thought to be the most abundant of all animals?
    a. Copepods
    b. Millipedes
    c. Nematomorphs
    d. Barnacles
    e. Apterygota
    *Answer: a*

*22. Which of the following is *not* one of the major recognized lineages of winged insects?
    a. Insects that cannot fold their wings back against the body
    b. Insects that can fold their wings and that undergo incomplete metamorphosis
    c. Insects that do not undergo metamorphosis
    d. Insects that can fold their wings and that undergo complete metamorphosis

    e. All of the above are recognized lineages.
    *Answer: c*

23. Members of which of the following phyla do *not* have a circulatory system?
    a. Nematoda
    b. Chelicerata
    c. Crustacea
    d. Nematomorpha
    e. Both a and d
    *Answer: e*

24. Which of the following is *not* a characteristic or purpose of spider webs and threads?
    a. They can function as a home.
    b. They are produced by modified abdominal appendages connected to internal secretory glands.
    c. They are composed primarily of carbohydrates.
    d. They function as a snare for catching prey.
    e. They provide safety lines for climbing.
    *Answer: c*

*25. You have found a small animal in marine sand and are examining it under a microscope. It is about 0.3 mm in length and appears to have fleshy unjointed legs. To which phylum does this organism most likely belong?
    a. Tardigrada
    b. Onychophora
    c. Chelicerata
    d. Crustacea
    e. Hexapoda
    *Answer: a*

26. Members of which of the following phyla have a pseudocoel?
    a. Chaetognatha and Nematomorpha
    b. Nematomorpha and Nematoda
    c. Nematomorpha, Nematoda, and Chaetognatha
    d. Nematoda, Chelicerata, and Chaetognatha
    e. Nematomorpha, Nematoda, and Uniramia
    *Answer: b*

27. Members of which of the following phyla have a complete gut?
    a. Chaetognatha and Nematomorpha
    b. Chaetognatha and Nematoda
    c. Chaetognatha, Crustacea, Hexapoda, and Nematoda
    d. Chaetognatha, Crustacea, Hexapoda, and Chelicerata
    e. Chaetognatha, Crustacea, Hexapoda, and Nematomorpha
    *Answer: d*

*28. The wingless insects belong to which group?
    a. Apterygota
    b. Pterygota
    c. Myriapoda
    d. Arachnida
    e. Merostomata
    *Answer: a*

29. Organisms of which phylum are probably most similar to ancestral arthropods?
    a. Chelicerata
    b. Nematoda
    c. Crustacea
    d. Onychophora
    e. Priapulida
    *Answer: d*

*30. Butterflies belong to which order of winged insects?
    a. Lepidoptera
    b. Coleoptera
    c. Diptera
    d. Trichoptera
    e. Isoptera
    *Answer: a*

■31. A firm exoskeleton has protective and supportive advantages, but it poses a problem for insects. What is this problem?
    a. The animal must consume large amounts of food to support the growth of the exoskeleton.
    b. The exoskeleton prevents the animal from moving rapidly.
    c. The exoskeleton cannot grow as the animal body inside it grows.
    d. The exoskeleton attracts predators.
    e. None of the above
    *Answer: c*

*32. Which of the taxa below contains fewer than 100 species?
    a. Copepoda
    b. Arthropoda
    c. Priapulida
    d. Nematomorpha
    e. Nematoda
    *Answer: c*

# 34 Deuterostomate Animals

## Fill in the Blank

1. The ray-finned fishes are capable of regulating their buoyancy because of an organ called the _____.
   *Answer: swim bladder*

2. During the Devonian the partly terrestrial _____ arose from an ancestor they shared with lungfishes.
   *Answer: Amphibia*

3. The first major animal group that could live completely out of water is the class _____.
   *Answer: Reptilia*

4. Part of the extraordinary success of the reptiles can be attributed to the evolution of the _____, which made development possible outside of the water.
   *Answer: amniote egg*

5. The evolutionary lineage leading to birds is nested within the more inclusive monophyletic group, _____, which also contains the Ornithischians and Saurischians.
   *Answer: Dinosauria*

6. The class _____ includes monotremes, therians (marsupials), and eutherians.
   *Answer: Mammalia*

7. The muscular pharynx of an acorn worm (Hemichordata) opens to the outside with a number of _____ through which water can exit.
   *Answer: pharyngeal slits*

8. In the evolutionary lineage leading to the chordates, a dorsal supporting structure called the _____ evolved.
   *Answer: notochord*

9. Members of the phylum _____ are bilaterally symmetrical animals that possess pharyngeal slits, a dorsal hollow nerve cord, a ventral heart, a tail that extends beyond the anus, and a notochord at some stage of their life cycle.
   *Answer: Chordata*

10. Structures called placentas, which nourish developing embryos, evolved in _____.
    *Answer: mammals*

11. The evolution of primitive _____ in the sister group of the ray-finned fishes set the stage for the invasion of land because they provided an alternative method of gas exchange.
    *Answer: lungs (or lunglike sacs)*

12. In the evolutionary lineages leading to dinosaurs and birds and to mammals, two factors were important in enabling these organisms to run at fast rates for long periods of time: a more vertical position assumed by the legs, and special _____ muscles that could operate independently of locomotory muscles.
    *Answer: ventilatory*

13. The heavily armored fishes that evolved jaws are known as the _____.
    *Answer: placoderms*

14. The expanded mental abilities of humans are largely responsible for the development of _____, the process by which knowledge and traditions are passed from one generation to the next.
    *Answer: culture*

15. The network of calcified hydraulic canals that functions in gas exchange, locomotion, and feeding in echinoderms is called a _____.
    *Answer: water vascular system*

16. Vertebrates have a jointed, dorsal _____ that replaced the notochord as their primary support.
    *Answer: vertebral column*

17. The cartilaginous fishes control their movement with pairs of unjointed appendages called _____.
    *Answer: fins*

18. Females of the group _____ have ventral pouches in which they carry and feed their offspring.
    *Answer: Marsupialia*

19. The lineages leading to modern reptiles began to diverge around _____ million years ago.
    *Answer: 250*

## Multiple Choice

1. Which of the following is *not* a characteristic of echinoderms?
   a. Radial symmetry as adults
   b. An external skeleton
   c. Lack of a brain
   d. Bilateral symmetry as larvae
   e. An extensive fossil record
   *Answer: b*

2. Which of the following phyla has a water vascular system?
   a. Mollusca
   b. Chordata
   c. Annelida
   d. Echinodermata
   e. Hemichordata
   *Answer: d*

*3. A major change in the evolutionary lineage leading to the chordates was
   a. evolution of the water vascular system.
   b. the development of a dorsal hollow nerve cord.
   c. calcification of an internal skeleton.
   d. development of the lophophore as an adaptation to predatory life.
   e. the ability to filter feed.
   *Answer: b*

4. Which of the following is *not* found in the phylum Chordata?
   a. Bilateral symmetry
   b. A dorsal hollow nerve cord
   c. An external skeleton
   d. Pharyngeal slits at some stage during development
   e. A notochord at some stage during development
   *Answer: c*

5. Which of the following is *not* a unique characteristic of the vertebrate body plan?
   a. A rigid internal skeleton with the vertebral column
   b. Two pairs of appendages
   c. A dorsal hollow nerve cord
   d. An anterior skull with a large brain
   e. Both c and d
   *Answer: c*

6. The important evolutionary novelty that evolved from the gill arches of jawless fishes and is retained by Placoderms, Chondrichthyes, ray-finned fishes, lobe-finned fishes, lungfishes, and tetrapods is
   a. heavily armored skin.
   b. the jaw.
   c. fins.
   d. the ability to swim.
   e. the vertebral column.
   *Answer: b*

7. The _____ are likely the most recent common ancestors of all deuterostomes.
   a. yunnanozoans
   b. echinoderms
   c. vertebrates
   d. arthropods
   e. sponges
   *Answer: a*

8. The cartilaginous fishes, including sharks, skates and rays, and chimaeras,
   a. belong to the class Agnatha.
   b. have heavy external armor.
   c. have an open circulatory system.
   d. have a few bones in their skeletons.
   e. have less external armor and are faster swimmers than their ancestors.
   *Answer: e*

9. Which of the structures below is (are ) *not* part(s) of the amniote egg?
   a. Extraembryonic membranes
   b. Yolk
   c. Shell
   d. Bone
   e. Chorion
   *Answer: d*

10. Which of the following is a major difference between the classes Chondrichthyes and Actinopterygii (ray-finned fishes)?
    a. Only Chondrichthyes have a lung or swim bladder.
    b. Chondrichthyes have a cartilaginous skeleton, whereas ray-finned fishes have a bony skeleton.
    c. Only ray-finned fishes have paired fins.
    d. Chondrichthyes evolved in fresh water, whereas ray-finned fishes evolved in salt water.
    e. Only Chondrichthyes have true jaws.
    *Answer: b*

11. The swim bladders of bony fishes
    a. evolved from lunglike sacs that supplemented the gills in respiration.
    b. are used for respiration in most contemporary species.
    c. are organs of buoyancy that help the fishes control their depth in the water column.
    d. prevented them from existing in a marine environment.
    e. Both a and c
    *Answer: e*

12. Lobe-finned fishes had several adaptations that were instrumental in the transition to life on land. Adaptations in lobe-finned fishes that were important in the evolution of the amphibians included
    a. primitive lungs.
    b. jointed fins with strong muscular support.
    c. watertight skin.
    d. Both a and b
    e. a, b, and c
    *Answer: d*

13. Most amphibians breathe air by which means?
    a. Gills and swim bladders
    b. Gills and thin skin
    c. Lungs only
    d. Lungs and thin skin
    e. Thin skin only
    *Answer: d*

14. Reptiles were the first group of animals to be completely liberated from a need to return to water for some portion of their life cycle. Adaptations first found in reptiles that were important to this shift include which features?
    a. A hard-shelled (amniote) egg
    b. Parental care
    c. Watertight skin
    d. The ability to breathe and run simultaneously
    e. Both a and c
    *Answer: e*

15. The most unique characteristic of birds is the
    a. scales on their legs.
    b. ability to lay eggs that will not dry out.
    c. presence of feathers.
    d. enormous amount of parental care they provide their young.
    e. None of the above
    *Answer: c*

16. Aves
    a. descended most recently from an amphibian ancestor.
    b. should be included in the class Mammalia.
    c. should include some modern reptiles.
    d. should include all modern reptiles.
    e. are in the monophyletic group that includes dinosaurs and crocodiles.
    *Answer: e*

17–19. Match the groups of organisms from the list below with their most important evolutionary novelty.
    a. Reptiles
    b. Birds
    c. Ray-finned fishes
    d. Chondrichthyes
    e. Echinoderms

17. Swim bladders
    *Answer: c*

18. Powered flight
    *Answer: b*

19. Jaws
    *Answer: d*

**20. In several important ways, deuterostome and protostome evolution are similar. Which of the following is *not* true of both protostome and deuterostome evolution?
    a. Both lineages exploited the abundant food supplies buried in soft marine substrates.
    b. Many groups in both lineages developed elaborate structures for extracting prey from water.
    c. In both lineages, a coelomic cavity evolved and became divided into compartments that allowed better control of body shape and movement.
    d. Both lineages evolved locomotor abilities.
    e. Both groups invaded land and evolved into very large terrestrial animals.
    *Answer: e*

21. The oldest fossil remains of members of our genus, *Homo*, suggest that early relatives of humans lived
    a. near China, where food is plentiful throughout the year.
    b. near Siberia, in an area where they were protected from predation.
    c. in the midwestern United States, where there is some of the most fertile soil in the world.
    d. in the American tropics, where there are long growing seasons and many species of fruits and berries.
    e. in dry African savannas, where they ate roots, bulbs, tubers, and animals.
    *Answer: e*

22.–24. From the list below, match the correct class of echinoderms with the descriptions that follow.
    a. Crinoidea
    b. Ophiuroidea
    c. Asteroidea
    d. Echinoidea
    e. Holothuroidea

22. To which class of echinoderms do the brittle stars belong?
    *Answer: b*

23. Which class of echinoderms has tube feet on arms (from five to hundreds of them) that extend into the current for feeding?
    *Answer: a*

24. Which class of echinoderms is herbivorous and has no arms?
    *Answer: d*

25. From the following list of echinoderm taxa, select the most inclusive taxon that was much more abundant in the past than it is today.
    a. Crinoidea
    b. Pelmatozoa
    c. Asteroidea
    d. Eleutherozoa
    e. Holothuroidea
    *Answer: b*

26. Which feature does *not* characterize animals in the phylum Echinodermata?
    a. Bilaterally symmetric larvae
    b. A network of water-filled canals leading to tube feet
    c. Presence of a water vascular system
    d. An internal skeleton of calcified plates
    e. An external skeleton of chitin
    *Answer: e*

27. The chordates
    a. all have a bony backbone.
    b. include some animals without a nervous system.
    c. pass through a developmental stage with pharyngeal slits.
    d. are poorly represented in the fossil record.
    e. are all filter feeders
    *Answer: c*

28. Which of the following is *not* unique to all members of the phylum Chordata?
    a. A notochord
    b. A ventral heart
    c. Vertebrae
    d. A tail that extends beyond the anus
    e. A hollow dorsal nerve cord
    *Answer: c*

29. Primates likely descended from small arboreal _____ mammals.
    a. insectivorous
    b. omnivorous
    c. carnivorous
    d. frugivorous
    e. clawed
    *Answer: a*

*30. Evolution of jaws first occurred in the group of fishes known as the
    a. ostracoderms.
    b. placoderms.
    c. cartilaginous fishes.
    d. lobe-finned fishes.
    e. ray-finned fishes.
    *Answer: b*

31. The sharks, skates, and rays are members of the vertebrate class known as
    a. Tetrapoda.
    b. Agnatha.
    c. Placodermi.
    d. Chondrichthyes.
    e. Urochordata.
    *Answer: d*

32. Dinosaurs dominated the earth for _____ million years.
    a. 60
    b. 215
    c. 150
    d. 100
    e. 500
    *Answer: c*

33. Which of the following statements about the lobe-finned fishes is *false*?
    a. They are in the class Actinistia.
    b. *Latimeria chalumnae* is a well-known living species.
    c. They have lungs, but no gills.
    d. The bones in their pectoral and pelvic limbs have joints.
    e. The lineage is the sister taxon of the lungfishes and tetrapods.
    *Answer: c*

34. Which of the following statements about the amphibians is *false*?
    a. The class Amphibia was more abundant in the past than it is today.
    b. Waterproof coverings on the skin of amphibians permit them to be terrestrial.
    c. Most amphibians must reproduce in or near water.
    d. Amphibian eggs are very sensitive to drying.
    e. Living amphibians belong to three major classes.
    *Answer: b*

35. Which of the following make up the sister taxon to the snakes?
    a. Birds
    b. Tuataras
    c. Crocodiles
    d. Pterosaurs
    e. Dinosaurs
    *Answer: b*

36. Which of the following subclasses of the class Reptilia has the fewest living members?
    a. Squamata
    b. Sphenodontida
    c. Chelonia
    d. Crocodylia
    e. Aves
    *Answer: b*

37. Which group of modern reptiles includes entirely carnivorous species?
    a. Lizards
    b. Turtles and tortoises
    c. Snakes
    d. Caecilians
    e. None of the above
    *Answer: c*

38. Which of the following statements comparing *Archaeopteryx* with modern birds is *false*?
    a. Unlike *Archaeopteryx*, modern birds lack teeth.
    b. Unlike modern birds, *Archaeopteryx* had a long tail.
    c. Unlike modern birds, *Archaeopteryx* had clawed fingers on its forearms.
    d. Although *Archaeopteryx* could fly, it lacked true feathers.
    e. None of the above
    *Answer: d*

39. Which of the following characteristics is unique to the marsupial mammals?
    a. They are the only egg-laying mammals.
    b. Their mammary glands have no nipples.
    c. They have no placentas.
    d. Gestation is short, and the young complete development outside the uterus in a special pouch.
    e. They are found only in Australia.
    *Answer: d*

40. Which of the following hominids is oldest?
    a. *Homo habilis*
    b. *Homo erectus*
    c. *Homo sapiens*
    d. Neanderthals
    e. Cro-Magnons
    *Answer: a*

*41. Which of the following was *not* a change that accompanied the transition from the australopithecines to *Homo habilis*?
    a. Increase in brain size
    b. Change in diet
    c. Increase in body size
    d. Use of tools
    e. Dispersal of populations to Europe and Asia
    *Answer: e*

*42. Which of the following evolutionary themes did *not* occur in both the protostomes and deuterostomes?
    a. Evolution of structures for filtering food from water
    b. Evolution of wormlike burrowing forms
    c. Evolution of adaptations needed for invasion of the land
    d. Evolution of large terrestrial species
    e. Evolution of jointed appendages for improved locomotion
    *Answer: d*

43. Which of the following is *not* true of deuterostomes?
    a. Radial cleavage
    b. Three germ layers including the ectoderm, mesoderm, and endoderm
    c. A blastopore that becomes the anus
    d. Greater numbers of species than among the protostomes
    e. In most of them, well-developed coelomic body cavities
    *Answer: d*

44. Acorn worms and pterobranchs have a body plan consisting of a
    a. head and trunk.
    b. head, trunk, and abdomen.
    c. proboscis, collar, and trunk.
    d. head, proboscis, collar, and trunk.
    e. head, proboscis, and trunk.
    *Answer: c*

■45. The principal reason we consider tunicates similar to the ancestor of all chordates is that
    a. the body plan of adult tunicates parallels that of chordates.
    b. tunicate larvae possess the synapomorphies of adult chordates and thus reveal close evolutionary relationships with chordates.
    c. tunicates have a lophophore-style mouth.

d. tunicate larvae are primitive in all of their features.
e. tunicate adults are very similar to the ancestors of cephalochordates and vertebrates.
*Answer: b*

■46. We sometimes refer to birds as "feathered dinosaurs" because
a. birds are descendants of a lineage of dinosaurs.
b. all dinosaurs had feathers as well as scales.
c. birds lay amniote eggs just as dinosaurs did.
d. dinosaurs had jointed appendages just as birds do.
e. birds are triploblastic, just as dinosaurs were.
*Answer: a*

47. The calcification of an internal skeleton is characteristic of the phylum
a. Urochordata.
b. Echinodermata.
c. Hemichordata.
d. Brachiopoda.
e. Phoronida.
*Answer: b*

**■48. The extant species *Latimeria chalumnae* belongs to a taxon that is an important link in evolution because it was the first group of
a. fishes with lungs.
b. fishes with paired fins.
c. fishes with lobed fins.
d. bony fishes.
e. fishes with cartilage.
*Answer: c*

■49. A difference between amphibians and reptiles is that
a. amphibian eggs can survive out of water and reptile eggs cannot.
b. amphibians have thin skins whereas reptiles have thick skins.
c. amphibians have gills and lungs whereas reptiles have only lungs.
d. Both a and b
e. a, b, and c
*Answer: b*

*50. The brain of *Homo sapiens* reached its modern size approximately _____ years ago.
a. 160 thousand
b. 160 million
c. 16 million
d. 60 million
e. 600 million
*Answer: a*

51. Which of the following is *not* a general characteristic of all deuterostomate animal phyla?
a. Three embryonic germ layers
b. Closed circulatory systems
c. Coelomic body cavities
d. Origin of the anus in the blastopore
e. Both b and c
*Answer: b*

*52. Which of the following is *not* a major trait distinguishing the primates from other mammals?
a. Dexterous hands with opposable thumbs that can grasp branches and manipulate food
b. Nails rather than claws
c. Maternal care of the young
d. Eyes on the front of the face that provide good depth perception
e. Both b and c
*Answer: c*

53. Early in its evolutionary history, the primate lineage split into two main branches: the _____ and the _____.
a. australopithecines; *Homo* species
b. ardipithecines; australopithecines
c. prosimians; monkeys
d. prosimians; anthropoids
e. the humans; apes
*Answer: d*

54. The hominids—the evolutionary lineage that led to humans—separated from the other ape lineages about _____ mya in Africa.
a. 1
b. 3
c. 6
d. 30
e. 60
*Answer: c*

55. Members of which group are probably most similar to the ancestors of all chordates?
a. Urochordata
b. Cephalochordata
c. Vertebrata
d. Amphibia
e. Reptilia
*Answer: a*

*56. Which of the following represents the probable order in which human ancestors made their appearance?
a. *Homo habilis; Australopithecus afarensis; Homo erectus; Australopithecus garhi*
b. *Australopithecus garhi; Australopithecus afarensis; Homo habilis; Homo erectus*
c. *Australopithecus garhi; Australopithecus afarensis; Homo habilis; Homo erectus*
d. *Australopithecus afarensis; Australopithecus garhi; Homo habilis; Homo erectus*
e. *Australopithecus afarensis; Australopithecus garhi; Homo erectus; Homo habilis*
*Answer: d*

57. Which of the following is *not* accurate in describing the Old World primates?
a. A prehensile tail
b. Arboreal species
c. Terrestrial species
d. Species that live and travel in large groups
e. Both a and b
*Answer: a*

# 35 The Plant Body

## Fill in the Blank

1. Unlike eudicots, monocots have their vascular bundles in a scattered arrangement in their stems and possess a central region called the _____ inside the xylem layer of their roots.
*Answer: pith*

2. When a tree is cut down, the cut surface of the stump often shows _____ due to variations in the size of the vessels.
*Answer: annual rings*

3. The most common type of undifferentiated cells in plant bodies are _____ cells.
*Answer: parenchyma*

4. Just inside the epidermis of the root is a region of unspecialized cells called the _____.
*Answer: cortex*

5. Sclerenchyma cells have thickened _____ cell walls and function when dead.
*Answer: secondary*

6. The _____ is a layer of root cells that contain the waterproofing substance suberin in their cell walls.
*Answer: endodermis*

7. Some vascular bundles contain a single layer of actively dividing cells known as the vascular _____.
*Answer: cambium*

8. In the epidermis of a leaf are pores called _____ whose opening is controlled by the action of a pair of _____ surrounding the pore.
*Answer: stomata; guard cells*

9. Division of cells that will form all of the plant's organs occurs from regions called _____.
*Answer: meristems*

10. Photosynthesis takes place in the leaf tissue called

_____.
*Answer: mesophyll*

11. Potatoes are a portion of the _____ of the plant, and their eyes contain lateral _____.
*Answer: stem; buds*

12. The younger, paler, actively conducting portion of the secondary xylem is called _____.
*Answer: sapwood*

## Multiple Choice

1. Unlike primary cell walls, secondary cell walls have
   a. plasmodesmata.
   b. deposits of lignin or suberin.
   c. deposits of cellulose.
   d. pit pairs.
   e. permeability to small molecules.
   *Answer: b*

2. Of vessel elements and sieve tube elements, only sieve tube elements
   a. are dead at maturity.
   b. are stacked end-to-end.
   c. transport substances through the plant.
   d. often have companion cells.
   e. occur in all plant organs.
   *Answer: d*

3. Sieve tube elements are unusual cells in that they lack
   a. cell walls.
   b. cytoplasm.
   c. water.
   d. nuclei.
   e. plasma membranes.
   *Answer: d*

4. Grasses and other flowering plants with parallel-veined leaves are examples of
   a. monocots.
   b. gymnosperms.
   c. eudicots.
   d. magnoliids.
   e. Both b and c
   *Answer: a*

5. Roses and other flowering plants with net-veined leaves are examples of
   a. monocots.
   b. grasses.
   c. magnoliids.
   d. eudicots.
   e. gymnosperms.
   *Answer: d*

6. A(n) _____ is the point at which a leaf attaches to a stem.
   a. internode
   b. bud
   c. node
   d. apical bud
   e. petiole
   *Answer: c*

*7. Heartwood differs from sapwood in that it
   a. has a lighter color.
   b. stores resin.
   c. conducts water and minerals.
   d. is a younger wood.
   e. has knots.
   *Answer: b*

*8. Cacti thorns result from a modification of the same plant organ that produces
   a. coconut trunks.
   b. maple leaves.
   c. strawberry runners.
   d. potato tubers.
   e. corn adventitious roots.
   *Answer: b*

9. Tracheids, vessel elements, and sclereids are similar in that they all
   a. lack secondary cell walls.
   b. conduct water and minerals.
   c. function when dead.
   d. have open ends.
   e. Both a and b
   *Answer: c*

10. A _____ is an embryonic shoot.
   a. node
   b. petiole
   c. blade
   d. bud
   e. root
   *Answer: d*

11. A root is called adventitious if it
   a. forms a mycorrhizal association.
   b. belongs to a fibrous root system.
   c. originates from a stem or leaf.
   d. is modified for storage.
   e. is actively growing.
   *Answer: c*

12. The widening of a tree trunk is mostly due to the activity of its
   a. apical meristem.
   b. secondary phloem.
   c. phelloderm.
   d. vascular cambium.
   e. primary xylem.
   *Answer: d*

13. In each vascular bundle, the tissue nearest the center of the stem is
   a. collenchyma.
   b. endodermis.
   c. phloem.
   d. vascular cambium.
   e. xylem.
   *Answer: e*

14. Moving from the center of a tree trunk outward, which of the following represents the correct order of vascular tissues?
   a. Primary xylem, secondary xylem, vascular cambium, secondary phloem, primary phloem
   b. Secondary xylem, primary xylem, vascular cambium, primary phloem, secondary phloem

   c. Primary xylem, primary phloem, secondary xylem, secondary phloem, vascular cambium
   d. Primary xylem, primary phloem, vascular cambium, secondary phloem, secondary xylem
   e. Secondary xylem, secondary phloem, vascular cambium, primary xylem, primary phloem
   *Answer: a*

15. A _____ leaf has multiple blades arranged along an axis or radiating from a central point.
   a. simple
   b. compound
   c. star
   d. common
   e. complex
   *Answer: b*

16. The pull of gravity is detected by a root's
   a. apical meristem.
   b. cap.
   c. endodermis.
   d. pericycle.
   e. region of elongation.
   *Answer: b*

17. The collective term for phelloderm, cork cambium, and cork is
   a. pericycle.
   b. periderm.
   c. phloem.
   d. procambium.
   e. protoderm.
   *Answer: b*

18. One function of the root's pericycle is to
   a. give rise to lateral roots.
   b. control the access of water to the vascular tissue.
   c. store nutrients.
   d. elongate the root.
   e. interact with mycorrhizal fungi.
   *Answer: a*

19. Which of the following best describes a fibrous root system?
   a. Deep-growing
   b. Consists of thick roots
   c. Holds soil well
   d. Typical of many eudicots
   e. Food storage organ
   *Answer: c*

20. The meristem is
   a. the tip of the stem.
   b. the site on the stem where a bud forms.
   c. supporting tissue.
   d. growing tissue.
   e. the base of the leaves.
   *Answer: d*

21–25. Match the following descriptions with one of the cell types from the list below.
   a. Parenchyma cells
   b. Sclerenchyma cells
   c. Collenchyma cells

21. Cells in which photosynthesis occurs
   *Answer: a*

22. Cells that function when dead
   *Answer: b*

23. Stone cells of pears
*Answer: b*

24. Cells that support growing organs
*Answer: c*

25. Bulk of root cells
*Answer: a*

26. The xylem tissue of advanced angiosperms typically is distinguished by its
    a. elongate tracheids.
    b. short, stacked vessel elements.
    c. sieve tube elements.
    d. companion cells.
    e. thick-walled fiber cells.
    *Answer: b*

27. Stems function in support and transport. Which type of cell accomplishes most of this function?
    a. Collenchyma
    b. Companion
    c. Parenchyma
    d. Sclerenchyma
    e. Vessel elements
    *Answer: e*

28. The vascular tissue system of plants has the same function as what animal system?
    a. Circulatory
    b. Digestive
    c. Excretory
    d. Reproductive
    e. Respiratory
    *Answer: a*

29. A layer of cells that protects the plant is the
    a. cuticle.
    b. endoderm.
    c. epidermis.
    d. ground tissue.
    e. pericycle.
    *Answer: c*

30. The vascular cambium is located between the _____ and _____.
    a. phloem; cork cambium
    b. xylem; cork cambium
    c. phloem; bark
    d. xylem; phloem
    e. phloem; ground tissue
    *Answer: d*

31. Unlike primary growth, secondary growth
    a. involves growth in plant diameter.
    b. involves growth in plant height.
    c. is produced by meristems.
    d. involves growth by cell elongation.
    e. occurs in all eudicots.
    *Answer: a*

32. Which of the following is typical of cork cells?
    a. Location interior to cork cambium
    b. Waxy suberin
    c. Water storage
    d. Active cell division
    e. Abundance in monocots
    *Answer: b*

33. Root hairs are adaptations that
    a. increase surface area.
    b. defend the plant.
    c. reduce water loss.
    d. provide active growth.
    e. support the plant.
    *Answer: a*

34. Which of the following represents the developmental sequence of an individual root cell?
    a. Elongation, division, differentiation
    b. Differentiation, elongation, division
    c. Division, differentiation, elongation
    d. Elongation, differentiation, division
    e. Division, elongation, differentiation
    *Answer: e*

35. Pith occurs in
    a. both monocot and eudicot stems.
    b. both monocot and eudicot roots.
    c. monocot stems and eudicot roots.
    d. monocot roots and eudicot stems.
    e. both monocot and eudicot stems and roots.
    *Answer: d*

36. A root or stem increases in diameter when
    a. primary xylem cells divide.
    b. phloem cells divide.
    c. primary xylem cells elongate.
    d. vascular cambium cells divide.
    e. phloem cells elongate.
    *Answer: d*

37. Annual rings are seen in temperate-zone trees because
    a. heartwood cells alternate with sapwood cells.
    b. resin is deposited in rings in the stem.
    c. cork is deposited in rings in the stem.
    d. phelloderm cell size varies with season.
    e. xylem cell size varies with season.
    *Answer: e*

38. Lenticels are spongy regions on the surface of some woody stems which function in
    a. water uptake.
    b. water conservation.
    c. gas exchange.
    d. protection of growing layers.
    e. support of the plant.
    *Answer: c*

39. What is the advantage of the spongy arrangement of mesophyll cells in the lower leaf layer?
    a. Maximum absorption of sunlight for photosynthesis
    b. Maximum diffusion of $CO_2$ in the leaf
    c. Maximum movement of water to leaf cells
    d. Minimum water loss from the leaf
    e. Minimum exchange of $O_2$ within the leaf
    *Answer: b*

40. The veins of a leaf consist of
    a. mesophyll cells.
    b. guard cells.
    c. stomata.
    d. vascular cells.
    e. epidermal cells.
    *Answer: d*

41. Cacti are plants with stems modified for water storage. Which type of tissue is well developed in cacti for this function?
    a. Cork
    b. Xylem
    c. Phloem
    d. Parenchyma
    e. Epidermis
    *Answer: d*

42. Conducting cells called _____ elements are the part of xylem where water and minerals are transported.
    a. tracheary
    b. sieve tube
    c. sclerenchyma
    d. xylem
    e. phloem
    *Answer: a*

43. Unlike fibrous root systems, taproot systems
    a. maximize surface area.
    b. are used as food storage organs.
    c. anchor the plant.
    d. hold soil well.
    e. transport water and minerals to the stem.
    *Answer: b*

44. The region of cell division in a primary root is located
    a. in the root cap.
    b. in the apical meristem.
    c. in the region of elongation.
    d. in the area containing differentiated tissues.
    e. throughout the root.
    *Answer: b*

45. A common function of stems but not roots is
    a. anchorage.
    b. transport.
    c. storage.
    d. support.
    e. absorption.
    *Answer: d*

46. Which of the following is *least* likely to be formed from a lateral bud?
    a. Flower
    b. Branch
    c. Runner
    d. Leaf
    e. Root
    *Answer: e*

*47. Simple tissues in plants
    a. are composed of only one type of cell.
    b. have only one function.
    c. only produce primary cell walls.
    d. can be found only in apical meristems.
    e. are found only in primitive plants.
    *Answer: a*

48. "An organized group of plant cells, working together as a functional unit" best defines a(n)
    a. organism.
    b. organ.
    c. organ system.
    d. tissue.
    e. tissue system.
    *Answer: d*

49. Pit pairs allow plasmodesmata to travel through
    a. the primary cell wall.
    b. the secondary cell wall.
    c. both the primary and secondary cell walls.
    d. neither the primary nor the secondary cell walls.
    e. the secondary and sometimes the primary cell walls.
    *Answer: a*

50. Which of the following is *not* true of parenchyma cells?
    a. They are the most common cell in the plant.
    b. They may contain chloroplasts.
    c. They are commonly used for food storage.
    d. They help support leaves in nonwoody plants.
    e. They usually have thick cell walls.
    *Answer: e*

51. Compared to sclerenchyma, collenchyma cells
    a. have more secondary cell wall materials.
    b. are variously shaped.
    c. can be found in bundles.
    d. are used to support the plant.
    e. are more flexible.
    *Answer: e*

52. Unlike tracheids, vessel elements
    a. function when dead.
    b. are spindle-shaped.
    c. are found primarily in gymnosperms.
    d. lose part or all of the end walls.
    e. evolved to be progressively longer.
    *Answer: d*

53. In angiosperm phloem,
    a. both the sieve tube elements and the companion cells have nuclei.
    b. the sieve tube elements have nuclei but the companion cells do not.
    c. the companion cells have nuclei, but the sieve tube elements do not.
    d. neither the companion cells nor the sieve tube elements have nuclei.
    e. the sieve tube elements have nuclei, but the companion cells may or may not have nuclei.
    *Answer: c*

54. One function of the ground tissue in a plant is
    a. photosynthesis.
    b. to protect the plant.
    c. to anchor the plant.
    d. water conduction.
    e. conduction of sugars.
    *Answer: a*

55. The shoot epidermis secretes a layer of wax-covered cutin, the _____, which helps retard water loss from stems and leaves.
    a. lignin
    b. suberin
    c. cuticle
    d. stomata
    e. bark
    *Answer: c*

56. In the development of a root, the protoderm gives rise to the
    a. cortex.
    b. root hairs.
    c. endodermis.

d. xylem.
e. pith.
*Answer: b*

57. In a young root, xylem cells can be observed in the
    a. root cap.
    b. apical meristem.
    c. zone of cell division.
    d. zone of cell elongation.
    e. zone of cell differentiation.
    *Answer: e*

58. Vascular rays are composed of _____ cells and
    conduct materials in a _____ direction in an upright
    woody stem.
    a. fiber; horizontal
    b. tracheids; horizontal
    c. parenchyma; horizontal
    d. fiber; vertical
    e. parenchyma; vertical
    *Answer: c*

59. The _____ is the centermost tissue in a nonwoody
    eudicot stem.
    a. pith
    b. xylem
    c. phloem
    d. pericycle
    e. endodermis
    *Answer: a*

60. Branch roots arise from the
    a. epidermis.
    b. pericycle.
    c. endodermis.
    d. cortex.
    e. pith.
    *Answer: b*

**61. In which of the following states would you be *least* likely
    to find annual rings in eudicot tree trunks?
    a. Maine
    b. Washington
    c. Kansas
    d. Arizona
    e. Hawaii
    *Answer: e*

62. The periderm of a tree functions primarily to
    a. transport sugars.
    b. form branches.
    c. absorb water.
    d. protect the inner tissues.
    e. support the leaves.
    *Answer: d*

63. In a woody stem, gas exchange occurs through
    a. stomata.
    b. the waxy cuticle.
    c. lenticels.
    d. the cork cambium.
    e. bundle sheath cells.
    *Answer: c*

64. In a typical eudicot leaf, most of the chloroplasts are
    found in the
    a. upper epidermal cells.
    b. palisade mesophyll cells.
    c. bundle sheath cells.
    d. phloem cells.
    e. guard cells.
    *Answer: b*

65. The primary function of a typical leaf is
    a. photosynthesis.
    b. food storage.
    c. support.
    d. anchorage.
    e. absorption.
    *Answer: a*

66. Guard cells
    a. protect the plant from herbivores.
    b. secrete a waxy cuticle to prevent evaporation.
    c. contain chemicals that poison insects.
    d. control gas exchange.
    e. inhibit germination of fungal spores.
    *Answer: d*

67. The purpose of vascular rays is to transport
    a. nutrients through the phloem to storage cells.
    b. water from the roots to the xylem.
    c. water from the leaves to the phloem.
    d. nutrients from the sclerenchyma to the phloem.
    e. $CO_2$ from the leaves to the phloem.
    *Answer: a*

68. Growth in the diameter of the stems and roots, which is
    produced by vascular and cork cambia, is called
    a. outgrowth.
    b. primary growth.
    c. secondary growth.
    d. tertiary growth.
    e. ingrowth.
    *Answer: c*

69. Which of the following is *not* a function of cork
    cambium?
    a. Protection from microorganisms
    b. Minimizing of water loss
    c. Secondary growth of stems and roots
    d. Mineral uptake
    e. Breaking off and allowing expansion of tree trunks
    *Answer: d*

<br>

---

# 36 Transport in Plants

## Fill in the Blank

*1. The effect of increasing dissolved solutes is to make the solute potential (more/less) _____ negative.
*Answer: more*

2. Minerals taken up in their ionic form are moved into root cells against their concentration gradient by the process of _____ transport.
*Answer: active*

3. When osmotic _____ increases inside a cell, water moves into the cell passively by osmosis.
*Answer: potential*

4. A more negative water potential in the xylem of the root than in the soil causes water to be drawn into the xylem, generating a force called _____.
*Answer: root pressure*

5. Cell walls and spaces between cells make up a compartment called the _____; the continuous meshwork of living cells connected by plasmodesmata is called the _____.
*Answer: apoplast; symplast*

6. When water moves outside cells, the movement is (regulated/unregulated) _____.
*Answer: unregulated*

7. Evaporative water is lost through pores in the leaf called _____.
*Answer: stomata*

■8. Water in the xylem is pulled up to replace water lost by evaporation because of the _____ of water molecules to water molecules. This physical property of water is the result of _____ bonding between water molecules.
*Answer: cohesion; hydrogen*

■9. Small-scale tests have shown that the natural plant substance _____ is a good antitranspirant.
*Answer: abscisic acid*

10. Water moves toward the region of more _____ water potential.
*Answer: negative*

11. Parenchymal cells known as _____ cells help transport mineral ions from the symplast into the apoplast.
*Answer: transfer*

12. Water transport through the xylem results from _____ of water from the leaves and the subsequent _____ force in the xylem, which pulls water up.
*Answer: evaporation; tension*

13. Water enters a plant mainly through its _____.
*Answer: root hairs*

14. Endodermal cells are lined with waxy structures called _____ which prevent water and ions from moving between cells.
*Answer: Casparian strips*

15. The movement between living cells of a plant occurs through _____.
*Answer: plasmodesmata*

## Multiple Choice

1. The tendency for water to move toward greater solute concentration is an example of
   a. active transport.
   b. osmolarity.
   c. diffusion.
   d. reverse osmosis.
   e. passive transport.
   *Answer: c*

2. Pumping protons ($H^+$) out of a cell can trigger which movements of potassium ion ($K^+$) and chloride ion ($Cl^-$)?
   a. $K^+$ out of the cell and $Cl^-$ into the cell
   b. $K^+$ into the cell and $Cl^-$ out of the cell
   c. Both $K^+$ and $Cl^-$ out of the cell
   d. Both $K^+$ and $Cl^-$ into the cell
   e. $K^+$ into the cell and no movement of $Cl^-$
   *Answer: b*

3. Water tends to move into a cell that has a(n)
   a. high turgor pressure due to cell wall rigidity.
   b. high, positive water potential.
   c. interior solute concentration like that of distilled water.
   d. more negative water potential than its surroundings.
   e. low turgor pressure.
   *Answer: d*

4. The wilting of plant tissue occurs when _____ is _____.
   a. water potential; high
   b. turgor pressure; high
   c. interior solute concentration; high
   d. osmotic potential; high
   e. turgor pressure; low
   *Answer: e*

5. The Casparian strip is
   a. a layer of endodermal cells.
   b. a layer of epidermal cells.
   c. the apoplast.
   d. the symplast.
   e. the waxy layer between endodermal cells.
   *Answer: e*

■6. Endodermal cells differ from other cells in the root in that they
   a. lack a symplast region.
   b. are nonselective with regard to solute uptake.
   c. have a high rate of water transport.
   d. are completely surrounded by a waxy layer.
   e. prevent the movement of water and ions.
   *Answer: e*

7. Water enters the xylem tissue from surrounding root cells via
   a. active transport.
   b. facilitated diffusion.
   c. osmosis.
   d. pressure pumping.
   e. guttation.
   *Answer: c*

8. Guttation is most commonly observed under conditions of _____ atmospheric humidity and _____ soil water.
   a. high; plentiful
   b. low; plentiful
   c. high; little
   d. low; little
   e. varying; plentiful
   *Answer: a*

9. The phenomenon of guttation is related to
   a. active transport.
   b. osmosis.
   c. root pressure.
   d. transpiration.
   e. translocation.
   *Answer: c*

10. Water moves from the soil into the root by the process of
    a. active transport.
    b. passive transport.
    c. facilitated transport.
    d. simple diffusion.
    e. facilitated diffusion.
    *Answer: d*

11. Pure water under no applied pressure has a water potential of
    a. +10.
    b. +1.
    c. 0.
    d. −1.
    e. −10.
    *Answer: c*

■12. A plant cell placed in distilled water will
   a. expand until the osmotic potential reaches that of distilled water.
   b. become more turgid until the osmotic potential reaches that of distilled water.
   c. become less turgid until the osmotic potential reaches that of distilled water.
   d. become more turgid until the pressure potential of the cell reaches its osmotic potential.
   e. become less turgid until the pressure potential of the cell reaches the outside water potential.
   *Answer: d*

★■13. In which of the following water potential conditions will water move from the root hairs through the cortex to the xylem?
   a. Root hairs = 0, cortex = 0, xylem = 0
   b. Root hairs = −1, cortex = −1, xylem = −1
   c. Root hairs = −2, cortex = −1, xylem = 0
   d. Root hairs = 0, cortex = +1, xylem = +2
   e. Root hairs = 0, cortex = −1, xylem = −2
   *Answer: e*

14. Which of the following is part of the apoplast?
    a. Cell wall
    b. Plasma membrane
    c. Plasmodesma
    d. Cytoplasm
    e. Vacuole
    *Answer: a*

15. Cell walls impregnated with water-repellent suberin are found in the cells of the
    a. root hairs.
    b. cortex.
    c. endodermis.
    d. pericycle.
    e. tracheids.
    *Answer: c*

16. Per Scholander placed plant stems in pressure bombs in order to measure
    a. tension.
    b. cohesion.
    c. adhesion.
    d. osmosis.
    e. turgor.
    *Answer: a*

17. In most plants, the stomata at night are
    a. open to take in water.
    b. open to take in $CO_2$.
    c. closed to prevent water loss.
    d. closed to prevent $CO_2$ loss.
    e. closed to exclude $O_2$.
    *Answer: c*

18. In which of the following conditions do plants close their stomata?
    a. Bright sunlight
    b. Water stress
    c. High $CO_2$
    d. Lack of wind
    e. Warm temperatures
    *Answer: b*

19. According to the pressure flow model for translocation,
    a. sugar concentration is highest near the sink area.
    b. water enters the sieve tube by osmosis.
    c. sugar is transported out of the sieve tubes near the source area.
    d. osmosis accomplishes the bulk flow of water and nutrients.
    e. little ATP expenditure is required.
    *Answer: b*

20. The movement in plants caused by differences in pressure potential is called
    a. osmosis.
    b. passive transport.
    c. diffusion.
    d. active transport.
    e. bulk flow.
    *Answer: e*

21. On moderately dry, hot days, recently watered plants have _____ transpiration rates and _____ root pressures.
    a. high; low
    b. low; high
    c. low; low
    d. high; high
    e. moderate; moderate
    *Answer: a*

22. To initiate stomatal opening, $K^+$ ions
    a. passively diffuse into guard cells.
    b. passively diffuse out of guard cells.
    c. are driven into guard cells.
    d. are actively transported out of guard cells.
    e. bond to receptor sites on guard cell walls.
    *Answer: c*

■23. When a tree is girdled, the bark swells in the region _____ the girdle, and the _____ die before the rest of the tree.
    a. below; roots
    b. below; branches
    c. below; leaves
    d. above; roots
    e. above; branches
    *Answer: d*

★24. The active transport of sucrose molecules from a leaf's apoplast to its phloem requires that the sucrose be cotransported with
    a. $H^+$.
    b. amino acids.
    c. $K^+$.
    d. $Cl^-$.
    e. water.
    *Answer: a*

25. As a tree begins transpiring in the morning, tension pressure occurs first in
    a. the leaves.
    b. the branches.
    c. the trunk.
    d. the roots.
    e. All of the above
    *Answer: a*

■26. Mineral nutrients enter the plant body directly from the environment by means of uptake
    a. through the roots.
    b. by the leaves.
    c. from digested food molecules.
    d. into vascular tissue.
    e. through the stems.
    *Answer: a*

■27. When the ion concentration inside root cells is higher than the concentration in the soil solution, ions in the soil solution can enter root cells by means of
    a. active transport only.
    b. simple diffusion only.
    c. simple diffusion and facilitated diffusion.
    d. facilitated diffusion and active transport.
    e. simple diffusion and active transport.
    *Answer: a*

■28. The facilitated diffusion of ions from the soil solution into root cells requires
    a. that the concentration of ions at the root cells be the same outside and inside.
    b. that the concentration of ions outside the root cells be lower than the concentration inside.
    c. the expenditure of ATP.
    d. specific channel proteins in the membranes.
    e. cellular respiration.
    *Answer: d*

■29. What happens when a large amount of water enters a plant cell?
    a. Entry of water increases as the water potential increases.
    b. Entry of water is opposed by turgor pressure.
    c. Water moves toward the region of more positive water potential.
    d. Entry of water reduces the turgor pressure.
    e. Entry of water causes increased active transport into the cell.
    *Answer: b*

■30. Which of the following is true about the apoplast (the transport route through intercellular spaces)?
    a. Osmosis of water is involved.
    b. Movement of materials is regulated by membranes.
    c. Water and solutes can move by bulk flow.
    d. Plasmodesmata are involved.
    e. Water and solutes enter the stele via this channel.
    *Answer: c*

★31. The uptake of ions in plant cells is influenced by the
    a. electrical gradient.
    b. concentration gradient.
    c. ionic balance.
    d. pumping of $H^+$.
    e. All of the above
    *Answer: e*

32. In order to transport $K^+$ ions into their cells, plants
    a. pump $H^+$ out.
    b. use the sodium–potassium pump.
    c. pump $K^+$ in.
    d. pump water in.
    e. Both b and c
    *Answer: a*

33. Where bulk flow occurs in plants, the stream consists of
    a. water and minerals.
    b. water and organic molecules.
    c. dissolved minerals and organic molecules.
    d. water only.
    e. organic molecules only.
    *Answer: a*

34. To drive water into root cells, energy comes most directly from
    a. ATP.
    b. the sun.
    c. NADPH⁺.
    d. bulk flow.
    e. chloroplasts.
    *Answer: a*

*35. Per Scholander used a _____ to measure tension in xylem sap.
    a. balloon on a bell jar
    b. water column
    c. pressure gauge
    d. barometer
    e. pressure bomb
    *Answer: e*

36. Which technique allowed researchers to conclude that the fibrous proteins in sieve tube elements are normally dispersed and obstruct sieve plates only in response to cell damage?
    a. Analyzing phloem sap extruded from aphid stylets
    b. Freezing phloem tissue before cutting and examining it
    c. Watering a plant before cutting and examining it
    d. Using patch clamping to examine sieve plates
    e. Exposing plants to blue light before examination
    *Answer: b*

*37. To study the tension in limbs high up in trees, Scholander used
    a. lumberjacks to cut down trees.
    b. expert tree climbers.
    c. a ladder.
    d. aphids.
    e. sharp shooters.
    *Answer: e*

■38. Which of the following causes the root pressure that moves water upward in plant xylem?
    a. Negative water potential in the xylem sap
    b. High pressure potential of water in the soil
    c. Movement of water from root cells into the soil
    d. Active transport of minerals from soil to root cells
    e. High atmospheric humidity
    *Answer: a*

■39. Which force accounts for the movement of water upward through a narrow tube?
    a. Cohesion of water molecules via hydrogen bonding
    b. Negative water potential in the xylem
    c. Active transport of water molecules
    d. Passive osmosis of water following ion movement
    e. Pumping of water into the phloem
    *Answer: a*

40. The evaporation–tension–cohesion mechanism explains how
    a. water is lost from leaf openings.
    b. water is transported in the xylem.
    c. water and minerals enter the root.
    d. mineral ions move through the xylem.
    e. leaf epidermal cells minimize water loss.
    *Answer: b*

41. Bulk water flow is stopped by
    a. metabolic inhibitors.
    b. closed stomata.
    c. accumulation of K⁺.
    d. blue light.
    e. surface tension.
    *Answer: b*

42. Which of the following increases a plant's intake of $CO_2$ for photosynthesis?
    a. Thick waxy cuticle
    b. Loss of water vapor
    c. Darkness
    d. Electric imbalance
    e. Open stomata
    *Answer: e*

43. Stomata begin to open when K⁺
    a. becomes concentrated, water enters guard cells, and the cells become turgid.
    b. leaves guard cells and they become less turgid.
    c. enters guard cells and they become less turgid.
    d. leaves guard cells and they become turgid.
    e. reaches equilibrium in guard cells and their surroundings.
    *Answer: a*

44. Which structure of the leaf minimizes water loss?
    a. Stoma
    b. Epidermis
    c. Cuticle
    d. Phloem
    e. Xylem
    *Answer: c*

45. Most water moving through the apoplast from the soil into the stele cells first crosses a plasma membrane in the cells of the
    a. root hairs.
    b. cortex.
    c. endodermis.
    d. pericycle.
    e. tracheids.
    *Answer: c*

46. A transfer cell has knobby growths extending into the cell that facilitate movement of minerals from the _____ into the _____.
    a. cell wall; cytoplasm
    b. cytoplasm; cell wall
    c. cell wall; cell wall of the neighboring cell
    d. cytoplasm; next cell's cytoplasm
    e. cell wall; next cell's cytoplasm
    *Answer: b*

■47. Which process makes the water potential in a leaf more negative?
    a. The pressure placed on the leaf by the cuticle
    b. The evaporation of water from mesophyll cells
    c. The movement of water into the leaf by root pressure
    d. The increased K⁺ pumped out of guard cells
    e. The movement of water from the veins into the leaf
    *Answer: b*

48. The evaporative loss of water from the shoot is called
    a. translocation.
    b. transformation.
    c. transportation.
    d. transpiration.
    e. transcention.
    *Answer: d*

49. Cohesion is the tendency of water molecules to attract _____ molecules by means of _____ bonds,
    a. other water; covalent
    b. other water; hydrogen
    c. cellulose; covalent
    d. cellulose; hydrogen
    e. lignin; covalent
    *Answer: b*

■50. If a stem is cut, what will occur if the xylem sap is under tension?
    a. Xylem sap will spurt out.
    b. Xylem sap will stay at the cut surface.
    c. Air will be pulled into the xylem.
    d. The cut surface will form bubbles if placed under water.
    e. Xylem sap will run out if placed under water.
    *Answer: c*

■51. What is the process by which movement of $K^+$ ions initiates stomatal closing?
    a. Active transport into the guard cells
    b. Active transport out of the guard cells
    c. Active transport from one guard cell to another
    d. Diffusion into the guard cells
    e. Diffusion out of the guard cells
    *Answer: e*

52. Which of the following processes can take place even without the presence of living cells?
    a. Osmosis
    b. Transpiration
    c. Translocation
    d. Active transport
    e. Facilitated diffusion
    *Answer: b*

53. A good antitranspirant
    a. closes stomata.
    b. reduces water evaporation but not $CO_2$ uptake.
    c. is a leaf coating.
    d. is dry soil.
    e. increases $CO_2$ release.
    *Answer: b*

54. Cells in the leaf release abscisic acid in response to
    a. negative water potential.
    b. a reduction in $CO_2$.
    c. the reduction of light at night.
    d. an increase of $Cl^-$ in the stomata.
    e. an increase in $K^+$ in the stomata.
    *Answer: a*

55. Which of the following is a function of abscisic acid?
    a. Storage of $CO_2$ for daytime use
    b. Reduction of water loss
    c. Opening of stomata
    d. Increased $H^+$ pumping in guard cells
    e. Lowering of pH
    *Answer: b*

56. When sugars are actively transported into a cell, what happens to the turgor pressure inside that cell?
    a. There is no change; sugar concentration has no effect on turgor pressure.
    b. It increases, because sugar concentration directly affects turgor pressure.
    c. It increases, because water enters and affects turgor pressure.
    d. It decreases, because water exits and affects turgor pressure.
    e. It decreases, because sugar concentration directly affects turgor pressure.
    *Answer: c*

57. Sugars move from the sieve tubes into the plant's tissue
    a. by diffusion.
    b. via the apoplast.
    c. by active transport.
    d. by osmosis.
    e. via translocation.
    *Answer: c*

58. Transport through both the xylem and the phloem
    a. stops if the tissue is killed.
    b. requires ATP.
    c. can occur simultaneously in both directions.
    d. requires negative pressure (tension).
    e. involves long, thin channels.
    *Answer: e*

59. Plant physiologists can obtain pure phloem sap from individual phloem cells by
    a. using tiny drills and capillary pipettes.
    b. collecting material from aphid stylets.
    c. obtaining liquids oozing from cut stem surfaces.
    d. collecting droplets formed by guttation.
    e. gathering the mycorrhizal fungi for subsequent analysis.
    *Answer: b*

60. The pressure flow model of translocation depends entirely on the existence of mechanisms for loading sugars into phloem at the _____ regions and for unloading them at the _____ regions.
    a. sink; sink
    b. sink; source
    c. source; source
    d. source; sink
    e. source; source or sink
    *Answer: d*

61. According to the pressure flow model, during fruit development photosynthesizing leaves are the _____ and the fruit are the _____.
    a. sink; sink
    b. sink; source
    c. source; source
    d. source; sink
    e. source; source or sink
    *Answer: d*

62. Sugars pass from cell to cell in the leaf starting in the
_____ of the mesophyll; they move through the
_____ of other cells and finally pass into the _____
of the sieve tube element.
a. symplast; symplast; symplast
b. apoplast; apoplast; apoplast
c. symplast; apoplast; symplast
d. apoplast; symplast; symplast
e. apoplast; symplast; apoplast
*Answer: c*

*63. Normally, plasmodesmata allow molecules as large as
_____ daltons to pass.
a. 10
b. 100
c. 1,000
d. 10,000
e. 100,000
*Answer: c*

# 37 *Plant Nutrition*

## Fill in the Blank

1. Decomposed plant litter produces a dark-colored organic material called _____.
   *Answer: humus*

2. The source from which plants derive their carbon is the

   _____.
   *Answer: atmosphere*

3. Nitrogen fixation is catalyzed by the enzyme _____, which cannot function in the presence of the element oxygen.
   *Answer: nitrogenase*

4. Plants acquire their essential mineral nutrients from the

   _____.
   *Answer: soil*

5. Some essential elements that occur as positive ions in soils may be traded with H⁺ ions in soil solutions by the process of _____.
   *Answer: ion exchange*

6. Because of leaching and crop production, soils may become depleted of nutrients and require the addition of

   _____.
   *Answer: fertilizer*

7. The three elements most commonly added to agricultural soils are _____, _____, and _____.
   *Answer: nitrogen; phosphorus; potassium*

8. Nitrogen fixation, which is the conversion of atmospheric nitrogen to _____, is catalyzed by a single enzyme called _____.
   *Answer: ammonia; nitrogenase*

9. All nitrogen fixation is carried out by _____, some of which live in symbiosis with other organisms.
   *Answer: prokaryotes*

10. The process that is the opposite of nitrogen fixation is called _____.
    *Answer: denitrification*

■11. Some plants living in boggy areas of low pH have adaptations for carnivory in order to increase their intake of _____. These carnivorous plants are considered to be _____ because they acquire energy from photosynthesis.
    *Answer: nitrogen; autotrophs*

12. The large *Rhizobium* cells that live within plant root nodules are called _____.
    *Answer: bacteroids*

13. The minerals required by plants in concentrations of less than 100 µg per gram of dry matter are called _____.
    *Answer: micronutrients*

14. Nutrients that are necessary for plant growth and reproduction, are not replaceable by another nutrient, and are required by the plant directly are called _____.
    *Answer: essential elements*

15. Plants use the essential element _____ in the processing of hormonal and environmental cues.
    *Answer: calcium*

16. Most soils have at least _____ horizons.
    *Answer: two*

17. _____ are bacteria that both photosynthesize and fix nitrogen.
    *Answer: Cyanobacteria*

18. A stunted dark green plant with purple veins is showing signs of _____ deficiency.
    *Answer: phosphorus*

## Multiple Choice

1. Which four elements are found in highest concentration in plants?
   a. Phosphorus, calcium, hydrogen, carbon
   b. Magnesium, phosphorus, calcium, potassium
   c. Magnesium, iron, phosphorus, potassium
   d. Carbon, hydrogen, oxygen, nitrogen
   e. Carbon, hydrogen, oxygen, potassium
   *Answer: d*

2. Hydrogen in hydrocarbons enters living systems from
   a. the atmosphere.
   b. oil and coal.
   c. water.
   d. H₂S.
   e. All of the above
   *Answer: c*

★3. Carnivorous plants use insects as a source of
   a. calcium.
   b. carbon.
   c. energy.
   d. hydrogen.
   e. nitrogen.
   *Answer: e*

219

4. Which of the following serves as an energy source for chemosynthetic bacteria?
   a. Sugars
   b. $CO_2$
   c. $H_2S$
   d. Oxygen
   e. Protein
   *Answer: c*

5. A photoautotroph acquires its carbon from
   a. the soil.
   b. the air.
   c. water.
   d. the sun.
   e. carbon-fixing prokaryotes.
   *Answer: b*

6. Plants acquire minerals from the soil
   a. by recycling them.
   b. by growing.
   c. from rainwater.
   d. from soil microbes.
   e. All of the above
   *Answer: b*

*7. If a new nutrient were discovered, it would be a
   a. macronutrient.
   b. micronutrient.
   c. rare gas.
   d. small organism.
   e. None of the above
   *Answer: b*

8. The process of ion exchange is the means by which
   a. carbonic acid is added to soils.
   b. positive ion nutrients are replaced by $H^+$.
   c. negative ion nutrients are incorporated into soils.
   d. positive ions are replaced by negative ions.
   e. neutral atoms become ions in soils.
   *Answer: b*

*9. Plants in nature
   a. usually have adequate nitrogen supplies.
   b. usually have inadequate nitrogen supplies.
   c. usually have surplus nitrogen supplies.
   d. always display symptoms when nitrogen is deficient.
   e. have yellow older leaves when nitrogen is over-abundant.
   *Answer: b*

10. Nitrogen deficiencies cause _____ leaves to turn _____ in many crops.
    a. old; yellow
    b. young; yellow
    c. old; orange
    d. young; orange
    e. old; brown-spotted
    *Answer: a*

11. Iron deficiencies cause _____ leaves to turn _____.
    a. old; yellow
    b. young; yellow
    c. old; orange
    d. young; orange
    e. old; brown-spotted
    *Answer: b*

12. The form of nitrogen from the soil that most plants prefer is
    a. found in amino acids.
    b. ammonia.
    c. $N_2$.
    d. nitrate.
    e. nitrite.
    *Answer: d*

13. Which of the following nitrogen compounds is used directly by plants to build proteins?
    a. Ammonia
    b. $N_2$
    c. Nitrate
    d. Nitrite
    e. Nitrous oxide
    *Answer: a*

14. One example of a nutrient in reduced form is the
    a. carbon in $CO_2$.
    b. hydrogen in water.
    c. nitrogen in ammonia.
    d. phosphorus in phosphate.
    e. sulfur in sulfate.
    *Answer: c*

15. The most common gas in the atmosphere is
    a. $CO_2$.
    b. nitrogen.
    c. oxygen.
    d. ozone.
    e. water vapor.
    *Answer: b*

16. Bacteria that function as denitrifiers
    a. oxidize ammonium ions to nitrate.
    b. oxidize nitrate to nitrite.
    c. reduce $N_2$ to ammonia.
    d. reduce nitrates to ammonia.
    e. oxidize nitrate to $N_2$.
    *Answer: e*

17. Plant roots get oxygen
    a. from air spaces in the soil.
    b. from water in soil.
    c. from the leaves, by way of the phloem.
    d. from the leaves, which get oxygen from the air.
    e. All of the above
    *Answer: a*

18. Most plants continuously obtain new sources of mineral nutrients by
    a. breaking down organic matter.
    b. growing longer roots.
    c. shading the plants below them.
    d. evolving more elaborate photosystems.
    e. absorbing minerals through leaves.
    *Answer: b*

*19. One of the defining characteristics of an essential element is that it
    a. is necessary only for early growth of the seedling.
    b. can be replaced by another element.
    c. has a direct function in the plant.
    d. may function by relieving the toxicity of another element.
    e. is found in relatively high concentrations in the environment.
    *Answer: c*

20. Plants do *not* obtain which of the following elements from soil?
    a. Carbon
    b. Nitrogen
    c. Potassium
    d. Sulfur
    e. Zinc
    *Answer: a*

21. Which are the five component elements in most proteins?
    a. Carbon, oxygen, sulfur, phosphorus, potassium
    b. Carbon, hydrogen, oxygen, nitrogen, phosphorus
    c. Carbon, hydrogen, oxygen, nitrogen, sulfur
    d. Carbon, hydrogen, nitrogen, sulfur, potassium
    e. Carbon, hydrogen, oxygen, phosphorus, potassium
    *Answer: c*

22. Which of the following is a micronutrient in plants?
    a. Potassium
    b. Sulfur
    c. Calcium
    d. Iron
    e. Magnesium
    *Answer: d*

23. Soil scientists recognize _____ major zones: _____.
    a. two; A and B
    b. four; top, top middle, middle, and lower
    c. seven; 1, 2, 3, 4, 5, 6, and 7
    d. two; upper and lower
    e. three; A, B, and C
    *Answer: e*

24. The maximum diameter of a clay particle is _____ μm.
    a. 2,000,000
    b. 2,000
    c. 2
    d. 0.002
    e. 0.000002
    *Answer: c*

25. Most earthworms can be found in
    a. topsoil.
    b. subsoil.
    c. bedrock.
    d. Both a and b
    e. Both b and c
    *Answer: a*

26. Which of the following plants does *not* have nitrogen-fixing root nodules containing *Rhizobium*?
    a. Alfalfa
    b. Beans
    c. Clover
    d. Peas
    e. Rice
    *Answer: e*

27. Which mineral is deficient in a plant whose growth is stunted and whose oldest leaves turn yellow and die prematurely?
    a. Calcium
    b. Iron
    c. Magnesium
    d. Nitrogen
    e. Phosphorus
    *Answer: d*

28. Which three elements are most commonly added to agricultural soils in fertilizers?
    a. Nitrogen, phosphorus, iron
    b. Nitrogen, potassium, iron
    c. Potassium, sulfur, iron
    d. Nitrogen, potassium, phosphorus
    e. Nitrogen, sulfur, iron
    *Answer: d*

29. What is the advantage of adding organic fertilizers (as opposed to inorganic fertilizers) to soils?
    a. Organic fertilizers improve the physical properties of soil.
    b. Organic fertilizers can apply specific nutrient formulas to specific problems.
    c. Organic fertilizers allow for a more rapid increase in nutrients.
    d. Organic fertilizers leach the soil.
    e. Organic fertilizers add clay particles to the soil.
    *Answer: a*

30. Chemical weathering, an important part of soil formation, includes the
    a. splitting of clays.
    b. effects of freezing and thawing.
    c. hydrolysis of rock.
    d. crushing of rock.
    e. drying of soils.
    *Answer: c*

31. Which of the following is *not* a consequence of adding lime to soil?
    a. Raised soil pH
    b. Increased nutrient availability to plants
    c. Addition of calcium ($Ca^{2+}$) to the soil
    d. Release of hydrogen ions ($H^+$) by clay particles
    e. Increase in the soil's ability to retain water
    *Answer: e*

■32. Which one of the following is *not* essential to the process of nitrogen fixation?
    a. $N_2$ gas
    b. $O_2$ gas
    c. A strong reducing agent
    d. A great deal of energy
    e. The enzyme nitrogenase
    *Answer: b*

33–36. Match the mineral nutrient in the list below with the following description. Each item may be used once, more than once, or not at all.
    a. Magnesium
    b. Nitrogen
    c. Phosphorus
    d. Potassium
    e. Iron

33. When bonded to oxygen atoms, this mineral is important in many energy-storing and energy-releasing pathways.
    *Answer: c*

34. It is pumped into cells to create osmotic potential.
    *Answer: d*

35. This mineral is a constituent of both proteins and amino acids.
    *Answer: b*

36. This mineral is used as a cofactor by many enzymes.
    *Answer: a*

■37. The ionization of carbonic acid
    a. triggers the release of mineral ions from clay.
    b. lowers the pH of soil.
    c. reduces the leaching of phosphates and nitrates from soil.
    d. causes laterization of the A horizon of soil.
    e. increases the amount of soil carbon available to plants.
    *Answer: a*

38. Clay in soil
    a. provides a reservoir of cation nutrients.
    b. holds moisture.
    c. resists pH changes.
    d. Both a and b
    e. a, b, and c
    *Answer: e*

39. Which of the following elements does *not* reversibly attach to the surface of clay particles?
    a. Calcium
    b. Chloride
    c. Hydrogen
    d. Magnesium
    e. Potassium
    *Answer: b*

40. The best topsoils are composed of
    a. clay.
    b. sand.
    c. organic matter.
    d. Both a and b
    e. a, b, and c
    *Answer: e*

41. Most of the nitrogen on Earth exists in the form of
    a. ammonia.
    b. nitrate ions.
    c. $N_2$ gas.
    d. amino acids.
    e. proteins.
    *Answer: c*

42. Nitrogen fixers convert
    a. ammonia to $N_2$.
    b. $N_2$ to ammonia.
    c. ammonia to nitrate.
    d. nitrate to ammonia.
    e. $N_2$ to nitrate.
    *Answer: b*

43. All nitrogen fixers belong to the kingdom(s)
    a. Bacteria.
    b. Plantae.
    c. Bacteria and Plantae.
    d. Protista.
    e. Protista and Plantae.
    *Answer: a*

44. Cyanobacteria are known to fix nitrogen in association with all of the following *except*
    a. fungi in lichens.
    b. ferns.
    c. bryophytes.
    d. cycads.
    e. wheat.
    *Answer: e*

45. In nitrogen fixation, the nitrogenase enzyme binds a molecule of $N_2$ and a reducing agent transfers _____ hydrogen atoms before the product is released.
    a. two
    b. three
    c. four
    d. five
    e. six
    *Answer: e*

46. Nitrogenase enzymes are extremely sensitive to _____ molecules.
    a. hydrogen
    b. oxygen
    c. water
    d. $CO_2$
    e. calcium carbonate
    *Answer: b*

47. Nodules that are actively fixing nitrogen are pink, demonstrating the presence of
    a. iron.
    b. chlorophyll.
    c. leghemoglobin.
    d. anthocyanin.
    e. alkaloids.
    *Answer: c*

■48. Recombinant DNA technology is now being used in an attempt to "teach" new plants how to
    a. produce their own nitrogenase enzymes.
    b. fix nitrogen without using ATP.
    c. produce ammonia under anaerobic conditions.
    d. convert nitrates into ammonium ions.
    e. recycle their nitrogen instead of excreting it.
    *Answer: a*

49. Denitrifying bacteria are part of nature's nitrogen cycle, converting
    a. ammonia to $N_2$.
    b. $N_2$ to nitrate.
    c. ammonia to nitrate.
    d. nitrate to ammonia.
    e. nitrate to $N_2$.
    *Answer: e*

50. Haustoria are structures of
    a. heterotrophic plants.
    b. carnivorous plants.
    c. nitrate-reducing plants.
    d. denitrifying bacteria.
    e. nitrogen-fixing bacteria.
    *Answer: a*

51. The process of nitrogen fixation is the
    a. uptake of atmospheric nitrogen by plants.
    b. conversion of atmospheric nitrogen into ammonia.
    c. production of nitrogen-bearing compounds in plants.
    d. release of nitrogen into the atmosphere.
    e. release of ammonia into the atmosphere.
    *Answer: b*

52. Root nodules on plants of the legume family contain
    a. cyanobacteria.
    b. *Nitrosococcus* bacteria.
    c. *Rhizobium* bacteria.
    d. *Pseudomonas* bacteria.
    e. *Nitrobacter* bacteria.
    *Answer: c*

■53. Which of the following statements about the chemical process of nitrogen fixation in cells is true?
a. All three bonds between nitrogen atoms are broken simultaneously.
b. Hydrogen ions are added to $N_2$ in three successive pairs.
c. Very little energy in the form of ATP is required.
d. A different enzyme catalyzes each of the many reactions.
e. It is enhanced by high oxygen concentrations.
*Answer: b*

■54. How do nitrogen-fixing microbes first become symbiotic with their plants?
a. They are carried in the seed.
b. They are attracted by chemicals released by the root.
c. They move in via openings in the plant cell walls.
d. They move into the vascular system and multiply.
e. They enter root nodules previously produced by the plant.
*Answer: b*

55. In the legume–*Rhizobium* symbiosis, the _____ partner produces leghemoglobin to provide _____ to the _____ partner.
a. legume; oxygen; *Rhizobium*
b. *Rhizobium*; oxygen; legume
c. legume; nitrogen; *Rhizobium*
d. *Rhizobium*; nitrogen; legume
e. legume; sugars; *Rhizobium*
*Answer: a*

56. The industrial production of nitrogen-containing fertilizer is currently limited by
a. its high energy expense.
b. the inability to insert nitrogenase genes into plants.
c. the lack of nitrogenase for the industrial process.
d. the limited supply of $N_2$ gas.
e. the need to exclude free oxygen in the process.
*Answer: a*

57. The process that is the opposite of nitrogen fixation is
a. nitrification.
b. denitrification.
c. aerobic breakdown of amino acids.
d. release of ammonia.
e. nitrate reduction.
*Answer: b*

58. The products of nitrogen-fixing organisms can be oxidized to form nitrites and nitrates by
a. all living organisms.
b. most living organisms that utilize oxygen.
c. many types of microorganisms.
d. a few specific genera of soil bacteria.
e. only the nitrogen-fixing bacteria.
*Answer: d*

*■59. Although mistletoes are green, they are considered to be parasites because
a. they depend on other plants for water and minerals.
b. they cling to woody plants for physical support.
c. they capture and digest insects.
d. they have root nodules containing nitrogen-fixing *Rhizobium*.
e. their chlorophyll is not functional.
*Answer: a*

60. An advantage to using inorganic fertilizers is that
a. they also improve physical properties of the soil.
b. they can be taken up almost instantaneously.
c. their quality is better than that of organic fertilizers.
d. they contain minerals in their proper ionic form.
e. None of the above
*Answer: b*

61. Currently, commercial nitrogen-containing fertilizer is an imperfect solution to the worldwide problem of soil degradation because
a. its production requires a great deal of energy.
b. it pollutes lakes and streams.
c. soil applications do not always benefit the plants.
d. the cost of shipping offsets the benefits to agricultural production.
e. it can never be as good as organic sources.
*Answer: a*

62. The capture and digestion of insects allows carnivorous plants to
a. pollinate their flowers.
b. absorb nitrogen compounds.
c. disperse their fruits.
d. overcome insect parasitism.
e. neutralize acidic soils.
*Answer: b*

# 38 Regulation of Plant Growth

## Fill in the Blank

1. A germinating grass seedling embryo produces _____ that mobilize stored nutrients.
   *Answer: gibberellins*

2. The hormone _____ is responsible for the phenomenon of apical dominance in plants.
   *Answer: auxin*

3. It is thought that auxins control growth by cell _____, whereas cytokinins cause growth by cell _____.
   *Answer: elongation; division*

4. The hormone that generally shuts down or inhibits plant activity is _____.
   *Answer: abscisic acid*

5. Application of _____ to leaves can keep them green and delay senescence.
   *Answer: cytokinins*

6. Fruit shippers apply _____ to speed up ripening.
   *Answer: ethylene*

7. Growth of a shoot tip toward the light, which is called _____, is thought to be mediated by the actions of the hormone _____.
   *Answer: phototropism; auxin*

8. Bending toward the light occurs when the auxin moves to the side (toward/away from) _____ the light; it has the effect of _____ the cell walls in that region, causing asymmetric growth and thus bending.
   *Answer: away from; loosening*

9. The process of seed coat modification to increase ease of germination is called _____.
   *Answer: scarification*

10. Molecules that are active at small concentrations and regulate plant development are called _____.
    *Answer: hormones*

11. Eudicot seedlings form an apical _____ to protect the stem while it grows through the soil.
    *Answer: hook*

## Multiple Choice

1. Which of the following is *not* a factor in breaking seed dormancy?
   a. Fire
   b. Low oxygen levels
   c. Cycles of freezing and thawing
   d. Leaching of inhibitor molecules
   e. Soil microorganisms
   *Answer: b*

2. To make seedless grapes grow to normal size, they are
   a. sprayed with cytokinins.
   b. treated with NO.
   c. treated with ethylene.
   d. sprayed with auxin solution.
   e. sprayed with gibberellin solution.
   *Answer: e*

3. In shoot phototrophic responses, auxin is transported _____ and _____ from the apical meristem.
   a. down; asymmetrically
   b. down; symmetrically
   c. upward; symmetrically
   d. evenly; toward the roots
   e. upward; toward the tip
   *Answer: a*

*4. Production of auxin in response to light most likely involves
   a. phytochrome.
   b. photoreceptors.
   c. a mechanism that is currently unknown.
   d. phototropin.
   e. Both a and b
   *Answer: d*

5. Phototropin is classified as a(n)
   a. phytochrome.
   b. cryptochrome.
   c. gibberellin.
   d. cytokinin.
   e. abscisic acid.
   *Answer: b*

6. A protein receptor for auxin is
   a. auxin A.
   b. kinetin.
   c. ABP1.
   d. phototropin.
   e. chlorophyll.
   *Answer: c*

7. Abscisic acid closes stomata by
   a. stimulating proton pumping.
   b. triggering potassium pumps to remove potassium from the cytosol.
   c. releasing calcium into the cytoplasm.
   d. pumping water from the cell.
   e. means of the "stress effect."
   *Answer: c*

8. Steroid-like hormones that have some of the same effects as auxin are
   a. adinosteroids.
   b. estrogens.
   c. brassinosteroids.
   d. testosterones.
   e. Both a and b
   *Answer: c*

■9. As the old saying goes, "A rotten apple spoils the whole barrel (of apples)." This result occurs because the rotten apple emits the ripening hormone
   a. abscisic acid.
   b. auxin.
   c. cytokinin.
   d. ethylene.
   e. gibberellin.
   *Answer: d*

★10. Auxins cause elongation by
   a. triggering proton release.
   b. indirectly activating expansins.
   c. indirectly increasing plasticity.
   d. All of the above
   e. None of the above
   *Answer: b*

11. Plant growth substances generally
    a. have a single specific role.
    b. affect mainly the cells that produce them.
    c. are species-specific.
    d. are produced in many parts of the plant.
    e. elicit rapid responses.
    *Answer: d*

12. As a grass seed germinates, the embryonic plant secretes gibberellins that
    a. absorb light.
    b. cause elongation.
    c. mobilize stored foods.
    d. take up water.
    e. direct the shoot upward.
    *Answer: c*

13. The hormone responsible for phototropism is
    a. abscisic acid.
    b. auxin.
    c. ethylene.
    d. gibberellin.
    e. phytochrome.
    *Answer: b*

14. Gibberellins were discovered by scientists studying the "foolish seedling" disease of rice, in which seedlings
    a. grew unusually slowly.
    b. grew into tall, spindly plants.
    c. died after germination.
    d. produced seeds unusually early.
    e. had a harmful mutation.
    *Answer: b*

15. Which of the following suggests that plants produce gibberellin growth hormones?
    a. Genetically dwarf corn seedlings grow tall with gibberellin treatment.
    b. Genetically tall corn plants grow even taller with gibberellin treatment.
    c. Gibberellins are produced by tall and dwarf varieties of corn.
    d. Gibberellins are produced by dwarf varieties of corn.
    e. Different varieties of corn produce different gibberellins.
    *Answer: a*

16. In the experiments of Charles and Francis Darwin, which parts of the seedlings were sensitive to light?
    a. The entire seedling
    b. The entire shoot above the roots
    c. The entire coleoptile
    d. The coleoptile just below the tip
    e. The extreme tip of the coleoptile
    *Answer: e*

17. In plant tropisms,
    a. an imbalance in ethylene concentration causes curvature.
    b. roots grow toward the light.
    c. one side of the root or shoot grows more rapidly than the other.
    d. DNA is the light receptor or gravity receptor.
    e. auxin is the light receptor or gravity receptor.
    *Answer: c*

18. Leaf abscission is the
    a. separation of leaves from stems.
    b. orientation of a leaf toward the light.
    c. regeneration of a leaf after a wound.
    d. maturation of leaf tissue.
    e. initiation of growth of new tissue.
    *Answer: a*

19. The phenomenon of apical dominance is strengthened most by
    a. removal of the tip.
    b. auxin production.
    c. removal of leaves.
    d. production of fruits.
    e. removal of fruits.
    *Answer: b*

20. Cytokinins are formed primarily in which area of the plant?
    a. Tips of the shoot
    b. Leaves
    c. Stems
    d. Roots
    e. Lateral buds
    *Answer: d*

21. In the phenomenon of gravitropism, a plant part grows
    a. toward the center of Earth.
    b. in a direction determined by gravity.
    c. in a direction opposite to that of the main light source.
    d. in a direction opposite to that of the growth of the shoot.
    e. toward the darkest area.
    *Answer: b*

■22. Why are cell differentiation experiments often done with pith tissue cultures?
a. Pith cells are all unspecialized.
b. Only pith tissue grows rapidly.
c. Pith tissue of stem does not differentiate into root cells.
d. Pith tissue responds primarily to auxin.
e. Pith tissue is found in all plants.
*Answer: a*

23–26. Match the appropriate plant hormone from the list below with each description. Each hormone may be used once, more than once, or not at all.
a. Abscisic acid
b. Auxin
c. Cytokinin
d. Ethylene
e. Gibberellin

23. Stimulates lateral buds to grow into branches
*Answer: c*

24. A gas
*Answer: d*

25. Considered the plant's "stress hormone"
*Answer: a*

26. Has its highest concentrations in dormant (inactive) buds and seeds
*Answer: a*

27. Two different hormones have opposing effects on senescence: _____ promotes it, whereas _____ inhibits it.
a. auxin; cytokinin
b. ethylene; cytokinin
c. ethylene; auxin
d. cytokinin; auxin
e. cytokinin; ethylene
*Answer: b*

28. Bolting, or rapid stem elongation, is induced by _____ and can be inhibited by _____.
a. auxin; cytokinin
b. abscisic acid; ethylene
c. gibberellin; abscisic acid
d. ethylene; auxin
e. cytokinin; abscisic acid
*Answer: c*

29. The movement of auxin in plants is said to be "polar," which means that auxin
a. is a chemically polar molecule.
b. is produced only at one part of the plant.
c. is transported from the tip to the base of the plant.
d. moves away from the light.
e. cannot move through gelatin.
*Answer: c*

30. Which of the following processes is *not* increased by ethylene?
a. Breakdown of fruit cell walls
b. Stimulation of leaf abscission
c. Ripening of fruit
d. Inhibition of stem elongation
e. Change in leaf color from green to red or yellow
*Answer: e*

■31. If a shoot cutting is treated with auxin, which of the following is likely to result?
a. Extensive root production

b. Suppression of apical dominance
c. Growth of lateral buds
d. Bolting of the shoot
e. Nothing
*Answer: a*

32. Plants utilize which cue to detect the onset of winter?
a. Decreasing temperature
b. Increasing precipitation
c. Decreasing length of daylight
d. Increasing length of darkness
e. Height of the midday sun
*Answer: d*

33. In plants that germinate in response to a brief pulse of light,
a. green light is most effective in triggering germination.
b. far-red light is most effective in triggering germination.
c. far-red light reverses the effect of prior exposure to red light.
d. initiation of photosynthesis is the mechanism for germination.
e. a rise in temperature also triggers germination.
*Answer: c*

■34. A protein pigment called phytochrome is thought to monitor photoperiod because
a. it is converted between two forms by specific wavelengths of light.
b. a phytochrome that absorbs red light breaks down in darkness.
c. the photoperiod response depends on how much light it absorbs.
d. in darkness the pigment becomes inactive.
e. the pigment response can be observed only in intact living plants.
*Answer: a*

*35. Which of the following is most advantageous for a young plant seedling that has not yet been exposed to light?
a. Increased production of chlorophyll
b. Rapid elongation of the shoot
c. Increased production of phytochrome
d. Rapid photosynthesis
e. Increased uptake of water
*Answer: b*

36. Gibberellins were first discovered by a biologist studying the "foolish seedling" disease of
a. corn.
b. wheat.
c. rice.
d. barley.
e. millet.
*Answer: c*

37. Phinney reported the first evidence that gibberellins were produced by plants. In his studies with dwarf mutant strains of corn, he demonstrated that treatment with gibberellins caused the dwarf plants to _____, while the normal tall plants _____.
a. grow taller, also grew taller, but only slightly
b. grow taller, were virtually unaffected
c. stop growing immediately, stopped growing after a period of time
d. stop growing, were virtually unaffected
e. die, stopped growing
*Answer: b*

▪38. What is the function of the many gibberellins found in plant stems?
   a. Each is needed for a different process in plant growth and development.
   b. One gibberellin, $A_1$, controls stem elongation, and the others are intermediates in the production of $A_1$.
   c. They all play essential roles in the development of roots, leaves, and flowers.
   d. Both a and c
   e. None of the above
   *Answer: b*

*39. Spraying biennials such as cabbage with gibberellin causes them to bolt, a process first observed as an increase in
   a. leaf senescence.
   b. stem elongation.
   c. the number of flowers.
   d. the size of fruit.
   e. the number of seeds per fruit.
   *Answer: b*

40. The discovery of auxin is traced back to the work of Charles and Francis Darwin and their studies of
   a. photosynthesis.
   b. photorespiration.
   c. photophosphorylation.
   d. phototropism.
   e. photoperiodism.
   *Answer: d*

41. In the Darwins' experiment with grass coleoptiles, they observed that the photoreceptor was _____ and the bending took place _____.
   a. in the tip; at the tip also
   b. below the tip; below the tip also
   c. in the tip; below the tip
   d. below the tip; throughout the coleoptile
   e. at the tip; throughout the coleoptile
   *Answer: c*

*42. In a classic experiment, a gelatin block containing auxin was placed on one edge of a decapitated coleoptile. The result was that the coleoptile
   a. grew straight up.
   b. grew more on the side with the block.
   c. grew more on the side away from the block.
   d. grew a new coleoptile tip.
   e. did not grow.
   *Answer: b*

43. Auxin transport goes
   a. from apex to base.
   b. from base to apex.
   c. in either direction.
   d. primarily from apex to base, but sometimes in reverse.
   e. primarily from base to apex, but sometimes in reverse.
   *Answer: a*

44. Your cat knocks over the coleus plant you were keeping in a dark closet. After it spends a few days on its side, you notice that the shoots are growing upright again. You have just observed
   a. positive gravitropism.
   b. negative gravitropism.
   c. positive phototropism.

   d. negative phototropism.
   e. positive thigmotropism.
   *Answer: b*

45. Removal of the auxin source demonstrates that leaf abscission is _____ by auxin and apical dominance is _____ by auxin.
   a. promoted; also promoted
   b. inhibited; also inhibited
   c. promoted; inhibited
   d. inhibited; promoted
   e. promoted; unaffected
   *Answer: d*

46. An ideal herbicide to kill weeds in a wheat field would kill _____ and break down _____ in soil.
   a. monocots and eudicots; slowly
   b. monocots but not eudicots; slowly
   c. eudicots but not monocots; slowly
   d. monocots but not eudicots; rapidly
   e. eudicots but not monocots; rapidly
   *Answer: e*

47. A cell wall is a network of crystalline _____ molecules in a jelly-like matrix of _____ molecules.
   a. cellulose; starch
   b. starch; other polysaccharide
   c. cellulose; other polysaccharide
   d. starch; nonpolysaccharide
   e. cellulose; nonpolysaccharide
   *Answer: c*

48. The growth of a plant cell is driven primarily by the
   a. breakdown of ATP to ADP.
   b. uptake of water into the vacuole.
   c. strengthening of cell wall components.
   d. deposition of new cell wall materials on the outside of the cell wall.
   e. forces in transpirational pull.
   *Answer: b*

*49. The "wall-loosening factor" from the cytoplasm is now believed to be
   a. auxins.
   b. cellulose-digesting enzymes.
   c. starch-digesting enzymes.
   d. expansins.
   e. calcium ions.
   *Answer: d*

50. Cytokinins are believed to form primarily in the plant's
   a. roots.
   b. stems.
   c. leaves.
   d. vegetative apical meristem.
   e. floral apical meristem.
   *Answer: a*

51. Unlike other plant hormones, ethylene
   a. is not produced by all plants.
   b. exerts a number of effects.
   c. inhibits plant development.
   d. acts either as an inhibitor or a promoter.
   e. is a gas.
   *Answer: e*

52. Leaf senescence is important for the survival of a plant. Which of the following statements about senescence is *false*?
    a. The delicate leaves can be a liability during winter.
    b. Amino acids in the leaves are exported to the stems.
    c. It is a reversible process.
    d. It is promoted by the hormone ethylene.
    e. Controlled leaf abscission costs the plant little and benefits it greatly.
    *Answer: c*

53. Abscisic acid concentrations are _____ in some dormant seeds and _____ in buds during winter dormancy.
    a. high; high
    b. high; low
    c. low; high
    d. low; nonexistent
    e. low; low
    *Answer: a*

54. Which of the following statements about phytochrome $P_r$ is true?
    a. It absorbs green light.
    b. It looks red in a test tube.
    c. It is the active form of phytochrome.
    d. It converts spontaneously to the other form in the dark.
    e. It controls a variety of plant responses.
    *Answer: e*

55. When seeds germinate below the soil surface, the young etiolated seedlings
    a. grow very slowly.
    b. produce chlorophyll.
    c. elongate so that the apical meristem is the first part of the shoot to break through the soil surface.
    d. have small, unexpanded leaves.
    e. keep the cotyledons enclosed in a coleoptile.
    *Answer: d*

56. Which of the following does *not* serve to break dormancy and initiate seed germination?
    a. Scarification by abrasion
    b. Scarification by fire
    c. Exposure to water
    d. Abscisic acid
    e. Growth promoters
    *Answer: d*

57. A shift away from rapid vegetative growth often accompanies which of the following phenomena?
    a. Repeated mitotic division
    b. Production of new leaves
    c. Production of flowers
    d. Increased root development
    e. Increased rate of photosynthesis
    *Answer: c*

58. Which of the following is *not* a way to break a seed's dormancy?
    a. Soaking it with water
    b. Singeing it with a flame
    c. Passing it through an animal's gut
    d. Tumbling it among stones
    e. Infusing it with $CO_2$
    *Answer: e*

59. Seed dormancy is usually an adaptation to ensure that
    a. the embryo is mature.
    b. germination occurs at a favorable time.
    c. seeds germinate near the parent plant.
    d. levels of abscisic acid are high enough.
    e. plenty of other seeds are ready to germinate.
    *Answer: b*

60. All of the following occur during the earliest stages of germination *except*
    a. intake of water.
    b. activation of enzymes.
    c. cell division.
    d. lengthening of the root.
    e. mobilization of food reserves.
    *Answer: c*

61. The adaptive advantage of seed dormancy is that it may
    a. result in germination at a favorable time.
    b. increase the probability of a seed's germinating in the right place.
    c. be a way to avoid competition.
    d. result in an increase in the likelihood of dispersal.
    e. All of the above
    *Answer: e*

62. During the initial stages of seed germination, all of the following increase *except* for
    a. cell size.
    b. respiration.
    c. RNA synthesis.
    d. DNA synthesis.
    e. protein synthesis.
    *Answer: d*

63. Which of the following enhances energy storage in a small space?
    a. Protein
    b. Lipid
    c. Starch
    d. Sugar
    e. Amino acid
    *Answer: b*

64. As a seed germinates, DNA synthesis begins
    a. when gibberellins are secreted by the embryo.
    b. when the embryonic root, or radicle, begins to grow.
    c. as imbibition takes place.
    d. when the endosperm starts metabolizing starches, proteins, and lipids.
    e. when the aleurone layer assembles enzymes, proteases, and ribonucleases.
    *Answer: b*

65. Treatment of some plants with gibberellin or auxin causes parthenocarpy; this is the
    a. formation of flowers with petals in multiples of four.
    b. division of one shoot into two separate shoots.
    c. formation of fruit with only one seed instead of many.
    d. formation of fruit without fertilization.
    e. formation of an embryo without fertilization.
    *Answer: d*

■66. What is the difference between the cytokinins kinetin and zeatin?
  a. Kinetin is found only in aged DNA, whereas zeatin is found only in fresh DNA.
  b. Kinetin is the active form of zeatin.
  c. Zeatin is the active form of kinetin.
  d. Only zeatin is a naturally occurring plant cytokinin.
  e. Zeatin is a synthetic cytokinin, whereas kinetin is natural.
  *Answer: d*

■67. Florists use silver thiosulfate to
  a. delay abscission of petals caused by the action of ethylene.
  b. delay abscission of leaves caused by the action of abscisic acid.
  c. keep the petals from turning brown because of the action of ethylene.
  d. keep the leaves green longer.
  e. promote flower fertilization.
  *Answer: a*

68. Abscisic acid promotes the formation of bud scales, which
  a. help the plant retain water in dry areas.
  b. waterproof the leaf primordia and stem during winter.
  c. promote cell elongation at the apical meristem.
  d. assist in leaf drop during autumn.
  e. trap $CO_2$ in the leaves.
  *Answer: b*

69. As a first step in seed germination, the seed must
  a. dehydrate.
  b. be placed in the ground.
  c. imbibe.
  d. be activated by germisin.
  e. pump protons.
  *Answer: c*

70. Which one of the following is *not* a plant hormone?
  a. Abscisic acid
  b. Brassinosteroid
  c. Cytokinin
  d. Gibberellin
  e. Phytochrome
  *Answer: e*

71. The physiologically active form of phytochrome is
  a. $P_{fr}$.
  b. $P_r$.
  c. G protein.
  d. cryptochrome A.
  e. 730 nm.
  *Answer: a*

# 39 Reproduction in Flowering Plants

## Fill in the Blank

1. Angiosperm plants are characterized by double fertilization, in which one sperm nucleus fertilizes the _____ to begin formation of the embryo and the other fertilization results in production of _____ tissue.
   *Answer: ovum; endosperm*

2. The male gametophyte in seed plants is the _____; the mature female gametophyte is an embryo sac with _____ haploid nuclei.
   *Answer: pollen grain; eight*

3. The process of transfer of pollen grains to the stigma is called _____.
   *Answer: pollination*

4. The major role of the fruit of a flowering plant is to facilitate _____.
   *Answer: seed dispersal*

5. In general, the production of progeny all having identical genotypes to the parent is called _____ reproduction, whereas _____ reproduction is the modification of a part of the plant to produce new individuals.
   *Answer: asexual; vegetative*

6. The asexual production of seeds is called _____.
   *Answer: apomixis*

7. One agricultural industry in which grafting is an important technique is _____ production.
   *Answer: fruit*

8. Instead of germinating immediately after release, many seeds undergo an inactive period of _____.
   *Answer: dormancy*

9. Experiments on the effect of light cues on flowering have shown that the significant cue that is sensed or measured by plants is _____.
   *Answer: night length*

10. The physiological mechanism by which plants measure photoperiod involves a pigment called _____. This pigment alternates between two forms, one of which absorbs _____ light and the other of which absorbs _____ light.
    *Answer: phytochrome; red; far-red*

11. Most plants are _____ plants; that is, they do not rely on light cues for the induction of flowering.
    *Answer: day-neutral*

12. Pollination before the flower bud opens is an example of _____.
    *Answer: self-fertilization*

13. Angiosperms are unique in that the number of chromosome sets in the endosperm is _____.
    *Answer: triploid*

14. The production of flowers from floral meristems is an example of _____ growth.
    *Answer: determinate*

15. A northern tree that is placed too far south will not flower well because it will fail to undergo _____.
    *Answer: vernalization*

## Multiple Choice

1. Lifestyle categories for flowering plants include all the following *except*
   a. annual.
   b. biennial.
   c. perennial.
   d. axial.
   e. Both a and c
   *Answer: d*

2. An apomictic seed contains an embryo that is
   a. produced when two sperm fertilize one egg.
   b. developed from one egg alone.
   c. the result of parental self-fertilization.
   d. genetically identical to its parent.
   e. homozygous for most genetic traits.
   *Answer: d*

3. A plant's transition to a flowering state is often marked by
   a. an increased rate of photosynthesis.
   b. a decrease in vegetative growth.
   c. an increase in root development.
   d. a decreased rate of respiration.
   e. an increase in lateral bud growth.
   *Answer: b*

4. After pollination, which of the following events is crucial for fertilization?
   a. Sperm must swim to the egg and the polar nuclei.
   b. Petals must close around the reproductive parts.
   c. Meiosis must occur within the pollen grain.
   d. A pollen tube must grow from the stigma to the ovule.
   e. An insect must deliver pollen to the stigma.
   *Answer: d*

5. Vegetative parts that are modified for asexual reproduction are short, vertical stems called _____ and _____.
   a. rhizomes; tubers
   b. tubers; corms
   c. bulbs; corms
   d. bulbs; rhizomes
   e. corms; rhizomes
   *Answer: c*

6. Flowering is triggered by _____, although this substance has not yet been isolated and characterized.
   a. florigen
   b. NO
   c. ethylene
   d. abscisin
   e. phytochrome
   *Answer: a*

7. After fertilization of the egg, the integument of the megasporangium develops into the
   a. cotyledons.
   b. embryo.
   c. endosperm.
   d. fruit.
   e. seed coat.
   *Answer: e*

■8. In some eudicots, no distinct endosperm can be seen because the
   a. embryo has digested the endosperm.
   b. cotyledons have absorbed the endosperm.
   c. seeds never produced endosperm.
   d. endosperm has become the seed coat.
   e. fruit has incorporated the endosperm.
   *Answer: b*

9. The loss of water from a developing seed causes it to
   a. produce a root.
   b. die.
   c. be released from the fruit.
   d. become dormant.
   e. be protected from animal predators.
   *Answer: d*

★10. When a short-day plant (SDP) and a long-day plant (LDP) are grafted together and the SDP is exposed to a photoperiod that causes it to flower, the LDP flowers as well. This experiment supports the theory that a specific flower-initiating hormone is produced in plants. The proper control for this experiment would be to repeat the same design and
   a. omit the grafting.
   b. omit the LDP.
   c. omit the SDP.
   d. omit the SDP photoperiod.
   e. include an LDP photoperiod.
   *Answer: a*

11. Which of the following is a gametophyte of a flowering plant?
   a. Flower
   b. Egg
   c. Pollen grain
   d. Anther
   e. Entire plant
   *Answer: c*

12. The megagametophyte of flowering plants consists of the
   a. pollen grain.
   b. pollen tube.
   c. eight-nucleate embryo sac.
   d. ovule.
   e. megasporangium and cells within it.
   *Answer: c*

★13. From megasporocyte to egg cell, which processes are required?
   a. Meiosis followed by mitosis
   b. Mitosis followed by meiosis
   c. Several meiotic divisions only
   d. Several mitotic divisions only
   e. Several nuclear fusion events
   *Answer: a*

14. Within which of the following structures does meiosis occur?
   a. Petal
   b. Ovule
   c. Stigma
   d. Sepal
   e. Pollen grain
   *Answer: b*

★15. The advantage of self-fertilization in plants is
   a. increased genetic recombination.
   b. that meiosis can occur.
   c. greater efficiency of pollination.
   d. that no flowering is needed.
   e. that only asexual reproduction is necessary.
   *Answer: c*

16. The advantage of cross-fertilization in plants is
   a. increased genetic recombination.
   b. that meiosis can occur.
   c. greater efficiency of pollination.
   d. that no flowering is needed.
   e. that only asexual reproduction is necessary.
   *Answer: a*

17. Plants with wind-pollinated flowers tend to have
   a. colorful petals.
   b. smooth stigmas.
   c. large quantities of pollen.
   d. large quantities of nectar.
   e. pollination before the bud opens.
   *Answer: c*

18. Where does fertilization occur in flowering plants?
   a. Where pollen lands on the stigma
   b. Where the pollen tube germinates
   c. Inside the pollen tube
   d. At the base of the embryo sac
   e. Inside the seed
   *Answer: d*

19. Winter wheat can be planted in the spring for fall harvest if it is
   a. sprayed with gibberellins.
   b. soaked in water.
   d. planted under a full moon.
   d. stored in the dark for 50 days.
   e. vernalized.
   *Answer: e*

20. The egg can be fertilized by
    a. one tube nucleus.
    b. one sperm nucleus.
    c. two sperm nuclei.
    d. one generative nucleus.
    e. two synergid nuclei.
    *Answer: b*

*21. What is the fate of the seven cells of the embryo sac?
    a. All but one disintegrate upon fertilization.
    b. Two become fertilized; the others disintegrate.
    c. Two become fertilized; the others fuse to form endosperm.
    d. All are involved in nuclear fusion events.
    e. They all become part of the seed tissue.
    *Answer: b*

22. Which nuclei fuse to form the endosperm?
    a. The egg nucleus and the sperm nucleus
    b. The egg nucleus and two sperm nuclei
    c. The synergid nuclei and the sperm nucleus
    d. Polar nuclei and the generative nucleus
    e. Polar nuclei and the sperm nucleus
    *Answer: e*

23. A seed consists of a _____ embryo, a _____ endosperm, and _____ seed coats.
    a. diploid; triploid; diploid
    b. haploid; triploid; diploid
    c. diploid; triploid; haploid
    d. diploid; diploid; diploid
    e. haploid; diploid; haploid
    *Answer: a*

24. Which of the following is the "suspensor"?
    a. The pollen tube
    b. The organ supporting the ovule
    c. The base of the flower
    d. The narrow part of the embryo
    e. The stalk of the stamen
    *Answer: d*

25. The fruit generally develops from which part of the flower?
    a. Petals
    b. Sepals
    c. Ovary
    d. Stamens
    e. Pedicel
    *Answer: c*

26. Early in development, _____ initiates cell specialization in the embryo.
    a. mitotic division of the zygote
    b. uneven distribution of cell contents
    c. absorption of food from the endosperm
    d. seed germination
    e. maturation of the fruit
    *Answer: b*

27. Which of the following is a distinguishing characteristic of all angiosperms?
    a. Double cotyledons
    b. Fleshy cotyledons
    c. Seeds with nutrients
    d. Double fertilization
    e. Pollen production
    *Answer: d*

28. The nutritious flesh of many fruits has the function of
    a. nourishing the embryo.
    b. attracting seed eaters.
    c. attracting pollinators.
    d. attracting seed dispersers.
    e. ensuring that the seeds fall close to the parent plant.
    *Answer: d*

29. Asexual reproduction is the best strategy for plants
    a. that are well adapted to a stable environment.
    b. as winter approaches.
    c. when new genes must be introduced.
    d. that have underground stems.
    e. that have low seed production during a particular season.
    *Answer: a*

30. What is necessary for successful grafting?
    a. Each section must be able to form roots.
    b. The grafted section must be able to form seeds.
    c. Fusion of the two vascular tissues must occur.
    d. Fusion of the two cambial tissues must occur.
    e. Each section must be from the same species.
    *Answer: d*

**31. Fruit-eating bats tend to feed extremely rapidly on fruits and have relatively inefficient digestion (sometimes defecating seeds as early as an hour after feeding on them). Why are they good seed dispersal agents?
    a. Seed survival in bat guts is low.
    b. Undigested seeds are deposited in a heap at the bats' roost site.
    c. Undigested seeds are deposited at various bat feeding sites.
    d. Undigested seeds are deposited near the same plant that produced them.
    e. Digested seeds are dispersed in bat waste products.
    *Answer: c*

32. Which of the following is a characteristic of wind-dispersed seeds?
    a. Abundant pollen
    b. Hooked extensions
    c. Fleshy fruit
    d. Air chambers
    e. Flat, winged extensions
    *Answer: e*

33. Bamboo reproduces by means of
    a. rhizomes.
    b. tubers.
    c. corms.
    d. stolons.
    e. wind-dispersed seed germination.
    *Answer: a*

34. Navel oranges are produced by
    a. cross-hybridization.
    b. forced seed germination.
    c. apomixis.
    d. cloning.
    e. vernalization.
    *Answer: d*

■35. Quaking aspen trees usually reproduce asexually in Colorado but sexually in New England. From this information, what can one hypothesize about the stability of the two environments in which the aspens grow?
   a. No hypothesis can be developed from this information.
   b. The Colorado aspens grow in a more stable environment.
   c. The New England aspens grow in a more stable environment.
   d. Both environments are stable.
   e. Neither environment is stable.
   *Answer: b*

36. The process of grafting involves which of the following?
   a. Allowing a piece of one plant to grow onto the root of another
   b. Allowing cross-fertilization between two plants
   c. Preparing several cuttings from a plant, each of which will grow into an individual plant
   d. The production of xylem and phloem from the same cambium layer
   e. Interbreeding of two species of plants
   *Answer: a*

37. In a flower, the microsporangia are found in the
   a. anther.
   b. filament.
   c. stigma.
   d. ovule.
   e. ovary.
   *Answer: a*

38. Which is the correct order of events for female gametophytes?
   a. Megagametophyte, megasporocyte, megaspore
   b. Megagametophyte, megaspore, megasporocyte
   c. Megasporocyte, megaspore, megagametophyte
   d. Megaspore, megasporocyte, megagametophyte
   e. Megaspore, megagametophyte, megasporocyte
   *Answer: c*

39. In the mature embryo sac, the cells closest to where the pollen tube enters the ovule (the micropyle) are the
   a. polar nuclei.
   b. synergids.
   c. eggs.
   d. egg and polar nuclei.
   e. egg and synergids.
   *Answer: e*

40. In flowering plants, pollen is transferred to the
   a. stigma.
   b. style.
   c. ovary.
   d. ovule.
   e. micropyle.
   *Answer: a*

41. A flower that is wind-pollinated would be least likely to
   a. have numerous anthers.
   b. have sticky or feathery stigmas.
   c. produce large numbers of pollen grains.
   d. have a colorful corolla.
   e. have smooth wall sculpturing on its pollen.
   *Answer: d*

42. Pollination is the
   a. fusion of the egg and sperm nuclei.
   b. transfer of pollen from the anther to the stigma.
   c. development of the two-celled pollen grain.
   d. growth of the pollen tube after pollen germination.
   e. division of the generative nucleus to produce two sperm nuclei.
   *Answer: b*

43. The three nuclei in a mature pollen grain are formed by
   a. one meiotic division and one mitotic division.
   b. two meiotic divisions and one mitotic division.
   c. one meiotic division and two mitotic divisions.
   d. two meiotic divisions and two mitotic divisions.
   e. one meiotic division in which one of the four cells degenerates.
   *Answer: c*

44. The "embryo sac" is also called the
   a. megaspore.
   b. megasporangium.
   c. megasporocyte.
   d. megasporophyll.
   e. megagametophyte.
   *Answer: e*

45. Double fertilization results in the formation of
   a. two diploid embryos.
   b. one diploid embryo and a diploid endosperm.
   c. two diploid embryos and a haploid endosperm.
   d. one diploid embryo and a triploid endosperm.
   e. two diploid embryos and a diploid seed coat.
   *Answer: d*

46. The integuments of the ovule develop into the _____, and the carpels ultimately become the wall of the _____.
   a. cotyledons; endosperm
   b. seed coats; fruit
   c. cotyledons; seed coats
   d. endosperm; seed coats
   e. cotyledons; fruit
   *Answer: b*

47. Coconut fruits are dispersed by
   a. monkeys.
   b. wind.
   c. fruit bats.
   d. water.
   e. birds.
   *Answer: d*

48. Self-pollination results in progeny that
   a. are identical to the parent.
   b. are somewhat different from the parent because mutations are common.
   c. may express a recessive gene if the parent is heterozygous.
   d. may be heterozygous in a locus where the parent is homozygous.
   e. may be as varied as progeny resulting from cross-pollination.
   *Answer: c*

49. A clone of white potatoes may be derived from underground
    a. stolons.
    b. tubers.
    c. rhizomes.
    d. bulbs.
    e. root suckers.
    *Answer: b*

50. The ability of plants to measure night length was determined by experiments in which
    a. plants were grown in 12 hours of darkness alternating with 12 hours of light.
    b. plants were grown in continuous light.
    c. dark periods were interrupted with brief pulses of light.
    d. flowering was measured in plants of different ages that had been placed in the same light–dark schedule.
    e. plants were grown in a 24-hour cycle in which the length of the light period was increased gradually.
    *Answer: c*

51. Which of the following photoperiods would induce flowering in a short-day plant with a critical day length of 15 hours?
    a. 12 hours of light alternating with 12 hours of darkness
    b. 16 hours of light alternating with 8 hours of darkness
    c. 14 hours of light alternating with 8 hours of darkness
    d. 8 hours of light alternating with 8 hours of darkness
    e. 15 hours of light alternating with 9 hours of darkness, interrupted by one short burst of white light
    *Answer: a*

52. Plants that require vernalization must experience which of the following before flowering can occur?
    a. Exposure to cold
    b. Availability of soil calcium
    c. Minimal day length
    d. Sufficient moisture for a minimum period
    e. One full year of growth
    *Answer: a*

53. Short-day annuals usually flower
    a. in the spring.
    b. in midsummer.
    c. in late summer.
    d. in midsummer and late summer.
    e. throughout the summer.
    *Answer: c*

54. Technically, short-day plants flower when the _____ period _____ a critical period.
    a. light; exceeds
    b. light; is less than
    c. light; equals
    d. dark; exceeds
    e. dark; is less than
    *Answer: d*

55. Long-day plants will not flower if exposed to
    a. a long day.
    b. a long day interrupted by a dark period.
    c. a long day interrupted by a prolonged period of red light.
    d. a long night interrupted by a period of far-red light.
    e. a long day interrupted by a brief period of red light.
    *Answer: b*

56. Circadian rhythms
    a. in nature are always 24-hour cycles.
    b. are found only in multicellular organisms.
    c. have a period that is remarkably sensitive to temperature.
    d. do not persist when the plant is placed in complete darkness.
    e. can be made to coincide with the light–dark regime.
    *Answer: e*

57. Although the mysterious "flowering hormone" has never been discovered, there is evidence that it is synthesized in the
    a. floral meristem.
    b. vegetative apical meristem.
    c. stems.
    d. leaves.
    e. roots.
    *Answer: d*

58. A benefit of sexual reproduction in plants is
    a. the greater number of progeny that results.
    b. ease of pollination.
    c. the improved ability of plants to adapt to new environments.
    d. that the haploid plant becomes diploid.
    e. farther dispersal of progeny.
    *Answer: c*

59. Long-short-day plants bloom
    a. in the spring.
    b. in the fall.
    c. in the summer.
    d. only after a cold winter.
    e. only after the second year of life.
    *Answer: b*

60. The production of French wine grapes (*Vitis vinifera*) that are resistant to plant lice was accomplished by
    a. self-fertilization of resistant plants.
    b. grafting the scion onto resistant plants' roots.
    c. recombinant DNA techniques using the resistance gene.
    d. planting seeds in California, where there are no plant lice.
    e. inserting cuttings, or slips, of the plants into California soil, where there are no plant lice.
    *Answer: b*

61. In the seed, nutrients for the seedling are generally stored as
    a. monomers in solution.
    b. monomers in fat storage.
    c. macromolecules.
    d. cellular enzymes.
    e. cellular organelles.
    *Answer: c*

62. As a pollen tube grows into the female organ, the nucleus that enters the synergid first is called the
    a. sperm nucleus.
    b. generative nucleus.
    c. tube nucleus.
    d. pollen nucleus.
    e. microspore.
    *Answer: c*

63. For plants that flower in response to photoperiodic stimuli, the most critical determinant in the light–dark cycle is the
    a. temperature.
    b. length of the dark period for short-day plants and the length of the light period for long-day plants.
    c. length of the light period for short-day plants and the length of the dark period for long-day plants.
    d. length of the uninterrupted dark cycle.
    e. length of the uninterrupted light cycle.
    *Answer: d*

# 40 Plant Responses to Environmental Challenges

## Fill in the Blank

*1. Plants limit the spread of viral pathogens by blocking their plasmodesmata with _____.
*Answer: polysaccharides*

2. Plants adapted to dry environments are called _____.
*Answer: xerophytes*

3. Plants with long dormant periods interrupted by short periods of rapid growth and reproduction typically live in _____ environments.
*Answer: dry*

4. When cells accumulate the amino acid _____, their water potential becomes more negative.
*Answer: proline*

5. Some swamp plants have root extensions called _____ that grow into the air and deliver _____ to the rest of the root system.
*Answer: pneumatophores; oxygen*

6. The major stress encountered by plants living in water-logged soil is lack of soil _____.
*Answer: oxygen*

7. In cacti, leaves are modified to form _____, and photo-synthesis is carried out in the _____ region.
*Answer: spines; stem*

8. Globally, the toxic substance that most restricts plant growth is _____.
*Answer: sodium chloride*

9. Halophytes accumulate chloride and sodium ions and transport these substances to their _____.
*Answer: leaves*

10. Plants with fleshy, water-storing leaves are called

_____.
*Answer: succulents*

11. Plants produce small molecules called _____ within hours of infection by fungi or bacteria.
*Answer: phytoalexins*

12. Substances that are not used for basic cellular metabolism, but rather for special functions such as inhibiting other organisms, are called _____ metabolites.
*Answer: secondary*

13. Large molecules called _____ proteins are produced as a defense against infection and in cleanup operations.
*Answer: pathogenesis-related*

14. Cold weather affects the _____ fluidity of plants.
*Answer: membrane*

## Multiple Choice

1. Phytoalexin production is triggered by
a. high temperature.
b. high salt concentration.
c. oligosaccharins.
d. suberin.
e. heat shock protein.
*Answer: c*

2. The secondary plant products that prevent the normal development of insect herbivores are
a. alkaloids.
b. flavonoids.
c. phenolics.
d. quinones.
e. steroids.
*Answer: e*

■3. Some plants produce the amino acid canavanine, which is toxic to many insects because
a. insects lack a tRNA specific for canavanine.
b. canavanine is a component of alkaloids.
c. canavanine prevents cells from synthesizing the amino acid arginine.
d. insects lack large vacuoles for the storage of secondary products.
e. insect proteins that incorporate canavanine function poorly.
*Answer: e*

4. In defense against tissue damage caused by pathogens,
a. animals repair damaged tissues and plants seal off damaged tissues.
b. animals seal off damaged tissues and plants repair damaged tissues.
c. both plants and animals sometimes seal off and sometimes repair damaged tissues.
d. both plants and animals repair damaged tissues.
e. both plants and animals seal off damaged tissues.
*Answer: a*

5. Phytoalexins
a. are always present in plants.
b. occur in uniform concentration throughout a plant.
c. are toxic to many fungi and bacteria.
d. have no effect on viral infections.
e. cause plants to seal off areas of damaged tissue.
*Answer: c*

6. Which of the following is *not* a hypersensitive reaction of a plant to infection?
   a. Production of phyoalexins by cells around the infection
   b. Long-term resistance to the infective agent
   c. Synthesis of pathogenesis-related proteins
   d. Death of both infected cells and cells near the infection
   e. Transport of phytoalexins to all parts of the plant.
   *Answer: e*

7. Salicylic acid in plants does not
   a. increase resistance to pathogens.
   b. trigger the production of pathogenesis-related proteins.
   c. cause herbivorous insects to produce defective proteins.
   d. protect against tobacco mosaic virus.
   e. play a role in the hypersensitivity response.
   *Answer: c*

8. Non-water-soluble (hydrophobic) poisons are stored in a plant's
   a. chloroplasts.
   b. epidermal waxes.
   c. Golgi bodies.
   d. mitochondria.
   e. vacuoles.
   *Answer: b*

**9. Plants can produce the respiratory poison cyanide without poisoning themselves because plants
   a. do not respire.
   b. store a cyanide precursor in one compartment and store activating enzymes in another.
   c. store water-soluble cyanide in laticifers.
   d. possess enzymes that are unaffected by cyanide.
   e. also produce proteins that bind and inhibit cyanide.
   *Answer: b*

*10. Which evidence best supports the hypothesis that the presence of toxic latex in leaves deters insects from feeding on a plant?
   a. Many insects do not feed on latex-producing plants.
   b. Latex-producing plants release milky latex when their leaves are damaged.
   c. Beetles that drain latex out of part of a leaf can then feed on that part.
   d. Beetles that cut veins in the leaves can then feed on the latex released.
   e. Latex-producing plants have high survival rates.
   *Answer: c*

11. Plants can be treated with _____ to stimulate the production of pathogenesis-related proteins.
   a. salicylic acid
   b. PR inducer
   c. phytoalexin
   d. cellulose
   e. willowgen
   *Answer: a*

12. Once secondary compounds are formed by plants, where are these defensive compounds usually stored?
   a. Vacuoles
   b. Nuclei
   c. Cell walls
   d. Cytoplasm
   e. Bound to membrane proteins
   *Answer: a*

*13. According to some data, a unique feature of oil of wintergreen is that it can stimulate production of PR protein in
   a. leaves.
   b. roots.
   c. flowering regions.
   d. a plant harboring an infection.
   e. plants neighboring the infected plant that produced it.
   *Answer: e*

14. In order for an *R* gene to confer resistance, a predator must have a corresponding
   a. *R* gene.
   b. tRNA.
   c. virus.
   d. *Avr* gene.
   e. bacterium.
   *Answer: d*

15. Which of the following are "secondary products"?
   a. Proteins
   b. Lipids
   c. Alkaloids
   d. Carbohydrates
   e. Nucleic acids
   *Answer: c*

16. Steroids produced by plants may
   a. attract pollinators and animals that disperse seeds.
   b. affect nervous systems of animals.
   c. inhibit fungal action.
   d. prevent normal development of insects.
   e. impair growth of competing plants.
   *Answer: d*

17. Polysaccharides serve to
   a. store water in plants.
   b. defend against pathogens.
   c. repel predators because they are toxic to animals.
   d. act as salt glands.
   e. strengthen cell walls to form a barrier against invasion of a pathogen.
   *Answer: e*

18. Laticifers are
   a. specialized cells for containing sodium ions.
   b. latex-containing tubes for storing hydrophobic products.
   c. waxy cells in the epidermis.
   d. cells that produce poisons such as alkaloids.
   e. cells in roots that take up water in dry environments.
   *Answer: b*

19. In the gene-for-gene resistance mechanism, if a plant has a dominant resistance gene and a pathogen has a dominant avirulence gene,
   a. the pathogen's gene overrides the plant's gene and infects the plant.
   b. epistasis occurs and the plant is infected.
   c. epistasis occurs and the plant is resistant to the pathogen.
   d. the plant is resistant to all pathogens whether or not they have the dominant avirulence gene.
   e. None of the above
   *Answer: c*

20. The plant substance canavanine is toxic to most insects because it
   a. inhibits respiration.

b. causes defects in protein structure and function.
c. interferes with protein digestion in the gut.
d. burns tissues due to its high acidity.
e. inhibits the synthesis of reproductive hormones.
*Answer: b*

21. The nicotine in tobacco is an example of a plant
    a. alkaloid.
    b. flavonoid.
    c. glycoside.
    d. steroid.
    e. tannin.
    *Answer: a.*

22. Grazing can increase photosynthesis production by
    a. causing increased branching.
    b. removing old or dead leaves.
    c. providing increased root nutrients to the remaining leaves.
    d. All of the above
    e. None of the above
    *Answer: d*

23. The difference between systemin and jasmonate is that systemin is a _____, whereas jasmonate is a _____.
    a. nucleic acid; steroid
    b. steroid; nucleic acid
    c. fatty acid derivative; polypeptide
    d. nucleic acid; steroid
    e. polypeptide; fatty acid derivative
    *Answer: e*

24. Plants protect themselves from toxins they produce by all of the following methods *except*
    a. compartmentalizing them.
    b. building up a tolerance to them.
    c. adjusting the timing of toxin production.
    d. storing them in waxes.
    e. storing them in vacuoles.
    *Answer: b*

25. To convert *Arabidopsis* from nonhalophyte to halophyte, plants were altered to
    a. have salt glands.
    b. not express $Na^+/H^+$ symport.
    c. not express $Na^+/H^+$ antiport
    d. overexpress $Na^+/H^+$ antiport.
    e. overexpress $Na^+/H^+$ symport.
    *Answer: d*

26. Which of the following is *not* an adaptation to a dry environment?
    a. Thick cuticle
    b. Vertically hanging leaves
    c. Stomata in sunken cavities
    d. Epidermal hairs
    e. Salt glands in leaves
    *Answer: e*

27. The typical environment for annual plants with a brief growing period and seeds capable of long dormant periods is a
    a. desert.
    b. salt marsh.
    c. freshwater marsh.
    d. environment contaminated with heavy metals.
    e. grazed field.
    *Answer: a*

28. Corn and related grasses roll up their leaves in response to
    a. excess water.
    b. lack of water.
    c. excess salt.
    d. heavy metals.
    e. herbivores.
    *Answer: b*

■29. Compared to plants in moderate environments, xerophytes carry out photosynthesis more
    a. slowly, due to interference from the amino acid proline.
    b. slowly, because transpiration occurs more quickly.
    c. slowly, because their adaptations minimize carbon dioxide uptake.
    d. quickly, because their stems are also photosynthetic.
    e. quickly, because they have short periods of intense growth.
    *Answer: c*

30. Swamp plants typically have root systems that
    a. grow quickly.
    b. penetrate deeply into the soil.
    c. can carry out alcoholic fermentation.
    d. alternate periods of growth with periods of die-back.
    e. accumulate the amino acid proline.
    *Answer: c*

31. The pneumatophores of swamp plants are modified
    a. flowers.
    b. leaves.
    c. roots.
    d. spines.
    e. stems.
    *Answer: c*

32. Leaf parenchyma tissue with large spaces between cells _____ in _____.
    a. provide buoyancy; aquatic plants
    b. decrease transpiration rates; aquatic plants
    c. store water; desert plants
    d. form succulent leaves; desert plants
    e. excrete salt; halophytes
    *Answer: a*

33. By accumulating the amino acid proline, plants
    a. become toxic to most herbivores.
    b. can carry out alcoholic fermentation.
    c. can extract more water from the soil.
    d. can prevent toxic effects from sodium.
    e. attract animals that disperse seeds.
    *Answer: c*

■34. If a nonhalophyte and a halophyte are both placed in a salty environment, which will accumulate more sodium internally, and why?
    a. The halophyte, because the nonhalophyte will not absorb much sodium
    b. The halophyte, because it requires sodium as a nutrient
    c. The halophyte, because as a succulent, it has more internal storage
    d. The nonhalophyte, because it cannot excrete sodium after absorption
    e. The nonhalophyte, because it requires sodium to create a negative water potential
    *Answer: a*

35. Which of the following adaptations is *not* found in both xerophytes and halophytes?
    a. High root-to-shoot ratios
    b. Sunken stomata
    c. Large air spaces in the leaf parenchyma
    d. Reduced leaf area
    e. Thick cuticles
    *Answer: c*

■36. The reason that certain plants can grow in soils contaminated with high levels of heavy metals is that they
    a. do not take up the heavy metals.
    b. excrete the heavy metals.
    c. have a genetic tolerance to the heavy metals.
    d. use the heavy metals for normal biochemical functions.
    e. are toxic to herbivores due to the heavy metals.
    *Answer: c*

37. Which of the following is *not* true of plants that tolerate heavy metals?
    a. They take up the heavy metals.
    b. Different populations of plants have an equal capacity to tolerate the heavy metals.
    c. Tolerant populations can evolve rapidly.
    d. A plant's tolerance is determined by its genotype.
    e. They usually experience little competition.
    *Answer: b*

■38. Grazing increases photosynthetic rates in certain plant species because
    a. it allows more light to reach younger, more active leaves.
    b. the remaining leaves can transport sugars more slowly to the roots.
    c. it allows the roots to take up more nitrogen.
    d. older, dying leaves are sugar sinks.
    e. it reduces competition for atmospheric carbon dioxide.
    *Answer: a*

39. The ability of grasses to grow from the base of the shoot and leaf is an adaptation to
    a. dry environments.
    b. soil fungi.
    c. heavy metals.
    d. grazing.
    e. saline environments.
    *Answer: d*

40. Grazed plants may exhibit increased productivity in all of the following ways *except*
    a. faster photosynthetic rates.
    b. growth of more stems.
    c. greater seed distribution.
    d. growth later into the season.
    e. growth of more roots.
    *Answer: e*

41. Secondary products
    a. are essential for basic biological reactions.
    b. are similar in all plants.
    c. occur more often in animals than in plants.
    d. attract or inhibit other organisms.
    e. are usually of high molecular weight.
    *Answer: d*

42. Which of the following is *not* a special adaptation of leaves to dry environments?
    a. Modification into spines

b. Stomata in sunken cavities
    c. Dense epidermal hairs
    d. Fleshy leaves
    e. Horizontal leaves
    *Answer: e*

43. The presence of pneumatophores in plants is an adaptation for success in which type of habitat?
    a. Desert
    b. Mountain
    c. Grassland
    d. Seashore
    e. Swamp
    *Answer: e*

44. Some halophytic plants have salt glands that
    a. accumulate salt in their roots.
    b. serve as barriers to salt intake.
    c. maintain high salt concentrations in the plant.
    d. secrete salt onto the leaf surface.
    e. increase water loss from the plant.
    *Answer: d*

45. Which combination of adaptations is often observed in plants found in saline environments?
    a. Salt glands and succulence
    b. Salt glands and broad leaves
    c. CAM metabolism and succulence
    d. Spines and thin cuticles
    e. Dense stomata and thick cuticles
    *Answer: a*

46. Certain plants concentrate the harmless amino acid proline in their cells. What effect does this have on the plant?
    a. Increased negative water potential
    b. Increased rate of transpiration
    c. Increased positive water potential in leaves
    d. Decreased salt loss
    e. Decreased water uptake
    *Answer: a*

47. When some leaves are removed from a plant, what typically happens?
    a. Less light is available to the leaves.
    b. Less nitrogen is obtained from the soil.
    c. The plant dies.
    d. The rate of photosynthesis in the remaining leaves increases.
    e. The transport of sugar from the remaining leaves decreases.
    *Answer: d*

48. In conditions where water is plentiful but oxygen scarce, plants respond by
    a. rapidly growing roots that penetrate deeply into the soil.
    b. fermenting sugars to lactic acid.
    c. inhibiting the production of ATP.
    d. forming aerenchyma tissue.
    e. producing oxygen from water.
    *Answer: d*

49. All of the following are adaptations to saline environments *except*
    a. accumulation and transport of sodium ions.
    b. sequestering of sodium ions in roots.
    c. salt glands in leaves.

d. fleshy, gummy leaves.
e. smaller leaves with smaller cells.
*Answer: b*

50. One way that grazing increases the productivity of a plant is by
    a. supporting food chains in nature.
    b. reducing the rate of photosynthesis in remaining leaves.
    c. increasing the number of sinks for absorbed nitrogen.
    d. shading younger leaves.
    e. causing the production of more replacement stems.
    *Answer: e*

51. Many halophytes accumulate the amino acid
    a. proline.
    b. arginine.
    c. glycine.
    d. methionine.
    e. chlorine.
    *Answer: a*

52. High temperatures trigger heat shock proteins such as
    a. nucleases.
    b. kinases.
    c. chaperonins.
    d. antifreeze proteins.
    e. proline.
    *Answer: c*

**\*\***53. An RNA virus attack on a plant triggers the production of interference RNA (RNAi), which is derived from the RNA of the _____ and causes the plant to _____.
    a. virus; die from a viral infection
    b. plant; become immune to the virus
    c. plant; form mechanical barriers
    d. virus; become immune to the virus
    e. virus; form mechanical barriers
    *Answer: d*

54. Plant defensive steroids affect insect
    a. nervous systems.
    b. digestive systems.
    c. life cycles.
    d. protein synthesis.
    e. appetites.
    *Answer: c*

55. Which element is *not* a heavy metal?
    a. Cadmium
    b. Chromium
    c. Lead
    d. Mercury
    e. Sulfur
    *Answer: e*

# 41 Physiology, Homeostasis, and Temperature Regulation

## Fill in the Blank

1. The four types of tissues are epithelial, muscle, nervous, and _____.
*Answer: connective*

★2. In certain "hot" fish such as bluefin tuna, heat is exchanged between vessels carrying blood in opposite directions. This adaptation is called _____.
*Answer: countercurrent heat exchange*

3. The metabolic rate of a resting animal at a temperature within the thermoneutral zone is called the _____.
*Answer: basal metabolic rate*

4. Small endotherms can extend the period over which they can survive without food by dropping body temperature. This adaptive hypothermia is called _____.
*Answer: daily torpor*

5. The maintenance of more or less constant physiological conditions within an organism is called _____.
*Answer: homeostasis*

6. _____ feedback amplifies a response.
*Answer: Positive*

7. Animals whose temperature fluctuates to match that of the environment are termed _____.
*Answer: poikilotherms*

8. Animals that can affect their body temperature by generating metabolic heat are called _____.
*Answer: endotherms*

9. In vertebrates, the thermostat is located in the _____.
*Answer: hypothalamus*

10. During the condition called _____, when an endotherm's body temperature is far below normal, metabolic rates become _____. Although this condition can be harmful to endotherms, it is a useful strategy for some overwintering mammals called _____.
*Answer: hypothermia; lower; hibernators*

11. In mammals, most nonshivering heat production occurs in the adipose tissue called _____.
*Answer: brown fat*

12. The linings of many hollow organs consist of _____ tissue.
*Answer: epithelial*

13. The most abundant protein in our bodies is _____, found in connective tissue.
*Answer: collagen*

14. Cells with the same characteristics form _____, and different tissues form discrete structures with specific functions called _____.
*Answer: tissues; organs*

★15. _____, derived from bacteria or viruses that invade the body, cause fever.
*Answer: Pyrogens*

## Multiple Choice

1. Which of the following statements about tissues is *false*?
   a. The protein elastin is found in tissues that are regularly stretched.
   b. An organ is usually composed of a single type of tissue.
   c. Epithelial tissue forms a boundary between the inside and outside of the body.
   d. Adipose tissue is a form of connective tissue.
   e. None of the above
   *Answer: b*

2. Which of the following statements about heat exchange is *false*?
   a. Conduction is the direct transfer of heat between two objects of different temperatures that have come into contact.
   b. Evaporation of water from the surface of the body heats the body.
   c. Ectotherms and endotherms can change the rate of heat exchange between their bodies and the external environment by changing blood flow to the skin.
   d. Animals may lose heat by convection when they are exposed to a wind with a temperature below that of their body surface.
   e. None of the above
   *Answer: b*

■3. Homeostasis
   a. favors a relatively constant internal physiological environment regardless of the changes in the external environment.
   b. keeps vital organs working at their maximum potential.
   c. keeps all cells working at the same metabolic rate.
   d. keeps the body's metabolic rate constant in varying environmental temperatures.
   e. keeps the body's temperature constant in varying environmental temperatures.
   *Answer: a*

4. Which term refers to organisms that depend largely on external sources of heat to maintain body temperature?
   a. Homeothermic
   b. Endothermic
   c. Heterothermic
   d. Ectothermic
   e. None of the above
   *Answer: d*

5. In response to a 10°C rise in environmental temperature, an endotherm's body temperature will
   a. rise at a constant rate.
   b. fall at a constant rate.
   c. fall to a point, then become stable.
   d. rise to a point, then become stable.
   e. remain relatively constant.
   *Answer: e*

■6. Which of the following adaptations would *not* favor an increase in an ectotherm's body temperature?
   a. Muscle contractions
   b. Cluster or huddling behavior
   c. Decreased surface-to-surface contact with the cold environment
   d. Circulatory changes to maintain core or internal temperatures greater than the animal's peripheral temperatures
   e. Metabolic compensation
   *Answer: e*

7. Readjustment of an organism's metabolic rate to compensate for seasonal thermal change is termed
   a. homeostasis.
   b. negative feedback.
   c. metabolic compensation.
   d. acclimatization.
   e. regulation.
   *Answer: c*

8. Which term best describes an organism that maintains a constant body temperature?
   a. Ectotherm
   b. Homeotherm
   c. Heterotherm
   d. Poikilotherm
   e. None of the above
   *Answer: b*

■9. After cool nighttime temperatures have changed, a desert lizard crawls slowly from its burrow and positions itself on a rock warmed by the mid-morning sun. Which one of the following is most likely *not* one of the lizard's thermoregulatory responses?
   a. Orientation to the sun to maximize exposure to solar radiation
   b. Increased peripheral circulation via dilation of surface blood vessels
   c. Increased metabolic heat production
   d. Increased physical contact between the warm rock and the relatively cool lizard
   e. Orientation of the lizard on the rock into more suitable microenvironments to avoid cold air currents
   *Answer: c*

10. The principal components of the endocrine system are
    a. hormones.
    b. neurons.
    c. glial cells.
    d. ductless glands.
    e. skeletal muscle cells.
    *Answer: d*

11. _____ are cells of the nervous system that communicate via electrochemical signals.
    a. Hormones
    b. Neurons
    c. Glial cells
    d. Ductless glands
    e. Skeletal muscle cells
    *Answer: b*

12. The chemical messages of the endocrine system are sent in the form of
    a. hormones.
    b. neurons.
    c. glial cells.
    d. ductless glands.
    e. skeletal muscle cells.
    *Answer: a*

13. _____ are cells of the nervous system that do not conduct signals.
    a. Hormones
    b. Neurons
    c. Glial cells
    d. Ductless glands
    e. Skeletal muscle cells
    *Answer: c*

14. What (is) are the normal value(s) for most biological $Q_{10}$'s?
    a. 1
    b. 1–2
    c. 2–3
    d. 2–10
    e. 0–45
    *Answer: c*

15–21. Use the following answer choices to indicate the probable metabolic response of a mammal exposed to a 3–5°C environmental temperature change.
    a. Increased metabolic rate
    b. Decreased metabolic rate
    c. No change in the metabolic rate
    d. Death

■15. The environmental temperature increases above the upper critical temperature.
    *Answer: a*

■16. The environmental temperature increases above the lower critical temperature.
    *Answer: c*

■17. The environmental temperature decreases far below zero.
    *Answer: d*

■18. The environmental temperature decreases below the lower critical temperature.
    *Answer: a*

■19. The environmental temperature decreases below the upper critical temperature.
    *Answer: c*

■20. The environmental temperature fluctuates between the upper and the lower critical temperatures.
    *Answer: c*

■21. The environmental temperature increases above zero.
*Answer: b*

*22. An endotherm's thermoneutral zone can be described variously. Which of the following does *not* describe the limits of the thermoneutral zone?
   a. A range of environmental temperatures between the upper critical and lower critical temperature
   b. A range of environmental temperatures over which the organism exhibits a basal metabolic rate
   c. A range of body temperatures at which the metabolic rate is maximum
   d. A range of environmental temperatures over which the organism's metabolic rate neither increases nor decreases for thermoregulation
   e. A range of environmental temperatures over which the organism produces minimal metabolic activity to support itself
*Answer: c*

23. Increased heat for thermoregulation or thermogenesis is produced either by shivering or by nonshivering mechanisms. Nonshivering thermogenesis is dependent upon all of the following *except*
   a. brown fat.
   b. thermogenin.
   c. production of ATP.
   d. a rich blood supply.
   e. increased mitochondria.
*Answer: c*

24. The hypothalamus serves in part as an integrated thermoregulatory center defining an organism's response to changes in its thermal environment. Because the hypothalamus normally serves to produce metabolic changes to reverse the direction of environmental temperature change, its control is termed
   a. positive feedback.
   b. metabolic compensation.
   c. negative feedback.
   d. feedforward.
   e. metabolic torpor.
*Answer: c*

■25. Which of the following is *not* true of hibernation?
   a. It may be interrupted by brief returns to normal body temperature.
   b. It is a form of regulated hypothermia.
   c. Body temperature is turned down to a low level.
   d. Metabolic rate is reduced to only a fraction of the basal metabolic rate.
   e. It is found in more species of birds than in species of mammals.
*Answer: e*

26. Which of the following is an organ system involved in communication within the body?
   a. Urinary system
   b. Nervous system
   c. Endocrine system
   d. Both b and c
   e. a, b, and c
*Answer: d*

■27. Most physiological processes
   a. occur more rapidly at higher temperatures.
   b. occur less rapidly at higher temperatures.

   c. are not temperature-sensitive.
   d. first increase in rate and then decrease in rate at high temperatures.
   e. None of the above
*Answer: a*

28. _____ is the process of physiological and biochemical change that an animal undergoes in response to seasonal changes in climate.
   a. Acclimatization
   b. Homeostasis
   c. Sublimity
   d. Metabolic compensation
   e. Hybridization
*Answer: a*

29. Adaptations used by endotherms to reduce heat loss include which of the following?
   a. Rounder body shapes
   b. Shorter appendages
   c. Increased thermal insulation
   d. Both b and c
   e. a, b, and c
*Answer: e*

30. The vertebrate thermoregulatory center ("thermostat") is located within the central nervous system in the
   a. pons.
   b. cerebellum.
   c. hypophysis.
   d. medulla.
   e. hypothalamus.
*Answer: e*

31. Homeostasis refers to the tendency to keep the body systems
   a. matched to the external environment.
   b. the same relative to one another.
   c. at a steady state over time.
   d. under the control of the brain.
   e. at the same specific temperature.
*Answer: c*

32. Which type of tissue is responsible for secreting digestive enzymes?
   a. Connective
   b. Epithelial
   c. Matrix
   d. Muscle
   e. Nervous
*Answer: b*

33. Bone is which type of tissue?
   a. Connective
   b. Epithelial
   c. Matrix
   d. Muscle
   e. Nervous
*Answer: a*

34. Blood is which type of tissue?
   a. Connective
   b. Epithelial
   c. Matrix
   d. Muscle
   e. Nervous
*Answer: a*

35. Connective tissues differ from one another mostly in their
    a. cellular structure.
    b. function in support.
    c. matrix properties.
    d. location in the body.
    e. cell packing.
    *Answer: c*

*36. In regulatory systems, the phenomenon of negative feedback
    a. is the least common type of feedback mechanism.
    b. stimulates a return to set point.
    c. amplifies a response.
    d. disrupts homeostasis.
    e. None of the above
    *Answer: b*

**37. Which of the following would serve to increase the set point of a regulatory system?
    a. Negative feedback
    b. Positive feedback
    c. Feedforward
    d. Insensitivity to information
    e. None of the above
    *Answer: c*

38. Cellular functions are generally limited to what temperature range (in °C)?
    a. 20–100
    b. 20–45
    c. 0–20
    d. 0–45
    e. 0–100
    *Answer: d*

39. The upper temperature limit at which cells can function is determined by the
    a. boiling point of water.
    b. melting point of water.
    c. melting point of fats.
    d. denaturation point of nucleic acids.
    e. denaturation point of proteins.
    *Answer: e*

40. The $Q_{10}$, which describes the sensitivity of a reaction to temperature, is calculated as the
    a. rate of a process at a certain temperature divided by its rate at 10°C.
    b. rate of a process at a certain temperature divided by its rate at a temperature 10°C lower.
    c. rate of a process at a certain temperature divided by its rate at a temperature 10°C higher.
    d. temperature at which the rate of a certain process doubles.
    e. temperature at which the rate of a certain process becomes insignificant.
    *Answer: b*

41. The rate of a particular biological function is X at 15°C. Which of these statements about the rate of that function at 25°C is true?
    a. If the $Q_{10}$ were 1, the rate would be X.
    b. If the $Q_{10}$ were 1, the rate would be 25.
    c. If the $Q_{10}$ were 2, the rate would be X.
    d. If the $Q_{10}$ were 3, the rate would be 2X.
    e. If the $Q_{10}$ were 3, the rate would be 30.
    *Answer: a*

42. If an animal exhibits metabolic compensation,
    a. it adapts its physiology to local environmental conditions.
    b. its metabolic rate is always slightly higher in summer.
    c. its $Q_{10}$ stays constant despite seasonal temperature change.
    d. its body temperature and its metabolic rate are independent.
    e. it uses the same reaction pathways at all temperatures.
    *Answer: a*

43. Poikilothermic animals are those whose body temperature
    a. is maintained at a constant level some of the time.
    b. is maintained at a constant level all of the time.
    c. fluctuates with environmental temperature.
    d. is kept at a higher temperature than that of the environment.
    e. is kept at a lower temperature than that of the environment.
    *Answer: c*

44. Which of the following animals are endotherms?
    a. Fishes
    b. Amphibians
    c. Birds
    d. Mammals
    e. Both c and d
    *Answer: e*

45. Which of the following animals would be most successful with the part-time temperature regulation strategy of heterothermy?
    a. A desert lizard that basks in the sun
    b. A deep ocean fish in an environment with little temperature change
    c. An insect that is warm-blooded in flight and cold-blooded at rest
    d. An amphibian that is sometimes in water and sometimes on land
    e. A hummingbird that drops its body temperature at night
    *Answer: e*

46. Which of the following animals is behaving as an endotherm in order to warm its body?
    a. A moth that shivers its wings before flight
    b. A black beetle that absorbs solar radiation
    c. A snake that lies on a warm blacktop road
    d. A fish that moves to a warm, shallow part of a pond
    e. An insect that positions its body for maximum exposure to sunlight
    *Answer: a*

47. A certain desert lizard thermoregulates as an ectotherm. Which of the following phenomena would *not* be a part of its thermoregulatory behavior?
    a. Staying in a burrow when the surface temperature is below 10°C
    b. Basking in the sun during the early morning hours
    c. Moving into the shade during the midday hours
    d. Pressing its body to the ground during the midday hours
    e. Consuming prey that is ectothermic
    *Answer: d*

■48. In which case would a poikilothermic animal have a higher body temperature than a homeotherm would have?
a. When resting in an underground burrow
b. At midday on the desert floor
c. When swimming in the cold ocean
d. When basking in the early morning sun
e. When the poikilotherm is much larger
*Answer: b*

49. Which of the following exemplifies adaptive thermoregulatory behavior in humans?
a. A cowboy wearing a broad-brimmed hat
b. An Eskimo wearing lightweight white clothing
c. A desert home with large windows facing the sun
d. Swimming in winter weather
e. None of the above
*Answer: a*

50. Which physiological control mechanism is a response to a rise in body temperature?
a. Slower heart rate
b. Increased blood flow to the skin
c. Constriction of blood vessels in the skin
d. Contraction of muscles
e. Retention of water
*Answer: b*

51. What is the adaptive advantage of honeybee workers' clustering to maintain warm temperatures in the hive?
a. Protection of the queen bee
b. Protection of the comb structure
c. Optimal pollen collection
d. Optimal digestion of honey
e. Optimal brood development
*Answer: e*

*52. Within a range of environmental temperatures called the thermoneutral zone, the metabolic rate of an endotherm is
a. constant.
b. low and independent of temperature.
c. high and independent of temperature.
d. near the basal metabolic rate.
e. dependent upon the temperature.
*Answer: b*

**■53. In a mammal, metabolic rate is highest when the temperature is _____ the thermoneutral zone and the animal is _____.
a. within; at rest
b. below; active
c. within; active
d. above; at rest
e. below; at rest
*Answer: b*

54. Which of the following statements about brown fat is *false*?
a. It contains abundant mitochondria.
b. It provides the most energy for shivering.
c. It allows metabolic fuels to be consumed without producing ATP.
d. It uses thermogenin to uncouple proton movement from ATP production.
e. It is more abundant in hibernating animals.
*Answer: b*

*55. The mechanism of heat production in brown fat depends on
a. inefficient use of ATP in metabolism.
b. rapid breakdown of protein.
c. rapid breakdown of fatty acids.
d. the shivering of skeletal muscles.
e. uncoupling proton movement from ATP production.
*Answer: e*

56. Which of the following is an important adaptation of animals to cold climates?
a. Increased tendency to shiver
b. Thinner layers of body fat
c. Reduced density of fur or feathers
d. Reduced surface area-to-volume ratio
e. Increased flow of blood to surface
*Answer: d*

57. The elephant is better adapted to tropical habitats than to cold climates because of its
a. sparse hair.
b. large size.
c. stocky appendages.
d. vegetarian diet.
e. thick skin.
*Answer: a*

58. Why is evaporative cooling used only as a last resort by animals in dry environments?
a. It is ineffective at dissipating heat.
b. Water may be a limiting resource.
c. Sweating requires little energy expenditure.
d. It requires an insulating layer in the skin.
e. Resetting the thermostat is required.
*Answer: b*

59. When the temperature of the hypothalamus rises, which thermoregulatory response results?
a. Increased metabolic heat production
b. Resetting of the thermostat higher
c. Dilation of blood vessels in the skin
d. Overall increase in body temperature
e. Initiation of shivering movements
*Answer: c*

**60. Which of the following accurately describes the thermoregulatory set point?
a. The set point for shivering is the same as the set point for panting.
b. In a given individual, the set points are relatively constant.
c. The set points are the same for all members of the same species.
d. Information on skin temperature can change the metabolic set point.
e. Temperature information serves as a positive feedback signal.
*Answer: d*

61. Which of the following statements about thermoregulation is *false*?
a. Thermoregulatory set points are higher during sleep.
b. Entrance into hibernation begins with a decrease in metabolic rate.
c. During a fever, aspirin can lower the hypothalamic set point.
d. Hypothermia may result from starvation.
e. None of the above
*Answer: a*

62. Single-celled and simple multicellular animals meet their needs by having every cell exposed to the environment; more complex animals have an internal environment. Which of the following is *not* an advantage of an internal environment?
    a. Each cell is capable of performing every function.
    b. Cells can be specialized for one function.
    c. Cells can be more efficient.
    d. It can be maintained independently of the external environment.
    e. Complex animals can occupy otherwise inhospitable habitats.
    *Answer: a*

63. Which of the following statements about muscle tissue is *false*?
    a. It consists of cells that can contract.
    b. It is the most abundant tissue in the body.
    c. It is a form of connective tissue.
    d. It uses a lot of energy when animals are active.
    e. It includes skeletal, smooth, and cardiac muscle.
    *Answer: c*

64. When scientists removed fish from a lake and measured their $Q_{10}$ in the lab, their prediction of the metabolic rate differed from the observed rate because
    a. the fish had acclimated to the temperature changes that occur in the lake.
    b. metabolic rate varies in different parts of the fish.
    c. fish do not produce metabolic compensation.
    d. in the lab, the internal temperature of the fish is higher than the surrounding environment.
    e. the $Q_{10}$ of a poikilotherm is often inconsistent.
    *Answer: a*

65. "Hot" fish, such as bluefin tuna, keep a higher temperature difference between their body and the surrounding water than cold fish do because they
    a. are endotherms.
    b. use a countercurrent heat exchange system of veins and arteries.
    c. use shivering to create more heat.
    d. have many brown fat tissues.
    e. have a large dorsal aorta that keeps them warmer.
    *Answer: b*

66. Your body responds to an infection by producing _____, which cause many of the symptoms of sickness such as fever and body aches.
    a. pyrogens
    b. interleukins
    c. prostaglandins
    d. Both a and b
    e. Both b and c
    *Answer: a*

67. As the environmental temperature increases, up to 25°C, the metabolic rate of an ectotherm _____ and that of an endotherm _____.
    a. increases; increases
    b. increases; decreases
    c. decreases; increases
    d. decreases; decreases
    e. stays the same; decreases
    *Answer: b*

# 42 Animal Hormones

## Fill in the Blank

1. In an emergency, the adrenal glands produce _____ immediately and then _____ for prolonged response to stress.
   **Answer: *epinephrine; cortisol***

2. _____ is a hormone that is produced from cholesterol in skin cells.
   **Answer: *Vitamin D***

★3. _____ mediates inflammation and is an example of a paracrine hormone.
   **Answer: *Histamine***

4. Hormonal and other systems in which production is shut off when the product becomes abundant are known as _____ feedback systems.
   **Answer: *negative***

5. _____ is a pituitary hormone that stimulates the production of milk.
   **Answer: *Prolactin***

6. A lobed gland located around the front of the windpipe is called the _____.
   **Answer: *thyroid***

7. Glands that lack ducts are called _____ glands.
   **Answer: *endocrine***

8. Overproduction or underproduction of _____ in children can cause gigantism or dwarfism.
   **Answer: *growth hormone***

9. Pituitary peptide hormones that affect other target glands are called _____ hormones.
   **Answer: *tropic***

10. Local or circulating hormones are capable of acting on target cells only if target cells are equipped with response components termed _____.
    **Answer: *receptors***

11. The hormone in *Rhodnius* that is produced by the corpora allata and prevents metamorphosis into an adult is called

    _____.
    **Answer: *juvenile hormone***

12. The _____ is an important endocrine gland derived embryonically from an outpocketing of the mouth region of the digestive tract and a downgrowth of the floor of the brain.
    **Answer: *pituitary gland***

13. The three classes of steroid hormones produced by the adrenal cortex are glucocorticoids, _____, and _____.
    **Answer: *mineralocorticoids; sex steroids***

14. Protein hormones do not cross the cell membrane, but _____ hormones do.
    **Answer: *steroid***

★15. When a hormone's receptor is on the secreting cell, it is said to have an _____ function.
    **Answer: *autocrine***

16. When people fail to get adequate amounts of iodine in their diet, they may develop goiter. The hormone that is nonfunctional is called _____.
    **Answer: *thyroxine***

17. If the circulating levels of calcium are high, _____ stimulates the osteoblasts to resorb it.
    **Answer: *calcitonin***

18. The steroid hormones are synthesized from _____.
    **Answer: *cholesterol***

## Multiple Choice

1. Which of the following is a local hormone?
   a. Adrenaline
   b. Estrogen
   c. Histamine
   d. Insulin
   e. Thyroxine
   **Answer: *c***

■2. The hormones of invertebrates
   a. are never secreted from glands in the head.
   b. are secreted only in adulthood.
   c. play no role in molting and metamorphosis.
   d. have different functions from those in vertebrates.
   e. require large quantities to have an effect.
   **Answer: *d***

■3. Which of Wigglesworth's observations of *Rhodnius* bugs helped describe the role of insect hormones?
   a. A blood meal triggers molting in these bugs.
   b. If decapitated immediately following a blood meal, the bug will molt.
   c. When two bugs are connected, they molt simultaneously.
   d. When two bugs are connected, the feeding status of one can trigger molting in the other.
   e. When two bugs are decapitated and connected, they will never molt into adults.
   **Answer: *d***

4. Insect brain hormone serves to
   a. stimulate the prothoracic gland to release the hormone that stimulates molting.
   b. stimulate the corpora cardiaca to release molting hormone.
   c. directly stimulate molting if food reserves are adequate.
   d. inhibit molting until the insect is a certain size.
   e. have a general inhibitory effect on insect growth.
   *Answer: a*

5. Which of the following hormones directly stimulates an insect larva to molt?
   a. Brain hormone
   b. Ecdysone
   c. Juvenile hormone
   d. Moltin
   e. Prolactin
   *Answer: b*

■6. Juvenile hormone be used by humans as an effective control of insect populations because it causes
   a. juvenile insects to die.
   b. the insects to fail to molt.
   c. the insects to fail to develop into adults.
   d. even tiny insects to pupate.
   e. insects to molt too quickly.
   *Answer: c*

7. The neurohormones vasopressin (antidiuretic hormone) and oxytocin are produced by the
   a. anterior pituitary and released by the posterior pituitary.
   b. hypothalamus and released by the posterior pituitary.
   c. pituitary and signal to the hypothalamus.
   d. hypothalamus and signal to the brain.
   e. pituitary and signal to the reproductive organs.
   *Answer: b*

8. Which of the following is an effect of oxytocin?
   a. Stimulation of uterine contractions at birth
   b. Increased reabsorption of water in the kidney
   c. Stimulation of tropic hormone release
   d. Increased productivity of the hypothalamus
   e. Increased rate of ovulation in the ovary
   *Answer: a*

★9. Which of these pairs correctly matches a pituitary hormone with its target organ?
   a. Oxytocin–oviducts
   b. Melanocyte-stimulating hormone–kidney
   c. Endorphins–brain
   d. Growth hormone–skin and hair
   e. Luteinizing hormone–thyroid
   *Answer: c*

10. The best source of pituitary hormones for medical uses today is
    a. slaughtered sheep or cattle.
    b. human cadavers.
    c. amino acids synthesized in the laboratory.
    d. genetically engineered bacteria.
    e. human blood samples.
    *Answer: d*

■11. Under which of these conditions would a mammal need to increase thyroxine levels?
    a. Following childbirth in a female
    b. During illness and fever
    c. When blood glucose levels are high
    d. During sleep and rest
    e. When exposed to cold
    *Answer: e*

12. Which of the following would signal a reduction in thyrotropin release?
    a. Increased levels of thyrotropin
    b. Decreased levels of thyrotropin
    c. Increased levels of thyroxine
    d. Decreased levels of thyroxine
    e. Decreased activity of the thyroid
    *Answer: c*

13. Which type of goiter (enlarged thyroid) can be reduced by addition of iodine to the diet?
    a. Hyperthyroid goiter, in which the thyroxine does not turn off the pituitary
    b. Hyperthyroid goiter, in which the thyroid is activated
    c. Hypothyroid goiter, involving low functional thyroxine and high thyrotropin
    d. Hypothyroid goiter, involving high functional thyroxine and low thyrotropin
    e. All types of goiter
    *Answer: c*

14. In addition to thyroxine, the mammalian thyroid gland produces
    a. adrenaline.
    b. calcitonin.
    c. iodine.
    d. prolactin.
    e. thyrotropin.
    *Answer: b*

15. The parathyroid glands are involved in regulation of blood levels of
    a. calcium.
    b. glucose.
    c. iodine.
    d. sodium.
    e. thyroxine.
    *Answer: a*

16. A lack of insulin cause diabetes mellitus because insulin is required for
    a. excretion of glucose.
    b. glucose breakdown.
    c. glucose uptake by cells.
    d. converting glucose to glycogen.
    e. synthesizing glucose.
    *Answer: c*

17. Which of the following is *not* produced by the pancreas?
    a. Cortisol
    b. Glucagon
    c. Insulin
    d. Somatostatin
    e. Digestive enzymes
    *Answer: a*

18. The adrenal medulla develops from nervous tissue and produces the hormone
    a. epinephrine.
    b. adrenocorticotropin.
    c. aldosterone.
    d. cholesterol.
    e. cortisol.
    *Answer: a*

19. Which of the following is *not* a function of cortisol?
    a. Metabolizing fats for energy
    b. Mediating response to stress
    c. Slowing down metabolism of glucose
    d. Stimulating the immune response
    e. Metabolizing proteins for energy
    *Answer: d*

20. Muscle-building anabolic steroids are known to cause all of the following effects *except*
    a. irregular menstrual periods in women.
    b. increased body and facial hair in women.
    c. increased body and facial hair in men.
    d. breast enlargement in men.
    e. kidney disease.
    *Answer: c*

21. The hormones secreted by the gonads are synthesized from
    a. complex carbohydrates.
    b. amino acids.
    c. cholesterol.
    d. hemoglobin.
    e. nucleic acids.
    *Answer: c*

22. What determines whether a developing mammalian gonad will become an ovary or a testis?
    a. Any gonad with cells containing the Y chromosome will become a testis.
    b. A steady high level of estrogens causes a gonad to become an ovary.
    c. The absence of estrogens causes a gonad to become a testis.
    d. The absence of androgens causes a gonad to become an ovary.
    e. Presence of a Y chromosome causes the release of androgens at a critical fetal stage, resulting in the development of testes.
    *Answer: e*

■23. Which of the following is the earliest event in puberty?
    a. The pituitary produces more gonadotropins.
    b. The level of circulating androgens rises in males.
    c. The menstrual cycle is initiated in females.
    d. The gonads differentiate into testes or ovaries.
    e. Subcutaneous fat increases in males.
    *Answer: a*

24. Vitamin D
    a. is synthesized by skin cells.
    b. becomes active after passing through the liver and kidneys.
    c. is lipid-soluble and thus can enter cells.
    d. is involved in the regulation of calcium.
    e. All of the above
    *Answer: e*

★25. Growth factors
    a. are paracrine hormones.
    b. can be found in blood plasma and tissue extracts.
    c. may act as autocrine messages during negative feedback.
    d. a, b, and c
    e. None of the above
    *Answer: d*

26. Which of the following statements about hormones is *false*?
    a. The same hormone can cause different responses in different types of cells.
    b. Hormone structure evolves more rapidly than hormone function does.
    c. The receptor for a hormone may be the secretory cell itself.
    d. Pheromones are chemical messages that influence other individuals of the same species.
    e. None of the above
    *Answer: b*

★■27. Which of the following statements about hormones is *false*?
    a. Target cells have the appropriate receptors to bind a particular hormone.
    b. The secretion, diffusion, and circulation of hormones is much slower than the transmission of nerve impulses.
    c. Hormones often control long-term physiological processes.
    d. All hormones travel in the blood to target cells.
    e. Both a and b
    *Answer: d*

★28. Steroid hormones
    a. initiate second messenger activity.
    b. bind with membrane proteins.
    c. alter gene expression.
    d. activate enzyme pathways.
    e. bind with membrane phospholipids.
    *Answer: c*

29. A _____ stimulus releases hormones that produce and coordinate major developmental, physiological, and behavioral changes in the male cichlid fish of Lake Tanganyika. The hormone transforms "wimpy" males into big, brightly colored, aggressive, sexually attractive "macho" males.
    a. chemical
    b. physical
    c. behavioral
    d. Both a and b
    e. a, b, and c
    *Answer: c*

30. Hormones are secreted by
    a. endocrine glands, such as the thyroid gland.
    b. individual cells, such as those lining portions of the digestive tract.
    c. cells in the nervous system (neurohormones).
    d. Both a and c
    e. a, b, and c
    *Answer: e*

■31. Which of the following is the chronological order of events in the molting of *Rhodnius*, as determined by Sir Vincent Wigglesworth?
    a. Blood meal, brain hormone release, ecdysone release
    b. Blood meal, ecdysone release, brain hormone release
    c. Ecdysone release, blood meal, brain hormone release
    d. Brain hormone release, blood meal, ecdysone release
    e. None of the above
    *Answer: a*

32. Hormones belong to a number of distinct chemical groups. Which of the following is *not* one of these groups?
    a. Steroids
    b. Proteins
    c. Amino acids and peptides
    d. Carbohydrates
    e. Modified fatty acids
    *Answer: d*

33. Which of the following hormones is *not* produced within the islets of Langerhans of the pancreas?
    a. Somatostatin
    b. Insulin
    c. Glucagon
    d. Calcitonin
    e. Both a and b
    *Answer: d*

34. Somatostatin
    a. is released when glucose and amino acids rise rapidly in the blood.
    b. inhibits the release of insulin and glucagon.
    c. slows digestive activity.
    d. is produced in the pancreas and hypothalamus.
    e. All of the above
    *Answer: e*

*35. What was the target for the brain hormone demonstrated by Sir Wigglesworth's experiments with *Rhodnius*?
    a. Prothoracic gland
    b. Pupal stage
    c. Larval stage
    d. Corpora cardiaca
    e. Corpora allata
    *Answer: a*

36. Hormones released by the posterior pituitary are produced in the
    a. anterior pituitary.
    b. hypothalamus.
    c. pineal gland.
    d. thymus.
    e. thyroid.
    *Answer: b*

37. Which of the following hormones is (are) produced by the posterior pituitary gland?
    a. Prolactin
    b. Oxytocin
    c. Endorphins
    d. Growth hormone
    e. Luteinizing hormone
    *Answer: b*

38. Tropic hormones control other endocrine glands. Which of the following is (are) *not* a tropic hormone?
    a. Thyrotropin
    b. Adrenocorticotropin
    c. Luteinizing hormone
    d. Enkephalins
    e. Follicle-stimulating hormone
    *Answer: d*

39. Which of the following is an effect of increased levels of vasopressin in the blood?
    a. Letdown of milk from mammary tissues
    b. Uterine contractions during birth
    c. Water conservation by the kidney
    d. Production of somatomedins by the liver
    e. Production of enkephalins by the hypothalamus
    *Answer: c*

40. Which of the following hormones is *not* paired correctly with its target organ?
    a. Oxytocin–adrenal glands
    b. Prolactin–mammary glands
    c. Endorphin–spinal cord neurons
    d. Vasopressin–kidneys
    e. Luteinizing hormone–gonads
    *Answer: a*

**41. Which of the following statements about growth hormone or its function is *false*?
    a. Growth hormone stimulates cells to take up amino acids.
    b. Growth hormone is not a tropic hormone.
    c. Pituitary dwarfism is now preventable.
    d. Neurohormones of the hypothalamus influence release of growth hormone.
    e. None of the above
    *Answer: b*

*42. The 1972 Nobel Prize in medicine was awarded to Roger Guillemin and Andrew Schally for discoveries relating to
    a. the cause of hypopituitary dwarfism.
    b. hypothalamic releasing factors.
    c. production of growth hormone using genetically engineered bacteria.
    d. the cause of acromegaly.
    e. the discovery of a cure for hypothyroidism.
    *Answer: b*

43. Malfunctioning of the thyroid gland in the human adult can result in
    a. cretinism.
    b. diabetes.
    c. dwarfism.
    d. goiter.
    e. hermaphroditism.
    *Answer: d*

44. Which of the following hormones is produced by the parathyroid glands?
    a. Parathormone
    b. Thyroxine
    c. Calcitonin
    d. Oxytocin
    e. Pitressin
    *Answer: a*

45–55. Match the hormones from the list below with the correct function or action.
    a. Insulin
    b. Glucagon
    c. Epinephrine
    d. Somatostatin
    e. Cortisol

45. First extracted by Frederick Banting and Charles Best
    *Answer: a*

46. Produced by the adrenal medulla
    *Answer: c*

47. Released in response to rapid rises of glucose and amino acids in the blood
    *Answer: d*

48. Responsible for the stimulation of cells to use glucose as a metabolic fuel and to convert excess glucose into fat and glycogen storage
    *Answer: a*

49. Responsible for the conversion of glycogen into glucose when serum glucose levels fall
    *Answer: b*

50. Inhibits the release of insulin and glucagon
    *Answer: d*

51. Slowly released in response to short-term stress
    *Answer: e*

52. Rapidly released in response to immediate stress
    *Answer: c*

53. Insufficient levels result in diabetes mellitus
    *Answer: a*

54. Inhibits the release of growth hormone
    *Answer: d*

55. Steroidal hormone synthesized from cholesterol
    *Answer: e*

**■56. Which of the following best describes the function of the corticosteroid hormones?
    a. Depresses the immune response
    b. Stimulates sexual and reproductive activity
    c. Influences blood glucose concentrations
    d. Influences ionic and osmotic concentration of the blood
    e. All of the above
    *Answer: e*

■57. Which of the following statements about water-soluble or lipid-soluble hormones is *false*?
    a. Lipid-soluble hormones pass readily through the plasma membrane.
    b. Lipid-soluble hormones act by stimulating the synthesis of new kinds of proteins through gene activation.
    c. Many water-soluble hormones normally require compounds such as cAMP and cGMP to function properly.
    d. Water-soluble hormones exert their effect by altering the activity of enzymes normally present in the target cell.
    e. Lipid-soluble hormone action is characterized by a cascade of regulatory steps resulting in an amplification of the effect of a single hormone molecule.
    *Answer: e*

58. Hormones produced by the adrenal cortex are
    a. water-soluble.
    b. proteins.
    c. steroids.
    d. carbohydrates.
    e. tropic hormones.
    *Answer: c*

59–62. Match the appropriate hormone from the list below with the correct gland, target organ, function, or action.
    a. Cholecystokinin
    b. Melatonin
    c. Aldosterone
    d. Testosterone
    e. Atrial natriuretic hormone

59. Increases sodium ion excretion
    *Answer: e*

60. Increases reabsorption of sodium ions
    *Answer: c*

61. An androgen
    *Answer: d*

62. Secreted by the pineal
    *Answer: b*

■63. Which of the following is *not* true of a hormone?
    a. One hormone can have different effects on different cells.
    b. Its actions are as fast as a neural impulse.
    c. It can act on the same cell that secretes it.
    d. It can enter a cell's nucleus.
    e. It is usually present in very small amounts.
    *Answer: b*

*64. Portal blood vessels connect the _____ to the _____.
    a. hypothalamus; brain
    b. hypothalamus; posterior pituitary
    c. hypothalamus; anterior pituitary
    d. anterior pituitary; posterior pituitary
    e. pancreas; liver
    *Answer: c*

65. The hormone responsible for releasing calcium from bone into the bloodstream is
    a. insulin.
    b. calcitonin.
    c. parathyroid hormone.
    d. thyroxine.
    e. somatostatin.
    *Answer: c*

66. Which of the following statements about sex steroids is *false*?
    a. Presence of the Y chromosome causes the embryonic gonads of mammals to produce androgens.
    b. Gonads produce hormones and gametes.
    c. Androgens are required to trigger male development in mammals and birds.
    d. The human embryo has the potential to develop into either a male or a female until about the seventh week of development.
    e. None of the above
    *Answer: c*

67. Which of the following statements about biological rhythms is *false*?
    a. Melatonin influences biological rhythms.
    b. Melatonin is released by the pineal gland during the day.
    c. Photoperiodicity is the phenomenon whereby seasonal changes in day length cause physiological changes in animals.
    d. Melatonin is inhibited by exposure to light.
    e. None of the above
    *Answer: b*

*68. Which of the following statements about regulation of hormone receptors is *false*?
    a. Upregulation of receptors is a negative feedback mechanism.
    b. Type II diabetes is characterized by a downregulation of insulin receptors.
    c. Downregulation is more common than upregulation.
    d. The abundance of receptors for a hormone can be under feedback control.
    e. None of the above
    *Answer: a*

69. In order for a cell to be responsive to a lipid-soluble hormone, it must have
   a. a specific cell surface receptor.
   b. a specific receptor in the cytoplasm or nucleus.
   c. cAMP.
   d. G protein.
   e. a specific DNA sequence to which the hormone binds.
   *Answer: b*

*70. Which of the following statements about responses to hormones is *false*?
   a. A dose-response curve quantifies response to a hormone.
   b. The half-life of epinephrine in the blood is a few minutes.
   c. Hormones are degraded by enzymes in the liver.
   d. Hormones that bind to carrier proteins have shorter half-lives than do those that circulate as free molecules.
   e. None of the above
   *Answer: d*

# 43 Animal Reproduction

## Fill in the Blank

1. A behavior that transfers sperm into the female's reproductive tract is _____.
   *Answer: copulation*

2. The uterus opens into the vagina at a muscular necklike region called the _____.
   *Answer: cervix*

3. A _____ consists of one egg cell and the surrounding ovarian cells.
   *Answer: follicle*

4. In humans, fertilization typically occurs when the egg is located in the _____.
   *Answer: oviduct*

5. Females of most mammals have a period of sexual receptivity called _____.
   *Answer: estrus*

6. The corpus luteum secretes the hormones _____ and _____.
   *Answer: progesterone; estrogen*

7. Following fertilization, human chorionic gonadotropin secreted by the _____ keeps the corpus luteum functional.
   *Answer: blastocyst (or embryo)*

8. The synthetic hormones in birth control pills cause the suspension of the ovarian cycle, which operates by means of feedback to the _____ and _____.
   *Answer: anterior pituitary; hypothalamus*

■9. A major evolutionary step allowing the first reptiles to succeed on land was the development of the _____.
   *Answer: amniote egg*

10. _____ is a reproductive technology that involves placing sperm in the correct location in the female reproductive tract for fertilization to occur. This technique is commonly used in the production of livestock.
    *Answer: Artificial insemination*

11. As opposed to viviparous animals, _____ animals lay eggs in the environment, and their offspring go through embryonic stages outside the body of the mother.
    *Answer: oviparous*

12. Very early in the life of a new mammalian embryo, a population of cells arises during the first few cell divisions. These special cells, called _____, ultimately end up in the region of the developing gonads, where they become gametes.
    *Answer: germ cells*

13. The _____ of the human male contains sperm and secretions from bulbourethral glands, seminal vesicles, and the prostate gland.
    *Answer: ejaculate (or semen)*

14. _____ is a specialized type of asexual reproduction in which offspring develop from unfertilized eggs.
    *Answer: Parthenogenesis*

15. Species with separate male and female members are described as being _____.
    *Answer: dioecious*

16. The male sexual response includes a _____ period following orgasm, in which he cannot achieve a full erection.
    *Answer: refractory*

17. The enzyme-containing cap at the front of the head of a sperm is the _____.
    *Answer: acrosome*

18. In mammals, the species-specific mechanism by which egg and sperm recognize one another is found in the _____.
    *Answer: zona pellucida*

## Multiple Choice

1. Which of the following is *not* an advantage of asexual reproduction?
   a. Only mitosis is necessary for cell division.
   b. Populations can grow until limited by resources.
   c. Single individuals can produce offspring.
   d. All the offspring are identical in a changing environment.
   e. No energy expenditure is required for mating and fertilization.
   *Answer: d*

■2. In order for an organism to reproduce by means of regeneration,
   a. each cell must be able to give rise to a new organism.
   b. a fragment must be broken off at a particular place.
   c. a fragment must contain all essential tissues.
   d. gamete formation must occur.
   e. Both a and b
   *Answer: c*

■3. Parthenogenetic reproduction involves
   a. meiosis but not fertilization.
   b. neither meiosis nor fertilization.
   c. fertilization but not meiosis.
   d. both meiosis and fertilization.
   e. identical copies of the same fertilized egg.
   *Answer: b*

■4. In populations that alternate between periods of asexual reproduction and periods of sexual reproduction, asexual reproduction is most advantageous
   a. when adults of both sexes are numerous.
   b. during the most environmentally harsh season.
   c. when there is a threat of hybridization.
   d. when conditions are favorable and stable.
   e. when the population is most dense.
   *Answer: d*

*5. In male gametogenesis, the second meiotic division produces four haploid
   a. germ cells.
   b. primary spermatocytes.
   c. secondary spermatocytes.
   d. spermatids.
   e. spermatogonia.
   *Answer: d*

*6. Unlike spermatogenesis, oogenesis in humans
   a. is continuous over the life of the woman.
   b. begins prenatally.
   c. produces four haploid gametes.
   d. occurs at a rapid rate.
   e. results in swimming cells.
   *Answer: b*

7. Hermaphrodites are organisms that
   a. possess both male and female reproductive systems.
   b. breed for a time as one sex, then breed as the other.
   c. develop offspring from unfertilized eggs.
   d. usually fertilize themselves.
   e. have abnormal reproductive organs.
   *Answer: a*

■8. Reproductive systems involving external fertilization are most common in
   a. terrestrial animals.
   b. populations with many more males than females.
   c. animals that are sessile.
   d. animals that produce few gametes.
   e. animals that are widely dispersed.
   *Answer: c*

9. For reproductive purposes, a penis is necessary for
   a. all male animals.
   b. all but hermaphrodites.
   c. species with external fertilization.
   d. species with internal fertilization.
   e. species in which courtship precedes mating.
   *Answer: d*

10. What is the physiological mechanism for erection of the penis in human males?
   a. Contraction of skeletal muscle
   b. Engorgement of spongy tissue
   c. Contraction of smooth muscles
   d. Emission of glandular fluid
   e. Accumulation of intercellular fluid
   *Answer: b*

*11. Which part of the testis produces the male sex hormones?
   a. The epididymis
   b. Leydig cells
   c. The scrotum
   d. The seminiferous tubules
   e. The vas deferens
   *Answer: b*

*12. Which of the following statements about mammals is *false*?
   a. All mammals have internal fertilization.
   b. Some species of mammals are oviparous.
   c. The uterine cycles of most mammals do not involve menstruation.
   d. Some species of mammals are ovoviviparous.
   e. None of the above
   *Answer: d*

13. A human female has the largest number of primary oocytes in her ovaries
   a. at birth.
   b. just prior to puberty.
   c. early in her fertile years.
   d. midway through her fertile years.
   e. at menopause.
   *Answer: a*

14. Following ovulation, what happens to the follicle cells if the egg is not fertilized?
   a. They move with the egg into the oviduct.
   b. They degenerate.
   c. They grow into a corpus luteum.
   d. They stop secreting hormones.
   e. They begin to develop a new egg.
   *Answer: b*

15. The egg is propelled through the oviduct by means of
   a. cilia on its surface.
   b. its flagellum.
   c. uterine contractions.
   d. cilia lining the oviduct.
   e. its amoeboid motion.
   *Answer: d*

■16. In what way does the human female differ from females of other mammalian species?
   a. She has fewer cycles per year.
   b. She reabsorbs the uterine lining following the completion of her uterine cycle.
   c. She must copulate in order to ovulate.
   d. She can be continuously sexually receptive.
   e. Her cycles are under hormonal control.
   *Answer: d*

17. Which two pituitary gonadotropins coordinate female reproductive events?
   a. Follicle-stimulating hormone and estrogen
   b. Estrogen and progesterone
   c. Follicle-stimulating hormone and progesterone
   d. Luteinizing hormone and progesterone
   e. Luteinizing hormone and follicle-stimulating hormone
   *Answer: e*

18. In the early half of the menstrual cycle, before ovulation, which hormone is at its highest level in the blood?
   a. Estrogen
   b. Follicle-stimulating hormone
   c. Luteinizing hormone

d. Progesterone
e. Testosterone
*Answer: a*

19. What event triggers ovulation and formation of the corpus luteum?
    a. A peak in estrogen
    b. A peak in luteinizing hormone
    c. A peak in progesterone
    d. Presence of sperm in the reproductive tract
    e. Readiness of the endometrial lining of the uterus
    *Answer: b*

*20. New follicles do not mature as long as the corpus luteum is maintained because
    a. the hormones released from the corpus luteum inhibit gonadotropin release.
    b. the hormones released from the corpus luteum inhibit ovarian secretion.
    c. the ovary receives negative feedback from the endometrial lining of the uterus.
    d. the ovary receives negative feedback as long as the egg is viable.
    e. new follicle development depends upon the hormones released from the corpus luteum.
    *Answer: a*

21. In the female sexual response, the structure that is most analogous to the male's penis is the
    a. breast.
    b. clitoris.
    c. labium.
    d. uterus.
    e. vagina.
    *Answer: b*

22. With a few exceptions, mammals are
    a. oviparous.
    b. ovoviviparous.
    c. viviparous.
    d. multiparous.
    e. marsupial.
    *Answer: c*

23. Which of the following statements about the prostate gland is *false*?
    a. Semen contains secretions from the prostate gland.
    b. Secretions from the prostate gland make the uterine environment more hospitable to sperm.
    c. Secretions from the prostate gland convert semen into a gelatinous mass.
    d. The prostate gland surrounds the urethra as it leaves the bladder.
    e. None of the above
    *Answer: e*

■24. Which sexually transmitted disease can be treated or cured with antibiotics?
    a. AIDS
    b. Hepatitis B
    c. Genital warts
    d. Genital herpes
    e. Chlamydia
    *Answer: e*

25. The intrauterine device is a means of birth control that prevents
    a. ovulation.

b. fertilization.
c. implantation.
d. ejaculation.
e. sperm from reaching the egg.
*Answer: c*

■26. Which of the following is *not* a part of asexual reproduction?
    a. Identical genetic makeup of parent and progeny
    b. Fertilization
    c. Parthenogenesis
    d. Regeneration
    e. Budding
    *Answer: b*

27. Sexual reproduction has an evolutionary advantage over asexual reproduction because it
    a. results in both males and females of a species.
    b. is a more lengthy process.
    c. promotes genetic variability to cope with changes in the environment.
    d. is controlled by many hormonal mechanisms.
    e. protects and nurtures the embryo.
    *Answer: c*

*28. Which of the following is the correct order of the structures through which sperm pass from the time they are produced to ejaculation?
    a. Vas deferens, seminiferous tubules, epididymis, urethra
    b. Epididymis, seminiferous tubules, vas deferens, urethra
    c. Seminiferous tubules, vas deferens, epididymis, urethra
    d. Seminiferous tubules, epididymis, vas deferens, urethra
    e. Vas deferens, urethra, epididymis, seminiferous tubules
    *Answer: d*

*29. Semen, which is the fluid matrix for sperm during emission and ejaculation, contains all of the following *except*
    a. seminal fluid from the seminal vesicles.
    b. glucose to serve as an energy source for the sperm.
    c. alkaline secretions from the prostate gland.
    d. hormone-like substances called prostaglandins.
    e. clotting enzymes.
    *Answer: b*

30. Which of the following is a probable advantage of hermaphroditism?
    a. For a given species, every organism is a potential mate.
    b. Simpler behavior patterns are required for mate selection.
    c. More rapid reproduction of genetically successful individuals is possible through asexual reproduction.
    d. Complex hormonal control and feedback mechanisms are not required for reproduction.
    e. Both b and c
    *Answer: a*

31–36. Choose the appropriate cell stage from the list below. There may be more than one correct answer.
    a. Oogonia
    b. Spermatogonia
    c. Spermatid
    d. Primary spermatocyte
    e. Secondary spermatocyte

*31. Germ cells from the female gonad
    *Answer: a*

*32. Germ cells of the male gonad
    *Answer: b*

*33. Cell resulting from the initial mitotic division of a spermatogonium
*Answer: d*

*34. First haploid germ cell produced in spermatogenesis
*Answer: e*

*35. Cell resulting from the second meiotic division
*Answer: c*

*36. At the end of spermatogenesis, the cell type that differentiates into a sperm cell
*Answer: c*

37–41. Choose the appropriate cell stage from the list below. There may be more than one correct answer.
  a. Second polar body
  b. Primary oocyte
  c. Secondary oocyte
  d. First polar body
  e. Ootid

*37. Stage during which the energy, raw materials, and RNA needed for the first cell divisions after fertilization are acquired
*Answer: b*

*38. First essential haploid germ cell produced during oogenesis
*Answer: c*

*39. Daughter cell to the secondary oocyte
*Answer: Both a and e*

*40. Largest haploid cell resulting from second meiotic division of oogenesis
*Answer: e*

*41. At the end of oogenesis, the cell type that differentiates into the mature ovum
*Answer: e*

42. After spermatogenesis, sperm cells are generally stored in the
  a. spermatophore.
  b. prostate gland.
  c. vas deferens.
  d. seminiferous tubule.
  e. epididymis.
  *Answer: e*

*43. Progeny inherit all the characteristics of a single parent through
  a. copulation.
  b. asexual reproduction.
  c. gametogenesis.
  d. sexual reproduction.
  e. fertilization.
  *Answer: b*

44. Generally, among vertebrates
  a. cells are haploid during embryological development and become diploid at birth.
  b. cells are diploid during embryological development and become haploid at birth.
  c. cells are diploid during adult life and are haploid as a blastocyst.
  d. only gametic cells are haploid.
  e. only gametic cells are diploid.
  *Answer: d*

45. Progesterone and estrogen are hormones produced by the
  a. anterior pituitary.
  b. posterior pituitary.
  c. Sertoli cells.
  d. hypothalamus.
  e. corpus luteum.
  *Answer: e*

46. The blastocyst normally implants within the
  a. vas deferens.
  b. ovarian follicle tissue.
  c. endometrium of the uterus.
  d. myometrium of the vagina.
  e. chorion.
  *Answer: c*

47. Which of the following describes the principal difference between menstrual and estrous cycles?
  a. The estrous cycle occurs if the female is fertilized, whereas the menstrual cycle occurs if the female isn't fertilized.
  b. The menstrual cycle occurs if the female is fertilized, whereas the estrous cycle occurs if the female isn't fertilized.
  c. The estrous cycle lacks menstruation.
  d. The menstrual cycle occurs in mammals, whereas the estrous cycle occurs in birds.
  e. Only the menstrual cycle is controlled by estrogen.
  *Answer: c*

48. Which of the following is the most effective form of birth control?
  a. Vasectomy
  b. Condom
  c. Intrauterine device
  d. Rhythm method
  e. Birth control pill
  *Answer: a*

49. A vasectomy is a minor operation that involves removing a small section of the _____ and tying off the loose ends.
  a. epididymis
  b. urethra
  c. oviduct
  d. vas deferens
  e. seminiferous tubules
  *Answer: d*

50. In most male mammals, which anatomical feature is shared by both the reproductive and excretory systems?
  a. Prostate gland
  b. Seminiferous tubules
  c. Vas deferens
  d. Urethra
  e. Epididymis
  *Answer: d*

51. In mammals, which is the correct sequence of structures through which a sperm passes before fertilizing an egg?
  a. Uterus, vagina, cervix, oviduct, ovary
  b. Vagina, cervix, uterus, oviduct
  c. Oviduct, uterus, vagina, ovary, body cavity
  d. Vagina, cervix, oviduct, uterus
  e. Vagina, uterus, oviduct, body cavity
  *Answer: b*

52. The hormone that normally triggers mammalian ovulation is
    a. progesterone.
    b. follicle-stimulating hormone.
    c. luteinizing hormone.
    d. estrogen.
    e. prolactin.
    *Answer: c*

*53. Which of the following statements about human reproduction is *false*?
    a. Women are most likely to be fertile from day 10 to day 20 of the ovarian cycle.
    b. Ovulation can be detected by tracking the woman's body temperature.
    c. An ovum remains viable for up to 36 hours after ovulation.
    d. Sperm may survive for up to 6 days in the female reproductive tract.
    e. None of the above
    *Answer: e*

54. Which birth control method prevents ovulation?
    a. Vasectomy
    b. Intrauterine device
    c. Condom
    d. Rhythm method
    e. Birth control pill
    *Answer: e*

55. Most methods of birth control focus on
    a. preventing oogenesis.
    b. preventing spermatogenesis.
    c. preventing fertilization.
    d. decreasing libido.
    e. decreasing testosterone levels.
    *Answer: c*

56. The end result of spermatogenesis is the production of four haploid spermatids. The end result of oogenesis is the production of
    a. four haploid ootids.
    b. two haploid ootids.
    c. one diploid ootid.
    d. two diploid ootids.
    e. one haploid ootid with two polar bodies.
    *Answer: e*

57. The terms "monoecious" and "dioecious" refer to
    a. patterns of inheritance.
    b. circulatory physiology.
    c. patterns of cleavage in developing embryos.
    d. the evolutionary origin of species.
    e. male and female reproductive systems.
    *Answer: e*

*58. At birth, a female has about one million primary oocytes in each ovary. During a woman's fertile years, about _____ of these oocytes will mature completely into eggs and be released at ovulation.
    a. 100
    b. 200
    c. 300
    d. 450
    e. None of the above
    *Answer: d*

59. Which of the following is *not* true of asexual reproduction?
    a. It occurs mostly in invertebrates.
    b. Asexually reproducing species tend to live in variable environments.
    c. Asexually reproducing species tend to be sessile.
    d. Asexually reproducing species live in sparse populations.
    e. One method of asexual reproduction is parthenogenesis.
    *Answer: b*

60. Which method of contraception also helps prevent the spread of sexually transmitted disease?
    a. Diaphragm
    b. Cervical cap
    c. Condom
    d. IUD
    e. Birth control pill
    *Answer: c*

61. The contragestational pill, RU-486, blocks _____ receptors.
    a. luteinizing hormone
    b. progesterone
    c. estrogen
    d. follicle-stimulating hormone
    e. gonadotropin-releasing hormone
    *Answer: b*

62. Pregnancy tests detect the presence of
    a. luteinizing hormone.
    b. follicle-stimulating hormone.
    c. human chorionic gonadotropin.
    d. estrogen.
    e. progesterone.
    *Answer: c*

*63. Which of the following methods of contraception prevents the egg from entering the uterus?
    a. Coitus interruptus
    b. Condom
    c. RU-486
    d. Diaphragm
    e. Tubal ligation
    *Answer: e*

*64. In which of the following procedures does fertilization occur within the female reproductive tract?
    a. In vitro fertilization
    b. Gamete intrafallopian transfer
    c. Intracytoplasmic sperm injection
    d. Both b and c
    e. None of the above
    *Answer: b*

65. Some species of mites and scorpions use indirect fertilization, excreting a gelatinous container of sperm called a(n)
    a. spermatophore.
    b. spermatogonium.
    c. spermatid.
    d. cloaca.
    e. amplexus.
    *Answer: a*

66. Which of the following represents the four phases of the human sexual response, in the correct order?
    a. Excitement, refractory, plateau, orgasm
    b. Plateau, excitement, orgasm, resolution
    c. Plateau, excitement, refractory, orgasm
    d. Excitement, plateau, orgasm, resolution
    e. Excitement, plateau, refractory, orgasm
    *Answer: d*

67. Gamete intrafallopian transfer is the process of
    a. removing eggs from a woman's oviduct and fertilizing them in a culture medium.
    b. moving eggs from the ovaries to the oviduct when the ovaries are blocked.
    c. injecting eggs and sperm into the oviduct when the oviduct entrance is blocked.
    d. injecting an embryo into the oviduct.
    e. removing an embryo from the oviduct when the oviducts are blocked and injecting it into the uterus.
    *Answer: c*

68. Which of the following statements about blocks to polyspermy in sea urchins is *false*?
    a. Blocks to polyspermy are triggered by fusion of the plasma membranes of the sperm and egg and entry of the sperm into the egg.
    b. Blocks to polyspermy prevent more than one sperm from entering an egg.
    c. The fast block causes release of the acrosomal enzymes.
    d. The fast block results from a change in electric charge difference across the plasma membrane of the egg.
    e. The slow block results from the release of calcium.
    *Answer: c*

# 44 Neurons and Nervous Systems

## Fill in the Blank

1. The part of the neuron specialized for receiving impulses is the _____.
   *Answer: dendrite*

2. The initial membrane event of an action potential is the flow of _____ ions across the membrane.
   *Answer: sodium*

3. Following an action potential, a neuron has a _____ during which it cannot be stimulated.
   *Answer: refractory period*

4. In myelinated axons of vertebrate neurons, breaks in the insulation occur at points called the _____.
   *Answer: nodes of Ranvier*

5. Deadly nerve gases inhibit the enzyme _____.
   *Answer: acetylcholinesterase*

★6. In vertebrates, the two most common inhibitory neurotransmitters are _____ and _____.
   *Answer: GABA; glycine*

7. The information that flows through the nervous system consists of chemical and _____ messages.
   *Answer: electrical*

8. When a neuron contacts another neuron, a muscle, or a gland, special junctions called _____ transmit the message carried by the incoming neuron.
   *Answer: synapses*

9. Special glial cells, called _____, surround the smallest, most permeable blood vessels in the brain, thereby participating in the formation of the blood–brain barrier.
   *Answer: astrocytes*

★10. The nicotinic receptors of acetylcholine are not metabotropic receptors but rather _____ receptors.
   *Answer: ionotropic*

11. The depolarization of a neuron must rise above the _____ before an action potential is achieved.
   *Answer: threshold*

12. In simple organisms such as the sea anemone, the nervous system consists of a _____.
   *Answer: nerve net*

★13. Neurotransmitters that depolarize the postsynaptic membrane bring about an _____ postsynaptic potential.
   *Answer: excitatory*

★14. Two gases, nitric oxide and _____, are used as intercellular messengers by neurons.
   *Answer: carbon monoxide*

## Multiple Choice

1. Neurons that transmit information from sensory cells to the central nervous system are part of the
   a. brain.
   b. peripheral nervous system.
   c. central nervous system.
   d. spinal cord.
   e. nerve net.
   *Answer: b*

2. The functions of glial cells include all of the following *except*
   a. supporting developing neurons.
   b. supplying nutrients.
   c. conducting nerve impulses.
   d. consuming foreign particles.
   e. insulating nerve tissue.
   *Answer: c*

3. About how many neurons are there in the human brain?
   a. 100 thousand
   b. 1 million
   c. 100 million
   d. 1 billion
   e. 100 billion
   *Answer: e*

■4. Electrical synapses
   a. do not integrate information well.
   b. are also called gap junctions.
   c. cannot be inhibitory.
   d. are a fast means of signal transmission.
   e. All of the above
   *Answer: e*

5. Which of the following describes the resting potential of the neuronal cell membrane?
   a. The inside is 60 millivolts more positive than the outside.
   b. The outside is 60 millivolts more positive than the inside.
   c. The inside is 30 millivolts more positive than the outside.
   d. The outside is 30 millivolts more positive than the inside.
   e. The inside has about the same charge as the outside.
   *Answer: b*

6. Which of the following can carry electric charges across the cell membrane?
   a. Electrons
   b. Protons
   c. Water
   d. Ions
   e. Proteins
   *Answer: d*

7. What generally maintains the electric charge across the neuronal membrane?
   a. Sodium–potassium pump
   b. Action potential
   c. Resting potential
   d. Voltage-gated channels
   e. Negative ion pump
   *Answer: a*

8. Which of the following describes the mechanism of voltage-gated channel proteins?
   a. If the membrane voltage reaches threshold potential, ions are pumped through.
   b. If the membrane voltage reaches threshold potential, ions can diffuse through.
   c. Ions can move through only if the overall membrane voltage stays the same.
   d. Ions are pumped through in order to maintain existing membrane voltage.
   e. When the gates close, membrane voltage changes.
   *Answer: b*

9. What happens if Na$^+$ channels open and sodium ions diffuse into the cell?
   a. The cell will become hyperpolarized.
   b. Other sodium ions will move out of the cell.
   c. Voltage-gated channels will remain closed.
   d. The charge across the nearby membrane will change.
   e. Action potentials will be triggered.
   *Answer: d*

10. Following depolarization, the neural membrane potential is restored when
    a. Na$^+$ ions rush outward through the membrane.
    b. K$^+$ ions rush outward through the membrane.
    c. Cl$^-$ ions rush inward through the membrane.
    d. a pump moves ions to their original concentrations.
    e. the membrane becomes freely permeable to many ions.
    *Answer: b*

•11. Which of the following would be recorded by an electrode as an action potential moves along an axon membrane?
    a. The inside membrane voltage becoming negative
    b. The membrane voltage changing permanently
    c. The action potential reversing its direction
    d. The action potential moving at a constant speed
    e. The height of the action potential remaining constant
    *Answer: e*

12. Nerves _____ have the most rapid action potentials.
    a. with the thinnest axon diameters
    b. with myelin sheaths
    c. of invertebrate animals
    d. with the greatest membrane potential
    e. with the most ion channels
    *Answer: b*

13. Saltatory conduction results when
    a. continuous propagation of the nerve impulse speeds up.
    b. a nerve impulse jumps from one neuron to another.
    c. the threshold for an action potential is suddenly increased.
    d. action potentials spread from node to node down the axon.
    e. the direction of an action potential suddenly changes.
    *Answer: d*

14. When an action potential arrives at the axon terminal, the voltage-gated calcium channels there
    a. release calcium into the synaptic cleft.
    b. actively transport neurotransmitter into the synaptic cleft.
    c. cause vesicles to release a neurotransmitter into the synaptic cleft.
    d. depolarize the membrane at the axon terminal.
    e. cause the membrane receptors to bind the neurotransmitter.
    *Answer: c*

15. When the neurotransmitter diffuses across the synaptic cleft,
    a. it automatically causes depolarization of the postsynaptic membrane.
    b. it can be excitatory or inhibitory, depending on the type of postsynaptic membrane.
    c. a single molecule is sufficient to trigger activation of the postsynaptic membrane.
    d. only a few molecules make it to the postsynaptic membrane.
    e. it must move through nodes in the myelin sheath.
    *Answer: b*

16. Neurotransmitters
    a. have multiple types of receptors.
    b. may be excitatory or inhibitory.
    c. may have different effects in different tissues.
    d. include dopamine and serotonin, which are monoamines.
    e. All of the above
    *Answer: e*

17. What is the effect of inhibiting acetylcholinesterase?
    a. Release of neurotransmitter from the presynaptic membrane is inhibited.
    b. Synthesis of neurotransmitter in cells is inhibited.
    c. Breakdown of neurotransmitter in the synapse is inhibited.
    d. Stimulation of the postsynaptic membrane is inhibited.
    e. Cholinergic receptors are inhibited.
    *Answer: c*

18. Some organisms, such as earthworms and squid, have clusters of neurons called
    a. the spinal cord.
    b. the central nervous system.
    c. the peripheral nervous system.
    d. ganglia.
    e. None of the above
    *Answer: d*

*19. Which of the following is a neurotransmitter that is *not* released synaptically?
    a. Glycine
    b. Norepinephrine
    c. Nitric oxide
    d. Adenosine
    e. None of the above
    *Answer: c*

20. Which of the following are the two primary cell types of the nervous system?
    a. Fibroblasts and chondrocytes
    b. Neurons and glial cells
    c. Epithelial cells and glandular cells
    d. Neurons and epithelial cells
    e. Neuromuscular cells and epithelial cells
    *Answer: b*

21. Synaptic clefts can be cleansed of neurotransmitters by means of
    a. enzymatic degradation.
    b. simple diffusion.
    c. active transport.
    d. Both a and c
    e. a, b, and c
    *Answer: e*

22. Which of the following ions is most responsible for generating an action potential?
    a. $Na^+$
    b. $K^+$
    c. $Cl^-$
    d. $H^+$
    e. $OH^-$
    *Answer: a*

23. Most nerve cells communicate with others by means of
    a. electric signals that pass across synapses.
    b. chemical signals that pass across synapses.
    c. bursts of pressure that "bump" the postsynaptic cell membrane.
    d. $Na^+$ ions as they are released from one cell and enter the next.
    e. None of the above
    *Answer: b*

24. The resting potential of a neuron is produced by
    a. voltage-gated channels in the membrane.
    b. chemically gated channels in the membrane.
    c. permanently open potassium channels in the membrane.
    d. the concentration difference in $Na^+$ across the membrane.
    e. blockage of the the sodium–potassium pump.
    *Answer: c*

25. The myelin sheath that surrounds some axons in the peripheral nervous system is formed by
    a. neurons.

    b. Schwann cells.
    c. bacteria that have invaded the nervous system.
    d. synapses.
    e. None of the above
    *Answer: b*

26. Which of the following limits the frequency at which a single neuron can "fire" action potentials?
    a. The number of synapses that the neuron forms
    b. The number of other cells that the neuron contacts
    c. The refractory period for the neuron's $Na^+$ channel
    d. The length of the axon of the neuron
    e. The number of dendrites on the neuron
    *Answer: c*

27. Which of the following can be said to be involved in triggering synaptic transmission?
    a. The action potential
    b. The opening of $Ca^{2+}$ channels at the synaptic terminal
    c. The entry of $Ca^{2+}$ into the presynaptic terminal
    d. Fusion of synaptic vesicles with the presynaptic membrane
    e. All of the above
    *Answer: e*

28. Neurons
    a. have a uniform shape throughout the nervous system.
    b. are more numerous than glial cells in the nervous system.
    c. are found in mammals and birds only.
    d. communicate with other cells at synapses.
    e. All of the above
    *Answer: d*

29. Dopamine is
    a. involved in schizophrenia.
    b. involved in Parkinson's disease.
    c. a neurotransmitter of the central nervous system.
    d. a monoamine.
    e. All of the above
    *Answer: e*

30. Narcotic drugs such as opium and heroin activate the receptors of
    a. substance P.
    b. endorphins.
    c. enkephalins.
    d. Both b and c
    e. a, b, and c
    *Answer: e*

*31. Long-term potentiation
    a. was discovered by neurobiologists working with brain slice preparations.
    b. involves an enhanced postsynaptic response.
    c. results from repeated stimulation of a presynaptic cell.
    d. involves activation of the NMDA receptor.
    e. All of the above
    *Answer: e*

32. The action potential
    a. begins with the membrane's increased permeability to potassium.
    b. returns to resting when the sodium channels open.
    c. can be triggered in very rapid succession, with no delay.
    d. involves voltage-gated channels in the membrane.
    e. propagates only because chloride ions move through the membrane.
    *Answer: d*

33. The action potential
    a. travels along all axons at the same speed.
    b. is slowed down if a nerve cell has myelin around it.
    c. is blocked at the nodes of Ranvier.
    d. causes a brief depolarization of the membrane potential.
    e. triggers a simultaneous change in potential along the entire axon.
    *Answer: d*

*34. Mice with modified NMDA receptors
    a. run through mazes more slowly than do normal mice.
    b. can remember mazes for longer periods of time than normal mice can.
    c. fail to learn tasks.
    d. remember mazes for shorter periods of time than normal mice do.
    e. a, c, and d
    *Answer: b*

■35. Choose the correct statement about the vertebrate nervous system.
    a. The nervous system consists of brain and spinal cord only.
    b. It can carry out many tasks at the same time.
    c. All nervous system functions are voluntary.
    d. Most information processing, storage, and retrieval occur outside the central nervous system.
    e. Ions are equally distributed across nerve cell membranes.
    *Answer: b*

36. Choose the statement about acetylcholine that is *false*.
    a. Acetylcholine is a neurotransmitter.
    b. Acetylcholine is found at mammalian neuromuscular junctions.
    c. Both smooth muscles and skeletal muscles respond to acetylcholine.
    d. Acetylcholine is degraded by acetylcholinesterase.
    e. Acetylcholine increases contractility of the heart.
    *Answer: e*

37. Which of the following statements about the nervous system is *false*?
    a. The nervous system is the most complex system of the human body.
    b. Oligodendrocytes cover the axons of neurons in the peripheral nervous system.
    c. Effectors are muscles or glands.
    d. Sensory cells transduce information into the electric signals that can be transmitted by neurons.
    e. Thousands of synapses impinge on most neurons.
    *Answer: b*

■38. Neurons
    a. fire action potentials only on the basis of the number of excitatory inputs they receive.
    b. sum excitatory and inhibitory postsynaptic potentials.
    c. make the "decision" to fire in the dendrites of the neuron.
    d. make a spatial summation, but not a temporal one.
    e. Both a and c
    *Answer: b*

39. Patch clamping
    a. fixes a break in a cell membrane.
    b. records electrical activity inside a cell.
    c. records ion movements through a single channel.
    d. records ion movements through the entire neuron.
    e. records an action potential through a single channel.
    *Answer: c*

40. When an action potential arrives at an axon terminal, it causes the opening of _____ channels, which triggers fusion of a neurotransmitter vesicle with the cell membrane.
    a. calcium
    b. sodium
    c. potassium
    d. chloride
    e. acetylcholine
    *Answer: a*

*41. Muscarinic receptors of acetylcholine
    a. are found in heart muscle.
    b. tend to be inhibitory.
    c. are an example of a metabotropic receptor.
    d. Both a and b
    e. a, b, and c
    *Answer: e*

*42. Long-term potentiation
    a. may be involved in learning and memory.
    b. involves a decreased postsynaptic response.
    c. involves AMPA receptors only.
    d. results from low frequency of stimulation.
    e. None of the above
    *Answer: a*

43. Which of the following neurotransmitters has been shown to be involved with sleep/wake cycles?
    a. Acetylcholine
    b. Norepinephrine
    c. GABA
    d. Serotonin
    e. Dopamine
    *Answer: d*

*44. Which of the following neurotransmitters is a peptide?
    a. Acetylcholine
    b. Norepinephrine
    c. Serotonin
    d. Glycine
    e. Endorphin
    *Answer: e*

45. A postsynaptic cell's processing of information from synapses at different sites is called
    a. excitatory postsynaptic potential.
    b. inhibitory postsynaptic potential.
    c. spatial summation.
    d. temporal summation.
    e. action potential.
    *Answer: c*

46. The most critical area in a neuron for "decision making" is the
    a. axon hillock.
    b. presynaptic terminal.
    c. postsynaptic terminal.
    d. cell body.
    e. synapse.
    *Answer: a*

*47. Some glial cells communicate with each other by means of
a. a myelin sheath.
b. axons.
c. dendrites.
d. gap junctions.
e. tight junctions.
*Answer: d*

48. Anesthetics and alcohol can permeate the blood–brain barrier because
a. they are small molecules.
b. they are water-soluble.
c. they are fat-soluble.
d. they pass through gated channels.
e. there are receptors for them on blood vessels.
*Answer: c*

49. The sodium–potassium pump
a. needs energy to work.
b. moves potassium ions to the inside of a neuron and sodium ions to the outside.
c. works against a concentration gradient.
d. Both a and b
e. a, b, and c
*Answer: e*

50. Some medications that elevate mood and relieve anxiety
a. enhance the activity of serotonin at the synapse.
b. slow the reuptake of serotonin.
c. increase endorphins.
d. Both a and b
e. a, b, and c
*Answer: d*

# 45 Sensory Systems

## Fill in the Blank

1. _____ are chemical signals used by individuals to communicate with other individuals of their species.
   *Answer: Pheromones*

2. Fish can detect pressure waves in water through their _____ sensory system.
   *Answer: lateral line*

3. The three tiny bones of the middle ear transmit vibrations from the _____ to the oval window.
   *Answer: tympanic membrane*

4. The inner ear is a long, coiled structure called the
   _____.
   *Answer: cochlea*

5. The molecule _____ is the basis for photosensitivity.
   *Answer: rhodopsin*

6. A dense layer of photoreceptor cells at the back of the vertebrate eye forms the _____.
   *Answer: retina*

7. Besides vertebrates, the group of animals that has eyes that form images like cameras is the _____ group.
   *Answer: cephalopod mollusk*

*8. Mammals can alter the shape of the lens by contracting the _____.
   *Answer: ciliary muscles*

*9. Just prior to transmission to the brain via the optic nerve, visual signals are processed by _____.
   *Answer: ganglion cells*

10. The phenomenon of emitting sounds and creating images from reflections of those sounds is called _____.
    *Answer: echolocation*

■11. _____ arriving in the visual cortex are interpreted as light; in the auditory cortex as sound; and in the olfactory cortex as smell.
    *Answer: Action potentials*

12. An important characteristic of many sensory cells is that they can stop being excited by a stimulus that initially caused them to be active. This ability to ignore background or unchanging conditions while remaining sensitive to changes or new information is called _____.
    *Answer: adaptation*

13. The snake's forked tongue fits into cavities in the roof of its mouth that are richly endowed with olfactory sensors. Thus the snake is really using its tongue to _____ its environment.
    *Answer: smell*

14. A common cause of _____ is cumulative and permanent damage to the hair cells of the organ of Corti.
    *Answer: nerve deafness (or deafness)*

15. The most photosensitive area of the retina is the _____.
    *Answer: fovea*

16. A change in the resting membrane potential of a sensory cell in response to a stimulus is called a _____ potential.
    *Answer: receptor*

17. Hair cells have a set of projections called _____ that look like organ pipes.
    *Answer: stereocilia*

18. Many animals can focus sounds by moving their ear _____ toward the sound.
    *Answer: pinnae*

19. The _____ in mammals is located on the septum dividing the two nostrils and functions in detecting pheromones.
    *Answer: vomeronasal organ*

## Multiple Choice

1. Sensory cells transduce physical or chemical stimuli
   a. from one form to another.
   b. into some type of membrane potential.
   c. from an action potential into a synaptic signal.
   d. by summing incoming action potentials.
   e. into different forms to be sent to the brain.
   *Answer: b*

■2. The magnitude of a receptor potential
   a. depends on the strength of the incoming action potential.
   b. remains high even after a long period of stimulation.
   c. is the same no matter what the type of stimulus.
   d. depends on the amount of neurotransmitter released.
   e. affects the frequency of resulting action potentials.
   *Answer: e*

3. Which of the following is an example of adaptation of sensory cells?
   a. Going into deep sleep
   b. Discriminating different colors
   c. Ignoring your shoes as you walk
   d. Ignoring a boring lecture
   e. Detecting sound and light simultaneously
   *Answer: c*

4. Pheromones are chemical signals that can signal, for example, from _____ to _____.
   a. one neuron; another
   b. the peripheral nervous; the central nervous system
   c. prey; predator
   d. parasite; host
   e. female; male within a species
   *Answer: e*

5. The chemosensory hairs that flies use to detect and taste food are located
   a. near the mouthparts.
   b. at the base of the wings.
   c. on the tip of the proboscis.
   d. on the feet.
   e. on the antennae.
   *Answer: d*

6. A male silkworm moth locates a female at a distance by
   a. flying toward a chemical signal.
   b. flying toward a sound signal.
   c. flying toward anything shaped like a female moth.
   d. emitting a sound as the female approaches.
   e. emitting a chemical as the female approaches.
   *Answer: a*

7. Sensitivity of the sense of smell is proportional to the
   a. amount of mucus in the nose.
   b. surface area of nasal epithelium.
   c. density of olfactory nerve endings.
   d. number of capillaries in the nose.
   e. typical body temperature.
   *Answer: c*

■8. The greatest intensity of perceived smell comes from the
   a. enzyme that binds with the most odorant molecules.
   b. odorant that binds to the most receptors.
   c. greatest variety of odorant molecules.
   d. greatest threshold of depolarization.
   e. greatest number of odorant molecules entering the cell.
   *Answer: b*

9. Which of the following are *not* mechanoreceptors?
   a. Stretch receptors
   b. Hair cells
   c. Pressure receptors
   d. Olfactory receptors
   e. Airflow receptors
   *Answer: d*

10. Action potentials are generated in a mechanoreceptor when
    a. ion channels close in response to membrane distortion.
    b. ion channels open in response to membrane distortion.
    c. receptors bind chemicals in response to pressure.
    d. sensitivity of the membrane to neurotransmitters increases.
    e. signals from other mechanoreceptors are summated.
    *Answer: b*

11. Which of the following is most descriptive of the Pacinian corpuscle?
    a. It has high two-point discrimination ability.
    b. It responds to steady pressure for a long time.
    c. It has concentric layers and is rapidly adapting.
    d. It has long, extensive dendritic processes.
    e. It is extremely sensitive to light touch.
    *Answer: c*

12. When the stereocilia of hair cells are bent, the hair cells then
    a. trigger muscle contraction.
    b. release neurotransmitter.
    c. undergo action potentials.
    d. contract their stereocilia.
    e. become less sensitive.
    *Answer: b*

*13. Stereocilia in hair cells in the canals of the fish lateral line
    a. are embedded in gelatinous material.
    b. are moved individually by pressure.
    c. are immobilized within a cupula.
    d. bend only after electrical stimulation.
    e. bend under the influence of gravity.
    *Answer: a*

14. Which of the following is sensed by elephants but not humans?
    a. Infrared light
    b. Electric fields
    c. Ultraviolet light
    d. Infrasound
    e. Color
    *Answer: d*

15. Vertebrates rely on information from which sensory structure to keep their balance?
    a. Eustachian tube
    b. Otoliths
    c. Semicircular canal
    d. Statocyst
    e. Tympanic membrane
    *Answer: c*

16. The auditory function of the middle ear is to convert _____ pressure waves into _____.
    a. air; fluid pressure waves
    b. fluid; air pressure waves
    c. air; nerve impulses
    d. fluid; nerve impulses
    e. air; hair cell movements
    *Answer: a*

17. Hair cells in the ear that give auditory information are concentrated in which of the following structures?
    a. Oval window
    b. Tympanic membrane
    c. Organ of Corti
    d. Semicircular canals
    e. Pinnae
    *Answer: c*

*18. What is the physiological basis for the auditory system's ability to distinguish different sound frequencies?
    a. The three bones of the middle ear respond differentially.
    b. The loops of the semicircular canals respond differentially.
    c. The oval window and round window respond differentially.
    d. Different sections of the basilar membrane respond differentially.
    e. Individual hair cells have different peak frequency responses.
    *Answer: d*

*19. The molecular mechanism by which light is absorbed into visual systems is
   a. shape change in the protein opsin.
   b. depolarization of the rhodopsin molecule.
   c. isomerization of the molecule retinal.
   d. oxidation of the rhodopsin molecule.
   e. None of the above
   *Answer: c*

*20. When an individual rod cell is stimulated with light, its membrane potential
   a. becomes more negative.
   b. becomes more positive.
   c. becomes more positive than other neurons are.
   d. begins to generate action potentials.
   e. begins to reduce membrane polarization.
   *Answer: a*

*21. The activation of a rhodopsin molecule sets off a chain reaction that leads to the
   a. opening of a sodium channel.
   b. closing of a large number of sodium channels.
   c. formation of an activated phosphodiesterase molecule.
   d. activation of a large number of transducin molecules.
   e. activation of other rhodopsin molecules.
   *Answer: b*

22. Arthropods have evolved compound eyes consisting of large numbers of
   a. retinas.
   b. cones.
   c. eye cups.
   d. ommatidia.
   e. pupils.
   *Answer: d*

23. The amount of light entering the eye is decreased when
   a. an optician puts in atropine drops.
   b. the shape of the lens changes.
   c. the autonomic nervous system opens the pupil.
   d. the light is focused.
   e. the iris constricts.
   *Answer: e*

■24. Which of the following does *not* affect the focus of an image on the retina?
   a. The shape of the lens
   b. The shape of the retina
   c. The ligaments suspending the lens
   d. The ciliary muscles
   e. The elasticity of the lens
   *Answer: b*

25. What causes the blind spot in the eye?
   a. An unusually high density of rod cells
   b. An unusually high density of cone cells
   c. The focal spot of incoming light
   d. Lack of photoreceptors where the optic nerve leaves the eye
   e. Saturated photoreceptors
   *Answer: d*

26. Which of the following is a difference between rods and cones?
   a. Cones are more sensitive at low light intensity.
   b. Rods are responsible for color vision.
   c. There are more cones than rods in the human retina.

   d. Strictly nocturnal animals have more cones than rods.
   e. Cones provide greatest acuity of vision.
   *Answer: e*

27. Which of the following best describes the visual processing that occurs within the retina?
   a. Impulses are processed within individual photoreceptors and then travel to the brain.
   b. Light signals are processed in retinal cells immediately in front of the photoreceptors.
   c. Signals pass directly from photoreceptor cells to ganglion cells.
   d. Action potentials in photoreceptor cells affect bipolar cells before reaching ganglion cells.
   e. Membrane potential changes in a network of retinal cells activate ganglion cells.
   *Answer: e*

28. Which of the following statements about echolocation is *false*?
   a. Porpoises and dolphins use echolocation.
   b. Humans cannot hear the pulses of sound emitted by echolocating bats.
   c. Bats have muscles in their middle ears that decrease auditory sensitivity during emission of echolocation cries.
   d. Bats use echolocation in foraging and navigation.
   e. None of the above
   *Answer: e*

*■29. Which of the following describes the receptive field of a retinal ganglion cell?
   a. There is no overlap with the receptive fields of neighboring ganglion cells.
   b. The ganglion cell receives input from one rod or cone cell.
   c. The receptive field depends on connections of horizontal and bipolar cells.
   d. A signal at the edge of the receptive field is more significant than one from the center.
   e. The receptive field is defined by signals from neighboring ganglion cells.
   *Answer: c*

30. Which of these statements about animal sensitivity is true?
   a. Humans and snakes can detect ultraviolet rays, but insects cannot.
   b. Humans and snakes can detect infrared rays, but insects cannot.
   c. Humans and insects can detect infrared rays, but snakes cannot.
   d. Humans cannot detect ultraviolet rays, but insects can.
   e. Humans cannot detect ultraviolet rays, but snakes can.
   *Answer: d*

31. In general, _____ are cells of the nervous system that transduce physical or chemical stimuli into signals that are transmitted to other parts of the nervous system for processing and interpretation.
   a. sensory cells
   b. effectors
   c. glial cells of the blood–brain barrier
   d. nuclei within the midbrain
   e. None of the above
   *Answer: a*

*32. Which of the following statements regarding the stimulation of receptor proteins within plasma membranes is true?
   a. The receptor proteins of mechanoreceptors, thermoreceptors, and electroreceptors are themselves the ion channels.
   b. The receptor proteins of chemoreceptors and photoreceptors initiate biochemical cascades that eventually open and close ion channels.
   c. Receptor proteins are integral to the sensory process.
   d. Both and b
   e. a, b, and c
   *Answer: e*

33. Receptor potentials produce action potentials in two ways: by generating action potentials within the sensory cells, or by causing the release of _____, which induces an associated neuron to generate action potentials.
   a. a hormone
   b. ATP
   c. interleukin
   d. a neurotransmitter
   e. glucagon
   *Answer: d*

34. Electroreceptors may be associated with the _____ system.
   a. lateral line
   b. visual
   c. auditory
   d. gustatory
   e. None of the above
   *Answer: a*

35. Which of the following statements about sensory cells is *false*?
   a. Most sensory cells are modified neurons.
   b. Specific sensory cells can respond to all types of stimuli.
   c. Changes in the stimulus strength lead to changes in the sensory cell's receptor potential.
   d. Sensory cells display a phenomenon called adaptation.
   e. The way in which information from sensory cells is interpreted depends on where the information is sent.
   *Answer: b*

36. Which of the following is *not* a necessary part of the definition of a sensory cell?
   a. It is specialized for detecting specific kinds of stimuli.
   b. It transduces energy into action potentials.
   c. It causes the opening of sodium channels in the membrane.
   d. It has a receptor potential.
   e. It can become insensitive to a source of continuous stimulation.
   *Answer: c*

37. As a male silkworm moth nears a female that is releasing bombykol,
   a. the female starts to release more bombykol.
   b. the action potentials in the antennal nerve increase.
   c. many bombykol-sensitive hairs are being stimulated per second.
   d. a larger number of the bombykol-sensitive hairs undergo adaptation.
   e. the receptor potential in bombykol-sensitive hairs is reduced.
   *Answer: b*

38. Which of the following events is triggered by the binding of an odorant molecule to a receptor protein?
   a. Opening of sodium channels
   b. Increase of second messenger in cytoplasm of sensory cell
   c. Activation of G protein
   d. Depolarization of sensory cell
   e. All of the above
   *Answer: e*

39. Which of the following statements about gustation is *false*?
   a. The taste buds of all vertebrates are confined to the oral cavity.
   b. Stimulated taste sensory cells respond by releasing neurotransmitter.
   c. Microvilli increase the surface area of taste sensory cells.
   d. Individual taste buds are replaced every few days.
   e. Taste sensory cells form synapses with dendrites of sensory neurons.
   *Answer: a*

40. Which of the following statements about mechanoreceptors is *false*?
   a. Mechanoreceptors transduce mechanical forces into changes in receptor potential.
   b. If a mechanoreceptor is subject to increased distortion, more ion channels within its membrane open.
   c. Usually, sensory nerves are not involved in circuits returning from mechanoreceptors.
   d. If the receptor potential of a mechanoreceptor rises above a threshold, an action potential is propagated.
   e. Stimulus strength determines the rate of generated action potentials.
   *Answer: c*

41. Which of the following receptors are located deep in the skin and are adapted specifically for sensing pressure?
   a. Meissner's corpuscles
   b. Pacinian corpuscles
   c. Expanded-tip tactile receptors
   d. Neuron-wrapped hair follicles
   e. Bare nerve endings
   *Answer: b*

42. Hair cells are associated with all of the following sensory systems *except* for the
   a. Golgi tendon organ.
   b. lateral line.
   c. skin sensation.
   d. vestibular apparatus.
   e. semicircular canals.
   *Answer: a*

*43. Which of the following statements about the lateral line system is true?
   a. The cupulae of the lateral line system each contain several hair cells.
   b. The lateral line system allows the fish to sense the presence of other fish.
   c. The lateral line system responds to pressure waves in the surrounding water.
   d. The canal of the lateral system has numerous openings to the external environment.
   e. All of the above
   *Answer: e*

44. Which of the following structures of the mammalian auditory system is responsible for signal amplification?
    a. Tympanic membrane
    b. Ear ossicles
    c. Oval window
    d. Cochlea
    e. Round window
    *Answer: b*

*45. Which of the following statements about the auditory functioning of the cochlea is *false*?
    a. The flexing of the round window follows the flexing of the oval window in a delayed fashion.
    b. The hair cells on the organ of Corti are moved against the rigid tectorial membrane.
    c. The intensity of the sound determines how many hair cells will be stimulated.
    d. The frequency of the sound determines which hair cells will be stimulated.
    e. Lower frequency sounds result in the stimulation of hair cells closer to the round window.
    *Answer: e*

46. Which of the following structures of the mammalian auditory system is involved in transduction of pressure changes into action potentials?
    a. Tympanic membrane
    b. Ear ossicles
    c. Oval window
    d. Organ of Corti
    e. Basilar membrane
    *Answer: d*

*47. Which of the following statements about rhodopsin is *false*?
    a. Rhodopsin consists of a protein and a light-absorbing molecule.
    b. When 11-*cis*-retinal absorbs light, it becomes all-*trans*-retinal.
    c. In vertebrate eyes, the retinal and the opsin never separate from each other.
    d. Rhodopsin is a transmembrane protein.
    e. All animals that can sense light do so using rhodopsin.
    *Answer: c*

48. Which of the following statements about the functioning of a rod cell is *false*?
    a. Rhodopsin is located in the stack of disks that is farthest from the light source.
    b. A rod cell is a modified neuron.
    c. The resting potential of a rod cell in the dark is less negative than a typical neuron.
    d. The membrane potential of a rod cell exposed to light becomes more positive.
    e. The plasma membrane of a rod cell is fairly permeable to $Na^+$ ions.
    *Answer: d*

*49. Which of the following statements about the molecular events of photoreception is *false*?
    a. A single photon can excite a rhodopsin molecule.
    b. In a well-lit setting, a rod cell has most of its sodium channels open.
    c. cGMP keeps the sodium channels open.

    d. Activated phosphodiesterase (PDE) catalyzes the reaction hydrolyzing cGMP into GMP.
    e. Activated transducin activates PDE.
    *Answer: b*

■50. If a flatworm is positioned relative to a stationary light source so that more light-sensitive cells in its right eye cup are stimulated than those in its left eye cup, the planarian will
    a. turn to the left.
    b. turn to the right.
    c. make a complete clockwise circle.
    d. make a complete counterclockwise circle.
    e. stop moving.
    *Answer: a*

51. Which of the following statements about the compound eyes of arthropods is *false*?
    a. Ommatidia are the optical units of compound eyes.
    b. The number of ommatidia per eye can vary greatly in different species.
    c. The eyes of cephalopods are similar to the compound eyes of arthropods.
    d. The compound eye communicates a relatively crude image to the central nervous system.
    e. Each ommatidium has a lens structure that directs light onto photoreceptors.
    *Answer: c*

52. Which of the following statements about the functioning of the vertebrate eye is *false*?
    a. The cornea is transparent so that it can transmit light.
    b. The size of the pupil varies with light levels.
    c. The iris is under control of the autonomic nervous system.
    d. The lens focuses the image on the retina.
    e. The fovea is the part of the retina with the lowest density of photoreceptors.
    *Answer: e*

**53. Which of the following statements about accommodation is *false*?
    a. Accommodation is the process by which objects from different portions of the visual field are focused on the retina.
    b. Vertebrates such as fish and reptiles accommodate by moving the lens relative to the retina.
    c. The suspensory ligaments keep the lens flattened.
    d. When a person attempts to focus on a distant object, the ciliary muscles contract.
    e. The ciliary muscles change the shape of the eye by counteracting the action of the suspensory ligaments.
    *Answer: d*

54. Which of the following statements about vertebrate vision is *false*?
    a. Visual acuity varies according to the density of photoreceptors in the retina.
    b. Some vertebrates have two foveas per eye.
    c. A blind spot is always located where the optic nerve leaves the eye.
    d. Unlike a camera, the vertebrate eye does not project inverted images on the retina.
    e. Two major types of photoreceptor cells are found in the retina.
    *Answer: d*

55. Which of the following statements about the vertebrate retina is *false*?
    a. There are more rods than cones in the retinas of humans.
    b. There is a higher proportion of cones than rods in the foveas of humans.
    c. Cones give us our highest visual acuity.
    d. Our peripheral vision involves more rod cells than cone cells.
    e. Some entirely nocturnal animals have only cones in their retinas.
    *Answer: e*

*56. Which of the following statements about color vision is *false*?
    a. There are three different types of cone cells in the human retina.
    b. The absorption spectra of cone cells differ because of molecular differences in the retinal molecules.
    c. Rods do not contribute to color vision.
    d. The different opsin molecules in the human retina differ according to the wavelength of light they absorb best.
    e. Nocturnal animals often have poor color vision.
    *Answer: b*

*57. Which of the following cell layers occurs farthest back in the retina?
    a. Horizontal
    b. Photoreceptors
    c. Bipolar
    d. Amacrine
    e. Ganglion
    *Answer: b*

58. Which of the following cell layers of the retina is responsible for producing action potentials that are sent to the brain?
    a. Horizontal
    b. Pigmented
    c. Bipolar
    d. Amacrine
    e. Ganglion
    *Answer: e*

■59. Which of these is the correct order of the flow of information?
    a. Stimulus, ion channel, action potential, receptor protein, neurotransmitter release
    b. Stimulus, neurotransmitter release, action potential, ion channel, receptor protein
    c. Stimulus, receptor protein, ion channel, neurotransmitter release, action potential
    d. Stimulus, action potential, neurotransmitter release, receptor protein, ion channel
    e. Stimulus, receptor protein, action potential, neurotransmitter release, ion channel
    *Answer: c*

60. Chemoreceptors
    a. are universal among animals.
    b. can cause strong behavioral responses.
    c. do not undergo adaptation.
    d. Both a and b
    e. a, b, and c
    *Answer: d*

61. Meissner's corpuscles
    a. adapt very slowly.
    b. are present uniformly on skin surfaces.
    c. sense pressure.
    d. have concentric layers of connective tissue.
    e. sense light touch.
    *Answer: e*

62. The Golgi tendon organ
    a. causes muscles to relax and protects against tearing.
    b. senses light touch.
    c. increases muscle contraction.
    d. is found in high densities on lips and fingertips.
    e. provides steady-state information about pressure.
    *Answer: a*

■63. Which of the following statements is *not* true of receptor potentials?
    a. They can cause the release of a neurotransmitter.
    b. They can cause an action potential.
    c. They are a change in membrane potential of the sensory cell.
    d. They can spread over long distances.
    e. They can be amplified.
    *Answer: d*

*64. Conduction deafness is caused by loss of function of
    a. the inner ear.
    b. the eustachian tube.
    c. the tympanic membrane and the middle ear ossicles.
    d. hair cells in the organ of Corti.
    e. the cochlea.
    *Answer: c*

# 46 The Mammalian Nervous System: Structure and Higher Functions

## Fill in the Blank

1. The basic functional unit of the brain is the _____.
*Answer: neuron*

2. The brain and spinal cord together are called the _____.
*Answer: central nervous system*

3. The _____ is made up of a network of nerves throughout the body.
*Answer: peripheral nervous system*

4. Vision, hearing, touch, and balance make up _____ information.
*Answer: afferent*

5. _____ information is sent from the brain to the muscles and glands.
*Answer: Efferent*

6. Efferent pathways that are involuntary are also called the _____ division.
*Answer: autonomic*

■7. The nervous system can engage in many tasks at the same time. This ability is called _____ processing.
*Answer: parallel*

8. A spinal _____ occurs when afferent information is converted to efferent information without involvement of the brain.
*Answer: reflex*

9. A group of neurons that is anatomically or neurochemically distinct is called a _____.
*Answer: nucleus*

10. The transfer of short-term memory to long-term memory is the function of the _____.
*Answer: hippocampus*

★11. The cortex is folded into ridges called _____ and valleys called _____.
*Answer: gyri; sulci*

★■12. A person who is blind in one eye will have difficulty discriminating _____.
*Answer: distance*

13. Sleep researchers use an _____ to measure electric potential differences between neurons.
*Answer: electroencephalograph*

14. The process by which experiences modify behavior is called _____; the ability of the brain to retain this information is called _____.
*Answer: learning; memory*

15. A response that comes about through the linking of two unrelated stimuli is called a _____ reflex.
*Answer: conditioned*

16. Remembering how to perform a motor skill, such as using a computer keyboard, involves _____ memory. This type of memory cannot be consciously recalled and described.
*Answer: procedural*

★17. A deficit in the ability to use or understand words is called an _____.
*Answer: aphasia*

18. The autonomic nervous system is crucial to the maintenance of _____ in the body.
*Answer: homeostasis*

19. The _____ is an evolutionarily more recent part of the telencephalon found in birds and mammals.
*Answer: neocortex*

20. In the visual cortex, cells that receive information from both eyes are called _____ cells.
*Answer: binocular*

## Multiple Choice

1. An exciting new finding about nervous systems concerns
a. the formation of new neurons in the brains of adult birds.
b. the formation of new neurons in the brains of adult mice.
c. the persistent decline in the number of neurons throughout life.
d. evidence that all neurons in the adult human brain are at the same maturational stage.
e. Both a and b
*Answer: e*

2. Afferent information flows _____ the CNS, and efferent information flows _____ the CNS.
a. to; to
b. to; from
c. from; to
d. from; from
e. from and to; to
*Answer: b*

3. The hindbrain develops into which structure?
   a. Medulla
   b. Pons
   c. Cerebellum
   d. All of the above
   e. None of the above
   *Answer: d*

4. The forebrain develops into which structure?
   a. Telencephalon
   b. Diencephalon
   c. Cerebellum
   d. Both a and b
   e. a, b, and c
   *Answer: d*

5. The thalamus and hypothalamus develop from which structure?
   a. Telencephalon
   b. Diencephalon
   c. Cerebrum
   d. Cerebellum
   e. Hindbrain
   *Answer: b*

*6. Select the correct direction of the flow of afferent information.
   a. Medulla, pons, midbrain, thalamus
   b. Medulla, pons, thalamus, midbrain
   c. Telencephalon, thalamus, pons, medulla
   d. Telencephalon, midbrain, pons, medulla
   e. Pons, medulla, thalamus, midbrain
   *Answer: a*

7. In general, the more autonomic functions are found in the _____, and the more complex functions are found in the _____.
   a. forebrain; hindbrain
   b. telencephalon; diencephalon
   c. thalamus; hypothalamus
   d. midbrain; hindbrain
   e. hindbrain; forebrain
   *Answer: e*

8. The largest difference between the brains of humans and the brains of fish is in the size of the
   a. medulla.
   b. cerebellum.
   c. cerebrum.
   d. diencephalon.
   e. thalamus.
   *Answer: c*

9. In the spinal cord, the gray matter contains the _____, and the white matter contains the _____.
   a. axons; cell bodies
   b. cell bodies; axons
   c. dorsal horn; ventral horn
   d. ventral horn; dorsal horn
   e. afferent information; efferent information
   *Answer: b*

*10. Efferent nerves leave the spinal cord through the
   a. ventral roots.
   b. dorsal roots.

   c. gray matter.
   d. interneurons.
   e. ventral and dorsal horns.
   *Answer: a*

*11. Interneurons are found in the
   a. midbrain.
   b. thalamus.
   c. white matter of the spinal cord.
   d. gray matter of the spinal cord.
   e. telencephalon.
   *Answer: d*

12. The function of the reticular system is to
   a. conduct impulses through the spinal cord.
   b. distribute information to its proper location in the forebrain.
   c. regulate the level of arousal of the nervous system.
   d. regulate physiological drives and emotion.
   e. transfer short-term memory to long-term memory.
   *Answer: c*

13. The reticular system is located in the
   a. spinal cord.
   b. midbrain.
   c. hindbrain.
   d. Both b and c
   e. a, b, and c
   *Answer: d*

14. The function of the limbic system is the
   a. regulation of instincts, emotions, and physiological drives.
   b. transfer of short-term memory to long-term memory.
   c. regulation of arousal levels.
   d. Both a and b
   e. a, b, and c
   *Answer: d*

15. What structure constitutes the largest part of the human brain?
   a. Telencephalon
   b. Diencephalon
   c. Medulla
   d. Pons
   e. Cerebellum
   *Answer: a*

16. The telencephalon is divided into two hemispheres covered by a sheet of gray matter called the
   a. hippocampus.
   b. reticular system.
   c. sulci.
   d. gyri.
   e. cerebral cortex.
   *Answer: e*

*17. Which part of the brain is involved in higher-order information processing?
   a. Association cortex
   b. Thalamus
   c. Limbic system
   d. Central sulcus
   e. Hippocampus
   *Answer: a*

18–21. From the list below, match the correct portion of the brain with the descriptions that follow.
 a. Temporal lobe
 b. Occipital lobe
 c. Parietal lobe
 d. Frontal lobe
 e. Cerebellum

18. If a person can see and hear but cannot recognize familiar faces, which part of the brain has most likely been damaged?
*Answer: a*

19. If a person cannot feel pressure applied to the hand even though the hand has not been injured, which part of the brain has most likely been damaged?
*Answer: c*

20. If a person cannot see motion even though the eyes are functioning normally, which part of the brain has most likely been damaged?
*Answer: b*

21. Which part of the brain has mostly like been damaged in a person who suffers from a personality disorder and cannot plan for future events?
*Answer: d*

*22. If a person exhibits contralateral neglect syndrome and ignores stimuli from the left side of the body, which part of the brain has most likely been damaged?
 a. The left parietal lobe
 b. The right parietal lobe
 c. The left frontal lobe
 d. The right frontal lobe
 e. The left temporal lobe
*Answer: b*

*23. The primary motor cortex is found in the _____ lobe and controls _____.
 a. parietal; the detection of touch or pressure
 b. parietal; movement
 c. temporal; movement
 d. frontal; movement
 e. frontal; the detection of touch or pressure
*Answer: d*

*24. The primary somatosensory cortex is located in the _____ lobe and controls _____.
 a. parietal; the detection of touch or pressure
 b. parietal; movement
 c. temporal; the detection of touch or pressure
 d. temporal; movement
 e. frontal; the detection of touch or pressure
*Answer: a*

25. _____ have the largest brains in the animal kingdom, whereas _____ have the largest brain-to-body size ratio.
 a. Dolphins; humans
 b. Elephants; whales
 c. Elephants; humans
 d. Dolphins; humans
 e. Humans; rodents
*Answer: c*

26. The fight-or-flight mechanisms are the function of the _____ branch of the autonomic nervous system.
 a. sympathetic
 b. parasympathetic
 c. contralateral
 d. efferent
 e. afferent
*Answer: a*

27. The parasympathetic division controls
 a. fight-or-flight response.
 b. increased heart rate and blood pressure.
 c. increased digestion and decreased heart rate.
 d. increased release of epinephrine and production of glucose.
 e. memory.
*Answer: c*

*28. In autonomic efferent pathways, preganglionic neurons use _____ as the neurotransmitter, whereas postganglionic neurons use _____ as the neurotransmitter.
 a. norepinephrine; acetylcholine
 b. acetylcholine; norepinephrine
 c. norepinephrine or acetylcholine; norepinephrine
 d. acetylcholine; norepinephrine or acetylcholine
 e. norepinephrine or acetylcholine; norepinephrine or acetylcholine
*Answer: d*

29. The anatomy of the sympathetic and parasympathetic divisions differs. For instance, the preganglionic neurons of the sympathetic division meet their postganglionic connections
 a. mostly in the brain stem.
 b. in the upper regions of the spinal cord.
 c. in ganglia arranged like chains along the spinal cord.
 d. near the target organs.
 e. in the midbrain.
*Answer: c*

30. Which is the correct pathway for the flow of visual information?
 a. Eye, thalamus, occipital lobe, optic chiasm
 b. Eye, thalamus, optic chiasm, occipital lobe
 c. Eye, optic chiasm, occipital lobe, thalamus
 d. Eye, thalamus, occipital lobe, optic chiasm
 e. Eye, optic chiasm, thalamus, occipital lobe
*Answer: e*

31. Complex cells in the visual cortex are stimulated by
 a. specific colors.
 b. bars of light with specific orientations and locations on the retina.
 c. bars of light with any orientation but at a specific location on the retina.
 d. bars of light with specific orientations at any location on the retina.
 e. any type of light flashed on the retina.
*Answer: d*

*32. Each retina sends _____ million axons to the brain; these are received by about _____ million neurons in the visual cortex.
 a. 1; 200
 b. 1; 2
 c. 100; 2
 d. 100; 200
 e. 200; 1
*Answer: a*

■33. In their studies on vision, David Hubel and Torsten Wiesel found that
  a. many areas of the retina can stimulate a single cell in the visual cortex.
  b. cells in the visual cortex respond to a receptive field on the retina.
  c. cats can see bars of light at specific orientations only.
  d. simple cells make connections to complex cells in the visual cortex.
  e. visual information crosses over the optic chiasm.
  *Answer: b*

34. During REM sleep, _____ occurs.
  a. dreaming
  b. loss of motor output by the brain
  c. sleepwalking
  d. Both a and b
  e. a, b, and c
  *Answer: d*

★35. On the cellular level, sleep occurs because of
  a. hyperpolarization of the cells of the thalamus and cortex.
  b. depolarization of the cells of the thalamus and cortex.
  c. increased synaptic input between axons and neurons in the thalamus and cortex.
  d. desynchronization of electrical impulses in the cortex.
  e. decreased opening of potassium and calcium channels in the membrane of cortical cells.
  *Answer: a*

36. The peripheral nervous system connects to the central nervous system via the
  a. spinal nerves.
  b. cranial nerves.
  c. hypothalamus.
  d. Both a and b
  e. None of the above
  *Answer: d*

★37. Long-term potentiation is
  a. increased sensitivity to an electrical stimulus.
  b. decreased sensitivity to an electrical stimulus.
  c. habituation to a stimulus.
  d. the application of high-frequency electrical stimulation.
  e. a decreased entry of calcium ions into the postsynaptic cell.
  *Answer: a*

38. Long-term depression is
  a. increased sensitivity to an electrical stimulus.
  b. decreased sensitivity to an electrical stimulus.
  c. the application of a continuous low-level stimulus.
  d. the inability of neurons to fire an electrical impulse.
  e. the inability to retain long-term memory.
  *Answer: b*

■39. The Russian physiologist Ivan Pavlov and his dog became famous for demonstrating
  a. how short-term memory converts to long-term memory.
  b. associative learning.
  c. long-term potentiation.
  d. the eye blink reflex.
  e. that muscles twitch during REM sleep.
  *Answer: b*

★40. In eye blink reflex studies, the conditioned reflex was localized to a region in the
  a. medulla.
  b. spinal cord.
  c. cerebellum.
  d. thalamus.
  e. frontal lobe.
  *Answer: c*

41. In an attempt to cure severe epilepsy, both sides of a person's hippocampus were removed. An unfortunate side effect of the surgery was that it caused loss of
  a. the ability to feel emotion.
  b. short-term memory.
  c. the ability to convert short-term memory into long-term memory.
  d. immediate memory.
  e. the ability to recognize faces.
  *Answer: c*

42. _____ memory is almost perfectly photographic.
  a. Immediate
  b. Short-term
  c. Long-term
  d. Declarative
  e. Procedural
  *Answer: a*

43. Short-term memory lasts about
  a. a few seconds.
  b. 10–15 minutes.
  c. 20–30 minutes.
  d. 1 hour.
  e. a few days.
  *Answer: b*

44. Roger Sperry won the Nobel Prize for his work on
  a. the hippocampus and memory loss.
  b. the conditioned reflex.
  c. eye puffs on rabbits.
  d. lateralization of language to the left hemisphere.
  e. REM sleep.
  *Answer: d*

45. The corpus callosum is the
  a. tract of gray matter connecting the two cerebral hemispheres.
  b. white matter in the spinal cord.
  c. gray matter in the spinal cord.
  d. tract of white matter connecting the two cerebral hemispheres.
  e. None of the above
  *Answer: d*

46. Damage to Broca's area results in loss of
  a. language comprehension.
  b. speech, or poor speech.
  c. the ability to recognize faces.
  d. the ability to read.
  e. hearing.
  *Answer: b*

47. Damage to Wernicke's area results in loss of
  a. language comprehension.
  b. speech, or poor speech.
  c. the ability to recognize faces.
  d. hearing.
  e. long-term memory.
  *Answer: a*

48–51. From the list below, match the correct structure with the following descriptions.
   a. Medulla
   b. Cerebellum
   c. Diencephalon
   d. Telencephalon

48. Contains the final relay station for sensory information going to the telencephalon
   *Answer: c*

49. Controls physiological functions such as breathing
   *Answer: a*

50. Orchestrates and refines motor commands
   *Answer: b*

51. Plays major roles in conscious behavior, learning, and memory
   *Answer: d*

52. Which of the following is *not* a function of the spinal cord?
   a. Generation of repetitive motor patterns
   b. Reflexes
   c. Conduction of motor impulses from the brain
   d. Refinement of motor and behavioral processes
   e. Conversion of afferent to efferent information
   *Answer: d*

53. The knee jerk reflex
   a. involves a sensory neuron that synapses with a motor neuron in the ventral horn of the spinal cord.
   b. can be checked readily by a physician.
   c. is a monosynaptic reflex.
   d. involves the leg extensor muscle.
   e. All of the above
   *Answer: e*

54. Norepinephrine _____ the heart rate, whereas acetylcholine _____ the heart rate.
   a. decreases; increases
   b. increases; decreases
   c. increases; increases
   d. decreases; decreases
   e. increases; does not affect
   *Answer: b*

*55. The human brain has about _____ neurons.
   a. 1 million
   b. 10 million
   c. 1 billion
   d. 10 billion
   e. 100 billion
   *Answer: e*

■56. Which of the following statements about nerves is *false*?
   a. A nerve is a bundle of axons.
   b. Some axons in a nerve may be carrying information to the central nervous system, while other axons in the same nerve are carrying information from the central nervous system to organs.
   c. A nerve is the axon of a single neuron.
   d. Both a and b
   e. None of the above
   *Answer: c*

*57. A comatose state
   a. is an enhancement of sensation.
   b. may result from damage to the brain stem below the reticular system.
   c. may result from damage to the midbrain or higher levels such that information from the reticular system cannot reach the forebrain.
   d. results when the spinal cord is damaged.
   e. None of the above
   *Answer: c*

■58. Ivan Pavlov
   a. discovered that if he rang a bell after presenting food to a dog, the dog would eventually salivate at the sound of the bell alone.
   b. discovered operant conditioning.
   c. discovered that if he rang a bell before presenting food to a dog, the dog would eventually salivate at the sound of the bell alone.
   d. discovered a form of learning in which two very similar stimuli become linked to the same response.
   e. None of the above
   *Answer: c*

# 47 Effectors: Making Animals Move

## Fill in the Blank

1. Microtubules are composed of subunits of _____ protein.
   *Answer: tubulin*

2. Intercalated discs join together cells in _____ muscle tissue.
   *Answer: cardiac*

3. Huxley and Huxley proposed the _____ theory of muscle contraction.
   *Answer: sliding filament*

4. An action potential spreading through the muscle fiber causes a minimal contraction known as a _____.
   *Answer: twitch*

5. _____ is the maximum tension of a muscle contraction.
   *Answer: Tetanus*

6. A simple _____ skeleton consists of a fluid-filled cavity controlled by muscle movement.
   *Answer: hydrostatic*

7. Cartilage tissue consists of _____ fibers in a rubbery matrix of proteins and polysaccharides.
   *Answer: collagen*

8. Cells that break down or resorb bone are the _____.
   *Answer: osteoclasts*

9. Muscles at joints are often arranged in antagonistic pairs of extensors and _____.
   *Answer: flexors*

10. Cellular effectors called _____ are fired like miniature missiles by jellyfish to capture prey.
    *Answer: nematocysts*

11. _____ are adaptations that animals use to respond to information that is sensed, integrated, and transmitted by their neural and endocrine systems.
    *Answer: Effectors*

12. _____ muscle provides the contractile force for most of our internal organs, which are under control of the autonomic nervous system.
    *Answer: Smooth*

★13. The _____ is a highly specialized network of intracellular membranes of striated muscle cells which enables them to sequester and release calcium for muscle contraction.
    *Answer: sarcoplasmic reticulum*

14. The living cells of bone that are responsible for the remodeling of bone structure are called _____.
    *Answer: osteoblasts*

15. The heartbeat is _____ (i.e., generated by the heart muscle itself).
    *Answer: myogenic*

16. _____, a nitrogen-containing polysaccharide, is found in the endocuticle of arthropods.
    *Answer: Chitin*

17. In humans, the _____ skeleton includes the skull, vertebral column, and ribs.
    *Answer: axial*

18. _____, a storage compound in muscles, is part of the immediate system for obtaining ATP.
    *Answer: Creatine phosphate*

## Multiple Choice

1. What mechanism is responsible for movements of cilia and flagella?
   a. Muscle contraction
   b. The microfilament system
   c. The microtubule system
   d. Crystal formation
   e. The skeletal system
   *Answer: c*

2. Which of the following does *not* rely on movement of cilia?
   a. Movement of an egg through an oviduct
   b. Movement of particles through the airways of lungs
   c. Feeding movements of a paramecium
   d. Movement of particles over the gills of a clam
   e. Amoeboid movement
   *Answer: e*

3. Which is the most important determinant of the proportion of fast- and slow-twitch fibers in skeletal muscle?
   a. Diet
   b. Exercise
   c. Genetic heritage
   d. Age
   e. None of the above
   *Answer: c*

*4. The glycolytic system by which muscles obtain ATP
   a. relies on creatine phosphate.
   b. is also referred to as the immediate system.
   c. is the most efficient means of producing ATP.
   d. produces lactic acid.
   e. completely metabolizes carbohydrates and fats.
   *Answer: d*

5. Anaerobic exercise such as weight lifting
   a. increases strength.
   b. can do minor tissue damage.
   c. induces the formation of new actin and myosin filaments in existing muscle fibers.
   d. produces bigger muscle fibers and hence bigger muscles.
   e. All of the above
   *Answer: e*

6. Which of the following statements about fast-twitch and slow-twitch fibers is *false*?
   a. The most important factor determining the proportion of fast- and slow-twitch fibers in skeletal muscles is genetic heritage.
   b. Aerobic training increases the oxidative capacity of fast-twitch fibers.
   c. A single muscle may contain fast- and slow-twitch fibers.
   d. Fast-twitch fibers fatigue slowly.
   e. None of the above
   *Answer: d*

7. The extension of a pseudopod in amoeboid movement occurs by means of
   a. the extension of actin microfilaments next to the membrane.
   b. contraction of the plasmagel next to the membrane.
   c. the changing of plasmagel to plasmasol, which expands and bulges outward.
   d. microfilament-generated cytoplasmic streaming of the plasmasol.
   e. rapid expansion of the cell membrane and the incorporation of plasmagel.
   *Answer: d*

*8. The resting potential of smooth muscle cells is
   a. not subject to forming action potentials.
   b. affected by stretching the cells.
   c. more negative than in most cells.
   d. unaffected by nearby potential change.
   e. nearly zero.
   *Answer: b*

9. The striated appearance of skeletal muscle is due to the
   a. dark color of myosin.
   b. multiple nuclei per fiber.
   c. regular arrangement of filaments.
   d. dense array of microtubules.
   e. dense packing of ATP molecules.
   *Answer: c*

10. An individual sarcomere unit consists of
    a. a stack of actin fibers.
    b. a stack of myosin units.
    c. overlapping actin and membrane.
    d. overlapping myosin and membrane.
    e. overlapping actin and myosin.
    *Answer: e*

■11. How do muscle fibers shorten during contraction?
   a. Individual protein filaments contract.
   b. More cross-bridges are formed between filaments.
   c. Arrays of filaments overlap each other.
   d. Protein filaments coil more tightly.
   e. Subunits of protein polymers detach.
   *Answer: c*

12. How do actin and myosin molecules interact?
   a. Globular myosin heads bind to actin filaments.
   b. Globular actin heads bind to myosin filaments.
   c. Other proteins connect the two.
   d. Myosin filaments bend to connect to actin.
   e. Actin filaments bend to connect to myosin.
   *Answer: a*

13. Why do muscles stiffen in rigor mortis when animals die?
   a. Without ATP, muscles cannot contract.
   b. Without ATP, actin and myosin cannot bind.
   c. Without ATP, actin and myosin cannot separate.
   d. ATP is required for synthesis of protein filaments.
   e. ATP forms cross-bridges between filaments.
   *Answer: c*

14. Vertebrate skeletal muscles are excitable cells because they
   a. can be stimulated by ATP.
   b. can be stimulated by an electric charge.
   c. can secrete neurotransmitter.
   d. possess voltage-gated sodium channels.
   e. can attain a high level of activity.
   *Answer: d*

*15. What is the role of the sarcoplasmic reticulum in muscle contraction?
   a. It stores $Ca^{2+}$ ions for release during contraction.
   b. It surrounds and protects the muscle filaments.
   c. It provides sites of ATP synthesis.
   d. It depolarizes when stimulated by an impulse.
   e. It synthesizes actin and myosin filaments.
   *Answer: a*

16. How does tropomyosin control muscle contraction?
   a. It provides a bridge between actin and myosin.
   b. It provides a site where ATP can be utilized.
   c. Changing its position exposes actin–myosin binding sites.
   d. It transmits electric charge to the filaments.
   e. Changing its shape opens membrane channels.
   *Answer: c*

■17. How can muscle fibers show a range of responses to different levels of stimulation?
   a. Each muscle fiber contraction is all or none.
   b. The availability of $Ca^{2+}$ sets an upper limit.
   c. A new contraction can occur only after the resting condition is reached.
   d. Following a stimulation, the fiber stays contracted.
   e. Individual twitches in the same fiber can summate.
   *Answer: e*

18. The legs of cross-country skiers and long-distance runners are likely to have
   a. almost all slow-twitch fibers.
   b. almost all fast-twitch fibers.
   c. about the same number of slow-twitch and fast-twitch fibers.

d. more slow-twitch fibers.

e. more fast-twitch fibers.

*Answer: d*

19. Fast-twitch skeletal muscle fibers, called white muscle, are characterized by
    a. a high concentration of myoglobin.
    b. abundant mitochondria.
    c. the rapid development of high tension.
    d. the ability to sustain activity for a long time.
    e. higher oxygen requirements than that of slow-twitch fibers.
    *Answer: c*

20. What is a feature of cardiac muscle that helps the heart withstand high pressures without tearing?
    a. The fibers branch and intertwine.
    b. The fibers are arranged in parallel.
    c. Each fiber has a single nucleus.
    d. The fibers have gap junctions between them.
    e. Some fibers have pacemaking functions.
    *Answer: a*

21. The exoskeleton of clams contains
    a. calcium carbonate.
    b. cartilage.
    c. chitin.
    d. collagen.
    e. cuticle.
    *Answer: a*

22. An advantage of an arthropod's exoskeleton is that it
    a. protects against abrasion.
    b. prevents water loss.
    c. provides attachment sites for muscles.
    d. bends at the animal's joints.
    e. All of the above
    *Answer: e*

23. An advantage of an endoskeleton over an exoskeleton is that it
    a. provides muscle attachment sites.
    b. provides protection.
    c. grows as the animal grows.
    d. supports the animal's weight.
    e. gives structure to the animal.
    *Answer: c*

24. What is the role of cartilage in the skeleton?
    a. To bear a heavy load
    b. To add flexibility
    c. To be lightweight
    d. To sustain vibrations
    e. To grow rapidly
    *Answer: b*

25. Bone tissue consists of _____ cells in a matrix of _____.
    a. osteocyte; polysaccharide
    b. osteocyte; collagen and calcium phosphate
    c. osteocyte; myosin and actin
    d. collagen; calcium phosphate
    e. collagen; polysaccharide
    *Answer: b*

26. How do osteoblasts cause bone to grow?
    a. They lay down new matrix until surrounded.
    b. They increase the number of cells within the matrix.
    c. They lay down new matrix on bone surface.
    d. They tear down old bone and deposit new bone.
    e. They tear down old cartilage and deposit bone.
    *Answer: c*

27. Haversian systems consist of
    a. osteocytes that connect different cavities.
    b. osteoblasts that will give rise to osteoclasts.
    c. a meshwork of cancellous bone tissue.
    d. cylindrical units surrounding a canal of blood vessels.
    e. a reinforced system of bone that resists fracturing.
    *Answer: d*

28. The role of tendons is to join
    a. two bones.
    b. two ligaments.
    c. bone and ligament.
    d. muscle and ligament.
    e. muscle and bone.
    *Answer: e*

▪29. Which of the following does *not* act as an effector?
    a. The poisonous secretion of pufferfish
    b. The mercaptan spray of skunks
    c. Pheromone secretion by butterflies
    d. Sensory neuron stimulation
    e. The secretion of saliva in the mouth
    *Answer: d*

30. Which of the following functions is *not* governed by microfilaments?
    a. Formation of daughter cells following mitosis
    b. Support of intestinal cell microvilli
    c. Movement of flagella
    d. Changes in cell shape
    e. Phagocytosis
    *Answer: c*

31. Whether a muscle contraction is strong or weak depends on how many motor neurons going to that muscle are firing and the rate at which they are firing. These two factors can be thought of, respectively, as spatial _____ and temporal _____.
    a. transmission; transduction
    b. transduction; transmission
    c. organization; summation
    d. summation; summation
    e. coordination; summation
    *Answer: d*

▪32. What is the function of the hydrostatic skeleton of invertebrates?
    a. Skeletal muscle contraction
    b. Skeletal muscle expansion
    c. Locomotion
    d. Propulsion
    e. b, c, and d
    *Answer: e*

33. Which of the following joints is found in humans?
    a. Ball-and-socket
    b. Hinge
    c. Pivot
    d. Saddle and ellipsoid
    e. All of the above
    *Answer: e*

34. _____ are a group of invertebrate effectors that enable animals to change color.
    a. Nematocysts
    b. Chromatophores
    c. Electroplates
    d. Poison glands
    e. Retractable stingers
    *Answer: b*

■35. Which of the following general statements about effectors is *false*?
    a. Any mechanism that an animal uses in order to respond is an effector.
    b. All effector action that involves movement is dependent on the interaction of microfilaments or microtubules.
    c. Muscle action depends on the movement of microtubules.
    d. Movement due to microfilaments and microtubules depends on the sliding action of long protein molecules.
    e. Effectors rely on energy made available by ATP.
    *Answer: c*

36. Aerobic exercise such as jogging increases
    a. myoglobin.
    b. the number of mitochondria in skeletal muscle cells.
    c. enzymes involved in energy utilization.
    d. the density of capillaries.
    e. All of the above
    *Answer: e*

■37. In which of the following locations would you *not* expect to find cilia?
    a. Within the air passages of the human lung
    b. Within the human female reproductive tract
    c. On the surface of intestinal cells in the human digestive tract
    d. On the surface of a clam gill
    e. On a filter-feeding animal
    *Answer: c*

38. Which of the following statements about the growth of long bones is *false*?
    a. Ossification starts at the center of long bones.
    b. Elongation occurs at epiphyseal plates.
    c. Secondary sites of ossification occur at the ends of long bones.
    d. Long bones begin as cartilaginous bones.
    e. Long bones form as scaffolding on a connective tissue membrane.
    *Answer: e*

■39. The three systems by which muscles obtain ATP
    a. differ in the capacity and rate at which they produce ATP.
    b. rely on the large amounts of ATP stored in muscle.
    c. produce similar amounts of ATP.
    d. rely on reactions within mitochondria.
    e. have similar time courses.
    *Answer: a*

40. The polymerization and depolymerization of tubulin is involved in helping to cause all of the following phenomena *except*
    a. the movement of chromosomes during mitosis and meiosis.

b. the growth of neurons during development.
    c. changes in cell shape.
    d. amoeboid movement.
    e. organelle movement.
    *Answer: d*

41. The dominant microfilament component in cells is
    a. actin.
    b. myosin.
    c. tubulin.
    d. dynein.
    e. troponin.
    *Answer: a*

42. Microfilaments are involved in helping to cause all of the following phenomena *except*
    a. the movement of chromosomes during mitosis and meiosis.
    b. the division of cytoplasm during animal cell mitosis.
    c. changes in cell shape.
    d. amoeboid movement.
    e. phagocytosis and pinocytosis.
    *Answer: a*

■43. Which of the following statements about amoeboid movement is *false*?
    a. Plasmasol is located in the interior of the cell.
    b. Plasmagel contains a network of actin microfilaments that interact with myosin.
    c. The cell moves by flowing into its pseudopodia.
    d. The plasmasol contracts during amoeboid movement.
    e. The pseudopod stops forming when the cytoplasm at the leading edge converts to gel.
    *Answer: d*

★44. Which of the following statements about bone is *false*?
    a. When astronauts spend long periods of time in zero gravity, they experience decalcification of their bones.
    b. Osteoclasts are derived from the lineage of cells that produce red blood cells.
    c. Most bones have compact and cancellous regions.
    d. Bones may become thicker or thinner depending on the amount of stress placed on them.
    e. None of the above
    *Answer: b*

45. Which of the following statements about smooth muscle is *false*?
    a. Smooth muscle is under the control of the autonomic nervous system.
    b. Smooth muscle cells are multinucleate.
    c. Smooth muscle appears unstriated when viewed with the microscope.
    d. Gap junctions are common in smooth muscle.
    e. Stretched smooth muscle will contract.
    *Answer: b*

★★■46. Which of the following events does *not* occur during muscle contraction?
    a. The distance between Z lines increases.
    b. The sarcomere shortens.
    c. The H zone is reduced.
    d. The I band is reduced.
    e. The area of actin and myosin overlap increases.
    *Answer: a*

47. Which of the following statements about the molecular arrangement of actin and myosin in myofibrils is *false*?
    a. A thin filament consists of actin and tropomyosin.
    b. Two chains of actin monomers are twisted into a helix.
    c. Two strands of tropomyosin lie in the grooves of the actin.
    d. Troponin forms the head of the myosin molecule.
    e. The myosin heads have ATPase activity and interact with the actin.
    *Answer: d*

■48. Starting with the arrival of an action potential at the neuromuscular junction, which of the following is the correct order of events?
    a. Calcium is released from the sarcoplasmic reticulum, an action potential travels down the T tubules, depolarization spreads through the T tubule, and myosin binds actin.
    b. An action potential travels down the T tubules, depolarization spreads through the T tubule, calcium is released from the sarcoplasmic reticulum, and myosin binds actin.
    c. An action potential travels down the T tubules, depolarization spreads through the T tubule, calcium is taken up by the sarcoplasmic reticulum, and myosin binds actin.
    d. An action potential travels down the T tubules, depolarization spreads through the T tubule, ATP binds to myosin, and myosin binds actin.
    e. A T tubule is depolarized, calcium is released from the sarcoplasmic reticulum, an action potential is created in the muscle cell, and myosin binds actin.
    *Answer: b*

49. When troponin binds calcium,
    a. it allows tropomyosin to bind actin.
    b. ion channels open and sodium rushes into the muscle cells.
    c. it changes conformation, twisting tropomyosin and exposing the actin–myosin binding site.
    d. it changes conformation, exposing the ATP binding site and allowing the actin–myosin bond to break.
    e. calmodulin binds to calcium and starts a cascade by activating myosin kinase.
    *Answer: c*

50. Muscle tone is the condition in which
    a. different muscles are taking responsibility for maintaining posture.
    b. the muscle has generated maximum tension.
    c. the muscle will eventually exhaust its supply of ATP.
    d. a small but changing number of motor units is active.
    e. the maximum number of action potentials is being received by the muscle.
    *Answer: d*

51. Fast- and slow-twitch fibers differ in all of the following ways *except* for the
    a. number of mitochondria.
    b. amount of myoglobin.
    c. amount of glycogen and fat.
    d. number of neuromuscular junctions.
    e. number of blood vessels.
    *Answer: d*

52. Which of the following phenomena does *not* involve a hydrostatic skeleton?
    a. Burrowing in earthworms
    b. Swimming in scallops
    c. Retraction of tentacles and body in the sea anemone
    d. Jet propulsion in squid
    e. Both b and d
    *Answer: b*

53. Which of the following statements about exoskeletons is *false*?
    a. During molting, the old exoskeleton is reabsorbed and a new exoskeleton forms.
    b. A clam shell is a exoskeleton.
    c. The outer layer of the arthropod cuticle prevents water loss in terrestrial species.
    d. The inner layer of the arthropod exoskeleton is used for muscle attachment.
    e. The arthropod exoskeleton is a continuous covering that is thinned at the joints.
    *Answer: a*

54. Which of the following statements about the endoskeletons of vertebrates is *false*?
    a. The muscles are attached to a living support structure.
    b. The rib bones are part of the appendicular skeleton.
    c. Molting is unnecessary in animals with endoskeletons.
    d. Endoskeletons consist of bone and cartilage.
    e. At least two bones are required to form a joint.
    *Answer: b*

55. Which of the following statements about cartilage and bone is *false*?
    a. Some vertebrates are composed entirely of cartilage and no bone.
    b. Some vertebrates are composed entirely of bone, with no cartilage.
    c. Bone has crystals of calcium phosphate.
    d. The endoskeleton in many vertebrates serves as a reservoir of calcium.
    e. The principal protein in cartilage is collagen.
    *Answer: b*

56. Long bones begin as _____ bones. At maturity the shaft of a long bone consists of a _____ of _____ bone.
    a. dermal; cylinder; compact
    b. dermal; solid rod; cancellous
    c. cartilage; cylinder; compact
    d. cartilage; cylinder; cancellous
    e. cartilage; solid rod; compact
    *Answer: c*

57. Which of the following statements about skeletal systems as a series of joints and levers is *false*?
    a. An extensor muscle moves a bone closer to the body.
    b. Antagonistic pairs include an extensor muscle and a flexor muscle associated with the same joint.
    c. Ligaments hold together bones that meet at joints.
    d. Tendons attach muscles to bones.
    e. The knee joint is a hinge joint.
    *Answer: a*

\*\*58. If you were designing a skeletal lever system for maximum power, you would
   a. insert both the muscle and the load force close to the joint.
   b. insert both the muscle and the load force far from the joint.
   c. insert the muscle close to the joint and place the load far from the joint.
   d. insert the muscle far from the joint and place the load close to the joint.
   e. use a large muscle.
   *Answer: d*

\*59. Which of the following mechanisms is *not* known to occur in chromatophore-mediated color adaptation?
   a. Rapid synthesis or destruction of pigment within chromatophores
   b. Dispersal or aggregation of pigment granules within chromatophores
   c. Shape change of chromatophores due to amoeboid movement
   d. Shape change of chromatophores due to interaction with muscle tissue
   e. All of the above
   *Answer: a*

■60. Glands are involved in all of the following, *except*
   a. intracellular communication.
   b. intercellular communication.
   c. communication between different individuals.
   d. defense against predators.
   e. prey capture.
   *Answer: a*

61. Which of the following statements about electric organs is *false*?
   a. In some species, production of electric fields is used for defense or for capture of prey.
   b. In some species, production of electric fields is used for orientation.
   c. Electric organs are derived from nervous tissue.
   d. Nerves, muscles, and electric organs use the same mechanisms to produce electric potentials.
   e. Large electric currents are achieved by coordinating the discharge of the cells in the electric organ.
   *Answer: c*

62–66. Match the correct muscle type below with the descriptions that follow. Each question may have more than one correct answer.
   a. Smooth
   b. Cardiac
   c. Skeletal

62. Which of these muscle types has gap junctions?
   *Answer: Both a and b*

63. Which of these muscle types is multinucleated?
   *Answer: c*

64. Which of these muscle types has the least regular arrangements of actin and myosin?
   *Answer: a*

65. Which of these muscle types uses calcium to trigger actin–myosin interactions for movement?
   *Answer: a, b, and c*

66. Which of these muscle types has a pacemaking function?
   *Answer: b*

■67. _____ complexes with calcium and acts as a second messenger in smooth muscle cells.
   a. Myosin
   b. Actin
   c. Troponin
   d. cAMP
   e. Calmodulin
   *Answer: e*

68. Which type of joint is found in the elbow?
   a. Ball-and-socket
   b. Pivotal
   c. Hinge
   d. Saddle
   e. Plane
   *Answer: b*

69. Which type of joint allows almost complete rotational movement?
   a. Ball-and-socket
   b. Pivotal
   c. Hinge
   d. Saddle
   e. Plane
   *Answer: a*

70. An annelid moves by
   a. stretching its longitudinal muscles and pushing the fluid-filled body cavity forward.
   b. forming pseudopods.
   c. ciliary movement.
   d. alternating contractions of longitudinal and circular muscles.
   e. contraction of circular muscles, which puts pressure on the hydrostatic skeleton and causes it to extend.
   *Answer: d*

# 48 Gas Exchange in Animals

## Fill in the Blank

1. External _____ are highly branched folds of the body surface involved in gas exchange.
   *Answer: gills*

2. The breathing tubes of insects are called _____.
   *Answer: tracheae*

3. The lungs of _____ are the most efficient lungs of vertebrates.
   *Answer: birds*

4. In fish gas exchange systems, blood flows through gill lamellae in a direction opposite to water flow, a phenomenon known as _____.
   *Answer: countercurrent exchange*

5. In the mammalian lung, the amount of air that is moved in normal breathing is called the _____.
   *Answer: tidal volume*

6. After extreme exhalation, the lungs and airways still contain a _____ of air.
   *Answer: residual volume*

7. The oxygen-binding molecule abundant in muscle is _____.
   *Answer: myoglobin*

8. Hemoglobin consists of _____ protein chains linked together.
   *Answer: four*

9. The notation $P_{O_2}$ stands for the _____ of oxygen.
   *Answer: partial pressure*

★10. The shape of the oxygen dissociation curve of hemoglobin is said to be _____.
   *Answer: sigmoid*

11. An unusual feature of the lungs of _____ is that they expand and contract relatively little during a breathing cycle. They also contract during inhalation and expand during exhalation.
   *Answer: birds*

12. Energy to carry out essential functions comes in the form of ATP, which can be sustained only where _____ gas is present.
   *Answer: $O_2$*

13. There are no active transport mechanisms for respiratory gases. Therefore, _____ is the only means by which respiratory gases are exchanged.
   *Answer: diffusion*

★14. In Fick's law of diffusion, the term $(P_1 - P_2)/L$ is a _____.
   *Answer: partial pressure gradient*

15. The respiration of birds is unique. In addition to lungs, birds have _____ at several locations in their bodies.
   *Answer: air sacs*

16. Mammalian lungs have two adaptive features: the production of mucus, and _____, a substance that reduces the surface tension of the liquid that lines the insides of the alveoli.
   *Answer: surfactant*

17. Breathing mechanisms _____ the environmental side of the gas exchange surface, while circulatory systems _____ the internal side of the surface.
   *Answer: ventilate; perfuse*

18. _____ is a genetic disease in which respiratory problems result from dehydrated, thick mucus.
   *Answer: Cystic fibrosis*

## Multiple Choice

1. Breathing provides the body with oxygen required to support the energy metabolism of all cells. Breathing also eliminates _____, one of the waste products of cell metabolism.
   a. $CO_2$
   b. CO
   c. carbon tetrachloride
   d. calcium carbonate
   e. carbonic acid
   *Answer: a*

2. Why can oxygen can be exchanged more easily in air than in water?
   a. The oxygen content of air is higher than that of water.
   b. Oxygen diffuses more slowly in water than in air.
   c. More energy is required to move water than air because water is denser.
   d. Both a and b
   e. a, b, and c
   *Answer: e*

■3. Aquatic _____ are in a double bind, because as the temperature of their environment increases, so does their demand for oxygen, but the oxygen content of water declines with increasing water temperatures.
 a. insects
 b. ectotherms
 c. endotherms
 d. plants
 e. None of the above
*Answer: b*

4. Which factor accounts for the efficiency of gas exchange in fish gills?
 a. Maximized surface area
 b. Minimized path length for diffusion
 c. Countercurrent flow of blood and water over opposite sides of the gas exchange surfaces
 d. Both a and b
 e. a, b, and c
*Answer: e*

5. Gas exchange in animals always involves
 a. cellular respiration.
 b. breathing movements.
 c. neural control of exchange.
 d. diffusion across membranes.
 e. active transport of gases.
*Answer: d*

6. The amount of gas exchange required to support an animal's metabolism increases with decreased
 a. $O_2$ in the air.
 b. temperature.
 c. body size.
 d. water breathing.
 e. body movement.
*Answer: a*

7. Rapid gas exchange is easier in air than in water because
 a. the $O_2$ content of water is higher than that of air.
 b. the $CO_2$ content of water is higher than that of air.
 c. oxygen diffuses more rapidly in water.
 d. water is more dense and viscous than air is.
 e. more energy is required to move air than to move water.
*Answer: d*

■8. Humans have difficulty breathing at high elevations because
 a. $O_2$ makes up a lower percentage of the air.
 b. the temperature is lower.
 c. the barometric pressure is higher.
 d. the partial pressure of $O_2$ is lower.
 e. the air is drier.
*Answer: d*

9. Small insects may take a bubble of air underwater when they dive. The bubble can serve as an air tank for some time because
 a. as $CO_2$ increases, the bubble inflates.
 b. as $O_2$ is consumed, air pressure in the bubble decreases.
 c. as $O_2$ is consumed, more $O_2$ diffuses into the bubble.
 d. the partial pressure of nitrogen remains constant.
 e. for each $O_2$ molecule used, a $CO_2$ molecule replaces it.
*Answer: c*

10. The respiratory system of insects consists of
 a. branched air tubes called spiracles that supply capillaries.
 b. branched air tubes called tracheae that supply capillaries.
 c. branching gill systems that end in openings called tracheae.
 d. branching gill systems that end in openings called spiracles.
 e. extensive layers of gas exchange tissue just under the exoskeleton.
*Answer: b*

*11. The delicate gills of fishes are supported by
 a. opercular flaps and gill arches.
 b. opercular flaps and gill filaments.
 c. gill filaments and gill arches.
 d. opercular flaps and a diaphragm.
 e. gill arches and a diaphragm.
*Answer: a*

12. Which of the following is *not* true about the air sacs of birds?
 a. They connect with each other.
 b. They occur in anterior and posterior pairs.
 c. They make the bird's respiratory system more efficient than a mammal's.
 d. They allow for one-way air flow through the lungs.
 e. They provide extra gas exchange surface.
*Answer: e*

13. Which of the following characterizes the lungs of mammals?
 a. Air sacs
 b. Tidal ventilation
 c. Complete emptying in exhalation
 d. Crosscurrent air flow
 e. Countercurrent air flow
*Answer: b*

14. Which of the following structures is the site of gas exchange in the lungs?
 a. Alveoli
 b. Bronchi
 c. Bronchioles
 d. Trachea
 e. Both a and b
*Answer: a*

15. Which factor maximizes the rate of gas exchange in mammals?
 a. Exceedingly high partial pressures of $O_2$ in the blood
 b. The larynx (voice box) opening into the trachea
 c. Enormous surface area for gas exchange
 d. The small increase in size during inhalation
 e. Production of mucus and surfactants
*Answer: c*

*16. Surfactant produced by cells lining the alveoli serves to
 a. increase surface tension.
 b. increase stretching of the alveolar walls.
 c. assist muscular movements of breathing.
 d. reduce cohesion of surface molecules.
 e. reduce ciliary movement.
*Answer: d*

17. One effect of cigarette smoking on the lungs is to
    a. remove mucus from the airways.
    b. immobilize cilia lining the airways.
    c. speed air flow through the airways.
    d. increase surfactant action.
    e. increase surface area in alveoli.
    *Answer: b*

18. The lower side of the thoracic cavity is formed by the
    a. diaphragm.
    b. esophagus.
    c. stomach.
    d. ribs.
    e. intercostal muscles.
    *Answer: a*

•19. The lungs expand in inhalation because
    a. the diaphragm contracts upward.
    b. the shoulder girdle moves upward.
    c. the volume of the thoracic cavity increases.
    d. lung tissue actively stretches.
    e. the lung tissue rebounds from exhalation.
    *Answer: c*

20. The process of exhalation is brought about by
    a. the contraction of intercostal muscles.
    b. the relaxation of muscles.
    c. the contraction of the diaphragm.
    d. complete collapse of the lung tissue.
    e. low pressure in the thoracic cavity.
    *Answer: b*

*21. If the thoracic wall is punctured, air leaking in will cause the lung to collapse. Thus in the normal, intact thoracic cavity,
    a. lung expansion is passive.
    b. air pressure is the same as it is on the outside.
    c. a slight suction keeps lungs inflated.
    d. breathing movements keep lungs inflated.
    e. Both a and b
    *Answer: c*

22. Which part of the blood is most efficient at carrying oxygen?
    a. Blood plasma solution
    b. Blood plasma proteins
    c. Blood platelets
    d. Membrane molecules of red blood cells
    e. Hemoglobin molecules of red blood cells
    *Answer: e*

23. Which of the following would increase the amount of oxygen diffusing from the lungs into the blood?
    a. Increasing the binding rate of $O_2$ to hemoglobin
    b. Decreasing the partial pressure of $O_2$ in the lung
    c. Increasing the partial pressure of $O_2$ in the blood
    d. Decreasing the red blood cell count of the blood
    e. Increasing the water vapor of air in the lungs
    *Answer: a*

24. Each molecule of hemoglobin, when fully saturated, carries how many molecules of oxygen?
    a. 1
    b. 2
    c. 4
    d. 20
    e. Nearly 100
    *Answer: c*

25. Hemoglobin delivers $O_2$ to body cells from the red blood cells
    a. until the hemoglobin is depleted of $O_2$.
    b. until the partial pressures of $O_2$ in the cells are equivalent.
    c. although the hemoglobin is never completely saturated.
    d. by releasing $O_2$ to cells with higher partial pressure of oxygen.
    e. until the fluid pressure in the red blood cells is lower.
    *Answer: b*

26. In which of the following $P_{O_2}$ environments will hemoglobin release its $O_2$ most easily?
    a. 30 mm Hg
    b. 50 mm Hg
    c. 70 mm Hg
    d. 90 mm Hg
    e. 110 mm Hg
    *Answer: a*

*27. What is the advantage of having myoglobin in muscle cells?
    a. It binds with $O_2$ just as easily as hemoglobin does.
    b. It contributes to the dark color of flight muscle in birds.
    c. It increases the effectiveness of muscles used in short bursts.
    d. It releases bound $O_2$ at lower $P_{O_2}$ conditions than hemoglobin does.
    e. It uses hemoglobin as a reserve source of oxygen.
    *Answer: d*

28. What benefit is provided by fetal hemoglobin?
    a. Fetal hemoglobin pumps $O_2$ from adult hemoglobin to the fetus.
    b. Fetal hemoglobin releases $O_2$ at a lower partial pressure than adult hemoglobin does.
    c. In the placenta, $O_2$ diffuses from adult hemoglobin to fetal hemoglobin.
    d. At low $O_2$ pressures, adult hemoglobin is more likely to pick up oxygen.
    e. Fetal hemoglobin occurs in higher density in red blood cells than adult hemoglobin does.
    *Answer: a*

29. In rapidly metabolizing, acidic tissues, how does hemoglobin respond in comparison to its action in less acidic environments?
    a. It releases more $O_2$.
    b. It releases less $O_2$.
    c. It releases more $CO_2$.
    d. It releases less $CO_2$.
    e. There is no difference in response.
    *Answer: a*

30. After humans acclimate to high altitudes, their hemoglobin
    a. is more often saturated with $O_2$.
    b. delivers more $O_2$ to the tissues.
    c. is more concentrated in red blood cells.
    d. shifts its $O_2$-binding curve to the left.
    e. resembles fetal hemoglobin in terms of $O_2$-binding.
    *Answer: b*

31. Most of the $CO_2$ in the blood is transported
    a. bound with hemoglobin.
    b. as dissolved gas.
    c. as $HCO_3^-$.
    d. as $H_2CO_3$.
    e. as calcium carbonate.
    *Answer: c*

32. Carbonic anhydrase is an enzyme in red blood cells that catalyzes a reaction between $CO_2$ and
    a. $HCO_3^-$.
    b. $H_2CO_3$.
    c. hemoglobin.
    d. $O_2$.
    e. water.
    *Answer: e*

33. Neural control of breathing is a function of the
    a. cerebrum.
    b. diaphragm.
    c. medulla.
    d. olfactory lobe.
    e. spinal cord.
    *Answer: c*

34. The breathing center initiates ventilation in response to
    a. a decrease in air pressure.
    b. a decrease in $O_2$.
    c. an increase in $CO_2$.
    d. the time since the last breath.
    e. the rate of gas exchange in the alveoli.
    *Answer: c*

35. Nodes that are sensitive to $O_2$ and blood pressure levels are located in the
    a. aorta and carotid arteries.
    b. brain stem capillaries.
    c. hypothalamus.
    d. pulmonary artery.
    e. pulmonary vein.
    *Answer: a*

36. Aquatic animals that lack an internal system for transporting $O_2$
    a. often have flat, leaflike bodies.
    b. are usually small.
    c. may have a very thin body built around a central cavity through which water circulates.
    d. have cells that are all close to the respiratory medium.
    e. All of the above
    *Answer: e*

37. Which of the following is *not* an expected response from a fish to a drop in water temperature?
    a. Increase in metabolism
    b. Increase in amount of $O_2$ available
    c. Decrease in blood flow
    d. Decrease in $O_2$ consumption
    e. All of the above
    *Answer: a*

**\*\*38.** A planet is discovered where the barometric pressure is 2,000 mm Hg and the air is 15 percent $O_2$. What is the partial pressure of $O_2$ at sea level on this planet?
    a. 15 percent
    b. 15 mm Hg
    c. 133 mm Hg
    d. 300 mm Hg

    e. 30 percent
    *Answer: d*

**\*39.** Which of the following statements does *not* explain why diffusion of $CO_2$ from an animal is not as great a problem as the diffusion of $O_2$ into the animal?
    a. $CO_2$ is more soluble in water than $O_2$ is.
    b. Cellular respiration produces less $CO_2$ than the $O_2$ that is consumed.
    c. The $CO_2$ content of air is less than the $O_2$ content.
    d. The atmospheric partial pressure of $O_2$ is greater than the atmospheric partial pressure of $CO_2$.
    e. There is a greater $CO_2$ concentration gradient from the cell to the atmosphere than is true for $O_2$.
    *Answer: b*

40. Which one of the following organisms does *not* require both external ventilation of its respiratory surfaces with the medium containing $O_2$ and internal ventilation of its respiratory surfaces with blood?
    a. Crayfish
    b. Rabbit
    c. Insect
    d. Squid
    e. Bird
    *Answer: c*

41. Which of the following statements about respiratory adaptations is *false*?
    a. Internalization of respiratory surfaces leads to the need for ventilation.
    b. External gills are found only in invertebrates.
    c. Some fish ventilate their gills by constantly swimming with their mouth open.
    d. In fish, water flows unidirectionally into the mouth, over the gills, and out from under the opercular flaps.
    e. Desiccation of the respiratory surface is more likely to occur in lungs than in gills.
    *Answer: b*

42. Which of the following statements about insect respiration is *false*?
    a. Spiracles are the air tubes that carry the respiratory gases.
    b. Because $O_2$ diffuses much faster in air than water, the insect respiratory system is highly efficient.
    c. Aquatic insects that carry air bubbles with them underwater are using the air bubble as a source of $O_2$.
    d. Some insects have external gills.
    e. None of the above
    *Answer: a*

43. Which of the following statements about the structure of the fish gill is *false*?
    a. Afferent vessels bring blood to the gills, whereas efferent vessels take blood away from the gills.
    b. Exchange of respiratory gases occurs within the lamellae of the gill filaments.
    c. The efferent and afferent vessels are the countercurrent flow system of the gill.
    d. The lamellae greatly increase the surface area for gas exchange.
    e. The opercular flaps enclose the gill chambers.
    *Answer: c*

44. Which of the following adaptations is *not* seen in fish gills?

a. A countercurrent exchange system
b. Bidirectional ventilation of the gills
c. Morphological features to increase the surface area available for gas exchange
d. Morphological features to decrease the path length for diffusion of the respiratory gases
e. Morphological features to maximize the efficiency of $O_2$ extraction
*Answer: b*

45. In the respiratory cycle of birds, air flows through the trachea, to the bronchi, and then
    a. into the parabronchi.
    b. back through the trachea.
    c. to the air capillaries.
    d. into the posterior air sacs.
    e. into the anterior air sacs.
    *Answer: a*

46. Bird and fish respiratory systems are similar because
    a. both employ a countercurrent exchanger.
    b. both have air sacs.
    c. both have unidirectional flow of the environmental medium over the gas exchange membranes.
    d. both are infoldings of the body.
    e. All of the above
    *Answer: c*

47. In the following equation, which quantity represents the volume of your normal breath?
    Total lung capacity = residual volume + expiratory reserve volume + inspiratory reserve volume + tidal volume
    a. Total lung capacity
    b. Residual volume
    c. Expiratory reserve volume
    d. Inspiratory reserve volume
    e. Tidal volume
    *Answer: e*

48. Which of the following is *not* an adverse consequence of tidal breathing as seen in mammals?
    a. Absence of countercurrent gas exchange
    b. Dead space
    c. Residual volume
    d. Short diffusion path length
    e. Limit in the $O_2$ concentration gradient between air and blood
    *Answer: d*

49. Carbon monoxide (CO)
    a. binds to hemoglobin with a higher affinity than $O_2$ does.
    b. may be released by faulty furnaces or the burning of charcoal or kerosene in unventilated areas.
    c. exposure may cause death from lack of $O_2$.
    d. All of the above
    e. None of the above
    *Answer: d*

50. Which of the following statements about surfactants in the lungs is *false*?
    a. Water coating the respiratory surfaces of the lungs creates surface tension.

b. Surfactants increase the cohesive forces between water molecules.
c. Certain cells in the alveoli produce surfactants.
d. Surface tension influences the amount of effort required to inflate the lungs.
e. Premature infants suffer respiratory distress syndrome if natural surfactants are not present within the lungs.
*Answer: b*

51. Which of the following statements about the mechanics of ventilation is *false*?
    a. If the pleural cavity is punctured, the lung may collapse.
    b. As the diaphragm relaxes, air is expelled from the respiratory system.
    c. As the diaphragm relaxes, the pressure within the pleural cavities increases.
    d. Less energy is expended during the exhalation phase than during the inhalation phase.
    e. There is a slight positive pressure within the pleural cavities between breaths.
    *Answer: e*

52. Which of the following statements about the transport of respiratory gases by the blood is *false*?
    a. The amount of $O_2$ that can dissolve directly in plasma is sufficient to support the resting metabolism.
    b. By binding $O_2$, hemoglobin helps to maintain a steeper $O_2$ concentration gradient.
    c. Internal ventilation of the gas exchange surfaces is dependent on blood flow.
    d. Each hemoglobin molecule can carry a maximum of four $O_2$ molecules.
    e. Molecules of $O_2$ are associated with the heme groups of hemoglobin.
    *Answer: a*

53. Which of the following statements about the binding of oxygen by hemoglobin is *false*?
    a. The percent of completely $O_2$-saturated hemoglobin increases as $P_{O_2}$ increases.
    b. The oxygen-binding curve is S-shaped.
    c. There is a narrow range of the $O_2$-binding curve where $O_2$ dissociation–association is rapid.
    d. It requires a very low $P_{O_2}$ to cause all four hemoglobin subunits to bind to $O_2$.
    e. If one subunit of a completely oxygen-saturated hemoglobin loses its $O_2$, the affinity of the remaining subunits for their $O_2$ increases.
    *Answer: d*

54. Which of the following statements about $O_2$-binding curves is *false*?
    a. It requires a smaller decrease in $P_{O_2}$ to go from 100 percent to 75 percent saturated hemoglobin than it does to go from 75 percent to 50 percent.
    b. During normal metabolism, the body is only using the upper one-fourth of its oxygen-binding curve.
    c. The blood normally carries an enormous reserve of $O_2$.
    d. The steepest part of the $O_2$-binding curve results in a great release of $O_2$ for only a modest drop in $P_{O_2}$.
    e. The $P_{O_2}$ of blood only goes below about 40 mm Hg during extreme exertion.
    *Answer: a*

55. Which statement about myoglobin is *false*?
    a. Iron in the myoglobin molecule can bind to $O_2$.
    b. Myoglobin is found in muscle cells.
    c. Myoglobin has a lesser affinity for $O_2$ than hemoglobin does.
    d. Diving mammals have high concentrations of myoglobin in their muscles.
    e. Skeletal muscles can differ in their myoglobin content.
    *Answer: c*

56. If you plot the $O_2$-binding curves of the following molecules, which curve will be to the left of all the others?
    a. Fetal human hemoglobin
    b. Adult llama hemoglobin
    c. Adult human hemoglobin
    d. Myoglobin
    e. Fetal llama hemoglobin
    *Answer: d*

*57. If you plot the $O_2$-binding curves of the following types of hemoglobin at the pH levels shown, which curve will be to the right of all the others?
    a. Fetal human hemoglobin at pH 7.6
    b. Adult human hemoglobin at pH 7.6
    c. Adult human hemoglobin at pH 7.2
    d. Fetal human hemoglobin at pH 7.2
    e. Adult llama hemoglobin at pH 7.2
    *Answer: c*

*58 If you plot the $O_2$-binding curves of adult hemoglobin subject to the following conditions, which curve will be to the right of all the others?
    a. pH 7.6, little 2,3 bisphosphoglyceric acid
    b. pH 7.6, much 2,3 bisphosphoglyceric acid
    c. pH 7.2, little 2,3 bisphosphoglyceric acid
    d. pH 7.2, much 2,3 bisphosphoglyceric acid
    e. pH 7.4, little 2,3 bisphosphoglyceric acid
    *Answer: d*

59. Most of the $CO_2$ is transported in the blood as _____, and it is located mainly in the _____.
    a. $CO_2$; plasma
    b. $H_2CO_3$; plasma
    c. $HCO_3^-$; plasma
    d. carboxyhemoglobin; erythrocytes
    e. $HCO_3^-$; erythrocytes
    *Answer: c*

*60. Due to the activity of carbonic anhydrase, within the lung there is a concentration gradient of $CO_2$ from the erythrocyte to the _____, and a concentration gradient of _____ from the plasma to the _____.
    a. plasma; $HCO_3^-$; plasma
    b. plasma; $HCO_3^-$; lung
    c. lung; $CO_2$; lung
    d. lung; $HCO_3^-$; erythrocyte
    e. lung; $HCO_3^-$; lung
    *Answer: d*

61. The breathing rhythm is generated in the _____ and is influenced by variation in levels of _____ in the blood.
    a. medulla; $CO_2$ and $O_2$
    b. medulla; $CO_2$
    c. medulla; $O_2$
    d. frontal lobe; $CO_2$ and $O_2$
    e. frontal lobe; $CO_2$
    *Answer: a*

62. Chemoreceptors on the surface of the medulla are sensitive to
    a. $O_2$.
    b. $N_2$.
    c. surfactant.
    d. $CO_2$ and pH of the cerebral spinal fluid.
    e. None of the above
    *Answer: d*

# 49 Circulatory Systems

## Fill in the Blank

■1. From fishes to amphibians to reptiles to mammals and birds, the complexity and number of chambers in the heart increases. An important consequence of this increased complexity is the gradual separation of the circulatory system into two independent circuits— _____ and _____.
*Answer: pulmonary; systemic*

2. The "lub-dub" sounds of the cardiac cycle are caused by the _____.
*Answer: shutting of the heart valves*

3. A blood clot that becomes established within the lumen of an artery is termed a thrombus. A piece of a thrombus breaking loose can travel through the arteries and eventually become lodged in a vessel of small diameter. If this occurs in the brain, the resulting condition is called a _____.
*Answer: stroke*

4. The interstitial fluid that accumulates outside the capillaries is returned to the heart by a separate system of vessels called the _____ system.
*Answer: lymphatic*

5. In a human being under normal conditions, the bone marrow can produce about 2 million _____ every second.
*Answer: red blood cells (or erythrocytes)*

6. In open circulatory systems of arthropods, fluid enters the _____ through holes called ostia.
*Answer: heart*

7. In humans, blood is pumped from heart to lungs and back through the _____ circuit.
*Answer: pulmonary*

8. The walls of large _____ have elastic fibers and can withstand high pressures.
*Answer: arteries*

9. The condition of _____ results from accumulation of interstitial fluids and tissue swelling.
*Answer: edema*

10. The primary pacemaker of the heart is the _____ at the junction of the superior vena cava and right atrium.
*Answer: sinoatrial node*

11. Blood cells form from _____ in the bone marrow.
*Answer: stem cells*

12. _____ are cell fragments responsible for blood clotting.
*Answer: Platelets*

13. The brain's cardiovascular control center is in the _____.
*Answer: medulla*

■14. The _____ slows the heart when the individual is plunged into cold water.
*Answer: diving reflex*

15. _____ are vessels that carry blood away from the heart, and _____ are vessels that carry blood back to the heart.
*Answer: Arteries; veins*

16. The lymphatic system empties intercellular fluid into the _____, which empty into large veins at the base of the neck.
*Answer: thoracic ducts*

17. In humans, contraction of the right ventricle pumps blood into the _____.
*Answer: pulmonary artery*

18. Snakes, lizards, and turtles can bypass the lung circuit and pump all blood around the body because of the incomplete division of their _____.
*Answer: ventricles*

## Multiple Choice

1. A cardiovascular system is *not* necessary in the hydra because it
   a. is an aquatic animal.
   b. has no skeleton.
   c. is only two cells thick.
   d. has tentacles to move water.
   e. does not move rapidly.
   *Answer: c*

2. In an open circulatory system,
   a. there is no heart.
   b. there is a gastrovascular cavity.
   c. there are no blood vessels.
   d. blood flows out of the body.
   e. there is no distinction between blood and tissue fluid.
   *Answer: e*

3. An example of an animal with an open circulatory system is the
   a. bat.
   b. earthworm.
   c. hydra.
   d. snail.
   e. lizard.
   *Answer: d*

4. In humans, which vessel empties into the right atrium?
   a. Pulmonary veins
   b. Inferior vena cava
   c. Superior vena cava
   d. Pulmonary artery
   e. Both b and c
   *Answer: e*

5. In vertebrates, exchange of substances between the blood and the interstitial fluids occurs in the
   a. arteries.
   b. arterioles.
   c. capillaries.
   d. veins.
   e. venules.
   *Answer: c*

6. In the fish circulatory system, blood
   a. moves from the muscular ventricle to the gills.
   b. entering the aorta is under high pressure.
   c. leaving the heart moves to body tissues.
   d. leaving the gills is under high pressure.
   e. is received from the body into a muscular atrium.
   *Answer: a*

■7. Circulatory systems of adult amphibians demonstrate which of the following advantages over those of fishes?
   a. A four-chambered heart
   b. Heart metamorphosis
   c. Partial separation of pulmonary and systemic circulation
   d. A pocket of the gut that serves as an air bladder
   e. Separation of oxygenated from deoxygenated blood
   *Answer: c*

■8. The advantage of having a heart with two atria is that
   a. oxygenated and deoxygenated blood are separated.
   b. blood is pumped directly from heart to tissues.
   c. blood can be slowed down before going to tissues.
   d. the body can support a higher blood pressure.
   e. there are two muscular regions for pumping.
   *Answer: b*

9. In which way are crocodilians different from other reptiles?
   a. They have a separate pulmonary circulation.
   b. They have lungs and no gills.
   c. They have an open circulatory system.
   d. They have more muscular atria.
   e. They have a four-chambered heart.
   *Answer: e*

10. In the human heart, blood is pumped from the left ventricle into the
    a. right ventricle.
    b. left atrium.
    c. right atrium.
    d. pulmonary circuit.
    e. systemic circuit.
    *Answer: e*

11. Which vessel transports oxygenated blood from the lung into the heart?
    a. Pulmonary artery
    b. Pulmonary vein
    c. Superior vena cava
    d. Inferior vena cava
    e. Coronary artery
    *Answer: b*

*12. In the cardiac cycle, blood pressure is at a maximum when the
    a. atria are contracting during systole.
    b. atria are contracting during diastole.
    c. ventricles are contracting during systole.
    d. ventricles are relaxing during systole.
    e. ventricles are relaxing during diastole.
    *Answer: c*

13. The heart beats because of cardiac muscle's unique properties, such as
    a. high resistance to outside stimulation.
    b. cell communication via gap junctions.
    c. a low level of electric activity.
    d. an external pacemaking system.
    e. rapid chemical communication.
    *Answer: b*

14. The specific location of the heart pacemaker is the
    a. sinoatrial node.
    b. atrioventricular node.
    c. Purkinje fibers.
    d. bundle of His.
    e. ventricular mass.
    *Answer: a*

*15. The timing of the spread of the action potential from atrium to ventricle is controlled by the
    a. sinoatrial node.
    b. atrioventricular node.
    c. Purkinje fibers.
    d. bundle of His.
    e. ventricular mass.
    *Answer: b*

16. The diameter of a capillary is about the same as that of a(n)
    a. arteriole.
    b. nerve.
    c. red blood cell.
    d. valve.
    e. venule.
    *Answer: c*

*17. As blood enters the capillaries from the arterioles, the blood pressure _____ and the osmotic potential _____.
    a. decreases; decreases
    b. decreases; increases
    c. increases; decreases
    d. decreases; remains steady
    e. increases; remains steady
    *Answer: b*

*18. Histamine causes swelling by
    a. making blood vessels expand and decreasing the permeability of capillaries.
    b. making blood vessels contract and decreasing the permeability of capillaries.

c. making blood vessels expand and increasing the pressure in capillaries.

d. decreasing the permeability of and increasing the pressure in capillaries.

e. increasing the permeability of and increasing the pressure in capillaries.

*Answer: c*

19. Lymph nodes are sites
    a. where bacteria can grow.
    b. where inflammation occurs.
    c. that regulate blood fluid.
    d. where new cells enter.
    e. of mechanical filtering.

    *Answer: e*

20. What causes blood to move in the veins toward the heart?
    a. Gravity
    b. The contraction of venous walls
    c. Pulsing movement from the heart
    d. The contraction of nearby muscles
    e. Venous capacitance

    *Answer: d*

*21. The Frank–Starling law suggests that during exercise the
    a. flow of blood through the heart speeds up.
    b. heart is stretched and contracts more forcefully.
    c. veins expand to handle increased blood flow.
    d. heart fills to capacity with blood at each beat.
    e. valves in veins can no longer prevent backflow.

    *Answer: b*

22. In normal humans the hematocrit, or percentage of the blood made up of cells, is about _____ percent.
    a. 10
    b. 25
    c. 40
    d. 70
    e. 90

    *Answer: c*

23. The most abundant cells in the blood are
    a. erythrocytes.
    b. leukocytes.
    c. phagocytes.
    d. platelets.
    e. None of the above

    *Answer: a*

24. The hormone erythropoietin is released by the kidney
    a. to remove old red blood cells from circulation.
    b. in response to high levels of oxygen in circulation.
    c. in response to low levels of hemoglobin.
    d. to stimulate production of red blood cells.
    e. to stimulate platelet formation.

    *Answer: d*

25. The lifetime of an individual red blood cell is approximately
    a. 4 hours.
    b. 4 days.
    c. 4 weeks.
    d. 40 days.
    e. 4 months.

    *Answer: e*

26. The mature red blood cells of humans
    a. are more than 75 percent hemoglobin.
    b. contain no nuclei or endoplasmic reticulum.
    c. have a nearly spherical shape.

d. have a relatively low surface area-to-volume ratio.

e. have a rigid shape and low flexibility.

*Answer: b*

27. A platelet is activated to initiate clotting
    a. during a histamine reaction.
    b. when leukocytes are activated.
    c. when collagen fibers are encountered.
    d. when blood flow rates drop.
    e. in the presence of prothrombin.

    *Answer: c*

28. How does the circulating protein fibrinogen contribute to blood clotting?
    a. It polymerizes to form fibrin threads.
    b. It acts as a catalyst for thrombin activation.
    c. It binds to red blood cells.
    d. It activates the platelets.
    e. It triggers phagocytosis by the leukocytes.

    *Answer: a*

29. Heart murmurs result from defective
    a. heart muscle.
    b. arterioles.
    c. ventricular contraction.
    d. blood supply to the heart.
    e. heart valves.

    *Answer: e*

30. Which of the following will lead to autoregulatory increase of blood flow to a tissue?
    a. High blood pressure
    b. High $CO_2$ concentration
    c. High $O_2$ concentration
    d. High ATP concentration
    e. High glucose concentration

    *Answer: b*

31. Which of the following statements about gastrovascular cavities is *false*?
    a. Organisms with gastrovascular cavities tend to be only a few cells thick.
    b. Gastrovascular cavities are always highly branched.
    c. Gastrovascular cavities are filled with interstitial fluid.
    d. The diffusion path length in organisms with gastrovascular cavities is usually short.
    e. Large organisms do not usually have gastrovascular cavities.

    *Answer: c*

32. Which of the following statements about open circulatory systems is *false*?
    a. A pump is usually present.
    b. Blood is pumped out of openings in the heart called ostia.
    c. The blood is the interstitial fluid.
    d. Tissue fluid bathes the tissues directly.
    e. Many mollusks and all arthropods have open circulatory systems.

    *Answer: b*

33. Which of the following structures is *not* part of the circulatory system of an earthworm?
    a. Contractile vessels
    b. A major dorsal vessel
    c. Capillary beds in the lungs
    d. Capillary beds in each segment
    e. A major ventral vessel

    *Answer: c*

34. The left ventricle in humans is more muscular than the right ventricle because
    a. resistance is higher in the systemic circuit.
    b. resistance is higher in the pulmonary circuit.
    c. it pumps more blood.
    d. it pumps more viscous blood.
    e. None of the above
    *Answer: a*

35. Which of the following vertebrates has a three-chambered heart with both pulmonary and systemic circuits?
    a. Fish
    b. Lungfish
    c. Amphibian
    d. Crocodile
    e. Both b and c
    *Answer: e*

36. Which of the following structures in a vertebrate with a four-chambered heart has blood with the lowest $O_2$ concentration?
    a. Pulmonary vein
    b. Pulmonary artery
    c. Left atrium
    d. Aorta
    e. Arteriole end of a capillary
    *Answer: b*

★37. When submerged, frogs receive most of their $O_2$ from capillaries within the skin. Which of the following structures would contain blood with the highest $O_2$ concentration in a submerged frog that is not able to breath using its lungs?
    a. Right atrium
    b. Left atrium
    c. Aorta
    d. Ventricle
    e. Pulmonary vein
    *Answer: a*

38. Which of the following carries oxygenated blood?
    a. Aorta
    b. Pulmonary vein
    c. Pulmonary artery
    d. Both a and b
    e. Both a and c
    *Answer: d*

39. A blood pressure represented as 140/100 means that the
    a. pressure during ventricular contraction is 140, while the pressure during ventricular relaxation is 100.
    b. pressure during ventricular contraction is 100, while the pressure during ventricular relaxation is 140.
    c. pressure during ventricular contraction is 140, while the pressure during atrial contraction is 100.
    d. pressure during atrial contraction is 140, while the pressure during ventricular contraction is 100.
    e. diastolic pressure is 140 and the systolic pressure is 100.
    *Answer: a*

★40. During the cardiac cycle, the blood pressure in the aorta is at a minimum
    a. at the end of systole.
    b. at the beginning of systole.
    c. at the beginning of diastole.
    d. when the physician hears the "dub" sound with the stethoscope.

e. None of the above
    *Answer: b*

★41. Which of the following heart regions is third in sequence of action potential propagation during a normal heartbeat?
    a. Bundle of His
    b. Atrioventricular node
    c. Purkinje fibers
    d. Sinoatrial node
    e. Ventricular muscle cells
    *Answer: a*

42. Which of the following statements concerning the heartbeat is true?
    a. Only the ventricles contract together.
    b. Only the atria contract together.
    c. The atria contract together, and the ventricles contract together.
    d. Only the atrium and ventricle on the right side of the heart contract together.
    e. Only the atrium and ventricle on the left side of the heart contract together.
    *Answer: c*

43. Gap junctions
    a. electrically isolate cardiac muscle cells.
    b. allow large numbers of cardiac muscle cells to contract in unison.
    c. provide enough strength to withstand the large pressure generated by the ventricles.
    d. are only found in the cardiac muscle cells of the ventricles.
    e. are especially abundant in the muscles between the atria and ventricles.
    *Answer: b*

44. The blood vessels with the greatest total cross-sectional area are the
    a. arteries.
    b. arterioles.
    c. capillaries.
    d. venules.
    e. veins.
    *Answer: c*

45. The _____ have blood with the lowest pressure.
    a. arteries
    b. arterioles
    c. capillaries
    d. venules
    e. veins
    *Answer: c*

★46. In the electrocardiogram (EKG) of a normal cardiac cycle, the wave called T corresponds to
    a. depolarization of the atria.
    b. depolarization of the ventricles.
    c. repolarization of the atria.
    d. repolarization of the ventricles.
    e. More than one of the above
    *Answer: d*

★47. Blood pressure is _____ than osmotic pressure at the arterial end of a capillary bed, and the process of _____ occurs there.
    a. greater; reabsorption
    b. greater; filtration

c. lesser; reabsorption
d. lesser; filtration
e. lesser; edema
*Answer: b*

48. Blockage of a coronary artery can cause
   a. heart attack.
   b. stroke.
   c. fainting.
   d. edema.
   e. anemia
   *Answer: a*

*49. Which of the following statements about the lymphatic system is *false*?
   a. The lymphatic system conducts lymph from the lymph capillaries to the thoracic duct, where it enters the circulatory system.
   b. Lymph is identical to blood except that it does not contain erythrocytes.
   c. Lymphoid capillaries are distributed throughout the body.
   d. Contraction of skeletal muscles propels the lymph.
   e. Lymph nodes are an important part of the defense system that combats infections.
   *Answer: b*

50. Which of the following is *not* a mechanism that facilitates return of venous blood to the heart?
   a. Gravity when veins are above the heart
   b. Contraction of skeletal muscles
   c. Breathing
   d. Contraction of smooth muscle in most of the veins of the body
   e. Valves in veins
   *Answer: d*

51. The blood vessels with the greatest capacity to store blood are the
   a. arteries.
   b. arterioles.
   c. capillaries.
   d. venules.
   e. veins.
   *Answer: e*

52. Which of the following statements about blood cells is *false*?
   a. There are at least 500 times as many red as white blood cells.
   b. Erythrocytes are anucleate in mammals.
   c. The biconcave shape of the erythrocyte creates a large surface area for gas exchange.
   d. Red blood cells are generated by stem cells in the spleen.
   e. Leukocytes destroy foreign cells.
   *Answer: d*

*53. Which of the following circulatory components is *not* normally found circulating in the plasma?
   a. Platelets
   b. Thrombin
   c. Albumin
   d. Fibrinogen
   e. Both a and c
   *Answer: b*

54. The purpose of precapillary sphincters is to
   a. increase blood pressure to the arteries.
   b. decrease blood pressure to the veins.
   c. shut off the supply of blood to the capillary bed.
   d. increase blood pressure to the capillaries.
   e. provide blood to smooth muscles surrounding capillary beds.
   *Answer: c*

55. Which of the following statements about veins is true?
   a. As much as 80 percent of total blood volume may be in the veins at any one time.
   b. Veins are capacitance vessels.
   c. The walls of veins are more expandable than the walls of arteries are.
   d. Veins are resistance vessels.
   e. a, b, and c
   *Answer: e*

56. Which one of the following responses characterizes the diving reflex?
   a. Reduction in the heartbeat rate
   b. Increase in systolic pressure
   c. Increase in diastolic pressure
   d. Tachycardia
   e. Dilation of arteries to most organs
   *Answer: a*

*57. Advantages of closed circulatory systems over open circulatory systems include which of the following?
   a. Exchange occurs more rapidly.
   b. Closed systems can direct blood to specific tissues.
   c. Cells and large molecules can be kept separate from the animal's intercellular material.
   d. Closed circulatory systems can support higher levels of metabolic activity.
   e. All of the above
   *Answer: e*

58. In the evolution of the vertebrate circulatory system, the _____ is the organism that reveals the transition step leading to separate pulmonary and systemic circuits.
   a. lungfish
   b. ancient ostracoderm
   c. bird
   d. amphibian
   e. reptile
   *Answer: a*

59. Systole and diastole describe the _____ and _____, respectively, of the _____ in mammals.
   a. relaxation; contraction; ventricles
   b. contraction; relaxation; ventricles
   c. relaxation; contraction; atria
   d. contraction; relaxation; atria
   e. None of the above
   *Answer: b*

60. Which of the following is *not* a risk factor for developing atherosclerosis?
   a. A high fat and high cholesterol diet
   b. Smoking
   c. High altitude
   d. Inactive lifestyle
   e. Hypertension
   *Answer: c*

61. Which of the following statements about valves is *false*?
    a. Defective heart valves produce sounds called heart murmurs.
    b. Veins and lymphatic vessels have two-way valves.
    c. Varicose veins result when valves within veins can no longer prevent backflow of blood.
    d. The pulmonary valve is located between the right ventricle and the pulmonary artery.
    e. Atrioventricular valves prevent backflow of blood into the atria when the ventricles contract.
    *Answer: b*

62. When a piece of thrombus breaks loose and lodges in a smaller vessel, blocking the flow of blood, it is called
    a. thrombosis.
    b. an infarction.
    c. an embolism.
    d. atherosclerosis.
    e. a hematocrit.
    *Answer: c*

63. When whole blood is centrifuged, it is separated into a liquid component called _____ and a bottom layer of _____.
    a. hematocrit; cells
    b. plasma; cells
    c. plaque; cells
    d. hematocrit; plasma
    e. packed cell volume; plasma
    *Answer: b*

64. The function of leukocytes is
    a. defense against infection.
    b. transport of respiratory gases.
    c. blood clotting.
    d. distribution of nutrients to tissues.
    e. production of the different types of blood cells.
    *Answer: a*

65. Which of these hormones causes constriction of the blood vessels?
    a. Epinephrine
    b. Vasopressin
    c. Angiotensin
    d. Both b and c
    e. a, b, and c
    *Answer: e*

66. Stretch receptors in the aorta and carotid arteries
    a. detect atherosclerosis.
    b. detect changes in blood pressure and relay them to the brain.
    c. constrict arteries when available $O_2$ is low.
    d. detect when a flexor muscle is stretched too far and then pump more blood to the muscle.
    e. None of the above
    *Answer: b*

67. Which of these factors contributes to increased blood flow to tissues?
    a. Low $O_2$ concentration
    b. Increased concentration of lactate
    c. High concentration of carbon dioxide
    d. Both a and b
    e. a, b, and c
    *Answer: e*

# 50 Nutrition, Digestion, and Absorption

## Fill in the Blank

1. The crowns of the teeth of mammals are covered with a hard material called _____.
   *Answer: enamel*

2. _____ enzymes break down macromolecules into their monomeric units by adding water.
   *Answer: Hydrolytic*

★3. Many digestive enzymes are produced in an inactive _____ form.
   *Answer: zymogen*

4. A wave of smooth muscle contraction called _____ pushes food along the digestive tract.
   *Answer: peristalsis*

5. In addition to serving as an endocrine gland, the pancreas secretes the digestive enzyme precursor _____.
   *Answer: trypsinogen*

6. Many species of _____ live in the human large intestine.
   *Answer: bacteria*

7. The initial digestion of protein in the vertebrate digestive tract begins in the _____.
   *Answer: stomach*

8. The _____ is the organ that manages most of the balance between circulating and storing nutrients.
   *Answer: liver*

9. The liver converts pyruvate and other molecules to glucose by the process of _____.
   *Answer: gluconeogenesis*

10. The regulation of blood glucose is accomplished mainly by hormones produced by the _____.
    *Answer: pancreas*

■11. Most plants, some bacteria, and some protists are autotrophs. Animals, however, are _____ because they derive both their energy and their structural molecules from food.
    *Answer: heterotrophs*

★12. Most animals process their food through _____ digestion in a digestive cavity called a gut.
    *Answer: extracellular*

13. If 1 kcal = 4.184 joules, then the number of kcal found in a diet consisting of 13,398 joules is _____ kcal.
    *Answer: 3,200*

14. We can synthesize almost all the lipids required by the body from acetyl units obtained from carbohydrates or fats, but we must have a dietary source of essential _____, notably linoleic acid.
    *Answer: fatty acids*

15. _____ prevents fat droplets from aggregating, or clumping together, so that the maximum surface area is exposed to lipase action.
    *Answer: Bile*

16. Crabs and earthworms that feed actively on dead organic matter are called _____.
    *Answer: detritivores*

17. An unusual feature of the vertebrate digestive tract is that it has its own intrinsic _____.
    *Answer: nervous system*

18. _____ are organisms like protists and fungi that absorb nutrients from dead organic matter.
    *Answer: Saprobes*

19. The three sections of the small intestine are the _____, _____, and _____.
    *Answer: duodenum; jejunum; ileum*

20. Food intake is regulated by the region of the brain known as the _____.
    *Answer: hypothalamus*

## Multiple Choice

1. In which order are stored fuels utilized during starvation?
   a. Fats, glycogen, proteins
   b. Glycogen, proteins, fats
   c. Proteins, fats, glycogen
   d. Fats, proteins, glycogen
   e. Glycogen, fats, proteins
   *Answer: e*

2. Certain amino acids are called essential amino acids because they
   a. are required for making protein.
   b. cannot be made from other amino acids and must be obtained from food.
   c. are universally needed by all animals.
   d. are essential as an energy source.
   e. are required for making nucleic acids.
   *Answer: b*

3. Why are vitamins essential nutrients for cells?
   a. Vitamins are used as an energy source.
   b. Vitamins are used to digest foods.
   c. Vitamins function as coenzymes.
   d. Both a and b
   e. a, b, and c
   *Answer: c*

■4. Organisms that derive both their energy and molecular nutrients from other organisms are called
   a. autotrophs.
   b. herbivores.
   c. heterotrophs.
   d. photosynthetic.
   e. protists.
   *Answer: c*

5. The energy content of food is described in terms of calories because
   a. the amount of energy in food depends on the temperature.
   b. food heats up as it is being digested.
   c. the energy in food ultimately becomes heat.
   d. heat is the main product of digestion.
   e. heat is the main product of respiration.
   *Answer: c*

6. The major form of stored energy in animal bodies is
   a. protein, because it is a long-term energy storage form.
   b. glycogen, because it breaks down into readily usable carbohydrates.
   c. glycogen, because it is lightweight.
   d. fat, because it has the highest energy content per gram.
   e. fat, because it is readily stored with water.
   *Answer: d*

7. Which of the following statements about vitamin C is *false*?
   a. It functions in collagen production.
   b. It is important for healthy skin.
   c. It is an essential vitamin for humans.
   d. It is a fat-soluble vitamin.
   e. An excess of vitamin C is excreted.
   *Answer: d*

8. Vitamin D is obtained by different people in various ways. For example,
   a. dark-skinned people of high latitudes obtain it from sunlight.
   b. dark-skinned people of low latitudes obtain it from tropical sunlight.
   c. Inuit people obtain it from sunlight.
   d. some people obtain it from protein molecules that absorb sunlight.
   e. some people obtain it from water-soluble components of foods.
   *Answer: b*

9. The nutritional disease kwashiorkor results from a(n)
   a. protein deficiency.
   b. vitamin deficiency.
   c. calorie deficiency.
   d. overdose of fat-soluble vitamins.
   e. overdose of thyroxine.
   *Answer: a*

10. Which of the following diseases is *not* due to a vitamin deficiency?
    a. Pellagra
    b. Rickets
    c. Goiter
    d. Pernicious anemia
    e. Beriberi
    *Answer: c*

11. Which of the following does *not* represent a carnivore's feeding adaptation?
    a. Spider web
    b. Rattlesnake venom
    c. Bat echolocation
    d. Elephant's trunk
    e. Jellyfish tentacle
    *Answer: d*

12. A filter feeder acquires food items by
    a. using poison to restrain prey.
    b. using claws and jaws to restrain prey.
    c. ingesting mud and extracting particles.
    d. filtering food substances from blood.
    e. extracting particles suspended in water.
    *Answer: e*

■13. A mammal with a diet of grain and leaves would be expected to have which kind of teeth?
    a. Prominent canine teeth and small molars
    b. Prominent molars and small canines
    c. Prominent molars and canine teeth
    d. A balanced set of incisors, molars, and canines
    e. Prominent canine teeth and small incisors
    *Answer: b*

14. For most animals, digestion of food occurs
    a. intracellularly.
    b. in the gastrovascular cavity.
    c. in the coelomic body cavity.
    d. in the midgut.
    e. in the crop.
    *Answer: d*

*15. Most animals digest dietary proteins into their constituent amino acids and then synthesize new proteins because
    a. macromolecules like proteins cannot be readily transported through plasma membranes.
    b. protein function often varies with species.
    c. foreign proteins are considered invaders and are attacked by the immune system.
    d. All of the above
    e. None of the above
    *Answer: d*

■16. Which of the following is true about digestive systems in all animals?
    a. Food in the gut is digested extracellularly.
    b. There are two openings to the tract.
    c. Animals break food into smaller pieces before ingestion.
    d. All digestion occurs inside the gut.
    e. The digestive tract has specialized segments.
    *Answer: a*

*17. Which of the following structures does *not* serve to increase surface area for nutrient absorption?
    a. Human intestinal villi

b. An earthworm's infolding of the intestinal wall
c. A shark's spiral valve
d. A bird's gizzard
e. Human cell microvilli
*Answer: d*

18. An endolipase is an enzyme that breaks down
    a. carbohydrates.
    b. nucleic acids.
    c. proteins.
    d. fat molecules by cutting them in the middle.
    e. fat molecules by snipping at their ends.
    *Answer: d*

19. Most ulcers are caused by
    a. stress.
    b. an infectious bacterium.
    c. the oversecretion of digestive juices.
    d. old age.
    e. an acidic diet.
    *Answer: b*

*20. In the small intestine, the blood and lymph vessels that carry away absorbed nutrients lie in which layer?
    a. Microvilli
    b. Mucosa
    c. Submucosa
    d. Circular muscle layer
    e. Longitudinal muscle layer
    *Answer: c*

*21. Movement of food from the stomach into the esophagus is normally prevented by
    a. peristalsis.
    b. reverse peristalsis.
    c. the pyloric sphincter.
    d. a sphincter.
    e. the pharynx.
    *Answer: d*

22. The major enzyme produced by the stomach is
    a. amylase.
    b. chyme.
    c. mucus.
    d. pepsin.
    e. trypsin.
    *Answer:d*

23. What activates the inactive form of stomach enzymes?
    a. Activating enzymes
    b. ATP
    c. Low pH
    d. The appropriate substrate molecule
    e. The presence of water
    *Answer: c*

24. Most absorption of nutrients in the digestive tract takes place in the
    a. stomach.
    b. small intestine.
    c. large intestine.
    d. liver.
    e. pancreas.
    *Answer: b*

*25. Bile aids in the breakdown of lipids by
    a. hydrolyzing lipids.
    b. activating hydrolytic enzymes.
    c. aggregating droplets of lipids.

d. emulsifying lipids.
    e. making lipids water-soluble.
    *Answer: d*

*26. What neutralizes the acidic chyme in the small intestine?
    a. Bicarbonate from the pancreas
    b. Buffers from the jejunum
    c. Bile from the liver
    d. Trypsin activation
    e. A variety of zymogens
    *Answer: a*

27. How does sodium cotransport facilitate glucose absorption?
    a. Active transport of sodium aids in salt uptake.
    b. When sodium diffuses into cells, the carrier also binds a glucose.
    c. When sodium is pumped into cells, glucose moves out of the cells.
    d. Sodium and glucose both diffuse into cells from the gut.
    e. Glucose is actively transported into the gut.
    *Answer: b*

28. The major function of the colon or large intestine is
    a. the digestive breakdown of foods.
    b. to absorb nutrients from foods.
    c. to house parasitic bacteria.
    d. the secretion of bile and enzymes.
    e. to reabsorb water.
    *Answer: e*

29. The hormone secretin is a chemical message
    a. secreted by the pancreas.
    b. secreted by the stomach.
    c. whose release is stimulated by the nervous system.
    d. that triggers the intestine to release enzymes.
    e. that triggers pancreatic secretion.
    *Answer: e*

30. Insulin is released by the pancreas when blood
    a. glucose falls.
    b. glucagon falls.
    c. glucose rises.
    d. glucagon rises.
    e. insulin falls.
    *Answer: c*

31. The effects of glucagon include
    a. stimulating glucose uptake into cells.
    b. stimulating liver cells to break down glycogen.
    c. controlling blood glucose in the absence of insulin.
    d. stimulating cells to store energy as fat.
    e. None of the above
    *Answer: b*

32. Which of the following statements about energy storage is true?
    a. Carbohydrates are stored in the liver and in muscle as glycogen. The total glycogen stores are usually not more than the equivalent of a day's energy requirements.
    b. Fat is an important form of stored energy.
    c. Fat has the highest energy content per gram.
    d. Protein is the most important energy storage component.
    e. a, b, and c
    *Answer: e*

33. Digestive enzymes are classified according to the substances they hydrolyze and the site where the enzyme cleaves the given molecule. An endopeptidase hydrolyzes a _____ at an _____ site along the length of the molecule.
    a. carbohydrate; external
    b. protein; internal
    c. fat; internal
    d. carbohydrate; internal
    e. peptide; external
    *Answer: b*

34. The gut of an animal is often described as an elongated tube consisting of four layers of different cell types. Which of the following is the correct order of layers from innermost to outermost?
    a. Submucosa, cartilage, mucosa, endoplasmic reticulum
    b. Smooth muscle layers, submucosa, mucosa, cartilage
    c. Cartilage, smooth muscle layers, mucosa, circular muscle
    d. Mucosa, submucosa, circular muscle, longitudinal muscle
    e. Cartilage, mucosa, endoplasmic reticulum, submucosa
    *Answer: d*

35. _____ ingest both plant and animal food and process their food through _____ digestion.
    a. Predators; intracellular
    b. Omnivores; intracellular
    c. Carnivores; intracellular
    d. Herbivores; extracellular
    e. Omnivores; extracellular
    *Answer: e*

36. Which of the following statements about undernourishment is *false*?
    a. One-fifth of the world's population is undernourished.
    b. Several weeks of fasting are required to deplete glycogen reserves.
    c. Self-imposed starvation is called anorexia nervosa.
    d. The loss of blood proteins during starvation leads to edema.
    e. During starvation, fat reserves are metabolized before body protein is metabolized.
    *Answer: b*

*37. Herbivores require _____ essential amino acids than carnivores do and consequently are less likely to suffer from _____ than carnivores are.
    a. fewer; undernourishment
    b. fewer; malnourishment
    c. more; undernourishment
    d. more; malnourishment
    e. more; overnourishment
    *Answer: b*

*38. Which of the following statements about the essential amino acids is *false*?
    a. Most animals have some essential amino acids.
    b. Ingesting a surplus of one essential amino acid cannot compensate for a shortage of another.
    c. Some plant foods such as legumes supply all eight of the essential amino acids in humans.
    d. Acetyl groups can be combined with amino groups to produce many of the nonessential amino acids.
    e. As with amino acids, some fatty acids are essential.
    *Answer: c*

39. The enzyme lactase
    a. cleaves lactose into glucose and galactose.
    b. is produced by the small intestine.
    c. often is not produced by humans after four years of age.
    d. cleaves milk sugar.
    e. All of the above
    *Answer: e*

*40. A strictly vegetarian diet with no vitamin $B_{12}$ supplements can lead to
    a. beriberi.
    b. pellagra.
    c. pernicious anemia.
    d. scurvy.
    e. night blindness.
    *Answer: c*

41. The nutritional deficiency disease beriberi is caused by an inadequate supply of the water-soluble vitamin
    a. $B_1$ (thiamin).
    b. $B_2$ (riboflavin).
    c. $B_{12}$ (cobalamin).
    d. folic acid.
    e. C (ascorbic acid).
    *Answer: a*

42. The nutritional deficiency disease rickets is caused by an inadequate supply of the fat-soluble vitamin
    a. A (retinol).
    b. D (calciferol).
    c. E (tocopherol).
    d. K (menadione).
    e. biotin.
    *Answer: b*

43. The nutritional deficiency disease simple goiter is caused by inadequate supply of the micronutrient
    a. fluoride.
    b. iodine.
    c. chromium.
    d. zinc.
    e. copper.
    *Answer: b*

44. Which of the following statements about vitamins is *false*?
    a. Vitamins, like essential amino acids and fatty acids, are organic molecules.
    b. Most vertebrates require the very same vitamins.
    c. Vitamins function mostly as, or as parts of, coenzymes.
    d. Vitamins are required only in very small amounts.
    e. Humans require more water-soluble vitamins than fat-soluble vitamins.
    *Answer: b*

*45. Which of the following is the true stomach of a ruminant?
    a. Cecum
    b. Reticulum
    c. Rumen
    d. Omasum
    e. Abomasum
    *Answer: e*

46. The root of a typical tooth contains _____ and _____, but *not* _____.
    a. enamel; dentine; a pulp cavity
    b. enamel; a pulp cavity; dentine
    c. dentine; a pulp cavity; enamel
    d. dentine; enamel; a pulp cavity

    e. cement; dentine; a pulp cavity
*Answer: c*

47. Which of the following teeth are *less* prominent in animals with a diet consisting mainly of plants?
    a. Incisors
    b. Canines
    c. Premolars
    d. Molars
    e. Cheek teeth
*Answer: b*

48. Which one of the following structures of a tubular digestive tract is involved mainly with water and ion recovery?
    a. Gizzard
    b. Buccal cavity
    c. Stomach
    d. Hindgut
    e. Midgut
*Answer: d*

*49. In the stomach, pepsin activates pepsinogen molecules in a positive feedback process called
    a. emulsification.
    b. absorption.
    c. autocatalysis.
    d. rumination.
    e. excretion.
*Answer: c*

50. Which layer of the vertebrate gut shows adaptations for increasing absorptive surface area?
    a. Lumen
    b. Mucosa
    c. Submucosa
    d. Serosa
    e. Peritoneum
*Answer: b*

51. Which layer of the vertebrate gut is a fibrous tissue covering the abdominal organs?
    a. Circular muscle layer
    b. Mucosa
    c. Submucosa
    d. Serosa
    e. Longitudinal muscle layer
*Answer: d*

52. Which structure is *not* encountered by a food bolus as it is being swallowed?
    a. Epiglottis
    b. Larynx
    c. Soft palate
    d. Pharynx
    e. Esophagus
*Answer: b*

*53. Which of the following statements about movement of food in the gut is *false*?
    a. Three sphincters are found in the vertebrate gut.
    b. Peristalsis can move food in both directions.
    c. Stretched smooth muscle contracts.
    d. Peristalsis begins when food enters the glottis.
    e. The muscle in a sphincter is normally contracted.
*Answer: d*

54. The sweetness that you taste if you keep chewing a piece of bread is due to the enzyme

    a. maltase.
    b. sucrase.
    c. lactase.
    d. amylase.
    e. pepsin.
*Answer: d*

*55. Pepsinogen is converted into pepsin by
    a. low pH.
    b. chyme.
    c. enterokinase.
    d. trypsinogen.
    e. amylase from the salivary glands.
*Answer: a*

56. Which of the following is *not* caused by the HCl secreted in the stomach?
    a. Activation of the principal zymogen of the stomach
    b. Proper pH for the digestive enzyme of the stomach
    c. Breakdown of ingested tissues
    d. Formation of chylomicrons
    e. Death of ingested bacteria
*Answer: d*

*57. Which of the following statements about digestion in the small intestine is *false*?
    a. Most digestion occurs in the duodenum of the small intestine.
    b. Bile is produced by the liver and stored in the gall-bladder, and it emulsifies fat in the small intestine.
    c. Bile molecules have one end that is lipophobic and one end that is hydrophilic.
    d. The pancreas secretes zymogens and bicarbonate ions into the duodenum.
    e. Enterokinase is secreted by intestinal mucosal cells.
*Answer: c*

*58. Which one of the following proteases is produced by the small intestine?
    a. Dipeptidase
    b. Chymotrypsin
    c. Trypsin
    d. Carboxypeptidase
    e. Pepsin
*Answer: a*

59. Which of the following statements about the role of the large intestine in digestion is *false*?
    a. If too little water is reabsorbed from the feces, diarrhea results.
    b. The large intestine produces digestive enzymes.
    c. *Escherichia coli* is a normal inhabitant of the large intestine.
    d. Flatulence results from the metabolism of intestinal bacteria.
    e. The appendix is a vestigial cecum.
*Answer: b*

*60. Which of the following activities is *not* carried out by the liver in the regulation of fuel metabolism?
    a. Conversion of nutrients into glycogen and fat
    b. Gluconeogenesis
    c. Synthesis of plasma proteins from amino acids
    d. Production of high-density lipoproteins (HDL) for deposit in adipose tissue
    e. Processing of chylomicrons from the small intestine
*Answer: d*

*61. Which of the following statements about the hormonal control of fuel metabolism is *false*?
a. During the absorptive period, the liver converts glycogen into glucose.
b. During the postabsorptive period, the body cells preferentially use fatty acids for metabolic fuel.
c. Cells of the nervous system depend almost exclusively on glucose for metabolic fuel.
d. Glucagon plays a major role during the postabsorptive period by causing the liver to convert glycogen into glucose and stimulating gluconeogenesis.
e. Cells of the pancreas release insulin during the absorptive period.
*Answer: a*

62. People living in impoverished regions, as well as people with alcohol addiction, frequently have a niacin deficiency called
a. beriberi.
b. kwashiorkor.
c. pellagra.
d. scurvy.
e. rickets.
*Answer: c*

63. Primates should eat citrus fruit to prevent scurvy, a deficiency of
a. niacin.
b. thiamin.
c. calciferol.
d. ascorbic acid.
e. biotin.
*Answer: d*

*64. Ruminant animals, such as goats and cows,
a. produce enzymes in their guts that break down cellulose.
b. practice coprophagy.
c. have a four-chambered stomach that provides a large surface area for better absorption of nutrients.
d. produce hydrogen sulfide, an important greenhouse gas.
e. get much of their protein from the digestion of microorganisms.
*Answer: e*

65. One of the consequences of smoking is a lowered level of _____, which are beneficial molecules because they accept cholesterol and probably remove it from the tissues to the liver.
a. Chylomicrons
b. Very-low-density lipoproteins
c. Low-density lipoproteins
d. High-density lipoproteins
e. Cholecystokinin
*Answer: d*

66. Which of the following molecules contain mostly triglycerides and transport them to fat cells in tissues.
a. Chylomicrons
b. Very-low-density lipoproteins
c. Low-density lipoproteins
d. High-density lipoproteins
e. Cholecystokinin
*Answer: b*

67. Lipid-soluble compounds
a. dissolve in water.
b. are often stored for a long time.
c. metabolize quickly.
d. are easily filtered by the kidney.
e. are hydrophilic.
*Answer: b*

*68. Which of the following is an example of bioaccumulation?
a. A bird that eats fish, which eat invertebrates, which eat algae, which pick up a pesticide in a stream
b. Lead building up in a child's liver
c. A mouse eating poison in a mousetrap
d. A factory constantly dumping toxins into a river
e. A bird accidentally getting sprayed with pesticide and contaminating its offspring
*Answer: a*

69. The purpose of the cytochrome P450s is to
a. make energy in the electron transport chain.
b. break down cytochrome into simple diglycerides.
c. detoxify synthetic chemicals.
d. break down toxins into smaller, less harmful molecules.
e. break down glycogen into glucose in the liver.
*Answer: c*

70. Polychlorinated biphenyls are dangerous because they
a. cause hormones to shut off.
b. bioaccumulate and cause cognitive impairment.
c. bind vitamin $B_1$ and cause a nutrient deficiency.
d. cause a thinning of bird eggshells.
e. replace iron in blood and calcium in bones.
*Answer: b*

71. The protein leptin
a. is produced by cells of the hypothalamus.
b. signals the brain about the status of body fat reserves.
c. is found in lower-than-normal levels in most obese humans.
d. serves as a positive feedback signal to the brain to limit food intake.
e. is an environmental toxin that accumulates in body fat.
*Answer: b*

# 51 Salt and Water Balance and Nitrogen Excretion

## Fill in the Blank

1. Organisms whose tissue fluids can have the same osmolarity as the environment are called _____.
*Answer: osmoconformers*

2. An organism that maintains the osmolarity of its tissue fluids above that of the environment is known as a _____ regulator.
*Answer: hyperosmotic*

★3. Marine bony fishes actively excrete salt from their kidneys and gills, and nitrogenous wastes are lost as ammonia from the _____.
*Answer: gills*

4. Tadpoles living in fresh water excrete nitrogenous wastes in the form of _____.
*Answer: ammonia*

5. The Malpighian tubules of insects conserve water and excrete _____.
*Answer: uric acid*

6. In the nephron of the vertebrate kidney, the glomerulus is a dense knot of thin-walled _____.
*Answer: capillaries*

7. In humans, urine from the collecting ducts passes through the _____ to the urinary bladder.
*Answer: ureter*

8. Resorption in the renal tubules is facilitated by _____, which provide extra surface area in the cells lining the tubule.
*Answer: microvilli*

★■9. A normal kidney has several _____ mechanisms, which are adaptations to monitor and maintain kidney functions.
*Answer: autoregulatory*

10. High levels of antidiuretic hormone cause the collecting ducts to become (more/less) _____ permeable to water, and highly concentrated urine is produced.
*Answer: more*

11. The human kidneys filter about 180 liters of blood per day, but they produce about 2–3 liters of urine per day. Therefore, the percentage of fluid volume that ends up in urine is about _____.
*Answer: 1–2 percent (2–3 l/180 l)*

★12. Sharks and rays secrete salt from the _____.
*Answer: rectal gland*

13. Renin is a regulatory enzyme released by the kidney. It acts on a circulating protein to begin converting that protein into an active hormone called _____.
*Answer: angiotensin*

14. In addition to water and carbon dioxide, the metabolism of proteins and nucleic acids also produces nitrogenous waste found most frequently in the form of _____.
*Answer: ammonia (or $NH_3$)*

15. Birds that eat marine animals ingest excess salt, which they excrete through their _____ glands.
*Answer: nasal salt*

16. The hormone produced by the heart that influences kidney function is _____.
*Answer: atrial natriuretic hormone*

★17. Differences among tissues in water permeability relate to the presence or absence of proteins called _____.
*Answer: aquaporins*

18. High levels of atrial natriuretic peptide (ANP) in the blood may indicate the condition known as _____.
*Answer: congestive heart failure*

## Multiple Choice

1. The extracellular fluids in our bodies
   a. have the same composition as seawater.
   b. have a much higher osmotic concentration than fluids in cells have.
   c. contain no proteins.
   d. have a fixed water concentration.
   e. can supply water and nutrients to cells.
   *Answer: e*

2. Water moves into tissues as a result of all of the following processes *except*
   a. osmotic gradient.
   b. fluid pressure.
   c. active transport.
   d. principles of diffusion.
   e. the salinity gradient.
   *Answer: c*

3. Ammonia and urea are waste products derived from the metabolic breakdown of
   a. carbohydrates.
   b. lipids.
   c. sugars.
   d. proteins.
   e. salts.
   *Answer: d*

■4. Marine invertebrates in which the salinity of body fluids changes with the osmotic potential of their environments are known as
   a. osmoconformers.
   b. osmoregulators.
   c. osmoexcretors.
   d. hypotonic.
   e. hypertonic.
   *Answer: a*

5. Which of the following molecules is most toxic to cells?
   a. Water
   b. Sodium chloride
   c. Ammonia
   d. Urea
   e. Uric acid
   *Answer: c*

6. Which of the following animals is most likely to excrete mostly urea or uric acid instead of ammonia?
   a. Freshwater fishes
   b. Saltwater fishes
   c. Tadpoles
   d. Shrimp
   e. Seagulls
   *Answer: e*

*7. Which of the following substances is the *least* soluble in water?
   a. Ammonia
   b. Uric acid
   c. Urea
   d. Sodium chloride
   e. Amino acids
   *Answer: b*

■8. Organisms that are ionic regulators maintain an
   a. osmotic concentration that may differ from the environment.
   b. osmotic concentration that is the same as the environment.
   c. ion concentration that may differ from the environment.
   d. ion concentration that is the same as the environment.
   e. ion concentration by diffusion only.
   *Answer: c*

9. Which of the following operates by filtering fluids into a tube, and then secreting or resorbing specific substances?
   a. Flame cells of flatworms
   b. Metanephridia of annelid worms
   c. Malpighian tubules of insects
   d. Vertebrate nephrons
   e. All of the above
   *Answer: e*

*10. In which of the following are nitrogenous waste solutes eliminated via the gut?
   a. Flame cells of flatworms
   b. Metanephridia of annelid worms
   c. Malpighian tubules of insects
   d. Vertebrate nephrons
   e. All of the above
   *Answer: c*

11. The functional unit of the kidney is the
   a. Bowman's capsule.
   b. capillary.
   c. glomerulus.
   d. nephron.
   e. renal tubule.
   *Answer: d*

12. Which process drives the process of filtration from the capillaries into the glomerulus?
   a. Active transport
   b. Arterial blood pressure
   c. Venous blood pressure
   d. Osmotic pressure
   e. Secretion
   *Answer: b*

13. During filtration, which of the following does *not* enter Bowman's capsule from the bloodstream?
   a. Water
   b. Glucose
   c. Ions
   d. Amino acids
   e. Plasma proteins
   *Answer: e*

■14. Analysis of kidney function in simple vertebrates suggests that the earliest vertebrate nephron functioned mainly to
   a. remove excess fluid.
   b. remove excess salts.
   c. remove excess nutrients.
   d. filter nitrogenous wastes.
   e. conserve water.
   *Answer: a*

15. Marine bony fishes acquire excess salt when they drink seawater. They handle this salt load in all of the following ways *except* by
   a. producing very dilute urine.
   b. producing very little urine.
   c. excreting ions from gills.
   d. secreting salt from the renal tubules.
   e. not absorbing some ions from the gut.
   *Answer: a*

16. Cartilaginous fishes maintain an osmotic concentration in their tissue fluids that matches that of seawater by
   a. excreting large amounts of water.
   b. drinking large amounts of seawater.
   c. concentrating seawater salts in their blood.
   d. concentrating urea in their tissue fluids.
   e. excreting salts at their gills.
   *Answer: d*

17. Water conservation adaptations of reptiles include all of the following *except*
   a. scaly skin.
   b. excretion of uric acid.
   c. shelled eggs.
   d. internal fertilization of gametes.
   e. excretion of salts.
   *Answer: e*

18. Water conservation adaptations of amphibians living in dry environments include all of the following *except*
   a. waxy secretions on the skin.
   b. estivation during dry periods.
   c. production of concentrated urine.
   d. burrowing into the ground.
   e. large urinary bladders.
   *Answer: c*

19. Which structure is found in the inner medulla of the kidney?
    a. Bowman's capsule
    b. Convoluted tubule
    c. Glomerulus
    d. Loop of Henle
    e. Both a and d
    *Answer: d*

20. Of the fluid that is filtered into Bowman's capsule, approximately what percent is resorbed into the blood within the kidney?
    a. Less than 5
    b. About 25
    c. About 50
    d. About 75
    e. Over 95
    *Answer: e*

21. The cells lining the proximal convoluted tubule have numerous microvilli and mitochondria. Such structures indicate that _____ is conducted in these cells.
    a. rapid diffusion of water
    b. active transport
    c. conservation of water
    d. storage of salts
    e. production of urea
    *Answer: b*

22. Valuable molecules like glucose, amino acids, and vitamins are resorbed into the blood at which location in the nephron?
    a. Bowman's capsule
    b. Collecting duct
    c. Glomerulus
    d. Loop of Henle
    e. Convoluted tubule
    *Answer: e*

**23. The loops of Henle are considered to function as a countercurrent multiplier because
    a. ascending and descending limbs have different permeabilities.
    b. the amount of water conserved is related to the length of the loop.
    c. as urine flows in the loop, a concentration gradient is established in the medulla.
    d. as urine flows in the loop, it becomes more concentrated in the nephron.
    e. active transport in the ascending and descending limbs multiplies the concentration gradient.
    *Answer: c*

*24. Which of the following responses would *not* correct for a drop in glomerular blood pressure?
    a. Angiotensin elevating the body's blood pressure
    b. Angiotensin stimulating thirst in the brain
    c. Angiotensin constricting arterioles entering the kidney
    d. Stretch receptors triggering antidiuretic hormone release
    e. Aldosterone stimulating sodium resorption
    *Answer: c*

25. The effect of antidiuretic hormone is to
    a. reduce permeability in the loop of Henle.
    b. reduce permeability of the collecting ducts.
    c. reduce blood volume.
    d. increase urine volume.
    e. increase urine concentration.
    *Answer: e*

26. Antidiuretic hormone secretion increases when the hypothalamus is stimulated by
    a. angiotensin receptors.
    b. glucose receptors.
    c. osmoreceptors.
    d. renin receptors.
    e. stretch receptors.
    *Answer: c*

27. _____ produce urine that is more concentrated than their body fluids.
    a. Birds
    b. Catfish
    c. Crocodiles
    d. Frogs
    e. Sharks
    *Answer: a*

28. In adult humans, urination is controlled by
    a. a smooth muscle sphincter at the base of the urethra.
    b. the autonomic nervous system.
    c. a skeletal muscle sphincter at the base of the urethra.
    d. the voluntary nervous system.
    e. All of the above
    *Answer: e*

29. Which of the following is a function of excretory systems?
    a. They help regulate osmotic potential and the volume of extracellular fluids.
    b. They excrete molecules that are present in excess.
    c. They conserve molecules that are valuable or in short supply.
    d. They eliminate toxic waste products of phosphorus metabolism.
    e. a, b, and c
    *Answer: e*

30. Excretory systems control
    a. filtration.
    b. excretion.
    c. resorption.
    d. homeostasis.
    e. All of the above
    *Answer: e*

31. Which of the following is true of the human kidney?
    a. Functional units are called hepatocytes.
    b. Nephrons consist of proximal and distal convoluted tubules as well as collecting ducts and loops of Henle.
    c. Most of the substances and water filtered from the blood in the glomerulus return to the venous blood draining the kidney.
    d. Anything coming into the kidney has to come from the renal artery.
    e. b, c, and d
    *Answer: e*

*32. The kidneys help regulate acid–base balance by controlling the level of _____ in the blood.
    a. $CO_2$
    b. $H^+$
    c. $HCO_3^-$
    d. Both b and c
    e. All of the above
    *Answer: d*

33. The terms "ammonotelic," "ureotelic," and "uricotelic" are used to describe
    a. the actions of hormones on the excretory system.
    b. the types of nitrogenous waste produced by various classes of vertebrates.
    c. pathways of kidney evolution.
    d. modifications of kidney tubules to enhance excretion.
    e. modes of excretory system development.
    *Answer: b*

34. Organisms living in a freshwater environment normally
    a. excrete copious dilute urine and retain salts.
    b. excrete a small volume of dilute urine and retain salts.
    c. excrete copious concentrated urine.
    d. excrete small amounts of concentrated urine.
    e. conserve both water and salts.
    *Answer: a*

35. A ureotelic organism excretes most of its nitrogenous waste as
    a. ammonia.
    b. water.
    c. carbon dioxide.
    d. uric acid.
    e. urea.
    *Answer: e*

36. Terrestrial organisms must conserve water. The least amount of water is lost with the excretion of which nitrogenous waste product?
    a. Carbon dioxide
    b. Uric acid
    c. Ammonia
    d. Salt
    e. Urea
    *Answer: b*

37. Gout results from an accumulation of _____ in the joints.
    a. carbon dioxide
    b. water
    c. ammonia
    d. uric acid
    e. urea
    *Answer: d*

38–42. Choose the appropriate excretory system or component from the list below.
    a. Protonephridia
    b. Nephron
    c. Malpighian tubule
    d. Green gland
    e. Metanephridia

38. Annelid worms are associated with which type of excretory system?
    *Answer: e*

39. The vertebrate system is associated with which type of functional unit?
    *Answer: b*

40. Flatworms are distinguished by which type of excretory unit?
    *Answer: a*

41. Insects and most terrestrial arthropods have an excretory system composed of which type of functional unit?
    *Answer: c*

42. The primary excretory system of humans utilizes which type of functional unit?
    *Answer: b*

43. Which of the following is *not* part of a nephron?
    a. Podocyte
    b. Flame cell
    c. Renal corpuscle
    d. Glomerulus
    e. Bowman's capsule
    *Answer: b*

44. Blood enters a nephron's vascular component by way of the
    a. peritubular capillaries.
    b. glomerulus.
    c. efferent arteriole.
    d. afferent arteriole.
    e. renal vein.
    *Answer: d*

45–54. Choose the appropriate anatomical feature from the list below. There may be more than one correct answer.
    a. Renal pyramids
    b. Proximal convoluted tubule
    c. Distal convoluted tubule
    d. Loop of Henle
    e. Collecting duct

45. Reside(s) anatomically closest to Bowman's capsule
    *Answer: b*

46. Function(s) as a countercurrent multiplier system
    *Answer: d*

47. Make(s) up the internal core of the medulla
    *Answer: a*

48. Site of glucose and amino acid resorption
    *Answer: b*

49. Site of major water resorption
    *Answer: b*

50. Permeability is under the control of antidiuretic hormone
    *Answer: Both c and e*

51. The length is related to the potential maximum urine concentration
    *Answer: d*

52. Connect(s) the proximal and distal convoluted tubules
    *Answer: d*

53. Location of microvilli
    *Answer: b*

54. Located principally within the medulla
    *Answer: a, d, and e*

55. Which of the following is the proper sequence of the passage of urine from the kidney?
    a. Renal pyramid, ureter, bladder, rectal gland, urethra
    b. Ureter, renal pyramid, rectal gland, bladder, urethra
    c. Renal pyramid, urethra, bladder, ureter
    d. Renal pyramid, ureter, bladder, urethra
    e. Rectal gland, renal pyramid, ureter, bladder, urethra
    *Answer: d*

56. A person with diabetes insipidus fails to respond to ADH. Which of the following is a symptom of this condition?
    a. Glucose in urine
    b. Copious hyperosmotic urine
    c. Copious dilute urine

d. Small volume of concentrated urine

e. Failure to urinate

*Answer: c*

57. Beer contains ethyl alcohol, which inhibits ADH production. After consuming beer, the kidney produces

a. copious concentrated urine.

b. a small volume of concentrated urine.

c. a small volume of dilute urine.

d. copious dilute urine.

e. None of the above

*Answer: d*

58. A freshwater fish continuously needs to regulate its internal environment. The fish tends to _____ water because it is _____ to fresh water.

a. gain; hypoosmotic

b. gain; hyperosmotic

c. lose; hypoosmotic

d. lose; hyperosmotic

e. lose; isoosmotic

*Answer: b*

59. To osmoregulate in fresh water, a fish must _____ salts and produce _____ urine.

a. conserve; copious dilute

b. excrete; copious dilute

c. conserve; small volumes of dilute

d. excrete; small volumes of dilute

e. absorb; large volumes of concentrated

*Answer: a*

*60. The brine shrimp, *Artemia,* can live in an extremely high saline environment because it

a. uses active transport to move salt across its gill membranes.

b. uses active transport to move water across its gill membranes.

c. excretes salts through its nasal salt glands.

d. osmoconforms with its surroundings.

e. excretes concentrated uric acid, so it conserves water.

*Answer: a*

61. Why don't terrestrial animals excrete nitrogen as ammonia?

a. It takes more energy to convert nitrogen to ammonia than it does to convert nitrogen to urea.

b. They would lose more water by excreting ammonia.

c. Ammonia is very toxic to terrestrial animals.

d. Ammonia is not stable and quickly converts to urea.

e. Ammonia is insoluble in water and therefore it cannot be excreted as urine.

*Answer: c*

*62. Which molecule contains the most nitrogen?

a. Ammonia

b. Uric acid

c. Urea

d. Both b and c

e. All have the same amount.

*Answer: b*

63–66. Choose the correct structure(s) from the list below.

a. Protonephridia

b. Metanephridia

c. Malpighian tubules

d. Nephron

e. Aquaporin

63. This excretory system contains beating cilia and is found in flatworms.

*Answer: a*

64. This excretory system actively transports uric acid and ions into blind tubules and is found in insects.

*Answer: c*

65. This excretory system has capillary beds that come into close contact with podocytes, where there is a transfer of water and small molecules. It is found in amphibians.

*Answer: d*

66. This excretory system has ciliated openings in the coelom, called nephrostomes, which lead to a tubule that opens to the outside of the animal. It is found in annelids.

*Answer: b*

# 52 Animal Behavior

## Fill in the Blank

1. _____ are simple stimuli that elicit highly stereotyped, species-specific patterns of behavior.
*Answer: Releasers*

2. Scientists in the field of _____ usually study animals under natural conditions and focus on behavior that is highly stereotyped and species-specific.
*Answer: ethology*

3. Male songbirds learn, and later express, their songs under the influence of the hormone _____.
*Answer: testosterone*

4. The _____ dance is performed by honeybees when food is within 80 meters of the hive.
*Answer: round*

5. Learning a particular releaser during a critical period is the phenomenon of _____.
*Answer: imprinting*

6. A bird's song is said to be _____ when it has been learned in the form in which it will be sung thereafter.
*Answer: crystallized*

7. Konrad Lorenz investigated _____ behavior in hybrid ducks.
*Answer: courtship*

8. If behavioral patterns can be artificially selected for in laboratory populations, then behavior in that species must be at least partially under _____ control.
*Answer: genetic*

9. Many observations indicate that organisms have an internal, or _____, clock.
*Answer: endogenous*

10. A circadian rhythm can be reset with an environmental cue by a process called _____.
*Answer: entrainment*

11. The simplest form of navigation, using landmarks, is called _____.
*Answer: piloting*

★12. If the _____ gland of a bird is removed, the bird will no longer have circadian rhythms.
*Answer: pineal*

13. The regular departure and return of organisms seasonally is called _____.
*Answer: migration*

■14. Morphological structures, physiological apparatus, and behavior of organisms are all shaped by _____.
*Answer: natural selection (or evolution)*

15. Molecules used for chemical communication between animals are called _____.
*Answer: pheromones*

■16. _____ is the transmission of learned behavior through generations.
*Answer: Culture*

■17. _____ shapes the morphology, physiology, and behavior of animals.
*Answer: Natural selection*

## Multiple Choice

■1. Imprinting
a. must occur during a critical period in development.
b. always occurs between parents and offspring.
c. is genetically determined and requires no learning.
d. is an encoding of simple information.
e. occurs equally well with a variety of stimuli.
*Answer: a*

★2. Which outcome is expected when adult rats have their ovaries or testes removed and are subsequently treated with female sex steroids?
a. Both sexes will exhibit lordosis (female mating behavior).
b. Neither sex will exhibit lordosis.
c. Females will exhibit lordosis, and males will exhibit no response.
d. Females will exhibit lordosis, and males will exhibit mounting behavior (male mating behavior).
e. Females will exhibit no response, and males will exhibit lordosis.
*Answer: c*

★3. What outcome in terms of lordosis (female mating behavior) or mounting (male mating behavior) is expected when newborn rats have their ovaries or testes removed, are treated with testosterone as newborns, and are treated again with testosterone as adults?
a. Both sexes exhibit mounting behavior.
b. Neither sex exhibits mounting or lordosis behaviors.
c. Males exhibit mounting, and females exhibit no response.
d. Males exhibit mounting, and females exhibit lordosis.
e. Both sexes exhibit lordosis.
*Answer: a*

4. Experiments on songbirds determined that the reason female birds normally do *not* sing is that they
   a. lack the proper muscles.
   b. lack specific parts of the nervous system required to sing.
   c. lack testosterone.
   d. are unable to learn a song.
   e. have learned sex-specific behavior from a nonsinging mother.
   *Answer: c*

5. Anatomical comparisons of male and female songbirds reveal that _____ of male birds increase(s) in volume during periods of active singing.
   a. the throat muscles
   b. specific regions of the brain
   c. the pineal gland
   d. Both a and c
   e. a, b, and c
   *Answer: b*

*6. When purebred nonhygienic bees are crossed with purebred hygienic bees, they produce dihybrid nonhygienic offspring. If these offspring are backcrossed with the purebred hygienic bees, which type of behavior is *not* exhibited in the next set of offspring?
   a. Nonhygienic
   b. Nonhygienic, but the bees will remove pupae from uncapped cells
   c. Nonhygienic, but the bees will uncap cells of dead pupae
   d. Hygienic, but only when neighboring bees are hygienic
   e. Hygienic
   *Answer: d*

7. Male gypsy moths travel thousands of meters to reach female gypsy moths. Their behavior is a response to
   a. visual signals.
   b. auditory signals.
   c. pheromones.
   d. random search patterns.
   e. electric signals.
   *Answer: c*

8. The observation that potato washing spread through a population of Japanese macaques demonstrated that
   a. animals are capable of associative learning.
   b. animals other than humans are capable of transmitting learned behavior through generations.
   c. animals can optimally forage.
   d. behavior can be inherited.
   e. None of the above
   *Answer: b*

9. Chemical territory marking can convey all of the following information *except*
   a. height.
   b. direction of travel.
   c. individual identity.
   d. reproductive status.
   e. elapsed time.
   *Answer: b*

10. Which of the following is *not* a feature of visual communication?
   a. Ease of production
   b. Endless variety

   c. Effectiveness over long distances
   d. Possibility of rapid change
   e. Indications of the signaler's position
   *Answer: c*

11. When a food source is located in the direction of the sun, the direction of a honeybee's waggle dance on a vertical surface will be
   a. straight up.
   b. straight down.
   c. 90° to the right.
   d. 90° to the left.
   e. 45° to the right.
   *Answer: a*

12. The feature of a honeybee's waggle dance that indicates the distance to a food source is the _____ of the waggles.
   a. frequency
   b. speed
   c. direction
   d. size
   e. shape
   *Answer: b*

13. Insects use their antennae in which two forms of communication?
   a. Tactile and auditory
   b. Tactile and chemical
   c. Chemical and auditory
   d. Visual and tactile
   e. Visual and chemical
   *Answer: b*

14. Which of the following statements about the molecular genetics of fruit fly behavior is *false*?
   a. The gene *fruitless (fru)* controls male courtship behavior.
   b. A hierarchy of genes controls sexual differentiation and behavior.
   c. Fruit flies without the *transformer (tra)* gene develop into males anatomically and behaviorally.
   d. The *doublesex (dsx)* gene controls anatomical differentiation in males.
   e. Many single genes control complex behavior.
   *Answer: e*

15. Visual communication is better than auditory communication
   a. in dark environments.
   b. for conveying complex information.
   c. when the sender and the receiver are separated by a barrier.
   d. in terms of getting the attention of the receiver.
   e. over long distances.
   *Answer: b*

*16. Which of the following suggests that learning plays a role in a spider's web spinning?
   a. A newly hatched spider can spin a perfect web.
   b. Every web structure is the same.
   c. Web structure varies from species to species.
   d. A web size changes as an individual grows.
   e. A web is changed after contact with particular prey.
   *Answer: e*

■17. If behavior is under partial genetic control,
   a. no variability in the behavior will be seen.
   b. it is not subject to natural selection.
   c. it may be modified by experience.
   d. it is acquired only through learning.
   e. it will be species-specific.
   *Answer: c*

18. Which of the following is true of animal communication?
   a. It is shaped by natural selection.
   b. It may benefit the sender and the receiver.
   c. The particular channel of communication is shaped by the environment.
   d. Some species exploit the communication systems of other species.
   e. All of the above
   *Answer: e*

19. In most cases, the behaviors of animals
   a. are instinctive.
   b. are learned.
   c. have both genetic and learned components.
   d. are maladaptive.
   e. are not shaped by natural selection.
   *Answer: c*

20. A releaser
   a. is required in order for learning to occur.
   b. triggers stereotypic, species-specific behavior.
   c. triggers imitative learning.
   d. facilitates recognition of appropriate mates.
   e. causes an animal to remember the full stimulus.
   *Answer: b*

21. Experiments on interactions between herring gulls and their chicks showed that
   a. parental head shape is the releaser for begging by chicks.
   b. a red dot and/or a long thin bill release begging by chicks.
   c. chicks prefer parents with contrasting marks on their bills.
   d. appropriate sounds as well as visual stimuli are required to elicit begging behavior.
   e. parents give more food to visually discriminating chicks.
   *Answer: b*

22. Questions about the selective pressures that shaped a particular behavior are relevant for understanding the
   a. stimuli that elicit the behavior.
   b. underlying neural and hormonal mechanisms of the behavior.
   c. proximate causes of the behavior.
   d. ultimate causes of the behavior.
   e. All of the above
   *Answer: d*

23. Which of the following is critical to the production of normal song by adult male white-crowned sparrows?
   a. Auditory feedback after the song has crystallized
   b. Auditory feedback during the time the bird is matching its vocal output to memorized song
   c. Exposure to the song of his mother
   d. Exposure to the song of an adult male during a critical period
   e. Both b and d
   *Answer: e*

■24. Which of the following behaviors would be expected to be genetically programmed?
   a. Behavior that can be learned by trial and error
   b. Behavior that leads to successful mating
   c. Behavior that does not endanger an animal
   d. Behavior that involves social cues
   e. Behavior that is important in variable circumstances
   *Answer: b*

25. Which of the following statements about the singing behavior of birds is true?
   a. Male and female birds have song-control regions in their brains.
   b. Rising levels of testosterone cause song-control regions to decrease in size in the spring.
   c. Males isolated after their songs have crystallized produce abnormal song.
   d. Males deafened after their songs have crystallized produce abnormal song.
   e. None of the above
   *Answer: a*

26. What can behaviorists learn from hybridization experiments?
   a. A behavior is often coded by a single gene.
   b. Behavioral traits are usually all inherited from a single parent.
   c. A hybrid may show behavior that is intermediate between that of its parents.
   d. Hybrids show behavior patterns that are completely different from those of either parent.
   e. Only simple animals have genetically programmed behavior.
   *Answer: c*

27. One behavior that is inherited in a simple Mendelian manner is
   a. fruit fly courtship.
   b. lovebird nest-building.
   c. honeybee brood cleaning.
   d. dog retrieving behavior.
   e. rattlesnake prey capture.
   *Answer: c*

28. Circadian rhythms are characterized by
   a. seasonal changes.
   b. lack of synchrony with the environment.
   c. a tendency to lengthen with time.
   d. persistence in the absence of cues.
   e. changes with the animal's age.
   *Answer: d*

29. When a person changes time zones suddenly, as in jet travel, the circadian rhythm
   a. is destroyed temporarily.
   b. needs entrainment.
   c. adjusts within a few hours.
   d. becomes random.
   e. takes on a different periodicity.
   *Answer: b*

30. The circadian clock in mammals appears to be located in the
   a. hypothalamus.
   b. suprachiasmatic nuclei (SCN).
   c. pituitary.
   d. reticular activating system.
   e. visual cortex.
   *Answer: b*

31. A bird will be unable to entrain its circadian rhythms if light is blocked from reaching its
    a. eyes.
    b. optic nerve.
    c. pineal gland.
    d. pituitary.
    e. visual cortex.
    *Answer: c*

32. The adaptive advantage of circannual rhythms is
    a. that breeding is timed to coincide with peak resources.
    b. maximum in tropical species.
    c. that it entrains circadian rhythms.
    d. minimal in migratory birds.
    e. maximal in photoperiodic species.
    *Answer: a*

33. Homing pigeons released at a remote, unfamiliar site can find their way home by
    a. following visual landmarks.
    b. retracing the route along which they were carried.
    c. searching randomly until they locate home.
    d. following other birds.
    e. navigating without visual cues.
    *Answer: e*

34. Which of the following is an experimental method designed to test how migrating animals navigate?
    a. Food deprivation
    b. Testosterone injection
    c. Capture and displacement
    d. Captivity
    e. Induced migration at the wrong season
    *Answer: c*

*35. Biological rhythms play a role in the behavior of migratory species, as evidenced by
    a. their tendency to migrate over great distances.
    b. the phenomenon of migratory restlessness.
    c. their tendency to migrate at night.
    d. the phenomenon of orientation toward the sun.
    e. their tendency to form flocks.
    *Answer: b*

*36. Clock-shifting experiments revealed that birds determine direction from
    a. the sun.
    b. the stars.
    c. wind direction.
    d. sound cues.
    e. air pressure cues.
    *Answer: a*

37. Experiments with young songbirds in a planetarium suggested that birds can use star patterns of orientation if they
    a. are tested in a rotating star pattern.
    b. learned in a stationary star pattern.
    c. learned in a rotating star pattern.
    d. learned in the Northern Hemisphere.
    e. do not see the stars until adulthood.
    *Answer: c*

38. Glass knife fish, which live in murky waters and cannot see well, communicate by means of
    a. sound waves.
    b. chemical signals.
    c. tactile signals.
    d. electric signals.
    e. water vibrations.
    *Answer: d*

39. Which of the following is true concerning animal orientation systems?
    a. Each species has one sensory mechanism for orientation.
    b. All rely on light systems.
    c. All rely on bicoordinate navigation.
    d. They are based on redundant cues.
    e. An internal clock is usually one component.
    *Answer: d*

40. What is the scientific data suggesting that pigeons use magnetic cues for orientation?
    a. A neurophysiological magnetic transducer has been detected and described.
    b. Orienting behavior can be observed even on cloudy days.
    c. The attachment of small magnets disrupts pigeons' homing behavior.
    d. Pigeons seem to gain latitude information from magnetic field lines.
    e. Magnetite particles have been found in the cells of pigeons.
    *Answer: c*

41. Which of the following suggests that humans have some innate, unlearned motor patterns?
    a. All humans feel anger and fear.
    b. All cultures teach their young.
    c. Humans have a large capacity to learn.
    d. Blind infants do not smile normally.
    e. All cultures have similar facial expressions.
    *Answer: e*

*42. Genetically determined behaviors are expected to be common in species in all of the following circumstances *except* when
    a. there is no opportunity to learn.
    b. it is possible to learn the wrong behavior.
    c. environmental conditions are complex and changing.
    d. generations do not overlap.
    e. individual members of the species are dealing with dangerous prey.
    *Answer: c*

43. Human behavior is largely influenced by
    a. releasers.
    b. genetic traits.
    c. instinct.
    d. learning.
    e. orienting.
    *Answer: d*

44. Which of the following statements about web spinning in spiders is *false*?
    a. Juvenile spiders build a perfect web the first time.
    b. Instinctive behavior, such as web spinning, is not studied by ethologists.
    c. Web spinning is genetically determined.
    d. Web design and construction are species-specific.
    e. None of the above
    *Answer: b*

45. Which of the following is *not* part of the definition of instinctive behavior?
    a. It is stereotypic.
    b. It is genetically determined.
    c. It is unmodified by learning.
    d. It is inherited in a Mendelian fashion.
    e. It is species-specific.
    *Answer: d*

46. Which of the following is *not* an example of a stereo-typed, species-specific behavior?
    a. Web construction in orb-weaving spiders
    b. Food burying in squirrels
    c. Behaviors that are expressed during a deprivation experiment
    d. Song acquisition in the white-crowned sparrow
    e. Bill pecking in herring gull chicks
    *Answer: d*

47. If an animal shows a particular behavior during a depri-vation experiment, then the
    a. behavior may be instinctive.
    b. behavior has a strong learning component.
    c. correct releaser was present.
    d. behavior has substantial genetic determinants.
    e. a, c, and d
    *Answer: e*

48. Which of the following is *not* a releaser?
    a. The shape of the adult head that a herring gull chick sees
    b. The bill spot of the adult that a herring gull chick sees
    c. A tuft of red feathers seen by a male European robin
    d. The image of a duckling's parent to a duckling
    e. The presence of a nut for a tree squirrel
    *Answer: a*

49. Which of the following is true of cricket song?
    a. Cricket songs are species-specific.
    b. Only male crickets "sing."
    c. Song pattern is genetically determined.
    d. Female preference for song is genetically determined.
    e. All of the above
    *Answer: e*

50. Which of the following statements about imprinting is *false*?
    a. Imprinting consists of rapid learning that occurs during a critical period.
    b. Imprinting depends on either visual or auditory cues in mammals.
    c. Imprinting makes it possible for an animal to encode complex information in the nervous system.
    d. Imprinting occurs during song acquisition of some species of birds.
    e. Parent–offspring recognition is often dependent on imprinting.
    *Answer: b*

51. Which of the following statements about hygienic behav-ior in honeybees is *false*?
    a. Hygienic behavior in honeybees can be explained in a simple Mendelian manner.
    b. The strain showing hygienic behavior is resistant to a disease that kills larvae.
    c. In a cross of the hygienic and nonhygienic strains, all of the F₁ offspring are of one phenotype.

    d. The offspring from a backcross show a 3:1 ratio.
    e. The behavior of hybrids suggests that there are no separate genes for uncapping cells and for removing infected larvae.
    *Answer: e*

52. The *per* or "period" gene in fruit flies
    a. has mutations that cause flies to have short or long free-running circadian periods.
    b. is homologous to clock genes found in many other organisms.
    c. shows circadian rhythms during transcription.
    d. may form part of the molecular basis for circadian rhythms in this species.
    e. All of the above
    *Answer: e*

53. Which of the following scientists was an ethologist?
    a. Konrad Lorenz
    b. Niko Tinbergen
    c. Karl von Frisch
    d. All of the above
    e. None of the above
    *Answer: d*

54. Which of the following statements about circadian rhythms is *false*?
    a. Circadian rhythms are seldom exactly 24 hours long.
    b. Entrainment is the process of bringing two rhythms into phase.
    c. A rhythm that is shorter than 24 hours must be phase-advanced to remain in phase with the 24-hour cycle of light and dark.
    d. Animals in constant darkness exhibit circadian rhythms.
    e. The period of a rhythm is the length of one cycle.
    *Answer: b*

55. Which of the following statements about endogenous clocks is *false*?
    a. There are limits to the entrainment ability of endoge-nous clocks.
    b. The endogenous clock of a mammal is located in the suprachiasmatic nuclei.
    c. Endogenous clocks are found in virtually every animal group.
    d. Cells in the eyes of birds function as endogenous clocks.
    e. Many plants have endogenous clocks.
    *Answer: d*

*56. A blind bird with a circadian rhythm of 22 hours is fitted with a black cap that covers the top of its head. It is placed into an environmental chamber with a light–dark period of exactly 12 hours. After several days, the bird will
    a. become phase-advanced.
    b. become phase-delayed.
    c. lose all circadian rhythms.
    d. develop a free-running rhythm.
    e. be unaffected.
    *Answer: d*

57. Which of the following statements about piloting is *false*?
    a. Piloting is used by some wasps to find their nests.
    b. Piloting is useful only for short-distance migrations.
    c. Gray whales use piloting while moving from their wintering grounds to their summering grounds.
    d. Water currents can be used as piloting cues.
    e. Wind patterns can be used as piloting cues.
    *Answer: b*

58. Which of the following statements about homing is *false*?
    a. Simple piloting can explain many cases of homing.
    b. Homing pigeons fitted with frosted contact lenses are still able to home.
    c. Piloting is not sufficient to explain homing in the homing pigeon.
    d. Marine birds show many examples of homing over great distances.
    e. In most migratory bird species, young learn to home by following the adults.
    *Answer: e*

59. Studies of European starlings that have been transported to and released from different areas during their migratory period suggest that the starlings use the type of navigation called
    a. piloting.
    b. distance-and-direction navigation.
    c. bicoordinate navigation.
    d. random search navigation.
    e. homing.
    *Answer: b*

*60. Research has shown that in migratory species there is a correlation between the duration of _____ and the _____ the goal.
    a. migratory restlessness; distance to
    b. migratory restlessness; direction of
    c. homing; distance to
    d. homing; direction of
    e. circadian rhythms; distance to
    *Answer: a*

61. In bicoordinate navigation, an animal knows all of the following *except*
    a. its initial latitude and longitude.
    b. the latitude of its goal.
    c. the distance to its goal.
    d. the sun's position at its current location and at its goal.
    e. its initial longitude.
    *Answer: d*

62. A sun compass requires that an animal know
    a. its current location.
    b. its latitude and longitude.
    c. the current time.
    d. the sun's direction at the goal.
    e. the latitude and longitude of its goal.
    *Answer: c*

*63. A bird is trained to feed from the southeast end of a circular cage. At sunrise, a mirror is used to shift the sun 45° to the right of it real position. At which location in the cage will the bird attempt to feed?
    a. North
    b. East
    c. South

d. West
    e. Southwest
    *Answer: c*

*64. A bird is trained to feed from the west end of a circular cage. At sunrise, a fixed light is set up at the south end of the cage to mimic the appearance of the sun. Where will the bird attempt to feed at noon?
    a. North
    b. East
    c. South
    d. West
    e. Northwest
    *Answer: d*

*65. A bird is trained to feed from the south end of a circular cage. The bird is then placed into a light-controlled room and phase-advanced by 6 hours. If the bird is returned to the circular cage at noon, where will it search for food?
    a. North
    b. East
    c. South
    d. West
    e. Northwest
    *Answer: b*

*66. An animal is displaced such that the sun rises earlier and is lower in the sky than it would be at its home position. In which direction should the animal move in order to home?
    a. Southwest
    b. Southeast
    c. Northwest
    d. Northeast
    e. South
    *Answer: a*

67. A king snake must clamp down on the mouth of its prey, the rattlesnake, to avoid being bitten and killed by the rattlesnake. This type of behavior is
    a. learned.
    b. genetically programmed.
    c. imprinted.
    d. dependent upon hormone release.
    e. entrained.
    *Answer: b*

68. The fruit fly, *Drosophila*, makes a good subject for hybridization studies because
    a. it has a short life span.
    b. it produces large numbers of offspring.
    c. its behavior traits can be traced to simple Mendelian segregation.
    d. Both a and b
    e. a, b, and c
    *Answer: d*

69. Which of the following behaviors conveys information about the direction of a food source in honeybees?
    a. Round dance
    b. Odor on the body of a returning forager
    c. Angle of the straight run of the waggle dance
    d. Speed of the waggle dance
    e. Speed of the round dance
    *Answer: c*

# 53 *Behavioral Ecology*

## Fill in the Blank

1. Species that have social systems with sterile classes are termed _____.
   *Answer: eusocial*

2. Members of social groups often have _____ exposure to disease compared to solitary animals.
   *Answer: increased*

3. Members of social groups usually have _____ protection from predators compared to solitary animals.
   *Answer: increased*

4. _____ may cause traits that reliably signal male quality to become exaggerated through evolution.
   *Answer: Sexual selection*

5. _____ involves the defense of an area that contains food, nesting sites, or other resources.
   *Answer: Territorial behavior*

6. In white-fronted bee-eaters, about half the nonbreeding individuals help at the nests of other individuals. Helping relatives _____ fitness, and thus white-fronted bee-eaters usually help relatives rather than nonrelatives.
   *Answer: increases*

7. Ecologists often analyze their observations of animal behavior in terms of the costs and _____ to the performer.
   *Answer: benefits*

8. An important variable in determining the costs and benefits of altruism is the _____ between the performer and recipient.
   *Answer: relatedness*

9. The _____ cost of a behavior is the difference between the energy an animal would have expended had it rested, and the energy expended in performing the behavior.
   *Answer: energetic*

10. The _____ cost of a behavior is an animal's increased chance of being injured or killed as a result of performing it, as opposed to resting.
    *Answer: risk*

11. The _____ cost of a behavior is the sum of the benefits the animal forfeits by not being able to perform other behaviors during the same time interval. For example, an animal forfeits the possibility of feeding when it defends its territory.
    *Answer: opportunity*

## Multiple Choice

1. An elephant seal's environment is defined as all
   a. other elephant seals in the same area.
   b. other organisms that influence it.
   c. the physical characteristics of its area.
   d. other organisms of its food web.
   e. the physical and biological factors that influence it.
   *Answer: e*

2. Which of the following would be an appropriate study in the field of ecology?
   a. A comparison of vertebrate skeletal structures
   b. A study of the effect of adrenaline on human heart rate
   c. An investigation of the roles of hormones in plant growth
   d. A study of the effect of fire on forest animals' populations
   e. A classification of fungi into various groups
   *Answer: d*

3. _____ ecology is the study of how animals make decisions that influence their survival and reproductive success.
   a. Natural
   b. Social
   c. Behavioral
   d. Physical
   e. Environmental
   *Answer: c*

4. The term for the environment in which an organism lives is its
   a. community.
   b. ecology.
   c. habitat.
   d. niche.
   e. population.
   *Answer: c*

5. Red abalone larvae will settle on coralline algae substrates only after recognizing a chemical that the algae produce. As an example of a habitat selection cue, this chemical probably is
   a. required for abalone reproductive success.
   b. produced by the algae in order to attract the abalone.
   c. a good predictor of conditions suitable for abalone survival.
   d. a food source for the abalone.
   e. a waste by-product of the algae.
   *Answer: c*

6. Bluegill sunfish are put in two tanks with their prey, water fleas. Each tank contains an equal proportion of small, medium, and large water fleas. One tank has a low density of water fleas, and the other has a high flea density. According to foraging theory, the bluegill sunfish should eat
   a. an equal proportion of the three flea sizes at both densities.
   b. a greater proportion of large fleas at both densities.
   c. equal proportions of the three flea sizes at low density and mostly large fleas at high density.
   d. mostly large fleas at low density and equal proportions of the three flea sizes at high density.
   e. mostly medium and large fleas at low density and mostly large fleas at high density.
   *Answer: c*

7. Which of the following is *not* a necessary assumption of foraging theory?
   a. Efficient predators spend less time on predation than noneficient predators do.
   b. Superior predators can produce more offspring.
   c. Superior foraging ability can be genetically based.
   d. Predators choose prey in ways that maximize their energy intake.
   e. Efficient predators always choose the most abundant prey.
   *Answer: e*

8. Some species of animals may eat soil in particular areas. This behavior results from a need for _____ in the animal's diet.
   a. water
   b. carbohydrates
   c. minerals
   d. oxygen
   e. energy
   *Answer: c*

9. When wood pigeons draw together in a tight group in response to the presence of a hawk, they are exhibiting _____ behavior.
   a. altruistic
   b. flocking
   c. selfish
   d. spiteful
   e. territorial
   *Answer: b*

10. When a dog defends its food by attacking a second dog, the first dog is exhibiting _____ behavior.
    a. altruistic
    b. cooperative
    c. selfish
    d. spiteful
    e. territorial
    *Answer: e*

11. Kin selection is
    a. the mating of relatives.
    b. the recognition of relatives among societal animals.
    c. the adoption of young by generally unrelated adults.
    d. a process of forcing young males out of a society while keeping the females.
    e. a behavior that increases the survivorship of an individual's relatives.
    *Answer: e*

12. Altruistic acts are most likely to evolve into behavior patterns when the participants
    a. are capable of learning.
    b. have individual fitness.
    c. are genetically related.
    d. are largely nonsocial.
    e. are mating partners.
    *Answer: c*

■13. If a female animal who is reproductively mature rejects a courting male of the same species, she most likely determined that the male was
    a. monogamous.
    b. not from her social group.
    c. polygynous.
    d. not healthy.
    e. None of the above
    *Answer: d*

14. Although exceptions exist, a female animal generally _____ her reproductive success by mating with many males.
    a. reduces
    b. increases
    c. doubles
    d. does not change
    e. None of the above
    *Answer: d*

15. Together, individual fitness and kin selection determine the _____ fitness of an individual.
    a. social
    b. selective
    c. exclusive
    d. inclusive
    e. reciprocal
    *Answer: d*

16. Male courtship behavior accomplishes all of the following *except*
    a. inducing a female to mate with the male.
    b. displaying that the male is in good health.
    c. showing that the male is a good provider.
    d. conveying that the male has successfully mated in the past.
    e. signaling that the male has a good genotype.
    *Answer: d*

17. After copulation, a male may remain with a female for a time and prevent her from copulating with other males. Although this behavior has its benefits, it also carries a high _____ cost.
    a. energetic
    b. opportunity
    c. risk
    d. competitive
    e. altruistic
    *Answer: b*

■18. After copulation, many male insects are genetically programmed to stay near the female for a time. This behavior is adaptive because it
    a. guards the female against injury.
    b. prevents the female from copulating again until after fertilization.
    c. allows the male to recover from the energetic cost of mating.

d. defends a nesting territory.

e. creates a social bond between the male and female.

*Answer: b*

19. The most widespread social system among animals is called a
    a. family.
    b. pack.
    c. clan.
    d. flock.
    e. group.

    *Answer: a*

20. Females may choose a mate based on all of the following *except*
    a. the quality of the resources in his territory.
    b. his display of food items.
    c. his apparent good health.
    d. his physical features.
    e. the probability that his sperm will fertilize the eggs.

    *Answer: e*

21. Female reproductive success is improved by choosing a male with all of the following features *except* his
    a. high genetic quality.
    b. good health.
    c. interaction with other mates.
    d. high parental care.
    e. control of abundant resources.

    *Answer: c*

22. Parental care
    a. increases the chances of survival of the parent.
    b. increases the parent's ability to produce additional offspring.
    c. increases the chances of survival of the offspring.
    d. decreases as the number of offspring increases.
    e. decreases when costs to parents are lowered.

    *Answer: c*

■23. Which of the following characteristics is most likely to occur in a bird species having brightly colored males and dull-colored females?
    a. The males pursue the females during courtship.
    b. The species has a monogamous mating system.
    c. Offspring are raised by relatives of the parents.
    d. The species forms large nesting colonies.
    e. The relative size of the females is small compared to the males.

    *Answer: a*

24. The most widespread social system among animals is
    a. parents and offspring.
    b. peer grouping.
    c. several females and their offspring.
    d. several pairs and their offspring.
    e. a dominant pair, their offspring, and unrelated subordinates.

    *Answer: a*

25. Among mammals, which offspring, if any, tend to leave their parents' group?
    a. Both males and females leave.
    b. Females leave; males leave only if a new territory becomes available.
    c. Males leave; females do not.

d. Neither males nor females leave.

e. Males leave, but females leave only if a new set of offspring is born.

*Answer: c*

26. Large numbers of sterile individuals occur in all of the following *except*
    a. termites.
    b. ants.
    c. bees.
    d. beetles.
    e. wasps.

    *Answer: d*

27. _____ theory tries to answer questions related to how an animal looks for and acquires food.
    a. Storage
    b. Foraging
    c. Social
    d. Animal
    e. None of the above

    *Answer: b*

28. In species with diploid females and haploid males, which pair of relatives is most similar genetically?
    a. Mother and daughter
    b. Mother and son
    c. Father and son
    d. Full sisters
    e. Full brothers

    *Answer: d*

29. Which of the following groups exhibits eusocial behavior and has diploid females and haploid males?
    a. Penguins
    b. Bees
    c. Jackals
    d. Termites
    e. Naked mole rats

    *Answer: b*

30. W. D. Hamilton hypothesized that eusociality evolved because sterile worker females are genetically more related to their sisters than to any other relatives. According to this hypothesis, which one of the following would *not* be expected?
    a. Queens would produce equal numbers of sons and daughters.
    b. Workers would feed their sisters better than they feed their brothers.
    c. If the original queen is lost, the workers would not favor the new queen's daughters.
    d. Worker females would have the same father and mother.
    e. Both males and females would be diploid.

    *Answer: e*

31. Which of the following is *not* true of eusociality?
    a. Eusocial species form elaborate nests or burrows.
    b. Helping behavior may be a necessary prerequisite for the evolution of eusociality.
    c. Some eusocial species contain only diploid members.
    d. Eusociality may occur in one species and be lacking in a closely related species.
    e. Eusocial individuals typically live part of their lives away from their society.

    *Answer: e*

32. Large, hoofed, plant-eating mammals tend to
    a. feed preferentially on high-protein foods.
    b. live in herds.
    c. feed in forests.
    d. have monogamous mating patterns.
    e. have higher metabolic rates than smaller mammals do.
    *Answer: b*

33. Which is *not* true of social systems?
    a. A social system can be studied by asking how individuals in the system benefit from it.
    b. Social systems are dynamic; individuals' relationships with one another change constantly.
    c. Relationships within a social system are determined partly by genetic relatedness.
    d. Social systems have evolved primarily because societies have an increased success at obtaining food.
    e. Social systems continue to evolve in relation to animals' sizes, diets, and habitats.
    *Answer: d*

34. The benefits lost by an animal that is not able to feed while defending its breeding territory is an example of _____ cost.
    a. opportunity
    b. energetic
    c. risk
    d. feeding
    e. breeding
    *Answer: a*

35–38. Match the social acts in the list below with the correct descriptions.
    a. Altruistic act
    b. Eusocial act
    c. Kin selection
    d. Cooperative act

35. Willow leaf beetles from the same clutch feed together. Beetles in groups can initiate feeding on tough leaves more easily than lone beetles can, and there are group defenses against predators.
    *Answer: d*

36. Insects in a colony defend the colony against intruders.
    *Answer: b*

37. A bird assists its parents in raising offspring.
    *Answer: c*

38. Female mammals help other unrelated females raise their young.
    *Answer: a*

39. Which of the following is *not* a consequence of parental care?
    a. Increased chance of survival of the offspring
    b. Increased care of each offspring
    c. Increased chance of survival of the parent
    d. Increased fitness of the parent
    e. Decreased chance of production of additional offspring
    *Answer: c*

■40. Paternal care by male mammals is often less than that of birds. The reason for this difference may be that
    a. young mammals are less likely to be eaten by predators and thus need less care.
    b. paternity is more certain in birds than in mammals.

c. there is less cost to birds for paternal care.
    d. male mammals tend to leave, or are driven out of their social group.
    e. newborn mammals are more self-sufficient than newly-hatched birds are.
    *Answer: d*

41. Fitness for each sex is usually maximized in a mating cycle when males copulate _____ and females copulate _____.
    a. many times; once
    b. once; once
    c. many times; many times
    d. a few times; many times
    e. a few times; once
    *Answer: a*

42. Group living
    a. can result in foraging benefits.
    b. reduces exposure to disease.
    c. is usually advantageous when food is scarce.
    d. has no effect on reproductive success.
    e. usually decreases competition between individuals.
    *Answer: a*

43. To construct a hypothesis about how a foraging animal should behave, a scientist first specifies the objective of the behavior and then attempts to determine the behavioral choices that would best achieve that objective. This process is called _____ modeling.
    a. minimal
    b. optimality
    c. behavioral
    d. forager
    e. natural
    *Answer: b*

44. Parental care is an example of _____ behavior.
    a. selfish
    b. spiteful
    c. territorial
    d. altruistic
    e. reciprocal
    *Answer: d*

■45. Females are usually choosier than males with respect to mates because
    a. they typically invest more energy per offspring than males do.
    b. they typically have more total offspring than males do.
    c. males are more variable, so females have more to choose from.
    d. only females can nurse young in most animal species.
    e. they pass on more genes to their offspring than males do.
    *Answer: a*

46. Male mammals are most likely to provide care for young when
    a. fertilization is internal.
    b. the males are part of a mating pair.
    c. other males exhibit parental care.
    d. the males mate with many females.
    e. certainty of paternity is low.
    *Answer: b*

47. In some birds, individuals other than the parents help in rearing offspring. These helpers are usually
    a. unrelated to the parents.
    b. close relatives of the parents.
    c. unrelated females that have lost their own offspring.
    d. members of a different altruistic species.
    e. cuckolding males.
    *Answer: b*

48. When helpers participate in rearing young, the fitness of parental Florida scrub jays
    a. decreases due to competition with the helpers.
    b. decreases due to reduced interaction with their offspring.
    c. increases because the helpers often feed the parents.
    d. increases due to increased offspring survival.
    e. is usually unchanged.
    *Answer: d*

49. In many mammals, including primates,
    a. females are more likely than males to remain in natal groups.
    b. males are more likely than females to remain in natal groups.
    c. both males and females usually leave natal groups.
    d. both males and females typically remain in natal groups.
    e. natal groups rarely persist once young are independent.
    *Answer: a*

50. Baboons travel in groups that are defended from predators by adult males. This behavior of the males probably _____ the baboons' fitness because it _____.
    a. increases; increases their survival
    b. increases; increases survival of their offspring
    c. decreases; decreases their survival
    d. decreases; decreases survival of their offspring
    e. is unrelated to; has no effect on their survival
    *Answer: b*

51. An example of eusociality is
    a. female birds throwing each other's eggs out of the communal nest.
    b. the existence of a few mating females in an insect colony.
    c. male bees stealing one another's mates.
    d. male elephant seals defending harems against one another.
    e. parent birds feeding their offspring.
    *Answer: b*

52. Behavioral ecology studies
    a. how animals make decisions about their survival.
    b. how animals choose their mates.
    c. how animals select food and shelter.
    d. group interactions.
    e. All of the above
    *Answer: e*

53. Which of the following is *not* a criterion used by a predator in choosing its prey?
    a. How long it will take to capture the prey
    b. The energy it will receive from eating the prey
    c. The abundance of the prey
    d. Whether another predator is after the same prey
    e. Whether the prey contains the proper nutrients for the predator
    *Answer: d*

# 54 Population Ecology

## Fill in the Blank

1. The proportions of individuals in each age group in a population make up its _____.
*Answer: age distribution*

2. A group of individuals born at the same time is known as a _____.
*Answer: cohort*

3. The number of individuals of any particular species that a habitat or environment can support is that habitat's _____.
*Answer: environmental carrying capacity*

4. An interaction in which one of two participants is harmed but the other is unaffected is known as _____.
*Answer: amensalism*

5. An interaction in which both partners benefit is known as _____.
*Answer: mutualism*

6. An interaction in which two organisms share a resource that is insufficient to supply their combined needs is known as_____.
*Answer: competition*

7. The total number of individuals of a species per unit of area (or volume) is called the _____.
*Answer: population density*

8. When a single bacterium is placed in a culture vessel, the resulting population at first grows very quickly; however, as the population expands and uses up its resources, the growth rate eventually slows down and stops. This S-shaped growth pattern is best modeled mathematically as _____ growth.
*Answer: logistic*

9. Population regulation that works to decrease numbers when its population numbers are high and increase numbers when they are low is said to be _____.
*Answer: density-dependent*

10. If the death rate in a population is unrelated to the number of individuals in an area, the resulting change in population size is said to be _____.
*Answer: density-independent*

11. A short-term event that disrupts subpopulations more than it disrupts the entire population is called a(n) _____.
*Answer: disturbance*

12. The study of changes in the size and structure of populations is known as _____.
*Answer: demography*

13. A _____ consists of all the individuals of a species within a given area.
*Answer: population*

## Multiple Choice

1. Which one of the following characteristics of whales is *not* a factor in the whale population's slow response to management practices?
a. Whales have long prereproductive periods.
b. Whales produce one offspring at a time.
c. Whales have long life spans.
d. Whales have long intervals between births.
e. Whales are hunted by several nations.
*Answer: c*

2. Of the following factors that regulate population size, the most density-independent factor is
a. food supply.
b. predators.
c. disease.
d. availability of nesting sites.
e. sudden temperature changes.
*Answer: e*

3. If a population's birth rate is density- _____, the birth rate will _____ as the population increases.
a. dependent; increase
b. dependent; decrease
c. independent; increase
d. independent; decrease
e. dependent or -independent; increase
*Answer: b*

4. The relationship between a human and the fungus that causes athlete's foot is _____ when the fungus feeds only on dead skin cells, but becomes _____ if the fungus penetrates the skin and feeds on living cells.
a. amensalistic; commensalistic
b. commensalistic; amensalistic
c. a host–parasite interaction; commensalistic
d. amensalistic; a host–parasite interaction
e. commensalistic; a host–parasite interaction
*Answer: e*

*5. Agave and yucca plants appear similar and grow in the same environments, but agaves reproduce once and die, whereas yuccas reproduce several times before dying. Compared to agave, each yucca seed crop should use _____ energy and produce _____ seeds.
   a. more; more
   b. more; fewer
   c. equal amounts of; equal amounts of
   d. less; more
   e. less; fewer
   *Answer: e*

6. The fact that many tree species and mycorrhizal fungi cannot survive unless they are associated with one another indicates that their relationship is
   a. amensalistic.
   b. commensalistic.
   c. competitive.
   d. mutualistic.
   e. a parasite–host interaction.
   *Answer: d*

*7. A female fig wasp enters the syconium of a fig, pollinates the flowers, and lays eggs in the ovaries of some of the flowers. The young larvae grow up, eat (and kill) some, but not all, of the seeds, and complete their life cycle. The fig is completely dependent on fig wasps to pollinate its flowers, and the fig wasp requires figs to complete its life cycle. The interaction between figs and fig wasps has aspects of
   a. mutualism.
   b. commensalism and mutualism.
   c. amensalism and host–parasite interaction.
   d. predator–prey and host–parasite interaction.
   e. Both a and d
   *Answer: e*

*8. A different species of wasp lays its eggs in an already fertilized syconium of a fig. The wasp does this by inserting her long ovipositor through the wall of the syconium and laying her eggs in the developing seeds. The larvae of this wasp eat (and kill) the seeds of the fig. This wasp does not pollinate the fig or in any other manner benefit the fig. The interaction between figs and this wasp has aspects of
   a. mutualism.
   b. commensalism.
   c. amensalism.
   d. predator–prey or host–parasite interaction.
   e. competition.
   *Answer: d*

■9. If a certain age interval in a population contains ten individuals and has a death rate of 1.000, how many individuals will be in the next older age interval?
   a. None
   b. One
   c. Five
   d. Nine
   e. Ten
   *Answer: a*

10. The number of individuals in a population is *least* affected by the rate of
    a. births.
    b. deaths.
    c. mating.

d. immigration.
e. emigration.
*Answer: c*

11. A population of organisms consists of 25 individuals, and over the span of a year 5 individuals die. What can be inferred about the population?
    a. The organisms that died were weak.
    b. Survivorship of the population is 80 percent.
    c. Mortality in the population was high.
    d. Natural selection is at work.
    e. None of the above
    *Answer: b*

12. The barnacle *Chthamalus stellatus* is capable of growing at depths much lower than those at which it is actually found in nature. Experimental studies have determined that the lower limit of *Chthamalus* is determined by another barnacle, *Balanus balanoides*. The rapidly growing *Balanus* is able to smother, crush, or undercut the slower-growing *Chthamalus*. Because of this, *Chthamalus* cannot survive in the lower depths where *Balanus* is found. The interaction between these two barnacles is an example of
    a. predator–prey interaction.
    b. mutualism.
    c. commensalism.
    d. amensalism.
    e. competition.
    *Answer: e*

13. In northern Scandinavia, rodents called lemmings have irruptions, or periodic buildups of large populations. The most likely cause of these irruptions is a(n)
    a. increase in food supply.
    b. decrease in predator population.
    c. increase in favorable nesting sites.
    d. increase in birth rate.
    e. decrease in disease rate.
    *Answer: a*

*14. A population such as Sweden's, which has a low birth rate and a low death rate,
    a. will have a relatively even distribution of individuals of different ages.
    b. will have a population dominated by young individuals.
    c. will have a population dominated by old individuals.
    d. will have a population dominated by individuals of intermediate age, with relatively few young or old individuals.
    e. can have almost any age distribution. Birth and death rates do not affect the age distribution.
    *Answer: a*

15. Cattle egrets follow cattle around because the cattle disturb insects as they walk, making the insects easier for the egrets catch. There is no cost or benefit to the cattle from this interaction. This interaction is an example of
    a. commensalism.
    b. amensalism.
    c. mutualism.
    d. parasitism.
    e. competition.
    *Answer: a*

16. Which of the following is *not* a demographic process that determines the number of individuals in a population?
    a. Death

b. Birth
c. Migration
d. Immigration
e. Emigration
*Answer: c*

17. Legumes, such as soybeans, form root nodules that become infected by *Rhizobium* bacteria. These bacteria convert nitrogen into nitrates, a form that is usable by plants and is frequently limited in terrestrial environments. The plants benefit from the bacteria in this way, and the bacteria benefit as well because they receive nutrients and energy from the plants. This interaction is an example of
   a. commensalism.
   b. amensalism.
   c. mutualism.
   d. parasitism.
   e. predation.
   *Answer: c*

■18. In a life table, the number of individuals alive at the beginning of the 1-year to 2-year age interval is 800. During this interval 200 individuals die. What is the death rate for this interval?
   a. 0.25
   b. 200
   c. 800
   d. 0.2
   e. 0.8
   *Answer: a*

19. A single bacterium put in an environment with unlimited resources and no competition would
   a. reproduce logistically.
   b. reproduce exponentially.
   c. reproduce linearly.
   d. not reproduce.
   e. Both b and c
   *Answer: b*

20. _____ capacity is the maximum number of individuals of a species an environment can support.
   a. Habitat
   b. Growth
   c. Population
   d. Carrying
   e. None of the above
   *Answer: d*

21. Which of the following concerning exponential growth is true?
   a. No population can grow exponentially for long.
   b. Exponential growth slows down as the population nears its maximal size.
   c. Bacterial colonies have been observed to maintain exponential growth for over a month.
   d. Exponential growth is commonly observed in large, slow-growing species such as humans and elephants.
   e. Exponential growth includes a component of environmental resistance.
   *Answer: a*

22. The intrinsic rate of increase, $r_{max}$, is
   a. the number of individuals added to each generation in a growing population under optimal conditions.

b. the difference between the average per capita birth rate, $b$, and the average per capita death rate, $d$, under optimal conditions.
   c. the number of individuals added to each generation in a growing population under the conditions that are actually occurring.
   d. the average per capita birth rate, $b$, under optimal conditions.
   e. the difference between the average per capita birth rate, $b$, and the average per capita death rate, $d$, under the conditions that are actually occurring.
   *Answer: b*

23. The logistic growth equation describes a population that
   a. grows without limits.
   b. grows rapidly at small population sizes, but whose growth rate slows and eventually stops as it reaches the number the environment can support.
   c. rapidly overshoots the number the environment can support and then fluctuates around this number.
   d. grows very rapidly and then crashes when environmental resources are used up.
   e. is declining in size.
   *Answer: b*

24. Which of the following is *not* an assumption of the logistic growth equation?
   a. An individual exerts its effects immediately at birth.
   b. All individuals produce equal effects.
   c. Resources in the environment are limited.
   d. Each individual depresses population growth equally.
   e. There is a time lag between gathering resources and reproduction.
   *Answer: e*

25. Certain woodpecker-like African birds have become specialized for removing and eating ticks and parasitic insects from the bodies of large herbivores. The relationship between the birds and the ticks is an example of
   a. mutualism.
   b. parasitism.
   c. commensalism.
   d. predator–prey interaction.
   e. amensalism.
   *Answer: d*

26. In the example above, the relationship between the birds and the herbivores demonstrates
   a. mutualism.
   b. parasitism.
   c. commensalism.
   d. predator–prey interaction.
   e. amensalism.
   *Answer: a*

27. Elephants and other large herbivores trample many species of plants that are different from the plant species they use as food. The relationship between the elephants and the trampled plant species is an example of
   a. mutualism.
   b. parasitism.
   c. commensalism.
   d. predation.
   e. amensalism.
   *Answer: e*

28. If individuals frequently move between subpopulations, immigrants may prevent declining subpopulations from becoming extinct. This process is known as the _____ effect.
    a. emigration
    b. migration
    c. rescue
    d. dilution
    e. population
    *Answer: c*

*29. Which of the following is an example of density-independent population regulation?
    a. A contagious disease sweeps through a dense population of lemmings.
    b. Jaegers, which are predators of lemmings, search widely for places where lemmings are abundant and concentrate their hunting in those areas.
    c. An outbreak of lemmings leads to a depletion of the food supply. As a result, lemmings are underfed and produce few offspring.
    d. When lemmings are abundant, arctic foxes switch from a primary diet of mice to a primary diet of lemmings.
    e. An early cold spell kills 80 percent of the lemmings in a particular area.
    *Answer: e*

*30. If we wish to maximize the number of individuals in a deer population so that many deer can be harvested, we should manage the population so that the deer
    a. are at the number the environment can support to maximize the number of individuals.
    b. are far enough below the number the environment can support to have high birth and growth rates.
    c. are very rare and don't come into contact with one another.
    d. slightly exceed the number the environment can support so that the excess can be harvested.
    e. greatly exceed the number the environment can support to maximize the number of excess individuals.
    *Answer: a*

■31. Which would be the most effective way to minimize a rat population in an alley?
    a. Kill as many rats as possible by poisoning and trapping them.
    b. Clean up the alley so that the rats have no garbage to feed on.
    c. Lure the rats away to another site where they will be less harmful.
    d. Search out and kill very young pre-reproductive rats.
    e. Release cats into the alley.
    *Answer: b*

32. Preserving a rare, endangered species will usually be expensive unless we
    a. understand all aspects of its biology and reproduction.
    b. bring the species into zoos and manage them in a controlled setting.
    c. continually provide additional resources such as food.
    d. can make preservation efforts economically profitable by harvesting the species as a food source.
    e. provide the species with sufficient suitable habitat.
    *Answer: e*

33. Many tropical orchid species grow on the surface of tree branches without harming or benefiting the tree. This orchid–tree relationship is an example of
    a. mutualism.
    b. parasitism.
    c. commensalism.
    d. predation.
    e. amensalism.
    *Answer: c*

34. The endosymbiosis hypothesis for the evolution of eukaryotic cells, which states that modern mitochondria and chloroplasts descended from once free-living bacterial ancestors that infected larger cells, may describe a very early example of
    a. mutualism.
    b. parasitism.
    c. interspecific competition.
    d. predation.
    e. amensalism.
    *Answer: a*

*35. Which of the following relationships is *not* an example of mutualism?
    a. *Rhizobium* bacteria live in the protected environment of legume roots and provide the legumes with nitrogen.
    b. A prawn cleans a coral reef fish.
    c. A plant produces "hitchhiker" fruit that are dispersed by a passing animal.
    d. *Pseudomyrmex* ants live in the hollow spines of an *Acacia* tree.
    e. A plant produces fleshy fruits with digestion-resistant seeds that are eaten by an animal.
    *Answer: c*

36. What is the type of ecological relationship that can involve members of either the same or different species and harms both participants?
    a. Mutualism
    b. Parasitism
    c. Competition
    d. Predation
    e. Amensalism
    *Answer: c*

*37. A straight-line relationship with a negative slope between age and percent of survival describes a survivorship pattern in which
    a. most individuals die at about the same age.
    b. young individuals have the highest death rates.
    c. the probability of death is about the same for all ages.
    d. the majority of individuals survive to old age.
    e. death rates increase following reproduction.
    *Answer: c*

*38. Which pattern of survivorship is most likely for an organism such as the sea urchin, which produces large numbers of very small offspring and provides no parental care?
    a. Most of the offspring survive for most of their potential life span and then die at the same time.
    b. The probability of death is low throughout the life span.
    c. Survivorship is high early in life and declines slowly.
    d. Young individuals have a high probability of dying, but older individuals have a relatively low probability of dying.

e. Probability of death is equally high throughout the potential life span.
*Answer: d*

39. The growth rate of a population in an unlimited environment
    a. is described by an S-shaped curve.
    b. is greater than the intrinsic rate of increase.
    c. is equal to $(b - d) N$.
    d. depends on competition for resources.
    e. is infinite.
    *Answer: c*

40. Which of the following is *not* an assumption of the logistic model for population growth?
    a. Members of the population compete for resources at high density.
    b. All individuals have the same effect on population growth.
    c. Population growth rate increases as the size of the population approaches the carrying capacity.
    d. There is no time lag between birth and the time an organism begins to affect population growth.
    e. The environment imposes a carrying capacity on the population.
    *Answer: c*

41–44. Use the graph below to answer the following questions.

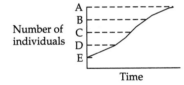

41. At what population size is the growth rate of this population greatest?
    a. A
    b. B
    c. C
    d. D
    e. E
    *Answer: c*

42. At what population size is the term $(K - N)/K$ largest?
    a. A
    b. B
    c. C
    d. D
    e. E
    *Answer: e*

43. At what population size is the rate of population growth beginning to slow down?
    a. A
    b. B
    c. C
    d. D
    e. E
    *Answer: b*

44. What would be the most effective practical means of managing this population in order to maximize the number of individuals harvested over a long period of time?
    a. Increase the carrying capacity
    b. Decrease the carrying capacity
    c. Harvest until the population size is equal to B

d. Harvest until the population size is equal to C
e. Increase $r$
*Answer: d*

45. Which of the following is *true* of the carrying capacity $(K)$?
    a. When $N = K$, the birth rate in a population is zero.
    b. The rate of population growth in an unlimited environment is proportional to $K$.
    c. $K$ is always determined by the amount of food in an environment.
    d. In a population at its K, the birth rate equals the death rate.
    e. $K$ changes over time for each population.
    *Answer: d*

46. Which of the following would be *unlikely* to contribute to large fluctuations in population size?
    a. Frequent disturbance
    b. A high death rate
    c. A density-dependent birth rate
    d. A high birth rate
    e. Seasonal variation in temperature
    *Answer: c*

47. A population of trees consists of large numbers of seedlings and saplings and a few reproducing adults. Which of the following is likely to be true for this population?
    a. The birth rate is density-dependent.
    b. The population is close to $K$.
    c. The death rate is low.
    d. Birth and death rates of young are high.
    e. Its members are evenly spaced.
    *Answer: d*

48. A graph showing the proportion of organisms of an initial group alive at different times throughout the life span is a(n)
    a. cohort.
    b. life table.
    c. survivorship curve.
    d. age structure diagram.
    e. population growth curve.
    *Answer: c*

49–51. Use the following life table for a hypothetical cohort of organisms to answer the questions below.

| Age interval (yr) | Number alive at beginning | Number dying | Death rate | Number of offspring |
|---|---|---|---|---|
| 0–1 | 200 | 150 | 0.75 | 0 |
| 1–2 | 50 | 10 | 0.20 | 0 |
| 2–3 | 40 | 20 | 0.50 | 3 |
| 3–4 | 20 | (?) | 0.50 | 9 |
| 4–5 | 10 | 10 | 1.00 | 4 |
| 5–6 | 0 | — | — | — |

49. During which of the following age intervals does this organism have the highest risk of mortality?
    a. 0–1
    b. 1–2
    c. 2–3
    d. 3–4
    e. Cannot tell from the information in the table
    *Answer: a*

50. How many individuals died during the interval 3–4?
    a. 5
    b. 10
    c. 20
    d. 40
    e. 50
    *Answer: b*

51. From the life table, you can conclude that
    a. individuals begin reproducing in their first year of life.
    b. the oldest individuals have the largest clutch size.
    c. on average, clutch size is larger for age interval 3–4 than for the interval 4–5.
    d. clutch size increases continuously with age.
    e. individuals of all ages produce about the same number of offspring.
    *Answer: c*

52. The intrinsic rate of increase is
    a. the rate of population growth in a limited environment.
    b. the maximum possible birth rate for a population.
    c. average birth rate minus the average death rate under optimal conditions.
    d. the average clutch size for individuals in a population in an unlimited environment.
    e. the rate at which a population colonizes a favorable new habitat.
    *Answer: c*

53. A population whose dynamics are dominated by periodic recovery from crashes will
    a. have a lower carrying capacity than a population not subject to crashes.
    b. be well below carrying capacity most of the time.
    c. live in a very stable environment.
    d. never reach exponential growth rates.
    e. None of the above
    *Answer: b*

■54. Many fish populations can be heavily harvested on a sustained basis because
    a. there are many "excess" individuals, since populations frequently overshoot their carrying capacity.
    b. they are not limited by the availability of suitable habitat.
    c. only a small number of females are needed to produce enough eggs to maintain population size.
    d. natural sources of mortality are low because their natural predators have gone extinct.
    e. each individual reproduces only once and would have died soon if they had not been harvested.
    *Answer: c*

■55. A population of seed-eating rodents infests your barn and consumes the corn stored there. Which of the following would be the most successful way to control the population of these pests?
    a. Leave corn out in the surrounding fields to lure the rodents away from the barn.
    b. Set large numbers of traps to kill as many rodents as possible.
    c. Introduce a voracious predator.
    d. Introduce a corn-eating competitor.
    e. Store grain in rodent-proof bins to reduce their access to corn and reduce their carrying capacity.
    *Answer: e*

56. Growth of the human population from prehistoric times to the present
    a. roughly fits an exponential curve.
    b. exhibits large oscillations around $K$.
    c. follows an S-shaped curve.
    d. has slowed considerably over the last 50 years.
    e. has been strongly influenced by density-dependent regulation.
    *Answer: a*

57. Earth's current carrying capacity for humans is set in part by
    a. limitations of food resources.
    b. the ability of Earth to absorb $CO_2$ produced by consumption of fossil fuel.
    c. disease.
    d. predation.
    e. space limitations.
    *Answer: b*

# 55 Communities and Ecosystems

## Fill in the Blank

1. The total amount of energy assimilated by photosynthesis is called _____.
*Answer: gross primary production*

2. The amount of energy assimilated by photosynthesis after the energy used by plants for maintenance and biosynthesis is subtracted is called _____.
*Answer: net primary production*

3. All organisms that get their energy from a common source (e.g., all herbivores) constitute a _____.
*Answer: trophic level*

4. The organisms that live together in a particular area constitute an _____.
*Answer: ecological community*

5. A set of linkages through which a plant is eaten by an herbivore, which in turn is eaten by a carnivore, and so on, is called a _____.
*Answer: food chain*

6. Organisms that reduce the remains of other organisms to mineral nutrients that can be taken up by plants are called _____.
*Answer: detritivores*

7. A network of linkages connecting a set of plants with a set of herbivores and carnivores is called a(n) _____.
*Answer: food web*

8. Two species of *Paramecium* are placed together in a test tube and one species drives the other to extinction. This phenomenon is an example of _____.
*Answer: competitive exclusion*

9. Growth on an area that at first supported no organisms but ends up supporting a mature forest is called _____.
*Answer: succession*

10. When the dead body of a freshly killed mouse decomposes, the change in the community of decomposers in the body is called _____.
*Answer: secondary succession*

11. Animals such as cows, which eat only plant tissues, are called _____.
*Answer: herbivores*

12. When a species has a larger influence on the community than its abundance would lead one to predict, it is called a _____ species.
*Answer: keystone*

13. Animals that eat other animals are called _____.
*Answer: predators*

## Multiple Choice

1. Which of the following changes would *not* result in an increase in net primary production?
   a. Increased precipitation in an arid area
   b. Increased soil fertility
   c. Increased latitude (moving from the equator toward the poles)
   d. Moving down a mountain to warmer temperatures
   e. Increasing the light in aquatic ecosystems
   *Answer: c*

2. Which of the following always has a "pyramidal" shape, that is, decreasing values at higher trophic levels?
   a. Pyramids of numbers
   b. Pyramids of biomass
   c. Pyramids of energy
   d. Both b and c
   e. a, b, and c
   *Answer: c*

■3. Which of the following statements about food chains and energy flow through ecosystems is *false*?
   a. A single organism can feed at several trophic levels.
   b. The lower the trophic level at which an organism feeds, the more energy is available.
   c. Detritivores feed at all trophic levels except the producer level.
   d. Food webs include two or more food chains.
   e. All organisms that are not producers are consumers.
   *Answer: c*

4. Assume that the energy transfer efficiency between trophic levels is 10 percent. How much grain would be required to produce 70 kg of human biomass if the grain is eaten by cows and the cows are eaten by humans?
   a. 210 kg
   b. 700 kg
   c. 2,100 kg
   d. 7,000 kg
   e. 70,000 kg
   *Answer: d*

5. Which characteristic of moose is most significant in determining that moose are a keystone species?
   a. A few moose have a large effect on forest succession.
   b. Moose are the largest animals in the ecosystem.
   c. Moose browse on a large variety of tree species.
   d. Moose use both terrestrial and aquatic habitats.
   e. Moose care for their young longer than any other species in the ecosystem.
   *Answer: a*

6. If the removal of a species from a community has a greater effect on the structure and functioning of the community than one would have predicted from the species' abundance, then that species is most likely a(n)
   a. keystone species.
   b. carnivore.
   c. herbivore.
   d. early successional species.
   e. mimic of another species.
   *Answer: a*

7. An inverted pyramid of _____ may occasionally be observed in _____ communities.
   a. energy; grassland
   b. energy; forest
   c. biomass; marine
   d. biomass; grassland
   e. biomass; forest
   *Answer: c*

8. In light, a plant fixes 0.12 ml of $CO_2$ per hour. However, in the dark the same plant releases 0.04 ml of $CO_2$ per hour. What is the estimated net primary production of this plant?
   a. 0.04 ml/hour
   b. 0.08 ml/hour
   c. 0.12 ml/hour
   d. 0.16 ml/hour
   e. 0.0048 ml/hour
   *Answer: b*

■9. The sea star *Pisaster ochraceous* is an abundant predator on the rocky intertidal communities on the Pacific coast of North America. The sea star feeds preferentially on the mussel *Mytilus californianus*. In the absence of the sea star, the mussel is a dominant competitor. Predation by the sea star, however, creates bare areas. For this reason,
   a. the number of species will be changed by the presence of the sea star, but the direction of change cannot be predicted.
   b. the variety of species will be unaffected by the presence or absence of the sea star.
   c. the variety of species will be greater when the sea star is present than when it is absent.
   d. the variety of species will be smaller when the sea star is present than when it is absent.
   e. *Mytilus* will go extinct when the sea star is present.
   *Answer: c*

10. Which of the following could *not* be used as an example of succession?
    a. A landslide exposes bare rock. Eventually a forest is reestablished on the site.
    b. An alien weed is introduced into the Hawaiian Islands. Eventually it displaces a native plant, driving the native plant to extinction.

c. A freshly killed mouse decomposes.
d. A tree falls in the forest. Eventually a new tree replaces it.
e. The leaf litter under a pine tree is in layers, with freshly fallen litter at the top and more decomposed layers deeper down.
*Answer: b*

11. Keystone species may influence the
    a. species richness of an ecosystem.
    b. flow of energy and materials in an ecosystem.
    c. environment of an ecosystem.
    d. Both a and b
    e. None of the above
    *Answer: d*

12. A community ecologist would most likely be concerned with
    a. energy flow through an ecosystem.
    b. population growth of a single species.
    c. interactions between individuals of the same species living together in a small area.
    d. interactions between individuals of different species living together in a small area.
    e. the cycling of matter through biotic and abiotic components of an area.
    *Answer: d*

13. Which characteristic is *not* true of ecological succession?
    a. The species composition of the community changes through time.
    b. The physical conditions of the community change through time.
    c. The rate of change is constant through time.
    d. Some species disappear quickly; others are more long-lived.
    e. The community becomes more complex through time.
    *Answer: c*

■14. Interactions within communities can control populations. Based on community studies involving white-footed mice, gypsy moths, and oak trees, scientists have determined that an increase in the mouse population leads to an increase in acorn production by the oak trees. If the mice prey on the larvae of the moths and the moth larvae eat the leaves of the oak, how can this be so?
    a. Gypsy moths do not influence acorn production.
    b. An increase in the mouse population leads to a decrease in the gypsy moth population.
    c. An increase in the mouse population leads to an increase in the gypsy moth population.
    d. Oak trees produce substances that deter the moth larvae from eating the leaves.
    e. Acorn production is not related to the mouse population.
    *Answer: b*

15. Solar energy is an important energy source in all of the following ecosystems *except*
    a. the tropical rainforest.
    b. temperate grassland.
    c. deserts.
    d. upper layers of the oceans.
    e. deep-sea hydrothermal vent systems.
    *Answer: e*

16. The greatest amount of energy loss in ecosystems occurs through
    a. photosynthesis.
    b. respiration.
    c. herbivory.
    d. excretion.
    e. digestion.
    *Answer: b*

■17. Average net primary production in the open ocean is low, but it makes up the largest percentage of net primary production on Earth. Why?
    a. Oceans can support only a small number of species.
    b. Sunlight is not a limiting factor in primary production in the open ocean.
    c. Oceans cover most of Earth, so the total percentage is the highest.
    d. Both a and b
    e. a, b, and c
    *Answer: c*

18. _____ exhibit _____ primary production due to a lack of moisture.
    a. Tropical rainforests; low
    b. Open oceans; high
    c. Deserts; low
    d. Tropical seasonal forests; high
    e. Savannas; high
    *Answer: c*

19. An example of species-specific coevolution is
    a. yucca plants and the single species of moth that pollinates them.
    b. egrets feeding near water buffalo.
    c. ants "milking" aphids for honeydew.
    d. tropical butterflies that look more similar to each other in each new generation.
    e. Both b and d
    *Answer: a*

■20. Species richness is influenced by primary productivity. Which of the following statements regarding this phenomenon is true?
    a. As species richness increases, primary production decreases.
    b. As species richness decreases, primary production increases.
    c. Species richness is highest at an intermediate level of primary production.
    d. As primary production increases, species richness continues to increase.
    e. Species richness is not influenced by primary production.
    *Answer: c*

21. Species richness has been observed to increase with primary productivity—to a point. At high levels of productivity, species richness decreases. Which of the following explains this phenomenon?
    a. Competitive exclusion
    b. Predation
    c. Herbivory
    d. Decomposition
    e. None of the above
    *Answer: a*

22. Photosynthetic plants are considered to be
    a. heterotrophs.
    b. consumers.
    c. autotrophs.
    d. detritivores.
    e. predators.
    *Answer: c*

23. Organisms that consume other organisms are called
    a. heterotrophs.
    b. detritivores.
    c. autotrophs.
    d. producers.
    e. photosynthesizers.
    *Answer: c*

■24. Biomass pyramids in the open ocean are inverted in comparison to most terrestrial ecosystems because
    a. primary production is lower in the open ocean.
    b. in the open ocean there are fewer consumers than there are producers.
    c. most of the biomass is found among the producers.
    d. the producers reproduce so rapidly that a smaller biomass of producers can support a larger biomass of herbivores.
    e. most of the biomass is found among the carnivores.
    *Answer: d*

25. Organisms that obtain their food from both primary consumers and another trophic level are called
    a. herbivores.
    b. omnivores.
    c. detritivores.
    d. producers.
    e. autotrophs.
    *Answer: b*

26. Organisms that eat herbivores are called
    a. tertiary consumers.
    b. primary producers.
    c. detritivores.
    d. secondary consumers.
    e. herbivores.
    *Answer: d*

★27. Why does a biomass pyramid for a forest show less biomass at the herbivore level than a grassland does?
    a. Forests contain fewer herbivores.
    b. Carnivores eat more herbivores in forests.
    c. Primary production of forests is higher than that of grasslands.
    d. Most of the biomass in a forest is tied up in material that is difficult to digest.
    e. Both a and b
    *Answer: d*

28. All of the following are detritivores *except*
    a. insects.
    b. fungi.
    c. wolves.
    d. insects.
    e. bacteria.
    *Answer: c*

29. Which of the following are examples of or causes of disturbances?
    a. Keystone species
    b. Hurricanes
    c. Fires
    d. Volcanic eruptions
    e. All of the above
    *Answer: e*

30. The intermediate disturbance hypothesis states that
    a. communities with high levels of disturbance have greater species richness.
    b. communities with low levels of disturbance have lower species richness.
    c. there is no relationship between species richness and the level of disturbance in a community.
    d. communities with intermediate levels of disturbance have greater species richness.
    e. communities with lower species richness have little or no competition.
    *Answer: d*

■31. A volcano has decimated a significant portion of the ecosystem around it. Following this disturbance, what type of process is likely to proceed?
    a. Secondary succession
    b. Tertiary succession
    c. Decomposition
    d. Primary succession
    e. All of the above
    *Answer: d*

32. _____ succession may begin in an area where dead parts of organisms remain.
    a. Secondary
    b. Primary
    c. Tertiary
    d. Quaternary
    e. Disturbance
    *Answer: a*

■33. During primary succession, what process is important for the reestablishment of the soil and the colonization of a plant community?
    a. Decomposition
    b. Predation
    c. Disturbance
    d. Secondary succession
    e. Nitrogen fixation
    *Answer: e*

34. _____ succession begins in areas that lack living organisms.
    a. Secondary
    b. Primary
    c. Tertiary
    d. Quaternary
    e. Disturbance
    *Answer: b*

35. The formation of the Central American land bridge 4 million years ago led to
    a. reduced dispersal of mammals between North and South America.
    b. mass extinction of most mammals in North and South America.
    c. increased dispersal of mammal species between North and South America.
    d. the establishment of a large number of South American mammals in North America.
    e. the establishment of a small number of North American mammals in South America.
    *Answer: c*

# 56 Biogeography

## Fill in the Blank

1. _____ is the science that documents and attempts to explain the patterns of distribution of populations, species, and ecological communities across Earth.
   *Answer: Biogeography*

2. The leeward side of a mountain where air descends, warms, and absorbs moisture rather than releasing it is called a _____.
   *Answer: rain shadow*

3. A species found only within a certain region is said to be _____.
   *Answer: endemic*

4. The appearance of a barrier that splits the range of a species is called a _____ event.
   *Answer: vicariant*

5. _____ and _____ are the most distinct biogeographic regions because they have been isolated from other continents for the longest time.
   *Answer: Australia; South America*

6. One theory of biogeography suggests that the number of species in an area is the result of a dynamic balance between _____ and _____.
   *Answer: extinctions; immigrations*

7. The major biome types are identified by their characteristic _____.
   *Answer: vegetation*

8. A particular community type, such as a hot desert, occurs where the _____ is suitable for that type of vegetation.
   *Answer: climate*

9. The _____ biome is found in the Arctic and high in the mountains at all latitudes.
   *Answer: tundra*

10. Arctic tundra is underlain by permanently frozen ground called _____.
    *Answer: permafrost*

11. A modern biogeographic technique that shows the distribution patterns of a taxonomic group is the _____.
    *Answer: area phylogeny*

12. Major ecosystems that vary according to the nature of their dominant vegetation are called _____.
    *Answer: biomes*

13. The ocean's _____ at some levels is the biggest barrier to colonization by a species.
    *Answer: temperature*

14. The amount of heat that an area of Earth receives depends on the _____ of the sun.
    *Answer: angle*

## Multiple Choice

1. An endemic species is one that
   a. is found only in one region.
   b. evolved in the region where it currently lives.
   c. occurs in two distinct areas separated by a barrier.
   d. has a good fossil record showing its evolution.
   e. disperses readily from one area to another.
   *Answer: a*

■2. Australia has a greater percentage of endemic species than other continents because it
   a. lies near several large chains of islands.
   b. has more varied habitats than other continents.
   c. has been isolated for the longest time due to continental drift.
   d. lacks a region near the equator.
   e. has received many immigrant species due to prevailing ocean currents.
   *Answer: c*

★3. A study of flightless weevils found in New Zealand showed that populations of the same species exist on both islands. What is the most likely explanation for this dispersal?
   a. A vicariant event split the population at some time in the past.
   b. Weevils, although flightless, managed to migrate from the South Island to the North Island.
   c. The same species of weevil evolved on both islands.
   d. Weevils were introduced to both islands.
   e. None of the above
   *Answer: a*

4. The island biogeographic model describes the relationship between the number of species on an island and its
   a. climate and vicariant events.
   b. size and distance from the mainland.
   c. species pool.
   d. rate of extinction.
   e. elevation and shape.
   *Answer: b*

■5. The poles are cooler than the equator because
a. the equator receives more hours of sunlight than the poles do.
b. the equator is closer to the sun than the poles are.
c. the angle at which the sun strikes Earth is shorter at the poles (the sun is lower on the horizon).
d. the long polar nights cool Earth much more than the short equatorial nights do.
e. winds tend to move heat from the poles toward the equator.
*Answer: c*

■6. It tends to rain more in the mountains than in adjacent lowlands because
a. as air travels up mountains, it cools. Cool air holds less moisture than warm air does, and the excess moisture is dropped as rain.
b. as air travels up the mountains, it is compressed. The mountains squeeze the water out of the air, and the water falls as rain.
c. of their great height. Mountains are often in the clouds, and precipitation is much greater in these clouds.
d. as air moves down mountains, it expands and does not form clouds, so no rainfall occurs.
e. the rainfall is equal in both mountains and lowlands; however, because the rain has farther to travel from clouds to the lowlands, more evaporates before reaching the ground in the lowlands.
*Answer: a*

7. Which of the following is *not* a modern barrier between major biogeographic regions?
a. Desert
b. Mountains
c. Waterways
d. High plateau–lowland boundary
e. Tundra
*Answer: e*

8. A common ecosystem on the leeward (away from the wind) side of mountains and also at 30° north or south latitude is
a. a coniferous forest.
b. a desert.
c. a rainforest.
d. a peatland.
e. tundra.
*Answer: b*

9. The geographic distribution of a biome may be affected by the seasonal variation of all of the following *except*
a. temperature.
b. latitude.
c. precipitation.
d. fire.
e. grazing.
*Answer: b*

10. A traveler starting out near Washington, D.C., on the East Coast of the United States and traveling toward the North Pole would encounter which of the following biomes, in which order?
a. Temperate deciduous forest, boreal forest, tundra
b. Chaparral, temperate deciduous forest, boreal forest
c. Savanna, boreal forest, chaparral
d. Temperate grassland, boreal forest, tundra
e. Boreal forest, tundra, chaparral
*Answer: a*

11. Forests in both the Northern and Southern Hemispheres are termed "boreal" forests when they are dominated by
a. evergreen trees with needles or small leaves.
b. cone-bearing gymnosperm trees.
c. deciduous beeches or related angiosperm trees.
d. a diverse assemblage of insect-pollinated trees.
e. evergreen trees adapted to frequent fires.
*Answer: a*

12. Which of the following biomes may lack seasonal temperature variations?
a. Tundra
b. Tropical evergreen forest
c. Boreal forest
d. Cold desert
e. Chaparral
*Answer: b*

13. Which of the following biomes lacks a pronounced seasonal variation in the amount of precipitation?
a. Tropical deciduous forest
b. Thorn forest
c. Tundra
d. Chaparral
e. Cold desert
*Answer: c*

14. The prevailing winds in the continental United States are from the west because of the
a. position of the sun.
b. location of the major mountain ranges.
c. Pacific Ocean's warmth in comparison to the Atlantic Ocean.
d. currents in the Atlantic and Pacific Oceans.
e. rotation of Earth.
*Answer: e*

**15. Biogeographers often apply the rule of parsimony when interpreting distribution patterns. Assume there are four organisms with the following similarities: Organisms A and B are very similar, and organisms C and D are very similar. Organisms A and B are similar to fossil E; organisms C and D are similar to a fossil F. Which of the following would be a parsimonious explanation for their evolutionary relationship?
a. B is more closely related to C than to A.
b. B is more closely related to C than to D.
c. B and A evolved from F; C and D evolved from E.
d. B and A evolved from E; C and D evolved from F.
e. A, B, C, and D evolved from F.
*Answer: d*

16. If a species lives in two distinct areas now separated by a barrier, and it was in those areas before the barrier existed, it is said to have a _____ distribution.
a. dispersal
b. vicariant
c. natural
d. migration
e. static
*Answer: b*

17. A vicariant distribution is one determined by
a. separation of formerly contiguous land masses.
b. dispersal.
c. extinction and recolonization.
d. adaptive radiation.
e. interbreeding of two similar species.
*Answer: a*

18–22. Refer to the following figure for the questions below.

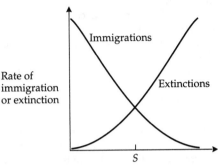

Number of species on island

18. The above figure suggests that the rate of extinction of species
    a. decreases as the number of species increases.
    b. increases as the number of species increases.
    c. is independent of the number of species.
    d. varies for different islands.
    e. is large when the number of species is small.
    *Answer: b*

19. The rate of immigration is high when the number of species is low because
    a. there is no competition.
    b. the first colonists to arrive are all "new" species.
    c. there are no predators.
    d. the island is newly formed and attracts species.
    e. there has not been enough time for extinctions to occur.
    *Answer: b*

20. *S* represents the
    a. point at which species are numerous enough to compete for resources.
    b. equilibrium composition of species: No new species can invade and none of the existing species will go extinct.
    c. minimum number of species that ever existed on the island.
    d. maximum number of species that ever existed on the island.
    e. equilibrium number of species: Extinctions will equal immigrations.
    *Answer: e*

21. If the number of species on the island is *S* + 3, we would expect to see
    a. on average, three more species go extinct than those that invade.
    b. three species go extinct and then no changes in the biota of the island.
    c. on average, three more species invade than those that go extinct.
    d. three species invade and then no changes in the biota of the island.
    e. Not enough information to determine
    *Answer: a*

22. If the island suddenly doubled in size, we would expect to see the
    a. extinction curve shift to the left and a decrease in *S*.
    b. immigration curve shift to the right and an increase in *S*.
    c. extinction curve shift to the right and a decrease in *S*.
    d. extinction curve shift to the right and an increase in *S*.
    e. immigration curve shift to the left and an increase in *S*.
    *Answer: e*

23. This biome is found in many climates, but all of them are relatively dry much of the year.
    a. Tundra
    b. Boreal forest
    c. Tropical evergreen forest
    d. Temperate grassland
    e. Chaparral
    *Answer: d*

24. This biome is characterized by cold temperatures, permafrost, and little rainfall. It is found in the Arctic and in high mountains at all latitudes.
    a. Tundra
    b. Boreal forest
    c. Tropical evergreen forest
    d. Temperate grassland
    e. Chaparral
    *Answer: a*

25. This is the richest of all biomes in species of both plants and animals, and it has the highest overall energy flow of all ecological communities. The soils in this biome are usually poor, however, as most of the nutrients are tied up in the vegetation.
    a. Tundra
    b. Boreal forest
    c. Tropical evergreen forest
    d. Temperate grassland
    e. Chaparral
    *Answer: c*

26. This biome is dominated by coniferous trees. It has Earth's tallest trees, and supports the highest standing biomass of wood of all ecological communities.
    a. Tundra
    b. Boreal forest
    c. Tropical evergreen forest
    d. Temperate grassland
    e. Chaparral
    *Answer: b*

27. This biome is found on the western sides of continents at moderate latitudes. The dominant vegetation is low-growing shrubs with evergreen leaves.
    a. Tundra
    b. Boreal forest
    c. Tropical evergreen forest
    d. Temperate grassland
    e. Chaparral
    *Answer: e*

*28. When the intertropical convergence zone is over an area,
    a. the trade winds prevail.
    b. the trade winds reverse direction.
    c. air sinks and the area becomes cool and dry.
    d. air rises and heavy rains fall.
    e. air stagnates and hot, dry conditions prevail.
    *Answer: d*

29. In general, ocean currents circulate
    a. clockwise in both the Northern and Southern Hemispheres.
    b. counterclockwise in both the Northern and Southern Hemispheres.
    c. clockwise in the Atlantic Ocean and counterclockwise in the Pacific Ocean.
    d. counterclockwise in the Northern Hemisphere and clockwise in the Southern Hemisphere.
    e. clockwise in the Northern Hemisphere and counterclockwise in the Southern Hemisphere.
    *Answer: e*

■30. North Dakota is known for its hot summers and cold winters. At the same latitude, western Washington has cooler summers and warmer winters. What is the main cause of this difference?
    a. Washington receives a steady westerly wind that leads to a more constant temperature. Because the westerly winds are blocked by mountains, North Dakota receives hot southern winds in the summer and cold northern winds in the winter.
    b. Western Washington is nearly always overcast. The cloud cover prevents the sun from heating the land in the summer and prevents excess heat loss during the winter. North Dakota is very dry, and it is heated and cooled more dramatically.
    c. Western Washington is near the Pacific Ocean, which moderates its climate. North Dakota is in the middle of a large continent and does not receive this buffering.
    d. Western Washington is at lower elevation than North Dakota. This difference in altitude accounts for the more extreme weather in North Dakota.
    e. Both a and b
    *Answer: c*

31. Rain shadows form on the _____ side of a mountain and result in _____ rainfall there than on the opposite side.
    a. windward; more
    b. windward; less
    c. leeward; more
    d. leeward; less
    e. eastward; more
    *Answer: d*

*32. As coastal mountains force a sea breeze to rise, a process occurs that is the similar to that occurring in the area of the _____, except that in the latter, the energy is provided by the _____.
    a. trade winds; Earth's spin
    b. westerlies; sun
    c. westerlies; Earth's spin
    d. intertropical convergence zone; sun
    e. intertropical convergence zone; Earth's spin
    *Answer: d*

33. Different regions on Earth with similar climates have _____ ecological communities with _____ species.
    a. similar; similar
    b. similar; identical
    c. identical; identical
    d. very different; very different
    e. similar; very different
    *Answer: e*

34. The direction of the westerlies and the trade winds is directly caused by
    a. the location of the intertropical convergence zone.
    b. the tilt of Earth's axis.
    c. variation in the rotational speed of Earth at different latitudes.
    d. differences in heat balance between the equator and the poles.
    e. global oceanic circulation.
    *Answer: c*

■35. Australia has been separated from other continents for about 65 million years. South America has been isolated from other continents for about 60 million years. North America and Eurasia have been joined for much of Earth's history. Which continents would you expect to have the most similar biota?
    a. Australia and South America
    b. North and South America
    c. North America and Eurasia
    d. Australia and both North and South America
    e. Australia and both North America and Eurasia
    *Answer: c*

*36. Two adjacent continents are separated by 150 km of ocean. On continent A there are ten species of organisms closely related to five species on continent B. By applying the principle of _____ you would conclude that the species moved from _____.
    a. parsimony; A to B
    b. parsimony; B to A
    c. evolution; A to B
    d. evolution; B to A
    e. island biogeography; A to B
    *Answer: a*

*37. Based on your knowledge of the history of continental drift on Earth, which of the following pairs of biogeographic regions should have the most similar fauna?
    a. Palearctic and Ethiopian
    b. Ethiopian and Neotropical
    c. Nearctic and Neotropical
    d. Australasian, Neotropical
    e. Oriental and Ethiopian
    *Answer: b*

38. Which of the following biogeographic regions has the greatest number of endemic taxa?
    a. Nearctic
    b. Neotropical
    c. Ethiopian
    d. Australasian
    e. Oriental
    *Answer: d*

39. If species arrive on an island at a constant rate, and there is no extinction, the rate of arrival of new species on the island should _____ as the total number of species on the island increases.
    a. increase
    b. decrease
    c. remain constant
    d. increase, then decrease
    e. decrease, then increase
    *Answer: b*

40. The number of species on an island at which the arrival rate of new species equals the extinction rate of species

already present is called the equilibrium species number. At this number, the species richness _____ and the species composition _____.
a. is constant; is constant
b. is constant; can be variable
c. can be variable; is constant
d. can be variable; can be variable
e. depends on the size of the island; is constant
*Answer: b*

41. Using the model of island biogeography developed by Robert MacArthur and E. O. Wilson, which of the following can be predicted?
a. Species composition on an island
b. Which species will become extinct
c. Which species will arrive
d. Species richness, given island size
e. Rate of evolution of endemic species
*Answer: d*

**\*\*42.** ISN and ISF are two small islands of the same size that are near and far respectively from a species pool, and ILN is a large island that is near the same species pool. Select the sequence in which these three islands are arranged from the smallest to the largest expected equilibrium species number.
a. ISF, ISN, ILN
b. ISN, ILN, ISF
c. ILN, ISF, ISN
d. ISF, ILN, ISN
e. ILN, ISN, ISF
*Answer: a*

43. Which of the following factors is *not* important in determining the equilibrium species number of an island?
a. Island size
b. Distance to the mainland
c. Distance to other islands
d. Size of the mainland
e. Direction of prevailing wind and currents
*Answer: d*

44. Which of the following is *not* true regarding biogeographic regions and biomes?
a. Generally, similar biomes are restricted to a single biogeographic region.
b. Biogeographic regions are based on similarities in biota.
c. Biomes are based on climate differences.
d. Many biomes are distributed latitudinally.
e. Both biomes and biogeographic regions exclude the oceans.
*Answer: a*

45. Select the biome that is found in dry regions with various temperatures and is characterized by much underground biomass.
a. Tundra
b. Boreal forest
c. Temperate deciduous forest
d. Temperate grassland
e. Cold desert
*Answer: d*

46. Select the biome that is dominated by coniferous, evergreen trees with short growing seasons.
a. Tundra
b. Boreal forest
c. Temperate deciduous forest

d. Temperate grassland
e. Cold desert
*Answer: b*

47. Select the biome that is characterized by the presence of permafrost and high latitude or altitude distribution.
a. Tundra
b. Boreal forest
c. Temperate deciduous forest
d. Temperate grassland
e. Cold desert
*Answer: a*

48. Select the biome that has a wide range of temperatures, ample precipitation year-round, and animal-dispersed pollen and fruit.
a. Tundra
b. Boreal forest
c. Temperate deciduous forest
d. Temperate grassland
e. Hot desert
*Answer: c*

49. Select the biome that often is located in rain shadows and in which most precipitation falls in winter.
a. Tundra
b. Boreal forest
c. Temperate deciduous forest
d. Temperate grassland
e. Cold desert
*Answer: e*

50. Select the biome that is found in areas with descending warm, dry air and in which most precipitation falls in summer.
a. Chaparral
b. Thorn forest
c. Tropical savanna
d. Tropical deciduous forest
e. Hot desert
*Answer: e*

51. Select the biome characterized by the greatest species richness and the highest energy flow.
a. Chaparral
b. Thorn forest
c. Tropical savanna
d. Tropical deciduous forest
e. Tropical evergreen forest
*Answer: e*

52. Select the biome that is found on the equatorial side of hot deserts and has a short, intense summer rainy season.
a. Chaparral
b. Thorn forest
c. Tropical montane forest
d. Tropical deciduous forest
e. Tropical evergreen forest
*Answer: b*

53. Select the biome that has cool, wet winters, hot, dry summers, and is dominated by low-growing shrubs with tough evergreen leaves.
a. Chaparral
b. Tropical savanna
c. Tropical montane forest
d. Tropical deciduous forest
e. Tropical evergreen forest
*Answer: a*

54. Select the biome characterized by a longer rainy season and taller trees than the neighboring thorn forest biome, and in which many trees flower in the dry season, when they are leafless.
    a. Boreal forest
    b. Temperate deciduous forest
    c. Tropical montane forest
    d. Tropical deciduous forest
    e. Tropical evergreen forest
    *Answer: d*

55. Select the biome with mostly wind-dispersed pollen and fruit and low species diversity.
    a. Boreal forest
    b. Temperate deciduous forest
    c. Tropical montane forest
    d. Tropical deciduous forest
    e. Tropical evergreen forest
    *Answer: a*

56. Which of the following pairs of biomes is maintained by grazing?
    a. Cold desert and hot desert
    b. Chaparral and thorn forest
    c. Temperate grassland and tropical savanna
    d. Temperate and tropical deciduous forests
    e. Cold desert and tundra
    *Answer: c*

57. Which of the following physical events is *not* a major influence on the distribution of organisms?
    a. Continental drift
    b. Earthquakes
    c. Sea level changes
    d. Glacial retreat
    e. Formation of mountains
    *Answer: b*

*58. _____ and _____ regions have more species richness than other regions do.
    a. Low latitude; flatland
    b. Low latitude; mountainous
    c. Tropical; peninsular
    d. High latitude; peninsular
    e. High latitude; mountainous
    *Answer: b*

# 57 Conservation Biology

## Fill in the Blank

1. The scientific study of how to preserve the diversity of life is called _____.
   *Answer: conservation biology*

2. The relationship between the area of a habitat patch and the number of species present is an example of a _____.
   *Answer: species–area relationship*

3. Nearly all of the mammals and birds native to Madagascar are found only on that island. Species with distributions like this are said to be _____.
   *Answer: endemic*

4. _____ are the phenomena that are influenced by adjacent habitats. They increase as patch size decreases.
   *Answer: Edge effects*

5. _____ are those that are in imminent danger of extinction over all or a significant part of their range.
   *Answer: Endangered species*

6. The subdiscipline of conservation biology that is concerned with converting disturbed areas back into natural areas (e.g., converting cropland into a native prairie) is called _____.
   *Answer: restoration ecology*

7. In the past (over a thousand years ago) the primary means by which humans caused the extinction of animals was _____.
   *Answer: overexploitation*

8. The most important cause of the endangerment of species is _____.
   *Answer: habitat loss*

9. Many species of birds became extinct in Hawaii due to _____ by humans.
   *Answer: overhunting*

10. Species that are likely to become endangered in the near future are labeled _____.
    *Answer: threatened species*

## Multiple Choice

1. Why should humans care about species extinctions?
   a. Humans derive pleasure and aesthetic benefits from interactions with other organisms.
   b. Humans depend on other species for food.
   c. A variety of species are necessary for the functioning of ecosystems.
   d. Biological diversity is the heritage of all humankind and should be passed on to future generations.
   e. All of the above
   *Answer: e*

2. In recent years, the number of species driven to extinction has increased dramatically. Which of the following is *not* a reason for this development?
   a. Overexploitation
   b. Habitat destruction
   c. Introduction of predators
   d. Natural predation
   e. Introduction of diseases
   *Answer: d*

3. When early Polynesian people first settled in Hawaii, they drove at least 39 species of endemic land birds to extinction because they
   a. introduced predators.
   b. overhunted.
   c. destroyed natural habitat.
   d. introduced diseases.
   e. competed with these birds for food.
   *Answer: b*

4. A species–area relationship is used by ecologists to
   a. determine the population density of a species in a certain habitat.
   b. examine how human populations are growing.
   c. estimate the numbers of species extinctions resulting from habitat destruction.
   d. produce a population model.
   e. None of the above
   *Answer: c*

5. Species benefit the functioning of ecosystems in all of the following ways *except* for
   a. maintenance of fertile soils.
   b. prevention of soil erosion.
   c. interference with the hydrological cycle.
   d. waste product recycling.
   e. plant pollination.
   *Answer: c*

6. Many species depend on particular patterns of _____ for maintenance of their populations.
   a. water temperature
   b. river flow
   c. rainfall
   d. disturbance
   e. wind
   *Answer: d*

7. Which parameter of a population do ecologists measure to assess its extinction risk?
   a. Genetic variation
   b. Behavior
   c. Physiology
   d. Both a and b
   e. a, b, and c
   *Answer: e*

■8. Biologists have measured the rate at which species richness tends to change with changes in patch or habitat size. These findings suggest that
   a. loss of habitat results in an increase in species richness.
   b. high loss of habitat results in significant loss of species richness in that habitat.
   c. habitat loss has no effect on species richness.
   d. humans have not caused habitat loss and thus have had no impact on species richness.
   e. species richness is completely unrelated to habitat loss or gain.
   *Answer: b*

9. Captive propagation in zoos and botanical gardens has been quite successful for several endangered species. Nevertheless, captive propagation is only a partial or temporary solution to the biodiversity crisis. Which of the following does *not* represent an inadequacy of zoos?
   a. There is not enough space in existing zoos and botanical gardens to maintain populations of more than a small fraction of rare and endangered species.
   b. Captive propagation projects in zoos have proven to be been useless in efforts to raise public awareness of the biodiversity crisis.
   c. A species in captivity can no longer evolve along with the other species in its ecological community.
   d. The preservation of endangered species cannot be accomplished simply by captive propagation. The habitat required to support wild populations must also be present for successful species preservation.
   e. Small, captive populations tend to have low genetic diversity.
   *Answer: b*

■10. Approximately 200 years ago, California condors ranged from Canada to Mexico. By 1978, the population had almost disappeared. Today condors are being reestablished because
   a. condors migrated into North America from other areas.
   b. nonbreeding birds began to pair when nesting conditions again improved.
   c. the climate in the western United States, which had gone through a warm period, began to cool making the climate suitable for condors once again.
   d. extensive natural forest areas have been restored on the eastern seaboard, allowing condors once again to nest successfully.
   e. captive-reared condors have been systematically released in the western United States.
   *Answer: e*

■11. Why are species extinctions a threat to humans?
   a. Many species are a human food source.
   b. Many pharmaceutical products are derived from natural products. Loss of species could mean loss of therapeutic drugs.

   c. Biodiversity is important in maintaining fertile soils, and in the prevention of erosion.
   d. Diverse species of animals and plants are important in the detoxification and recycling of nutrients.
   e. All of the above
   *Answer: e*

12. _____ are parts of habitats that allow animals to disperse from patch to patch.
   a. Windows
   b. Transects
   c. Corridors
   d. Pathways
   e. Portals
   *Answer: c*

*13. What effect, if any, would the extinction of a mutualistic pollinator have on the species of plant it pollinates?
   a. It would have no effect.
   b. Pollination of the plant species would be decreased, but would not cease altogether.
   c. The plant species would die out immediately.
   d. It would allow the plant species to be pollinated by another, competing pollinator.
   e. The plants of this species would not be pollinated, and thus the species would no longer be able to reproduce sexually.
   *Answer: e*

■14. Species native to islands such as Madagascar are particularly ravaged by habitat destruction because
   a. island populations of species tend to have more mutualistic relationships than mainland populations.
   b. many species found on islands are found nowhere else.
   c. habitats are more easily destroyed on islands than on continents.
   d. habitat fragmentation is more serious on islands than on continents because islands normally have a single continuous habitat.
   e. Both a and c
   *Answer: b*

15. A(n) _____ species is one that has spread widely and become unduly abundant, at a cost to native species.
   a. invasive
   b. endangered
   c. threatened
   d. endemic
   e. extinct
   *Answer: a*

*16. A lumber company proposes to clear-cut a large area of forest, but it aims to leave small patches of forest to provide habitat for forest animals. The problem with this method of conservation is that
   a. small patches cannot support populations of species that require large areas.
   b. small patches can harbor only small populations of the species that can survive there.
   c. due to edge effects, small patches have higher temperatures, stronger winds, and lower humidity levels than larger forest tracts.
   d. species that live in the clear-cut areas often invade the edges of the patches and compete with or prey on the species living there.
   e. All of the above
   *Answer: c*

17. _____ ecology is the field of conservation biology whose goal is to mend damaged and degraded ecosystems.
    a. Conservation
    b. Historical
    c. Restoration
    d. Renovation
    e. Redistribution
    *Answer: c*

18. The American chestnut, a formerly abundant tree in forests of the Appalachian Mountains, was virtually eliminated due to
    a. habitat loss.
    b. overexploitation.
    c. the introduction of a non-native pathogen.
    d. fire.
    e. illegal trade.
    *Answer: c*

19. Which of the following methods could be used to restore a population of animals from a few male and female individuals.
    a. Cross breeding
    b. Interbreeding
    c. Captive propagation
    d. Selective breeding
    e. Both a and b
    *Answer: c*

20. Which of the following programs would *not* be useful in the efforts to conserve species?
    a. International trade in pets
    b. Certification programs
    c. CITES
    d. Control of poaching
    e. The Forest Stewardship Council (FSC)
    *Answer: a*

21. Which of the following is *not* an essential service for humans that is provided by natural ecosystems, such as wetlands?
    a. Absorption of oxygen and release of carbon dioxide
    b. Reduction of erosion and water runoff
    c. Treatment and purification of wastewater
    d. Production of fish, waterfowl, and other wildlife
    e. Absorption of pollutants such as sulfates
    *Answer: a*

*22. Of the following organisms, which has the highest conservation priority?
    a. A plant found in North America and in Europe
    b. A plant endemic to Australia
    c. A plant found on the Galápagos Islands and in Brazil
    d. A plant found worldwide
    e. A plant found in North and South America
    *Answer: b*

23. Which of the following statements concerning the relationship between humans and the rest of the living world is true?
    a. Modern technology has made it so that we no longer depend on other living organisms.
    b. We are dependent on artificial ecosystems, such as agroecosystems, and we gain no benefit from natural ecosystems.
    c. We are dependent on natural ecosystems at present, but the technology exists to completely replace natural ecosystems so that we will no longer depend on them.
    d. Our survival is tightly linked to the survival of natural ecosystems throughout the world.
    e. Nonindustrial human societies are dependent on natural ecosystems, but industrial societies are not.
    *Answer: d*

24. In Southern California, 90 percent of the coastal wetlands have been destroyed. Efforts to restore these areas have included
    a. field experiments to examine the effects of species richness.
    b. introductions of non-native species.
    c. housing developments.
    d. human-initiated disturbance.
    e. All of the above
    *Answer: a*

**25. Hawaii has many species of long-lived *Lobelia* plants and many species of honeycreepers (birds) that pollinate the plants. Assume that *Lobelia* species A is pollinated exclusively by honeycreeper species A. If honeycreeper species A becomes extinct, what is the most likely fate of *Lobelia* species A?
    a. It will become extinct for lack of a pollinator.
    b. An empty niche will be created by the bird extinction, and another honeycreeper species will fill it and pollinate the plants.
    c. The plant flower shape will evolve toward a more generalized wind-pollinating form.
    d. The plants will evolve asexual methods of reproduction.
    e. The species will switch to self-crossing (i.e., the pollen produced by a flower will land on the female structures of the same flower) in order to survive.
    *Answer: a*

26. If global warming continues to cause average ocean temperatures to increase, the most likely effect on ocean coral reefs will be that
    a. they will lose their symbiotic dinoflagellates, a phenomenon called bleaching.
    b. they will grow uncontrollably.
    c. they will remain stable.
    d. coral reef reestablishment will be effective.
    e. None of the above
    *Answer: a*

27. The number of species that become extinct due to habitat destruction is greatest in _____ ecosystems.
    a. temperate
    b. tropical
    c. arctic
    d. desert
    e. savanna
    *Answer: b*

28. Major causes of human-induced extinctions of species include all of the following, *except*
    a. climate modification.
    b. overexploitation.
    c. habitat destruction.
    d. captive propagation.
    e. introduction of predators and diseases.
    *Answer: d*

29. Which of the following is *not* true of extinctions that occurred at least 2,000 years ago?
    a. Many exterminated species were large mammals.
    b. Humans had recently arrived in the area.
    c. Habitat destruction by humans was the cause.
    d. Many flightless birds were exterminated.
    e. The exterminated species were initially numerous.
    *Answer: c*

30. Which of the following is likely the most important cause of species extinctions?
    a. Habitat destruction
    b. Overexploitation
    c. Introduced predators
    d. Introduced disease
    e. Climate modifications
    *Answer: a*

31. Which of the following statements about captive propagation is *false*?
    a. Separating the organism from its ecological community can lead to difficulties.
    b. Captive propagation can play an important role in maintaining species during critical times.
    c. Successful captive propagation programs may make it possible to reduce the habitat needed for an endangered species.
    d. Reproduction in some captive endangered species has been successful.
    e. Captive propagation can be used until the external threat to the species is corrected.
    *Answer: c*

32. In forest ecosystems, edge effects within a patch will _____ and environmental variation will _____ with a decrease in habitat patch size.
    a. increase; decrease
    b. increase; increase
    c. increase; remain constant
    d. decrease; increase
    e. decrease; decrease
    *Answer: b*

33. Which of the following statements about the size effects of habitat fragmentation is *false*?
    a. Species with large home ranges and poor dispersal rates disappear as patch size decreases.
    b. Edge effects increase uniformly as patches become smaller.
    c. Death related to dispersal between suitable habitat patches increases as patches become smaller.
    d. The adverse effects of smaller habitat patches decrease if the patches are connected to larger habitat areas.
    e. All of the above
    *Answer: b*

34. Which of the following is *not* a reason to protect biodiversity?
    a. The aesthetic value
    b. Because of ecological relationships, whole communities could be endangered by the extinction of one species.
    c. Important medicinal compounds can be found only in certain species.
    d. Extinction is irreversible.
    e. None of the above
    *Answer: e*

35. Fynbos are the endemic shrubs of the hills of Western Cape Province in South Africa and the primary vegetation of the watershed that provides the region's water supply. During recent decades, fynbos have been invaded by alien plants that grow taller and faster, but a movement has begun to remove these aliens by cutting and digging them up. Which of the following is *not* a reason for removing them?
    a. It is cheaper to remove the aliens than to obtain water by another method.
    b. Tourism money comes into the country from people who want to see the fynbos.
    c. Removing the aliens is the most technologically advanced method of supplying water to the region.
    d. The fynbos can be sold as cut flowers.
    e. The intensity and severity of fires is lessened when the fynbos are present.
    *Answer: c*

# 58 Earth System Science

## Fill in the Blank

1. In temperate lakes, the depth at which the temperature abruptly changes from warm surface temperatures to cold deeper temperatures is the _____.
*Answer: thermocline*

2. The cycling of water between the oceans, atmosphere, and land is known as the _____.
*Answer: hydrological cycle*

3. The layer of the atmosphere in which humans live (the one closest to the ground) is the _____.
*Answer: troposphere*

4. The movement of an element through a living organism and the physical environment is called its _____.
*Answer: biogeochemical cycle*

5. The addition of nutrients, especially phosphorus, to a lake will result in blooms of algae that die and decompose, using all of the oxygen in the lake. This process is called _____.
*Answer: eutrophication*

6. With a very few exceptions, all of the energy available to living organisms ultimately comes from the _____.
*Answer: sun*

7. _____ is a new field of inquiry that focuses on Earth as a whole.
*Answer: Earth system science*

8. A _____ is a group of entities that interact to yield some product.
*Answer: system*

9. Waters in zones of upwelling are rich in _____.
*Answer: mineral nutrients*

10. The layer of the atmosphere that extends from the top of the troposphere up to about 50 km and contains very little water vapor is the _____.
*Answer: stratosphere*

11. Deposits of organic molecules that are eventually transformed into deposits of oil, gas, or coal are known as

    _____.
*Answer: fossil fuels*

12. _____ is the principal process that removes nitrogen from the biosphere and returns it to the atmosphere.
*Answer: Denitrification*

13. The rate of $CO_2$ movement to deep ocean waters depends on a circulation pattern called the _____.
*Answer: ocean conveyor belt*

14. A _____ is an area in a body of water that results from flows of nitrogen-enriched water.
*Answer: "dead zone"*

15. Human activity has most seriously disturbed the _____ biogeochemical cycle.
*Answer: carbon*

16. The troposphere is transparent to visible light and traps most outgoing infrared light. This property results in the _____ of Earth.
*Answer: warming*

## Multiple Choice

★1. Which of the following elements is *not* a major component of living systems?
a. Carbon
b. Nitrogen
c. Oxygen
d. Hydrogen
e. Potassium
*Answer: e*

★2. Erosion from poor farming practices leads to large losses of soil to the sea. What is the likely fate of the elements in the soil?
a. Most of the elements will be rapidly recirculated by ocean currents and reintroduced into terrestrial ecosystems.
b. Most elements will dissolve and eventually cycle into the atmosphere.
c. Most elements will dissolve in the soil and move into freshwater ecosystems, where they will eventually be reintroduced into terrestrial ecosystems.
d. Tidal forces will quickly return most of the soil to the shore, where the elements will collect.
e. Most elements will settle to the bottom of the oceans, where they will remain for millions of years until bottom sediments are elevated by movements of Earth's crust.
*Answer: e*

3. Excluding human-induced effects, most elements enter freshwater systems from
a. rainfall.
b. oceans (through tidal movements).
c. actions of animals.
d. weathering of rocks (via groundwater).
e. plants and plant parts decaying in streams, lakes, and rivers.
*Answer: d*

4. In a temperate zone lake in the middle of summer,
   a. most nutrients are near the surface of the lake, and $O_2$ is evenly distributed throughout all depths of the lake.
   b. most nutrients are near the bottom of the lake, and most $O_2$ is near the surface of the lake.
   c. nutrients and $O_2$ are evenly distributed throughout all depths of the lake.
   d. nutrients are evenly distributed throughout all depths of the lake, and $O_2$ is near the surface of the lake.
   e. most nutrients and $O_2$ are found near the surface of the lake.
   *Answer: b*

5. At which of the following temperatures is water most dense?
   a. 16°C
   b. 8°C
   c. 4°C
   d. 2°C
   e. 0°C
   *Answer: c*

6. Most of Earth's nitrogen is in
   a. the atmosphere.
   b. the oceans.
   c. freshwater systems.
   d. soil.
   e. organisms.
   *Answer: a*

7. The gas that is removed from the atmosphere by plants and algae is
   a. $N_2$.
   b. $O_2$.
   c. $CO_2$.
   d. methane.
   e. water vapor.
   *Answer: c*

■8. One difference between gases such as $O_2$ and $CO_2$ and minerals such as phosphorus and calcium that do not have a gas phase is that
   a. there is relatively little variation from site to site in the concentration of gases, whereas solids tend to remain where they are, creating local abundances and shortages of these minerals.
   b. gases such as $O_2$ and $CO_2$ are never found in a rock (solid) phase or dissolved in water, whereas minerals without a gaseous phase are frequently found as rocks or dissolved in water.
   c. gaseous elements (and compounds) are essential to living organisms, whereas minerals without a gaseous phase are not.
   d. gases are modified by living organisms, whereas minerals without a gaseous phase are not.
   e. the relative proportions of different gases in the atmosphere change radically over short periods of time, whereas the relative proportions of different solid minerals remain fairly constant.
   *Answer: a*

*9. Most of the world's carbon is found
   a. as $CO_2$ in the atmosphere.
   b. in living organisms.
   c. as bicarbonate and carbonate ions dissolved in the oceans.
   d. as carbonate minerals in sedimentary rock.
   e. in the decaying remains of dead organisms.
   *Answer: d*

10. Photosynthesis and respiration are central to which cycle?
    a. Nitrogen cycle
    b. Carbon cycle
    c. Phosphorus cycle
    d. Sulfur cycle
    e. Hydrological cycle
    *Answer: b*

11. For the past 150 years there has been a major new input to the carbon cycle. What is it?
    a. There are more humans releasing large quantities of $CO_2$ as respiration.
    b. Increased animal farming has resulted in greater $CO_2$ releases.
    c. Industrialization has resulted in the burning of fossil fuels such as oil and coal, which releases $CO_2$ into the atmosphere.
    d. Changes in ocean currents have lead to the release of large quantities of $CO_2$.
    e. Global warming has increased the release of $CO_2$ from carbonate minerals.
    *Answer: c*

■12. Why is nitrogen often in short supply in terrestrial ecosystems?
    a. There is very little free nitrogen in the air.
    b. Atmospheric nitrogen is primarily in the stratosphere and does not come into contact with terrestrial ecosystems.
    c. Atmospheric nitrogen cannot be used by most organisms. It needs to be converted to useful forms by bacteria and cyanobacteria.
    d. Nitrogen solubility in water is very low and therefore atmospheric nitrogen enters cells very slowly.
    e. Atmospheric nitrogen varies widely from location to location. As a result local shortages occur frequently.
    *Answer: c*

■13. The phosphorus cycle differs from the carbon cycle in that
    a. phosphorus does not enter living organisms, whereas carbon does.
    b. the phosphorus cycle does not include a gaseous phase, whereas the carbon cycle does.
    c. the phosphorus cycle includes a solid phase, whereas the carbon cycle does not.
    d. the primary reservoir of the phosphorus cycle is the atmosphere, whereas the primary reservoir for the carbon cycle is rock.
    e. phosphorus passes through living organisms many times, whereas carbon flows through living organisms only once.
    *Answer: b*

14. Acid precipitation
    a. occurs naturally when clouds are formed by evaporation over land.
    b. occurs naturally when clouds are formed by evaporation over oceans.
    c. occurs as a result of the burning of fossil fuels that contain sulfur and nitrogen compounds.
    d. occurs because of carbon monoxide emissions by automobiles.
    e. occurs when large forest tracts are logged or otherwise deforested.
    *Answer: c*

15. The two main human-induced causes of increased $CO_2$ levels are
    a. the burning of fossil fuels and the burning of forests.
    b. the burning of fossil fuels and the raising of cattle.
    c. modern agricultural fertilizers and the raising of cattle.
    d. pollution-induced mortality of oceanic algae and the burning of forests.
    e. modern agricultural fertilizers and pollution-induced mortality of oceanic algae.
    *Answer: a*

16. If $CO_2$ levels double, as they may by the middle of the next century, many serious consequences are likely. Which of the following is the one result that would *not* be expected?
    a. World mean temperature will increase 3–5°C.
    b. Droughts in the central regions of continents will become more common and more severe.
    c. Precipitation in coastal areas will increase.
    d. Polar ice caps will melt, raising sea levels and flooding coastal cities and agricultural areas.
    e. The destruction of the ozone layer by $CO_2$ will allow more ultraviolet light to reach Earth.
    *Answer: e*

17. In the movement of energy and nutrients through ecosystems, energy _____, and nutrients _____.
    a. flows; flow
    b. flows; cycle
    c. cycles; cycle
    d. cycles; flow
    e. flows; either cycle or flow
    *Answer: b*

18. Which of the following does *not* influence a terrestrial ecosystem?
    a. Solar energy fluxes
    b. Earlier successional stages
    c. Local species richness
    d. Global atmospheric circulation
    e. Eutrophication
    *Answer: e*

*19. The major differences between ocean and freshwater compartments of the global ecosystem exist because
    a. of the much greater depth of the oceans.
    b. the oceans turn over only once a year.
    c. freshwater systems receive input only from groundwater.
    d. temperature varies only with depth in freshwater systems.
    e. nutrient recycling is less rapid in freshwater systems.
    *Answer: a*

20. Which global ecosystem compartment is characterized by very slow movement of materials and exchange of gases, mostly via organisms?
    a. Oceanic
    b. Freshwater
    c. Atmospheric
    d. Terrestrial
    e. Both a and b
    *Answer: d*

*21. Which of the following comparisons of the troposphere and stratosphere compartments of the global ecosystem is true?
    a. Most water vapor resides in the troposphere.
    b. Vertical circulation of air occurs mostly in the stratosphere.
    c. Circulation of the stratosphere influences ocean currents more directly.
    d. The stratosphere represents most of the mass of the atmosphere.
    e. Most weather events occur in the stratosphere.
    *Answer: a*

*22. Which of the following statements regarding the hydrological cycle is *false*?
    a. Most input to the oceans occurs via runoff from rivers.
    b. The amount of water that evaporates from the surface of the oceans is greater than the amount that falls as rain over them.
    c. The amount of water that evaporates from the surface of the land is less than the amount that falls as rain over it.
    d. Water found in sedimentary rock is constantly exchanged with the ocean.
    e. The energy driving the hydrological cycle is heat from the sun.
    *Answer: d*

23. In biogeochemical cycles, elements that cycle the most quickly
    a. are found in organisms.
    b. are scarce.
    c. have a gaseous phase.
    d. do not become fixed into sediment.
    e. do not accumulate in the higher trophic levels.
    *Answer: c*

24. Which of the following biogeochemical cycles is characterized by a major reservoir that is gaseous, and a major inorganic form that can be utilized only by a small group of bacteria and cyanobacteria?
    a. Carbon cycle
    b. Nitrogen cycle
    c. Phosphorus cycle
    d. Sulfur cycle
    e. Oxygen cycle
    *Answer: b*

25. In which of the following biogeochemical cycles is the gaseous phase a major greenhouse gas?
    a. Carbon cycle
    b. Nitrogen cycle
    c. Phosphorus cycle
    d. Sulfur cycle
    e. More than one cycle
    *Answer: a*

26. Which of the following biogeochemical cycles lacks a gaseous phase?
    a. Carbon cycle
    b. Nitrogen cycle
    c. Phosphorus cycle
    d. Sulfur cycle
    e. Oxygen cycle
    *Answer: c*

27. Which of the following biogeochemical cycles has a gaseous phase made up of emissions from volcanoes and fumaroles?
    a. Carbon cycle
    b. Nitrogen cycle
    c. Phosphorus cycle
    d. Sulfur cycle
    e. Oxygen cycle
    *Answer: d*

■28. The land compartment of Earth is connected to the atmospheric compartment through
    a. runoff from rivers.
    b. organisms that process chemical elements.
    c. zones of upwelling.
    d. chemical elements that enter groundwater.
    e. None of the above
    *Answer: b*

29. In the human-induced condition called eutrophication, the main biogeochemical cycle that is altered is the _____ cycle; the effect is the creation of _____ conditions and a decrease in species diversity.
    a. hydrological; aerobic
    b. phosphorus; anaerobic
    c. hydrological; aerobic
    d. phosphorus; aerobic
    e. nitrogen; anaerobic
    *Answer: b*

★30. In the human-induced condition called acid precipitation, the main biogeochemical cycles that are altered are the _____ cycles; one effect in lakes is the _____ of populations of nitrifying bacteria.
    a. phosphorus and nitrogen; increase
    b. phosphorus and nitrogen; decrease
    c. nitrogen and sulfur; decrease
    d. nitrogen and sulfur; increase
    e. phosphorus and sulfur; decrease
    *Answer: c*

31. Which of the following occurrences is *not* considered to be a likely consequence of global warming?
    a. Melting of the polar ice caps
    b. Flooding and increased precipitation in coastal areas
    c. Increased ozone depletion
    d. Average worldwide temperature increase of 3–58°
    e. Droughts in central areas of continents
    *Answer: c*

32. Rain and snow become acidified when the burning of fossil fuels releases the acid forms of the elements
    a. phosphorus and sulfur.
    b. sulfur and nitrogen.
    c. nitrogen and chlorine.
    d. phosphorus and chlorine.
    e. nitrogen and phosphorus.
    *Answer: b*

■33. Primary production is higher in zones of upwelling because
    a. water in these zones conducts nutrients poorly.
    b. nutrient levels are lower in these zones.
    c. a thermocline exists in these zones.
    d. water is more dense at 4°C.
    e. the water in these areas has a greater concentration of nutrients.
    *Answer: e*

34. Which of the following gases is *not* part of Earth's atmosphere?
    a. $O_2$
    b. $N_2$
    c. Chlorine
    d. Water vapor
    e. $CO_2$
    *Answer: c*

35. In a generalized biogeochemical cycle, fluxes between _____ are slower than fluxes between _____.
    a. rocks and soils; organisms and the atmosphere
    b. oceans and the atmosphere; fresh water and the atmosphere
    c. organisms and the atmosphere; rocks and soils
    d. magma and igneous rock; metamorphic rock and magma
    e. nitrogen and phosphorus; carbon and water
    *Answer: a*

36. The process that redistributes nutrients in freshwater lakes is called
    a. upwelling.
    b. eutrophication.
    c. the hydrological cycle.
    d. turnover.
    e. denitrification.
    *Answer: d*